Phytochemistry and Nutritional Composition of Significant Wild Medicinal and Edible Mushrooms

Traditional Uses and Pharmacology

Phytochemistry and Nutritional Composition of Significant Wild Medicinal and Edible Mushrooms

Traditional Uses and Pharmacology

Edited by

Ajay Sharma
Chandigarh University, India
Email: sharmaajay9981@gmail.com

Garima Bhardwaj
Sant Longowal Institute of Engineering and Technology, India
Email: garimasliet@gmail.com

and

Gulzar Ahmad Nayik
Government Degree College Shopian, India
Email: gulzarnaik@gmail.com

ROYAL SOCIETY OF CHEMISTRY

Print ISBN: 978-1-83916-667-9
PDF ISBN: 978-1-83767-209-7
EPUB ISBN: 978-1-83767-210-3

A catalogue record for this book is available from the British Library

The Royal Society of Chemistry is a charity, registered in England and Wales, Number 207890, and a company incorporated in England by Royal Charter (Registered No. RC000524), registered office: Burlington House, Piccadilly, London W1J 0BA, UK, Telephone: +44 (0)20 7437 8656.

Visit our website at www.rsc.org/books

Printed in the United Kingdom by CPI Group (UK) Ltd, Croydon, CR0 4YY, UK

Dedication

This book is dedicated to my grandparents (the late Mr Ram Krishan Sharma and Mrs Tripta Devi) & my parents (Mr Ashwani Kumar and Mrs Anju Bala).

Phytochemistry and Nutritional Composition of Significant Wild Medicinal and Edible Mushrooms: Traditional Uses and Pharmacology
Edited by Ajay Sharma, Garima Bhardwaj and Gulzar Ahmad Nayik

Published by the Royal Society of Chemistry, www.rsc.org

Preface

Since ancient times, an unabated exploration and evaluation of health-promoting medicinal and edible mushrooms has been witnessed, which has led to an extensive use of several mushrooms with therapeutic properties for the treatment of a variety of ailments. Wild medicinal and edible mushrooms are a source of a large number of bioactive lead compounds and contain numerous health promoting minerals, vitamins, secondary metabolites, and some primary metabolites. Further, these properties of medicinal and edible mushrooms have been well supported by various scientific articles published in different reputed journals and books. Moreover, the knowledge of the nutritional potential, phytochemistry and health promoting benefits of wild medicinal and edible mushrooms is essential, not only for the discovery of new therapeutic agents, but also because such knowledge is necessary for the unveiling of new resources of bioactive leads. Additionally, the awareness of the phytochemistry and health promoting benefits of wild medicinal and edible mushrooms has also been useful in determining the real value of their folkloric potential. Therefore, the present book is going to be a boon in the identification of the role of health promoting wild medicinal and edible mushrooms in the improvisation and elevation of human health.

This book presents the traditional uses, nutraceutical composition, phytochemistry, pharmacology, mycology, cultivation, conservation status, toxicology and future prospectives of health-promoting wild medicinal and edible mushrooms. Further, the book will be very useful to Scientists/Pharmacognosists/Pharmacists/Chemists/Graduates/Undergraduates/Researchers in the fields of natural products, herbal medicines,

Phytochemistry and Nutritional Composition of Significant Wild Medicinal and Edible Mushrooms: Traditional Uses and Pharmacology
Edited by Ajay Sharma, Garima Bhardwaj and Gulzar Ahmad Nayik

Published by the Royal Society of Chemistry, www.rsc.org

ethnobotany, pharmacology, chemistry and biology, as well as teachers and students, who would like to learn more about health-promoting medicinal and edible mushrooms.

Ajay Sharma
Gulzar Ahmad Nayik
Garima Bhardwaj

Editor Biographies

Ajay Sharma is a Natural Product Chemist with a strong Organic Chemistry, Phytochemistry, Biochemistry, and Ethnobotany background. Presently, he is working as an Assistant Professor at the University Centre for Research and Development, Department of Chemistry, Chandigarh University, Mohali, India. He graduated (BSc Medical) from Govt. College Hoshiarpur (affiliated to Panjab University, Chandigarh), has a Master's Degree in Organic Chemistry (Natural Products) from Punjab Agriculture University Ludhiana and a PhD (Organic Chemistry – Natural product) from SLIET (Sant Longowal Institute of Engineering & Technology, Deemed University) Punjab. He has published over 55 peer-reviewed papers and book chapters in reputed international/national journals and books. Presently, his five edited books are under production in different international publication houses, such as Springer, Wiley, Taylor & Francis, Elsevier and the Royal Society of Chemistry. He has various honors/fellowships/awards to his credit, *i.e.* Role of honor from PAU Ludhiana (2011), Merit-cum-topper scholarship during MSc (PAU Ludhiana) (2009-10), MHRD fellowship Award for PhD (2012), Award of INSPIRE fellowship (JRF, 2013) from DST, Award of INSPIRE fellowship (SRF, 2015) from DST, 1st Prize in Poster Award at National Conference (NICS-2016) and Best Poster Award in (TSFS-2016). He conducts research in Organic Chemistry, Natural Products Chemistry, Phytochemistry, and Food Chemistry. His research activities have focused on the isolation and characterization of bioactive extracts from traditional medicinal plants and evaluation of their biological potential. Currently, his research group at Chandigarh University is investigating the field of more environmentally friendly syntheses of metal nanoparticles and carbon nanodots using plant extracts/plant-based biomass followed by assessing their biological, catalytic, and sensing potential. To date, he has successfully guided six MSc students through their dissertations.

Phytochemistry and Nutritional Composition of Significant Wild Medicinal and Edible Mushrooms: Traditional Uses and Pharmacology
Edited by Ajay Sharma, Garima Bhardwaj and Gulzar Ahmad Nayik

Published by the Royal Society of Chemistry, www.rsc.org

Gulzar Ahmad Nayik completed his master's degree in Food Technology from Islamic University of Science & Technology, Awantipora, Jammu and Kashmir, India, and his PhD at Sant Longowal Institute of Engineering & Technology, Sangrur, Punjab, India. He has published over 80 peer-reviewed research and review papers and 40+ book chapters, and he has edited 13 books with Springer, Elsevier, and Taylor & Francis. Dr. Nayik has also published a textbook on food chemistry and nutrition and has delivered several presentations at various national and international conferences, seminars, workshops and webinars. Dr. Nayik was shortlisted twice for the prestigious Inspire-Faculty Award in 2017 and 2018 from the Indian National Science Academy, New Delhi, India. He was nominated for India's prestigious National Award (Indian National Science Academy Medal for Young Scientists 2019–2020). Dr. Nayik also fills the roles of editor, associate editor, assistant editor and reviewer for many food science and technology journals. He has received many awards, appreciations and recognitions and holds membership at various international societies and organizations. Dr. Nayik is currently editing several book projects with Elsevier, Taylor & Francis, Springer Nature, the Royal Society of Chemistry, *etc.*

Garima Bhardwaj is an independent researcher. She earned her PhD in Organic Chemistry from SLIET Longowal, Punjab. The topic of her PhD research was on studies related to the utilization of oleo-chemical industry by-products for bio-surfactant synthesis. She has published her PhD work in various reputed international journals. After her PhD, she joined the Defence Institute of High-Altitude Research (DIHAR), Defence Research and Development Organization (DRDO) as a Research Associate in the Phytochemistry division. During her work at DIHAR, DRDO, she got the opportunity to work on the nutritional profiling of high-altitude vegetables and medicinal plants, where she developed her interest in Phytochemistry. She has published over 22 peer-reviewed papers and book chapters in reputed international/national journals and books. She has received the award of Research fellowships from Ministry of Human Resource Development (MHRD) and Ministry of Defence, DRDO, India, for her doctoral and post-doctoral research, respectively. She conducts research on Organic Chemistry, Biosurfactant synthesis, Natural Products Chemistry, Phytochemistry. During her PhD her research activities were focused on the isolation and characterization of novel strains to produce biosurfactants and their purification and structural elucidation using modern chromatographic and spectroscopic methods. In her post-doctoral research, she worked on the extraction and purification of bioactive compounds from high-altitude regions and their qualitative and quantitative analysis. She is currently working on the natural product chemistry of medicinal plants.

Contents

Phytochemistry and Nutritional Composition of Significant Wild Medicinal and Edible Mushrooms: Traditional Uses and Pharmacology
Edited by Ajay Sharma, Garima Bhardwaj and Gulzar Ahmad Nayik

Published by the Royal Society of Chemistry, www.rsc.org

CHAPTER 1

An Overview on Mushrooms

ABHISHEK KATOCH, MAHENDRA PAUDEL AND POOJA KATOCH*

Chandigarh University, University Institute of Agricultural Sciences, Mohali, City, 140301, India
*Email: pooja.agri@cumail.in

1.1 Introduction

A broad consensus-based classification of fungi has established seven phyla in the kingdom fungi, namely Chytridiomycota, Blastocladiomycota, Neocallimastigomycota, Microsporidia, and a subkingdom Dikarya, with the phyla Ascomycota and the Basidiomycota,[1] which also include mushrooms. Mushrooms or "fruits of fungus," are diverse filamentous achlorophyllous, spore-bearing, umbrella-shaped fleshy to leathery fungal fruit bodies (sporophores) produced under specific temperature, humidity, and nutritional conditions, and are epigeal (aboveground) or hypogeal (below ground) with a parasitic, saprophytic, or symbiotic mode of nutrition.

Historically, humans have foraged for edible mushrooms and used them for food and medicinal purposes.[2] However, in spite of macro fungi being the longest studied organism, many poisonous mushrooms are still consumed. A conservative estimate of macrofungal diversity on earth is around maximum of 110 000–140 000 species[3] and only 16–41% of macro fungi are estimated to be known.[4] To humans, a total of 2006 safe edible mushroom species in 99 countries have been deemed to be edible.[5] Moreover, around 90 mushroom species are cultivable (food or medicinal purposes),[6] but only 30 are grown commercially for food and consumed widely worldwide.[7]

As a superfood, mushrooms have many uses in nutritional supplementation, health care, and general well-being. Mushrooms have minimal fat,

Phytochemistry and Nutritional Composition of Significant Wild Medicinal and Edible Mushrooms: Traditional Uses and Pharmacology
Edited by Ajay Sharma, Garima Bhardwaj and Gulzar Ahmad Nayik

Published by the Royal Society of Chemistry, www.rsc.org

calories, and carbohydrate content while being abundant in fiber, protein, vitamins, and minerals.[8] Mushrooms are considered to be a substitute for meat and their nutritional value is comparable to several vegetables. Mushroom farming uses trash to grow fungi that are high in protein and nutritious; therefore, this is an excellent way of recycling agro-waste and providing vegetarians with better nutrition and medicinal benefits. India produces only 0.18 million tons of mushrooms compared to the 40 million tons produced worldwide by China, the United States, the Netherlands, Poland, Spain, France, Italy, Ireland, Canada, and the United Kingdom.[9] Even today, despite the fact that many mushrooms are now commercially grown, all highly prized uncommon and medicinal mushrooms are still harvested from their natural habitats because of their organoleptic properties.[10] Mushroom foraging from the wild constitutes the main source of revenue for inhabitants in forests ranging from modest woods and temperate forests to forests at high altitudes of the Himalayas to tropical miombo woodlands in south-central Africa.[11] Mushrooms are selected in accordance with their practicability in terms of size, shape, tenderness, and taste. However, traditional knowledge, such as ethnomycology/ethnomedicine that is passed down from generation to generation is now in danger of extinction due to a lack of recording or poor descriptions of it.

There have been substantial yield and quality improvements in mushroom production technology used in commercial cultivation, employing genetic improvement methods such as selection, hybridization, single spore evaluation, protoplasm fusion, and substrate-based studies. The exponential growth in mushroom output throughout the world as a consequence of changes in production technology and diversification has led to the income of mushroom producers doubling or tripling in a single year. Moreover, with an advancement in molecular techniques, more research areas are being explored for improving biological efficiency or bringing more mushrooms into cultivation. Molecular genetics have aided in resolving the phylogenetic link among many mushroom species.[12] Mushrooms are also increasingly being "value added" or "biofortified" to improve their quality and shelf life. As a result, mushroom farming has greatly expanded around the globe. In rural Indian communities and local tribal populations, mushroom foraging has proven to be viable for generating employment and ensuring the security of food, nutrition, and medicine.[9]

Rising knowledge of the health advantages of eating mushrooms and recent acknowledgements of mushrooms as a functional food have increased their consumption. Many cellular components and secondary metabolites of mushroom have been reported to impact the immune system and therefore might be utilized to treat disorders.[13] The market for metabolites from fungi has gained traction because of the high product yield and low cost of reproducing mushrooms. In the food market, mushrooms are recognized as a diverse group of fungi that serve as nutraceuticals, mycotherapeutics, cosmeceuticals, and are suitable for manufacturing functional foods with a considerable bioactive chemical content. Mushroom production also offers the possibility for the sustainable management of agro-industrial waste and creates a decent revenue with little input, which will help rural regions that are struggling economically.[14]

This chapter informs readers of the basic concepts, along with geographic distribution, Himalayan diversity, and climatic requirements with challenges faced for the sustainable usage of important mushroom species.

1.2 General Aspects of Mushroom Fungi

1.2.1 Mushroom Morphology

Mushrooms are macrofungi that are technically called basidiomata (sing. basidioma). These macro fungi live underground, either in wood or another species-specific substrate, wherein they form a fungal mycelium that is a network of hyphae (thread-like structures) that produces fruiting forms, in which spores are produced that help in the dispersal and dissemination of the fungi *via* various means.

Moreover, other structures (like sclerotia) are made up of organized hyphae compacted into pseudo tissues, and under the right conditions, proliferate into basidiomata (in basidiomycetes) and ascomata (in ascomycetes). Accurate descriptions are essential for identifying both ascomycetes and basidiomycetes since the ascomata and basidiomata have a diverse range of morphologies. The fertile part (hymenium) of ascomata or basidiomata is called the hymenial layer. In Ascomycota, ascospores are the sexual spores that are developed inside sac-like cells known as asci, whereas basidiospores are the reproductive or sexual spores of Basidiomycota, formed on top of the basidia, which are club-shaped cells that bear the sterigmata characteristic of the genus.

In most basidiomata, the pileus (also called the cap) and stipe (also called the stem) are well-shaped, and the hymenial layer is raised above the substrate (Figure 1.1a). In other cases, for better protection, the hymenium remains enclosed in the basidiomata until it is fully mature (this is the case with so-called *Gasteromycetes* or gasteroid fungi).

The ascomata in ascomycetes either have closed, tuber-like bodies that are born above the earth or beneath. An apothecium is the ascoma of a morel (*Morchella* spp.) that consists of a hard, stipe-like structure and spherical, spherically-elongated, or conical pileus with a honeycomb-like outer surface. The hymenial layer, which is located on the top surface and includes the structures where the spores are formed, is present (called asci). Truffles (Tuber spp.) have ascomata that are hypogenous (developed in the soil) and typically sub globose.

1.2.2 Mushrooms: Biology

Fungi may survive under a variety of conditions, including the driest, hottest deserts; the coldest, harshest ecosystems; the highest mountains; and the deepest seas. Ascomata and Basidiomata may create relationships with the

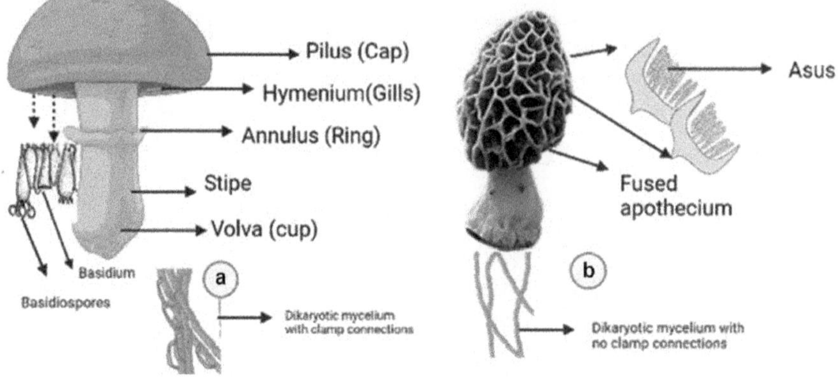

Figure 1.1 Morphological features of mushrooms from the subkingdom Dikarya (a) Basidiomycota and (b) Ascomycota.

nutrient substrate to crop up above or below the ground, in order to vary spore dissemination. Other mushrooms develop symbiotic relationships, mostly with woody or grassy plants, known as mycorrhizal association. The nutritional strategy of a fungus determines its life cycle and cultivation success as saprophytic fungi can be grown relatively easier than mycorrhizal fungi because these mycorrhizal fungi have a deep association with different tree species and other soil microbes. As a result of complex interactions that are not well known to the scientific community, great success has been found in growing mycorrhizal fungi, where truffles in artificial orchards have been effective.[15] Also, some pathogenic fungi in order to survive often flip their nutritional behavior from phytopathogenic to saprophytic or *vice versa*.

In addition to nutrition, geographic origin may impact the temperature shift that triggers the fruiting of mushrooms. Fungi produce mushrooms during their sexual reproductive cycles. In nature, temperature changes act as a stressor that initiate different morphogenesis steps either to develop vegetatively, stay dormant, or fruit. In temperate regions, mushroom species are suited to well-defined seasonal cycles, requiring the development of temperature recognition systems to initiate a new cycle, such as the autumn–winter period. In contrast, in tropical locations with undefined seasons and stable temperatures, mushrooms may be grown year-round without temperature-sensitive triggers, with rainfall triggering fruiting as fungi require a high relative humidity to fruit. Mycorrhizal mushrooms usually grow in the summer when high temperatures and fruiting bodies form after heavy rains. Although the basic steps in mushroom development have been determined, morphogenesis is still mostly unknown. The developmental stages ascertained are namely:[16]

- Hyphae knot formation
- Formation of initial aggregates
- Differentiation of the primordium with distinct cap and stipe tissues

- Fruit body maturation by cell extension
- Basidiospore formation

Coprinus cinereus (ink cap) and *Schizophyllum commune* (split gill) are used as model mushrooms to research the developmental processes in basidiomycetes because of their short life cycle and the ease with which they can be controlled in the laboratory;[17] they are also used as their mushroom body comprises different parts, namely a cap peel, a cap devoid of peel, and stipes (Figure 1.1a).

1.2.3 Taxonomy and Classification of Mushrooms/Basic Mycological Concepts

As taxonomy is a system for organizing and labelling living things and traditional mushroom taxonomists group mushrooms according to their outward features, with parts such as gills or pores considered to be conventional. Although this classification is important, it cannot be relied upon, therefore it is possibly outmoded as a method of fungal categorization. Several authors have made attempts to standardize disputes in scholarly journals; however, the application of modern scientific tools and techniques, such as chemical analysis, mating studies, and (most importantly) DNA sequencing, has led to significant shifts in mushroom categorization, prompting Ainsworth & Bisby's Dictionary of the Fungi to replace older approaches as the gold standard.[18] There are many distinct but often overlapping fungal species concepts such as morphological, ecological, phenetic, biological, recognition, evolutionary, genotypic cluster, and phylogenetic species concepts,[19] but disagreements about fungal species complex have prevailed that have led to a difference in the numbers of estimates of fungal species. This is attributed to historic mushrooms studies that have mostly relied on morphological traits, from color, shape, and size of different fruit bodies parts, tissues that produce spores, shapes and sizes of spores, to broad diverse morphological forms and different modes of reproduction and barriers. However, mushroom studies have recently moved towards multiple gene loci DNA sequencing[20] and virulence factors for pathogenic species, mating systems for model fungal species, mycotoxins or secondary metabolic producing fungi. Consequently, there is little known on genetic and environmental factors responsible for the variation in the morphological traits of closely related macro fungi. Moreover, a trait's phenotypic plasticity and environmental sensitivity consequences demand continuous revisions in fungal taxonomy.[21] The inconsistencies that currently exist within fungal taxonomy could be eliminated with the implementation of a genomic species concept, which would also assist in the integration of various fields of fungal biology.

Based on phylogenetic classification, a comprehensive study comprising different study group fungi classified mushrooms as a phylum. Ascomycota and Basidiomycota are included in a distinct dikaryotic subkingdom, *i.e.*, they exhibit a dikaryotic phase that occurs in the middle of their life cycles where two genetically distinct nuclei fuse to form a single, functional one before returning to a diploid state.[1] The members of the phylum, Ascomycota produce sexual spores within a sac, which is distinct from other fungi and release their spores only when the sac wall breaks down, with a few exceptions (*i.e.*, where members of the class Ascomycota simply do not reproduce sexually and do not produce the ascus which characterizes this taxonomic class). Ascomycota is the biggest phylum of fungi with 64 000 species. Taphrinomycotina, Saccharomycotina (Candida and Saccharomyces clades), and Pezizomycotina are the major subphyla (the largest subphylum, which includes Eurotiomycetes, Dothideomycetes, Sordariomycetes, and Leotiomycetes). In addition, Agaricomycotina (mushrooms), Pucciniomycotina (rusts), and Ustilaginomycotina are the three subphyla that make up the phylum Basidiomycota (smuts) where the subphylum Agaricomycotina includes mushrooms that are saprophytes or have a mycorrhizal symbiotic association with plants and other fungi that cause human diseases.[22] Here, Basidiomycota have a different structure called a basidium, where sexual spores are produced and dispersed into the environment (Figure 1.1). A summary on mushroom classification was given in 'Mushroom Taxonomy: The Big Picture', which elucidated and elaborated the entire fungal classification based on the Dictionary of Fungi.[23] When particular temperature, humidity, and nutritional conditions are met, these fungi form fruiting bodies, ascocarps or basidocarps, that are known as mushrooms.[24]

1.2.4 Categorization of Mushrooms

The categories of lamellate fungi that have gills in their fruiting bodies are agaricoid, macrolepiotoid, lentinoid, amanitoid, termitophilous, russuloid, pleurotoid and volvarioid. Non-lamellate mushrooms, *i.e.*, without gills, include chanterelles, false truffles, coral fungi, hedgehog mushrooms, cauliflower fungi, morels, truffles, jelly fungi, boletes, polypores, and a variety of other wild edible mushrooms, whereas Podaxis Desv. and puffballs are examples of mushrooms that belong to the gastromycetoid genus (spores are inside the fruiting body).

1.2.4.1 Ecological Niche[25]

In the wild, mushrooms may be found in many different habitats, namely:

- Aboveground or subterranean macro fungi
- Animal dung – coprophilous;
- Decaying leaf litter – humicolous;
- In symbiotic association with plant roots – mycorrhizal
- In living plants or logs – lignicolous
- In insects – entomogenous

1.2.4.1.1 Subterranean Macroscopic Fungi: Truffles (Phylum Basidiomycota and Ascomycota)

1.2.4.1.1.1 Aboveground Macroscopic Fungi

(a) From *Basidiomycota*:
- Gasteromycetes (puffballs, birdnest, stinkhorn, *etc.*)
- Mushroom-like fungus (with tubes (boletes) and with gills (*Agaricus*, *etc.*))
- Corals, tooth, and club fungi, *etc.*
- Polypores, crust, and jelly fungi on wood

(b) From *Ascomycota*: cup and flask shaped fungi.

1.3 Sexuality in Mushrooms

Fungi are capable of reproducing by asexual and sexual means. The processes of plasmogamy, karyogamy, and meiosis enable sexual reproduction, which comprise four developmental phases.[26] In general, fungi have developed two primary mechanisms that define mating behavior. Some species of fungus are capable of self-fertilization and are referred to be homothallic. Others, however, are incapable of mating using genetically similar cells and are self-sterile. In this latter fungus, known as heterothallic, mating occurs only between suitable partners that vary at the mating-type loci.[27,26] In heterothallic ascomycetes, the identification between female and male gametic cells is dependent on the interaction between pheromones and their associated receptors. The entire sexual reproduction in fungi involves initially establishing sexual identity based on the role pheromone precursor genes present in the specific genomic regions of fungi, *i.e.*, mating-type loci (MAT). These loci are the specialized genomic region that have implications in sex determination (pheromone precursor genes) in fungi.

In *Dikarya*, the fungus have pheromone-based communication systems and mating type-specific sexual pheromones (peptidic nature) enable the fungi to identify mating partners and determine compatibility that facilitates sexual reproduction among fungi that have different evolutionary origins and lifestyles. These mating partners then follow the mechanism of mate recognition, attraction, and mating signaling that leads to polarized cell growth and choosing of a mating partner.[28] In order to fuse and seize the male fertilizing nucleus, trichogynes, which are specialized hyphae, grow toward male gametes when they arise from ascogonial cells (Figure 1.2). These hyphae look like vegetative hyphae, but they are different from vegetative hyphae in two ways, as they may develop while the mycelium is in the stationary growth phase, which is when vegetative hyphal growth normally ceases, and their growth is affected by mating pheromones.[29]

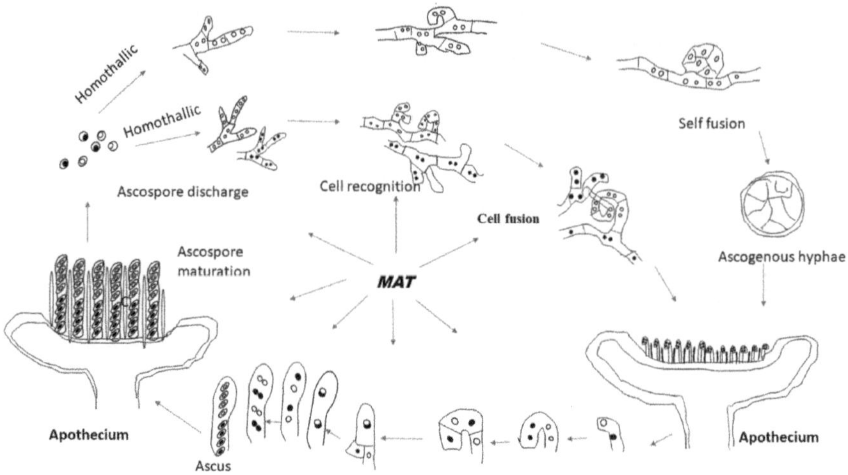

Figure 1.2 Generalized life cycle of filamentous pezizimycotina; homothallic (outer circle) and heterothallic (inner circle).

1.3.1 Plasmogamy

After mating partners have established recognition and attraction, their gametic cells come into contact, attach, and then conduct cell fusion. This allows nuclei that will undergo karyogamy to cluster in a single cell, *i.e.*, a diploid zygote. Plasmogamy needs carefully controlled cellular activities at the right place and time to keep cells whole and stop them from breaking apart during cell–cell fusion.

The absorptive behavior of fungi leads to high intracellular turgor pressure, which hinders plasmogamy. Successful plasmogamy involves synchronization between plasma membrane fusion and cell wall remodeling and fusion. Furthermore, cell wall remodeling takes place between the mating cell, wherein the process starts with the formation of a narrow channel that facilitates plasma membrane fusion. Later, cell wall remodeling for expansion of fusion by cell-wall degrading enzymes concentrated at the fusion site results in a widening of the narrow fusion channel that allows nuclear migration; however, cell integrity is preserved at the fusion site because of confined zones of cell wall degradation at the fusion site.[30] After cell wall dissolution, plasma membrane fusion takes place with the help of the pheromone-regulated membrane protein that is required for plasma membrane fusion during mating. Plasma membrane fusion is necessary for mating, wherein with the help of a protein called pheromone-regulated membrane protein, the fusion of plasma membrane occurs after the cell wall is broken down.[31]

1.3.2 Fruit Body Formation

1.3.2.1 Ascomycota

Sexual reproduction in filamentous ascomycetes (Pezizomycotina) is a complicated developmental process and involves the recognition of cells and their nuclei belonging to opposite mating types known as heterothallic (genetically self-incompatible) leading to outbreeding or genetically self-compatibility known as homothallic (Figure 1.3). These homothallic fungi exhibit hermaphrodite behavior, with a single strain alone having male and female hyphae and accomplishing nuclear fusion. The sexual cycle begins when protoperithecium leads to the formation of specialized hypha, trichogyne (female partner) and formation of conidium, antheridia or hyphal cell (male partner). These trichogyne are large coiled cells and antheridia are conidia-like cells. A male-cell nucleus entering the trichogyne (*i.e.*, fertilization) results in the production of ascogenous hyphae and ultimately fruiting bodies, known as a pseudothecium, perithecium, cleistothecium, apothecium, or gymnothecium, emerge. For instance, these fruit bodies in the case of morel, a highly prized mushroom, also known as *Morchella*, are a cluster of apothecia that have fused together to form a single, expansive structure.

1.3.2.2 Dikaryotic Phase in Ascomycetes

The dikaryotic stage in the majority of mycelial ascomycetes occurs within sexual fruiting bodies and is restricted to particular cell types. During all the stages of development MAT genes play an important role. Inside these fruiting bodies, the apical cells of dikaryotic ascogenous hypha normally differentiate into a hook-like crozier and dikaryotic nucleic simultaneously divide and resulting into multiple cells that result in the penultimate cell, the antepenultimate and the ultimate/apical cells (Figure 1.3). Of these cells, the penultimate reorganizes into a dikaryotic stage by choosing one nuclei from each simultaneously divided dikaryotic nucleus in the previous step, thereby resulting in ultimate/apical cell and antepenultimate uninucleates with each having a different nucleus. The further mechanism involves the production of two cells; one that will undergo meiosis. In most cases, mitosis occurs immediately after meiosis, and as a result, the mature ascus contains eight spores. However, in the antepenultimate and ultimate cell cases, the cells fuse and maintain the dikaryotic phase, which is aided by internuclear recognition facilitated by mating-type gene-dependence.[32]

1.3.2.3 Basidiomycota

Basidiomycota is another mushroom-producing phylum. Basidiomycetes are capable of both sexual and asexual reproduction, producing haploid and

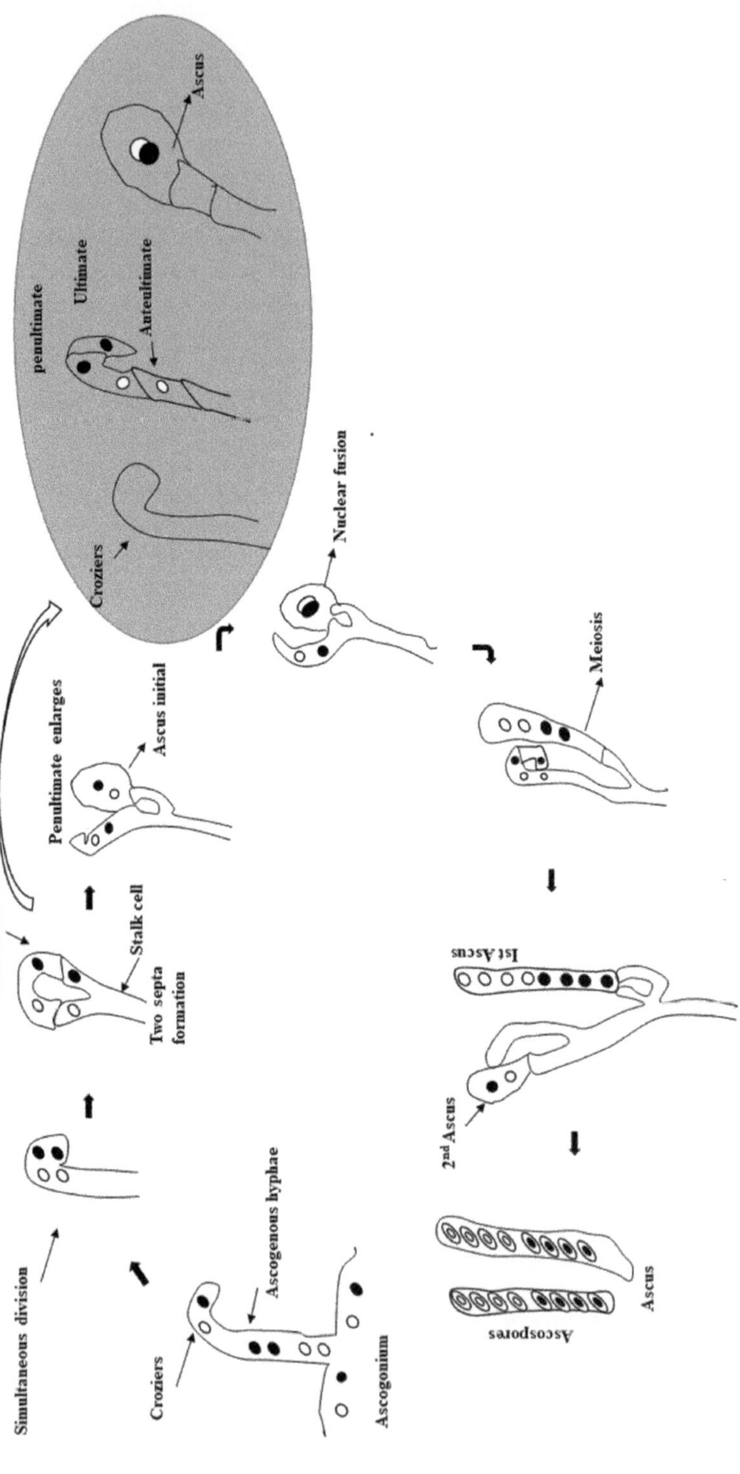

Figure 1.3 Illustration of ascospore formation from crozier (Ascomycota).

diploid mycelia. Mating-type genes control the establishment of the dikaryon and the appearance of fruiting bodies in basidiomycetes. The discovery of the first genes involved in mushroom growth marked the beginning of an understanding of fruiting body formation. The development of mushrooms also requires the participation of hydrophobins, lectins, haemolysins, and expansins.[33] However, the most common method of asexual reproduction utilizes chlamydospores, which are produced by secondary hypha and dissociates. Under optimal conditions these chlamydospores germinate into independent mycelium.

The most common method of asexual reproduction utilizes chlamydospores, which are produced by secondary hypha and dissociates. Under optimal conditions these chlamydospores germinate into independent mycelium.

1.3.2.4 The Dikaryotic Phase in Basidiomycota (Agaricomycotina)

In order to retain dikaryotic cells, basidiomycetes have developed several cellular mechanisms. The technique of forming clamp connections is fundamental to dikaryon formation, wherein in each hyphal compartment clamp connections separate a pair of nuclei that are compatible with one another (of an opposing mating type). The clamp cells grow backwards toward their parent hypha, and a nucleus migrates into the developing cell resulting in dikaryotic mycelium.

The union of hyphae that originate from vegetative monokaryotic haploid mycelia and result in haploid dikaryotic mycelium is typically the starting point for the process of cell anastomosis.[17] Each cell in this mycelium has two haploid nuclei, which will not fuse until formation of the basidium. The secondary mycelia will, in due time, give rise to a basidiocarp, which is a fruiting body that protrudes above the ground. Basidiocarp are large aerial sporophores that bear the developing basidia on the gills under their cap.

1.3.3 Karyogamy and Meiosis

In the basidium, the nuclei of two different mating strains fuse. This process, called karyogamy, creates a diploid zygote, which goes through meiosis in order to create sexual spores that contain basidiospores.

During meiosis, the chromosomes of two different mating partners exchange with each other, a process referred to as recombination, resulting in the formation of haploid nuclei. Fungi sexual development is a rapid process, where haploid nuclei move into four different chambers attached to the basidium and then become basidiospores. Each basidiospore grows into monokaryotic haploid hyphae when it germinates. This kind of mycelium is called a primary mycelium. These spores can be doubled or increased in number by one or more post-meiotic mitoses. Nevertheless, the process of mating in basidiomycetes is one that is rather complicated and asymmetrical in nature.[34] Even without the presence of meiosis, it is possible to produce recombinant genotypes in nature due to the tremendous genetic flexibility that exists therein. The vegetative cells of many fungal species have haploid nuclei that may merge to form diploid nuclei. An alternative to traditional sexual reproduction is the parasexual cycle, which increases genetic diversity and improves reproductive success.

1.4 Mushroom Diversity and Their Benefits

Agaricus bisporus, Flammulina velutipes, Volvariella volvacea, Boletus edulis, Grifola frondosa, Pleurotus spp., and *Lentinus edodes* are some of the most prized mushrooms that can be eaten. Medicinal mushrooms come from the *Ganoderma, Clitocybe, Trametes, Antrodia, Xerocomus, Pleurotus, Phellinus, Agaricus, Russula, Calvatia, Flammulina, Inonotus, Suillus, Inocybe, Lactarius, Funlia, Albatrellus, Cordyceps, Schizophyllum,* and *Fomes genera*. These mushrooms are interesting because they contain unusual compounds that are important in medicine, industry, and/or farming.[35]

Many mushrooms are the result of mycorrhizal or parasitic interactions with forest trees, which are significant both economically and environmentally. The diversity of woody plants is linked to the diversity of mushroom plants. The secondary metabolites of mushrooms, also called biochemicals, are very varied in their chemical composition and the biological processes they stimulate.[36,37] They often sprout at the start of the wet season and contribute much to the health of the forest environment.[38] Many of these mushrooms are extremely important wild-harvested items that have been utilized for culinary, medicinal, and pharmaceutical purposes, and have antioxidant, antibacterial, anti-inflammatory, and anticancer properties that have been studied for decades and continue to be to this day.[39,40] Numerous civilizations eat a variety of wild and cultivated fungi. In India, several types of medicine utilize the medicinal benefits of mushrooms (Figure 1.4). Due to their ecological value in bio-converting industrial and agricultural waste into edible biomass that can be utilised as a functional food or a source of drugs and pharmaceuticals, mushrooms are widely cultivated around the world.[41]

1.5 Cultivation of Mushrooms

Although mushrooms have been gathered from the wild and used by humans as a vital part of their diet and culture since prehistoric times, outdoor growing has become insufficient in meet the needs of the human population. To meet demands, cultivation technology to improve on traditional technology along with new innovative underwater liquid fermentation or bioreactors on solid substrate technology enable the production of large quantities of mycelium as a source of highly valuable protein, as well as secondary metabolites, from agricultural waste.[42] To simplify things, mushrooms may be placed into one of two categories.[6]

1. Mushrooms grown for profit, *i.e.*, commercially cultivated mushrooms
2. Mushrooms collected from the wild, *i.e.*, wild foraged

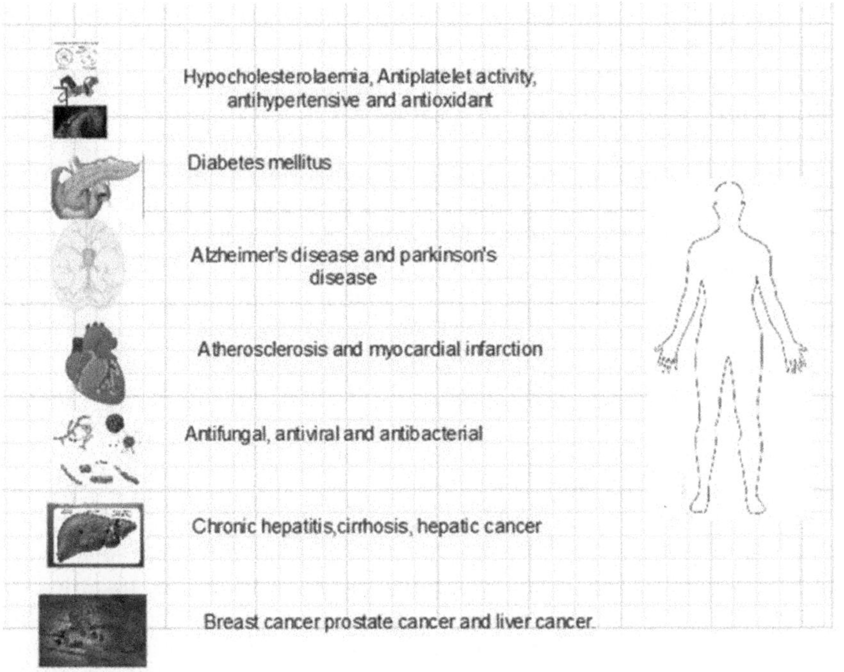

Figure 1.4. The therapeutic benefits of mushroom consumption by humans.

1.6 Common Edible Mushroom of the Himalayas

The Himalayas are a mountain range in Asia covering Bhutan, India, Nepal, China, and Pakistan and separating the lowlands of the Indian subcontinent from the Tibetan Plateau. The area is surrounded by large areas of hilly land and downwards sloping plains. The weather, altitude, topography, and soil of the Himalayas contribute to mushroom development.[43] The entire geographic region of the Himalayas is divided into three regions, *i.e.,* the north western Himalayas, central Himalayas, and the eastern Himalayas. The states of Himachal Pradesh, Uttarakhand, and Jammu and Kashmir, are located in the north-western area, whereas Assam Sikkim, Arunachal Pradesh, and West Bengal are located in the north-eastern region. These states include distinct ecological zones that act as major hotspots for endemism and ectomycorrhizal variety in the Himalayan region.[44]

1.6.1 Agaricus

The genus *Agaricus* comprises the oldest and most abundant basidiomacromycetes, with approximately 434 species worldwide that are reportedly consumed in 29 countries, making them a valuable source of food and medicine. Among these, the most popular species, the commonly named button mushroom, *i.e.*, *Agaricus Bisporus (A. bisporus)*, accounts for over 32% of total mushroom production.[45] *A. bisporus* is a temperate species that can be found in different parts of Europe, the UK, and the US.[46] *Agaricus bitorquis (A. bitorquis)* is a type of mushroom that grows well in subtropical climates but is also grown artificially. This species live in damp places and grow on dead plant matter and soil. Other species, namely, *A. campestris, A. arvensis* and *A. subrufescens* are also commercially cultivated with limited success, whereas *A. xanthodermus* and *A. litoralis* species are poisonous. The other known species in the genus *Agaricus* are: *A. volvatulus, A. comptulus, A. bingensis, A. subperonatus, A. aurantiacus, A. blazei, A. bulbillosus, A. nivescens, A. gennadii, A. purpurellus, A. goossensiae, A. maculatus, A. croceolutescens, A. silvaticus, A. semotus, A. subrutilescens. A. endoxanthus, A. erythrotrichus, A. micromegethus, A. placomyces, A. rodmani, A. sylvicola,* and *A. subedulis,*

1.6.2 Armillaria

The Basidiomycotan genus *Armillaria* is responsible for a wide variety of harmful plant diseases. White rot may be caused by *Armillaria* species on decaying wooden surfaces. Moreover, most species are parasitic, regarded as facultative necrotrophs. This genus comprises roughly 50–60 species that are usually called shoe string root-rot fungi or honey mushrooms, with most species of these fungi located in eastern Asia.[47] There have been different biological species reported from Europe (7), North America (10), Africa (5), Australia (5), and Asia (19); however, many remain unknown. Asia is recognized as the home of *Armillaria* biological species. Based on systematic DNA sequences, it is suggested that greater clarity prevails because of numerous variables that influence variety and speciation throughout biogeographic regions;[48] nonetheless, species overlap exists throughout geographic locations.

Armillaria s.s. and *Desarmillaria* are two species that belong to the basidiomycete genus *Armillaria* s.l. This genus has been reported to have different species on both the northern and southern hemispheres based on the distant clads in their phylogenetic analysis.[49] Common branches in the phylogeny suggest that some species are linked through shared characteristics, potentially from the human transmission of contaminated plant material.[50] *Armillaria mellea* (*A. mellea*) populations from several places, including Asia, western North America, eastern North America, and Europe, are genetically separated and classified into four distinct groups, which hints that they may be going through speciation.[51] This species is found across the globe and is critical to the dynamics of various woody habitats, including natural forests, tree plantations for lumber production, orchards, vineyards, and gardens. Diversity analyses conducted based on the translation elongation factor 1 gene, the ribosomal IGS- gene, and the ITS region have been used to consistently identify different species. In particular, based on the above technique, *A. mellea* is highly variable and yet phylogenetically distinct from other *Armillaria* species.

1.6.3 Flammulina

From late fall to early spring, this wood-destroying fungus may be seen growing on the trunks or stumps of broad-leaved trees, including aspens, willows, elms, and others. The fungus fructifies during winters and is commonly referred as a winter mushroom and reported to contain immunomodulatory, antitumor, and antibiotic substances.[52] It is found in China, Siberia, Asia Minor, Europe, Africa, North America, Australia, and in Japan, where the *Flammulina. velutipes (F. velutipes)* species has been harvested from the wild and consumed for millennia, but is currently produced under artificial conditions. This practice is gaining in popularity in Taiwan and Europe.[53] Today, it is the fifth most cultivated mushroom in the world, with 2.1 million tons produced annually.[54]

In India, *F. velutipes* has been reported in Jammu and Kashmir and the north eastern and north western Himalayas.[55] As a whole, the genus *Flammulina*, which includes winter mushrooms, has not had its phylogeny or species delimitations well resolved. In one study, at least four species, *F. rossica*, *Flammulina* sp., *F. velutipes* and *F. yunnanensis*, were collected and reported on in China.[56] However, one study has reported that all *F. velutipes* strains in east Asia are actually *F. filiformis* and a new species, *Flammulina finlandica*, was found in northern Europe.[57] Also, high genetic diversity among wild collections of *F. velutipes* has been reported in China. Core collections of Chinese isolates exhibiting high genetic diversity and other east Asian isolates were studied using SSR analysis.[58,59]

1.6.4 Fomes

Fomes fomentarius (L.) Fr. (syn. *Agaricus fomentarius* (L.) Lam., *Polyporus fomentarius* (L.: Fr.) Fr., Polyporaceae) can sometimes be referred to as tinder fungus, hoof fungus (as it is shaped like a horse's hoof), tinder conk, and tinder polypore.[60] Its cotton-like flesh is commonly known as "amadou" (tinder), and it has a variety of functions, including making fire, styptic treatment, clothing, fishing gear, ornamental uses, and medicines. *F. fomentarius* prefers a distinct kind of host called a polyphagus and produces big polypore fruit bodies. Fungus thrives on trees throughout Europe, Asia, and North America's temperate and boreal regions. Three widely distributed species of *F. fomentarius* have been recorded in the UK, as well as Himachal Pradesh, Jammu and Kashmir, Nagaland, and Meghalayan, states of India[61] Additionally, reports of its distribution have come from Bhutan (Thimphu), Nepal (Pokhra), Himachal Pradesh (Solan, Shimla), the Uttarakhand (Kumaun), Meghalaya (Khasi Hills), and West Bengal (Darjeeling), but *Fomes extensus* is exclusively found in Nagaland. Recently, the species has been reported to be non-homogeneous, comprising three distinct lineages/sublineages among the *F. fomentarius* strains based on internal transcribed spacer (ITS) sequence variability.[62] Three species of the genus Fomes, *Fomes fomentarius, Fomes extensus*, and *Fomes johnsonianus*, have been found in India. *Fomes fomentarius* can be found in Himachal Pradesh, Uttrakhand, and Jammu and Kashmir, whereas *Fomes johnsonianus* and *F. extensus* have been reported only in Himachal Pradesh and Uttarakhand. Also *Fomes fomentarius* is found in Bhutan, West Bengal, and Meghalaya, and *F. extensus* has been reported in Arunachal Pradesh and Meghalaya.[61]

1.6.5 Ganoderma

Karsten introduced the genus *Ganoderma* in 1881 with the species Ganoderma (*G.*) *lucidum* (Curtis). It is a saprophytic and pathogenic fungus that produces white rot on trees like oil palm, coconut, rubber, *Prosopis cineraria*, betelnut, tea, and *Acacia* species. Despite being pathogenic, their fruiting bodies are utilised in traditional medicine in China, Japan, and Korea, with several species having been described worldwide.[63] However, *Ganoderma* complex's chaotic taxonomy limits its further usage. Samples from Asia, Europe, and North America were evaluated morphologically and molecularly (ITS, tef1a, rpb1 and rpb2) and a multilocus phylogeny yielded 13 complicated species. Clade A was determined to include *G. flexipes, G. curtisii, G. multipileum, G. lingzhi, G. resinaceum, G. sichuanense, G. sessile,* and *G. tropicum*. Clade B was found to include *G. oregonense, G. lucidum,* and *G. tsugae*. Clade C *G. boninense* and *Ganoderma zonatum*.[64]

1.6.6 Lentinula

Lentinus edodes is the second most widely farmed fungus in the world. When the weather is warm and damp, *Lentinula* are found in groups and on the dead wood of broadleaf deciduous trees, particularly those in the family Fagaceae, such as maple, chestnut, sweet gum, beech, oak, alder, poplar, mulberry, and hornbeam.[65] In India, New Zealand, Bhutan, and Nepal, as well as in Indochina, China, Borneo, and New Guinea, shiitake mushrooms naturally grow on broad-leaved trees, often Fagaceae.[66] Genetic markers are used to differentiate between *Lentinula edodes, Lentinus squarrosulus, L. sajor-caju, L. tigrinus,* and one variety of *L. squarrosulus* in India (ITS, 18S rRNA and 28S rRNA genes in Nagaland, a north eastern state of India).[67]

1.6.7 Macrolepiota

Singer (Parasol Mushroom) is a basidiomycete fungus with a huge, visible fruiting body, an excellent flavor, a nice smell, a soft texture, and a slight nutty aroma originating from the cap. It is a good source of protein, vitamins, minerals, and fiber and has health benefits that make it a useful medicinal ingredient.[68] The stem of *Macrolepiota (M.)* is very tough and fibrous, so only the cap is consumed. This macrofungus has fruiting bodies that can be fried and eaten. The diameter of the cap ranges from 11 cm to 40 cm. It starts out round, but when matured, it opens up with a greyish-brown cap that is convex in the middle.[69] *M. procera* grows alone, in small groups, or grassy areas and mixed woods clumps on the ground,[70] and its fruiting bodies appear on the surface of soil in late summer or autumn in forests, meadows, parks, and gardens. The Indian subcontinent typically consumes the species of this genus, such as *M. dolichaula, M. procera*, and *M. rhacodes*, which are recognized for their edibility.[71] This mushroom has been recorded as being located in the Puliebzie forest range in Nagaland's Kohima District[72] and Assam.[69] Based on molecular and morphological studies, there is much species diversity in the genus *Macrolepiota*. Using ITS-based sequences, six distinct species have been determined—*M. dolichaula, M. detersa, M. orientiexcoriata, M. mastoidea, M. procera*, and *M. velosa*.[73] Moreover, the genus *Macrolepiota* has been classified into two major clades based on two (ITS and LSU rDNA) and morphological features. The first clade includes *M. Procera, M. mastoidea, M. clelandii*, and the other *M. globose* and *M. rachodes*.[74]

1.6.8 Stropharia

Stropharia is a member of the *Strophariaceae* family, which is found worldwide in northern temperature zones. This mushroom may be found in broadleaf woodlands in lawns or forests in solitary or gregarious forms, on fallen leaves or on heavily decayed wood as well as in certain cultivated places.[75] The fresh fruit bodies of *Stropharia (S.) rugosoannulata* are edible and may be farmed for use as a food source.[76] This delicious fungus is commonly farmed in many regions of the globe and has gained popularity in recent years (Delcaire, 1978). It has been recorded to be located most prevalently on the continent of Europe (519 cases), followed by the continents of North America (39), South America (2), Asia (4), and Oceania (1). However, there have been no reports of this fungus in Africa or Antarctica.[77]

In 1985, the fungus was detected for the first time in the Solan district of Himachal Pradesh (India); its crown is dark crimson to dark reddish-brown to wine red that fades to tan, greyish gills that mature to black, purple-black or black spores, and a grooved, segmented annulus.[75] Only 11 species of the genus *Stropharia* have been found in India.[78]

1.6.9 Volvariella

Volvariella is a member of the family *Pluteaceae* and phylum *Basidiomycota*, with a popular species being *Volvariella (V.) bombycina*, which is cultivated commercially and has a mild and pleasant taste.[79] *V. bombycina* has a high economic value worldwide and commercial production potential because of its high protein content and significant therapeutic benefits.[79]

The genus *Volvariella* has pink lamellae (whitish that become pink) and no annulus on its fruiting body stipe.[80] It prefers to grow in lignicolous environments, *i.e.*, rotting wood trunks or logs (maple, beech, elm, mango, *etc.*), leaf debris, rich agricultural soil or in coffee and palm plantations.[81]

The species has been reported to be located on all of the continents, except for Antarctica.[79] *V. bombycina* is a widespread but dispersed species in Europe, with few countries having placed this mushroom on red lists of fungi or referred to them as being in the highest category of endangered fungi. In India, total 28 taxa of *Volvariella* are known, where *V. bakeri, V. bombycina, V. diplasia, V. terastia, V. volvacea* var. *nigricans,* and *Volvopluteus gloiocephalus* have been reported to be edible taxa and the remaining taxa are known to form putative ectomycorrhizal associations.[81] *V. bombycina* shares the same lifecycle, hymenial cell type, cystidia, and basidia as *V. volvacea*. However, unlike *V. volvacea*, the mycelial development and fruiting of *V. bombycina* occur at 26–30 °C.[79] *V. bombycina* has been reported to grow under diverse environmental conditions, from the north western Himalayas on *Picea smithiana* stems[82] to Kerala on dead mango tree stumps[83] to humic soil in the Punjab and grasses in many different places.[81]

1.6.10 Aminata

Approximately 83 edible species have been documented in 31 countries, with *Aminata (A.) caesarea* being one of the most sought-after mushrooms in the world.[6] It is essential to correctly identify a target mushroom before eating it as some species have extremely high levels of toxicity, such as the *A. phalloides* plant, sometimes referred to as the "death cap" due to the large number of reported fatalities caused by its consumption.[84] More than half the instances of mushroom poisoning derive from members of the family *Amanitaceae* as a result of their mistaken identification as edible mushrooms, such as *Amanita, Galerina*, and *Lepiota. Amanitopsis, Amanita, Catatrama, Amarrendia, Saproamanita, Limacella,* and *Torrendia* are the eight genera that make up this family. The Indian Himalayan state of Uttarakhand has the highest species diversity (27), followed by Himachal Pradesh (24) and then Jammu and Kashmir (5) and Meghalaya (5).[85] Some of the edible species consumed worldwide are: *A. argentea, A. aurea, A. caesarea, A. bingensis, A. calyptratoides, A. calopus, A. calyptroderma, A. craseoderma, A. ceciliae, A. crassiconus, A. flammeola, A. crocea, A. gemmate, A. flavoconia, A. hemibapha, A. goossensiae, A. hovae, A. loosii, A. inaurata, A. muscaria, A. masasiensis, A. perphaea, A. robusta, A. rubescens, A. subviscosa, A. strobilaceovolvata, A. spissa, A. tuza, A. virgineoides, A. fulva, A. vaginata, A. xanthogala, A. umbonata*, and *A. zambiana*.[86,87] Recently, In the Himalayan state of Himachal Pradesh, *A. shorea* was found, while *A. chepangiana* has been reported as being located in Uttarakhand.[85]

1.6.11 Astraeus

The genus *Astraeus,* a Basidiomycota member fungi belonging to *Diplocystidiaceae* comprises highly-prized edible mushrooms with worldwide distribution found in the forests of tropical, sub-tropical and temperate regions in sandy/gravel/pebble-rich lateritic soils.[88–91] This gasteroid fungus (spores in the basidiocarps) grows slightly below the earth, and its fruiting bodies divide into star-like shapes. "Earthstar Fungi" are a delicacy with a high market value. This fungus is ectomycorrhizal and colonizes a wide variety of forest tree species such as the *Dipterocarpaceae, Fagaceae*, and *Pinaceae* families.[92] Early basidiomata of *A. hygrometricus* are frequently and commonly found in mycorrhizal associations with oak, oak–pine, oak–juniper–pine, low deciduous, subtropical scrub, and gallery forests.[93] To date, up to 10 species have been described worldwide: *Astraeus asiaticus, A. hygrometricus, A. koreanus, A. morganii, A. odoratus, A. pteridis, A. sirindhorniae, A. smithii, A. telleriae* and *A. thailandicus*.[94] *Astraeus* spp. are widespread in the Indian subcontinent (Himachal Pradesh, Jharkhand, Karnataka, Kerala, Orissa, Punjab, Uttarakhand, Uttar Pradesh, and West Bengal) in Southern India.[91]

1.6.12 Cantharellus

The *Cantharellus* genus, is a Basidiomycota that has a wide global distribution and is well known for its mycorrhizal properties, medicinal values, and edibility, with no toxic species yet known. To date, 22 edible species have been documented from boreal, temperate, and tropical forests all over the world.[95] Species of *Cantharellus* have been found to form ectomycorrhizal associations with a wide range of economically important host trees. Clarity on species diversity in this genus is evolving, with attempts having been made to resolve the *Cantharellus* species complex based on molecular phylogeny delimitation genera in six subgenera and there having been estimated to be 200 to 300 species worldwide. Moreover, clear distinguishing between *Craterellus* and *Cantharellus* has been established.[96,97]

Cantharellus (C.) cibarius, commonly referred to as yellow/golden chanterelles, is rich in antioxidants and this diverse species is noted for its delicious, apricot-like odor. The forests of the northern Himalayas are diverse and contain many families of ectomycorrhizal trees such as *Betulaceae, Fagaceae*, and *Pinaceae*. In India, *C. appalachiensis, C. cibarius, C. cinnabarinus, C. friesii, C. lateritius, C. lilloa, C. luteo, C. comus*, and *C. minor* are the species in this genus that have been documented as being located in India.[98] In another study, 13 species were reported overall in the north western Himalayas: *C. cibarius, C. appalachiensis, C. lateritius, C. minor, C. miniatescens Heinem, C. pseudoformosus*, with seven species, *C. applanatus, C. fibrillosus, C. elongatipes, C. indicus, C. himalayensis, C. umbonatus, C. natarajanii*[99] and *C. pseudoformosus* found in mixed evergreen forests under *Cedrus deodara*[100] that were new to scientists. Some edible species of this genus that are located across the world are: *C. cinereus, C. cibarius, C. congolensis, C. cinnabarinus, C. cyanoxanthus, C. cyanescens, C. eucalyptorum, C. densifolius, C. floccosus, C. formosus, C. floridulus, C. ignicolor, C. infundibuliformis, C. incarnatus, C. isabellinus, C. luteocomus, C. longisporus, C. luteopunctatus, C. madagascariensis, C. lutescens, C. miniatescens, C. odoratus, C. minor, C. platyphyllus, C. pseudofriesii, C. platyphyllus, C. pseudocibarius, C. rufopunctatus, C. ruber, C. splendens, C. subalbidus, C. symoensii, C. tenuis, C. tubaeformis*, and *C. subcibarius*.[101,102]

1.6.13 Craterellus

Craterelles or "horns of plenty" (*Craterellus*) are the thinner cousins of true chanterelles (*Cantharellus*), and most scientists place both genera in the same order (*Cantharellales*) and family (*Cantharellaceae*), where molecular phylogenetics data have been used to resolve taxonomic chaos,[96,103] These fungi are ectomycorrhizal in a small genus where only 20 species are known worldwide and the 140 total taxa have been reported[88] (http://www.index-fungorum.org). Although about nine species of *Craterellus* have been documented, this numbers is bound to increase with further molecular phylogenetic studies[103] as morphological characteristics alone are not sufficient to be able to differentiate between *Craterellus, Cantharellus* and *Pseudocraterellus* despite good knowledge of their hyphal anatomy, fruitbody development and pigment structure. Recently, two new species from the state of Kerala in India were identified as two unusual Craterellus species. The species from India include *C. inusitatus, C. shoreae, C. ignicolor, C. mussooriensis, C. cymatodermoides, C. albostrigosu, C. dubius, C. indicus, C. cornucopioides* var. *mediosporus, C. cinereus,* and *C. parvogriseus.*[104]

1.6.14 Leucopaxillus

Genus *Leucopaxillus*, a Basidiomycota commonly known as giant Leucopax/giant funnel is a saprobic fungus species belonging to *Tricholomataceae* and is a highly polyphyletic genus. It features a white or light cream crown and a funnel-shaped, gill-lined stem. Different species in the genus include: *L. albissimus, L. alboalutaceus, L. amarus, L. cerealis, L. eucalyptorum, L. gentianeus, L. giganteus, L. gracillimus, L. laterarius, L. lilacinus, L. cf. masakanus, L. mirabilis, L. monticola, L. nauseosodulcis, L. paradoxus, L. pulcherrimus,* and *L. tricolor*. These globally distributed fungi are delicious when young and grows in clusters or rings in forest clearings, grassy meadows, and roadside hedges. Moreover, they are known to possess several bioactive compounds with pharmaceutical possibilities, especially antitumor, antibacterial, antioxidative, and DNA-protective properties.[105,106]

1.6.15 Russula

The genus *Russula,* which belongs to the phylum Basidiomycota, class Agaricomycetes, the order *Russulales* and the family *Russulaceae* comprises 1100 species worldwide.[88] These are found in ectomycorrhizal association with trees of *Fagaceae, Leguminosae, Cistaceae, Salicaceae, Betulaceae, Dipterocarpaceae, Nothofagaceae, Pinaceae,* and *Myrtaceae.* These fungi are found in tropical to arctic zones and are the dominant ectomycorrhizal genus in the Indian Himalayas[107,108] and represent 158 taxa in India.[109] Over the last two centuries, various infrageneric classification systems have been suggested for the genus *Russula,* which has had a rich taxonomic history. Eight subgenera have been identified in this genus, according to recent molecular phylogenetic research based on a globally representative group. *Russula* is a mycorrhizal fungus, making it highly diverse and very difficult to cultivate. There have been reports of close to 128 edible *Russula* species from 28 different countries.[6] Some of the edible species include: *Russula (R.) atrovirens, R. alutacea, R. atropurpurea, R. cyclosperma, R. chamaeleontina, R. cyanoxantha, R. compressa, R. cellulata, R. congoana, R. diffusa* var. *diffusa, R. depallens, R. delica, R. emetic, R. erythropus, R. lepida, R. grisea, R. meleagris, R. hiemisilvae, R. minutula, R. olivacea, R. oleifera, R. pectinata, R. phaeocephala, R. pseudopurpurea, R. pseudostriatoviridis, R. roseostriata, R. roseoalba, R. rubra, R. striatoviridis, R. sesenagula, R. testacea, R. virescens, R. vesca, R. xerampelina*, and *R. viscida.*

1.6.16 Termitomycetes

Termitomyces heimii, a Basidiomycota *Lyophyllaceae* with gills are found in Malaysia, Thailand, Pakistan, India, and equatorial Africa.[110] There are a total of thirty different species that belong to the genus *Termitomyces*, all of which were included in the tenth edition of the Dictionary of Fungi.[88] This fungus is known to form pseudorhiza (pseudo root), which extend deep into the soil where the transport of spores is carried out by termites belonging to the order Isoptera. The fungus grows gregariously after heavy rains and it has a symbiotic relationship with termites. It is usually found near cocoa, oil palm and rubber plantation areas where termite mounds are present. It is recognized to be a revered genus with excellent nutritional benefits, such as *Termitomyces (T.) clypeatus,* which contains a substantial amount of nutrients.[111] *T. titanicus* is the world's largest edible species, with around 27 edible species recorded from over 35 countries.[6] They are more common in Africa and Asia, although their use is still inadequately documented. *T. clypeatus* is gathered and sold at local marketplaces in Tibet, Nepal, and northern India. Prominant edible species include: *T. aurantiacus, T. albuminosus, T. cylindricus, T. clypeatus, T. entolomoides, T. eurrhizus, T. globulus, T. fuliginosus, T. heimii, T. mammiformis, T. letestui, T. microcarpus, T. medius,* and *T. schim T. robustus.*[110] Two-thirds of the *Termitomyces* species found globally are found in six Western Ghats states and on the west coast of India, accounting for 19 species.[112]

1.6.17 Morchella

True morels (*Morchella* spp., *Morchellaceae*, Ascomycota) are a highly sought-after delicacy with significant commercial and scientific importance.[88] With the help of molecular phylogenetic studies, it has been found that there are at least 50 species of *Morchella* around the world.[113] A high degree of continental endemism has been observed within this genus; they are different and can be found everywhere. They live on land and can only be found in coniferous forests. Moreover, they live in the leaf litter of *Cedrus deodara, Cupressus torulosa, Picea smithiana*, and *Abies pindrow*, as well as in apple orchards.[114] Several north-western Himalayan yellow and black morel species have been discovered, *i.e., M. crassipes, M. spongiola* and *M. elata, M. angusticeps* and *M. gigas*.[115] In the north-western Himalayas, the species *M. esculenta* is predominant, *M. deliciosa* is predominant in Uttarakhand, and *M. elata* and *M. conica* are predominant in Himachal Pradesh. Also, *M. angusticeps, M. conica, M. elata, M. esculenta, M. hybrida, M. crassipes, M. rotunda, M. tomentosa*, and *M. semilibera* are often encountered in the forests of the Himalayas. Other species, namely *M. esculenta, M. angusticeps*, and *M. conica*, are in found mycorrhizal association with ferns, strawberry, and grasses.[116] Some of the edible species include: *M. conica, M. angusticeps, M. costata, M. conica* var. *rigida, M. deliciosa, M. crassipes M. elata, M. esculenta* var. *rotunda, M esculenta, M. intermedia, M. esculenta* var. *umbrina*, and *M. esculenta* var. *vulgaris*.[117]

1.6.18 Cordyceps

The *Cordyceps* genus is an ascomycota member of the family *Clavicipitaceae*, wherein all species are endoparasitoids to insects and arthropods. There are around 750 identified species that are globally distributed in Asia, Europe, and North America in tropical and humid temperate forests. However, there are only two species that are being researched and have received the greatest attention, namely *C. sinensis* and *C. militaris*. In the Himalayan region, *C. sinenses* is found on the Indian–Tibetan plateau, where it is collected by locals in alpine areas only in the months of May and July.[118] A total of thirty-five species of *Cordyceps* are reported to have medicinal value and contain valuable bioactive compounds. *C. sinenses* is amongst the most popular species used in Chinese–Tibetan medicine. This endoparasitic fungus is a rich source of cyclodepsipeptides, nucleosides and polysaccharides that exhibit high levels of immunomodulatory activity, as well as antioxidant, anticancer, cytotoxic, anti-inflammatory, anti-diabetic, anti-allergic, anti-HIV, antimalarial, analgesic, antifungal and antibacterial properties. *Cordyceps* species lie dormant in the ground until such time as they come into contact with a suitable host. They also exhibit morphological variety, in addition to cultivability and ecological relevance in the context of arthropod pathogens and fungi of the genus *Elaphomyces*.[119]

1.7 Himalayan Sustainability

It is clear from the history of human civilization that humans have relied on forests to provide food, fiber, and other necessities for survival. Melophilia has been the norm in human culture since the beginning of our existence. In ancient times, people from Greece, China, India, and Iran used mushrooms; there is evidence of India's entheomycological traditions, in which fungi are used in religious rituals to elicit spiritual experiences.[38,120] An increasing number of industrialized nations are turning to traditional medical practices such as Ayurveda, Chinese medicine, Unani, and other types of native medicine. This has prompted the World Health Organization (WHO) to publish a study on ethnomedicine.[121] Currently, the tribes and local people of India use 100 species of macrofungi belonging to 56 genera to treat common illnesses, serious diseases, personal ailments, and as herbal medicines.[122] Ethnic Indian groups have maintained ethnomedicinal techniques *via* religious and cultural traditions.[123]

Due to the rising economic prospects of nutraceuticals and food supplements, their consumption must be limited, and more research into cultivation technology and rationalization for sustainability is the need of the hour. Due to finite natural resources, sustainability is crucial for the human population, health, and the environment. The Food and Agriculture Organization of the United Nations (FAO) advocates the sustainable use of macrofungi for forest management, biodiversity protection, revenue creation, and food security. Conserving biodiversity ensures the survival of mushroom species and provides essential genes for the direct benefit and sustainability of human society. In this regard, in many areas, the commercial harvesting of mushrooms for consumption on a large scale has developed into an industry, and this trend is likely to continue in the foreseeable future. This raises a number of challenges, including preservation, ownership, and the long-term viability of supply sources. Effective and comprehensive management of the available forest resources seems to be the most important response.[124] Therefore, to be viable commercial harvesting requires the integration of conservation with mushroom and truffle hunting.[125] Reviews of the literature seem to demonstrate our ignorance rather than our proficiency on techniques used for cultivating wild Himalayan mushrooms. Therefore, emphasis must be placed on determining the causes of the loss of these wild mushrooms and finding strategies to safeguard them. The reasons can be attributed to:

1.7.1 Deforestation

The human destruction of natural habitats is jeopardizing the survival of specialized species that local mushroom hunters gather from dispersed places in mountainous settings. The main factor contributing to deforestation is the conversion of forest area to farmland and land for urban usage.[126] The survival of macrofungi is also seriously threatened by natural occurrences, such as floods, forest fires, or the invasion of the forest environment by parasites that kill trees.[125] A sustainable mushroom economy requires protecting the natural ecosystem; macroflora diversity must be protected.

1.7.2 Overexploitation

Many indigenous cultures across the globe rely on wild mushrooms for food and revenue. As these natural products are frequently prized for their rarity or unique appeal, such as flavor or taste, or both, wild-harvested mushrooms, including truffles with a unit price greater than that of gold, have enhanced the human appetite for economic profit.[125] In many tropical and temperate woods, they represent the main source of revenue.

1.7.3 Urbanization

Increased demand for lumber has led to the fast growth in the forestry industry, which has devastated many of the natural habitats of fungi, unbalancing forest economics and ecology. The contribution that mushrooms make to local lives and the environment of forests is undervalued; their impact is significant on economic, social, and cultural levels.

1.7.4 Lack of Awareness with Diminishing Ethnomedicine Knowledge

There are still a lot of indigenous species and many unchartered locations with high geographic variation that need to be investigated.[127] The eight states of Arunachal Pradesh, Manipur, Assam, Meghalaya, Nagaland, Mizoram, Tripura, and Sikkim in the north eastern Himalayan region of India are biodiversity hotspots, with little known about the fungi from these areas.[128] Despite this, there are records of wild plants being used for medicinal purposes by various ethnic groups.

1.7.5 Cultivation Technology: Geo Tagging and Restricted Usage

More research input is required for the identification of cultivation technology for wild mushrooms, with the adoption of a strategic approach for mass awareness about wild edible fungal flora and areas. Areas with high diversity should be geotagged and the restriction of mushroom hunting should give significant impetus for the sustained use of mushrooms. Recent research has shown various synergistic interactions between ecological and agricultural systems, and sustainable practices by farmers, farming communities, and agribusinesses have been extensively documented, which can play an essential part in sustainability.

1.8 Conclusions

More in-depth knowledge on ectomycorrhizal diversity is required, which is being negatively impacted by factors such as rising global temperatures, intensifying global trade, unsustainable land management practices, deforestation, and the spread of forest fires. These factors pose significant risks, and as a result, they require a heightened level of care and focus for the purposes of conservation and restoration in certain areas. There is also increasing concern to conserve the germplasm collection of important edible and poisonous mushrooms along with their chemical study in order to determine their properties. These properties hold promise for the treatment of various metabolic and viral disorders. Harvesting mushrooms can be a great way to tap into their vast potential while being mindful of our environment. By implementing eco-friendly methods and laws for their collection, we can reap numerous benefits and ensure a sustainable future for our planet.

References

1. D. S. Hibbett, M. Binder, J. F. Bischoff, M. Blackwell, P. F. Cannon, O. E. Eriksson, S. Huhndorf, T. James, P. M. Kirk and R. Lücking, *Mycol. Res.*, 2007, **111**, 509–547.
2. B. A. Wani, R. H. Bodha and A. H. Wani, *J. Med. Plants Res.*, 2010, **4**, 2598–2604.
3. D. L. Hawksworth, *Mycol. Res.*, 2001, **105**, 1422–1432.
4. G. M. Mueller, J. P. Schmit, P. R. Leacock, B. Buyck, J. Cifuentes, D. E. Desjardin, R. E. Halling, K. Hjortstam, T. Iturriaga, K.-H. Larsson, D. J. Lodge, T. W. May, D. Minter, M. Rajchenberg, S. A. Redhead, L. Ryvarden, J. M. Trappe, R. Watling and Q. Wu, *Biodiversity Conserv.*, 2007, **16**, 37–48.
5. H. L. Li, Y. Tian, N. Menolli, L. Ye, S. C. Karunarathna, J. Perez-Moreno, M. M. Rahman, M. H. Rashid, P. Phengsintham, L. Rizal, T. Kasuya, Y. W. Lim, A. K. Dutta, A. N. Khalid, L. T. Huyen, M. P. Balolong, G. Baruah, S. Madawala, N. Thongklang, K. D. Hyde, P. M. Kirk, J. C. Xu, J. Sheng, E. Boa and P. E. Mortimer, *Compr. Rev. Food Sci. Food Saf.*, 2021, **20**, 1982–2014.
6. E. R. Boa, *Wild edible fungi: a global overview of their use and importance to people*, Publishing Management Service, Rome, Italy, 2004.
7. S.-T. Chang and P. G. Miles, *Mushrooms: cultivation, nutritional value, medicinal effect, and environmental impact*, CRC Press, 2004.
8. R. H. Kurtzman, Jr., *Mycoscience*, 1997, **38**, 247–253.
9. M. P. Thakur, *Indian Phytopathol.*, 2020, **73**, 377–395.
10. C. Barroetaveña, and C. V. Toledo, *Wild Plants, Mushrooms and Nuts: Functional Food Properties and Applications*, 2016, pp. 65–81.
11. A. B. Cunningham and X. Yang, *Mushrooms in Forests and Woodlands: Resource Management, Values and Local Livelihoods*, Routledge, 2012.
12. C. P. Kala, *Int. J. Sustainable Dev. World Ecol.*, 2004, **11**, 205–210.
13. H. Rathore, S. Prasad and S. J. Sharma, *PharmaNutrition*, 2017, **5**, 35–46.
14. V. Ferraro, G. Venturella, L. Pecoraro, W. Gao and M. L. Gargano, *Plant Biosyst.*, 2022, **156**, 130–142.
15. E. S. Dias, and M. R. de Brito, *Edible and Medicinal Mushrooms: Technology and Applications,* 2017, pp. 15–33.
16. M. Nishihara, A. Watanabe and Y. J. Asada, *Mycoscience*, 2007, **48**, 109–116.
17. U. Kües and Y. Liu, *Appl. Microbiol. Biotechnol.*, 2000, **54**, 141–152.
18. G. C. Ainsworth, *Ainsworth & Bisby's dictionary of the fungi*, Cabi, 2008.

19. F. E. Zachos, *J. Genet.*, 2018, **97**, 811–815.
20. G. Wu, Y.-C. Li, X.-T. Zhu, K. Zhao, L.-H. Han, Y.-Y. Cui, F. Li, J.-P. Xu and Z. L. Yang, *Fungal Divers.*, 2016, **81**, 25–188.
21. J. Xu, *Genome*, 2020, **63**, 459–468.
22. M. A. Coelho, G. Bakkeren, S. Sun, M. E. Hood and T. Giraud, *Microbiol. Spectr.*, 2017, **5**(3), DOI: 10.1128/microbiolspec.FUNK-0046-2016.
23. M. Kuo, The Genus Scleroderma, 2015, http://www.mushroomexpert.com/scleroderma.html.
24. V. E. Rubatzky, and M. Yamaguchi, in *World Vegetables: Principles, Production, and Nutritive Values*, ed. V. E. Rubatzky, and M. Yamaguchi, Springer US, Boston, MA, 1997, pp. 723–744.
25. R. Singer, *The Agaricales in Modern Taxonomy*, Koeltz Scientific Books, Koenigstein, Germany, 1986.
26. L. Peraza-Reyes and F. Malagnac, in *Growth, Differentiation and Sexuality*, ed. J. Wendland, Springer International Publishing, Cham, 2016, pp. 407–455.
27. S. Pöggeler, M. Nowrousian, I. Teichert, A. Beier and U. Kück, in *Physiology and Genetics: Selected Basic and Applied Aspects*, ed. T. Anke and A. Schüffler, Springer International Publishing, Cham, 2018, pp. 1–56.
28. J. Wendland, *Growth, differentiation and sexuality*, Springer, 2016.
29. R. Debuchy, V. Berteaux-Lecellier, and P. C. Silar, *Cellular and Molecular Biology of Filamentous Fungi*, 2010, pp. 499–535.
30. O. Dudin, F. O. Bendezú, R. Groux, T. Laroche, A. Seitz and S. G. Martin, *J. Cell Biol.*, 2015, **208**, 897–911.
31. M. G. Heiman and P. Walter, *J. Cell Biol.*, 2000, **151**, 719–730.
32. N. D. Read and A. Beckett, *Mycol. Res.*, 1996, **100**, 1281–1314.
33. J. F. Pelkmans, L. G. Lugones and H. A. Wösten, in *Growth, differentiation and sexuality*, Springer, 2016, pp. 387–405.
34. B. P. Nieuwenhuis, S. Billiard, S. Vuilleumier, E. Petit, M. E. Hood and T. Giraud, *Heredity*, 2013, **111**, 445–455.
35. A. Dasgupta, S. Paloi and K. Acharya, *Res. J. Pharm., Biol. Chem. Sci.*, 2015, **6**, 943–948.
36. S. Ramesh, M. Majrashi, M. Almaghrabi, M. Govindarajulu, E. Fahoury, M. Fadan, M. Buabeid, J. Deruiter, R. Clark, V. Mulabagal, D. C. Agrawal, T. Moore and M. Dhanasekaran, in *Medicinal Mushrooms: Recent Progress in Research and Development*, ed. D. C. Agrawal and M. Dhanasekaran, Springer Nature, Singapore, 2019, pp. 103–141.
37. K. Kumar, R. Mehra, R. P. Guiné, M. J. Lima, N. Kumar, R. Kaushik, N. Ahmed, A. N. Yadav and H. Kumar, *Foods*, 2021, **10**, 2996.
38. K. Singha, S. Sahoo, A. Roy, A. Banerjee, K. C. Mondal, B. R. Pati and P. K. D. Mohapatra, *Int. J. Pharm. Sci. Res.*, 2020, **11**, 3160–3171.
39. A. Venkatachalapathi and S. Paulsamy, *Mycosphere*, 2016, 7, 118–130.
40. X.-M. Wang, J. Zhang, L.-H. Wu, Y.-L. Zhao, T. Li, J.-Q. Li, Y.-Z. Wang and H.-G. Liu, *Food Chem.*, 2014, **151**, 279–285.

41. S.-W. Chiu, S.-C. Law, M.-L. Ching, K.-W. Cheung and M.-J. Chen, *J. Gen. Appl. Microbiol.*, 2000, **46**, 269–282.
42. A. Klaus and W. A. A. Q. I. Wan-Mohtar, in *Wild Mushrooms*, CRC Press, 2022, pp. 23–65.
43. Himalayas, https://en.wikipedia.org/w/index.php?title=Himalayas&oldid=1157636224, accessed 30th May 2023.
44. D. Kumari, M. S. Reddy and R. C. Upadhyay, *Food Sci. Technol. Int.*, 2011, **17**, 557–567.
45. M. L. Largeteau, R. C. Llarena-Hernández, C. Regnault-Roger and J.-M. Savoie, *Appl. Microbiol. Biotechnol.*, 2011, **92**, 897–907.
46. M. L. Largeteau, P. Callac, A.-M. Navarro-Rodriguez and J.-M. Savoie, *Fungal Biol.*, 2011, **115**, 1186–1195.
47. R. A. Koch and J. R. Herr, *Front. Microbiol.*, 2021, **12**, 733159.
48. J. Cao, Z. Jin, W. Hu, Y. Zhang, S. Yao, X. Wang, Y. Zhang and Y. Tang, *Mar. Pet. Geol.*, 2010, **27**, 61–68.
49. R. Heinzelmann, C. Dutech, T. Tsykun, F. Labbé, J.-P. Soularue and S. Prospero, *Can. J. Plant Pathol.*, 2019, **41**, 1–23.
50. M. P. Coetzee, B. D. Wingfield and M. Wingfield, *Pathogens*, 2018, **7**, 83.
51. M. P. A. Coetzee, B. D. Wingfield, T. C. Harrington, D. Dalevi, T. A. Coutinho and M. J. Wingfield, *Mycologia*, 2000, **92**, 105–113.
52. V. P. Sharma and S. Kumar, Cultivation of Winter Mushroom (*Flammulina velutipes*), Directorate of Mushroom Research (Indian Council of Agricultural Research), Chambaghat, Solan Himachal Pradesh, India, https://dmrsolan.icar.gov.in/Folder_Flammulina_Mushroom__English_.pdf.
53. H. Tonomura, *The Biology and Cultivation of Edible Mushrooms*, 1978, pp. 409–421.
54. D. J. Royse, Proceedings of the 8th International Conference on Mushroom Biology and Mushroom Products (ICMBMP8), 2014.
55. V. P. Sharma, R. C. Upadhyay, S. Banyal, A. Barh and S. Kamal, *Mushroom Res.*, 2020, **29**, 37–45.
56. Z. Ge, Z. Yang, P. Zhang, P. Matheny and D. Hibbett, *Fungal Divers.*, 2008, **32**, 59–68.
57. P. M. Wang, X. B. Liu, Y. C. Dai, E. Horak, K. Steffen and Z. L. Yang, *Mycol. Prog.*, 2018, **17**, 1013–1030.
58. X. B. Liu, J. Li and Z. L. Yang, *Hereditas*, 2018, **155**, 3.
59. Q. Y. Wang, J. C. Zhang, C. Li, B. Wang, W. Y. Nong, Y. B. Bian and Y. Xiao, *Int. J. Med. Mushrooms*, 2018, **20**, 517–536.
60. Wikipedia, *In Wikipedia, The Free Encyclopedia*, 2022.
61. I. Prasher, *Wood-rotting non-gilled Agaricomycetes of Himalayas*, Springer, 2015.
62. J. Judova, K. Dubikova, S. Gaperova, J. Gaper and P. Pristas, *Fungal Biol.*, 2012, **116**, 155–160.
63. R. Mawar, L. Ram, Deepesh and T. Mathur, in *Beneficial Microbes in Agro-Ecology*, ed. N. Amaresan, M. Senthil Kumar, K. Annapurna, K. Kumar and A. Sankaranarayanan, Academic Press, 2020, pp. 625–649.

64. L.-W. Zhou, Y. Cao, S.-H. Wu, J. Vlasák, D.-W. Li, M.-J. Li and Y.-C. Dai, *Phytochemistry*, 2015, **114**, 7–15.
65. A. C. Campbell and M. Racjan, *Int. Biodeterior. Biodegrad.*, 1999, **43**, 101–107.
66. O. Shimomura, *J. Exp. Bot.*, 1992, **43**, 1519–1525.
67. T. Ao, C. R. Deb and S. R. Rao, *Mol. Biol. Rep.*, 2020, **47**, 621–630.
68. B. Kumari and N. S. Atri, *Int. J. Pharm. Pharm. Sci.*, 2014, **6**, 200–204.
69. K. Devi, J. Brahma and K. Shrivastava, *Int. J. Conserv. Sci.*, 2016, **7**, 709–718.
70. A. Pekşen and B. Kibar, *Acta Sci. Pol. Hortorum Cultus*, 2020, **19**, 11–20.
71. N. Atri, S. Saini, A. Gupta, A. Kaur, H. Kour and S. Saini, *Taxon. Ecol. Indian Fungi*, 2010, 161–169.
72. R. Kumar, S. Pandey, R. R. Rishi and K. Giri, *Asian J. Agric.*, 2017, **1**, 6–8.
73. Z. W. Ge, Z. L. Yang and E. C. Vellinga, *Fungal Divers.*, 2010, **45**, 81–98.
74. E. C. Vellinga, R. P. J. de Kok and T. D. Bruns, *Mycologia*, 2003, **95**, 442–456.
75. U. Singh, R. Bhatt, S. Stephenson, P. Uniyal and T. Mehmood, *Curr. Res. Environm. Appl. Mycol.*, 2017, **7**, 208–226.
76. W. Zhang, G. Tian, X. Geng, Y. Zhao, T. B. Ng, L. Zhao and H. Wang, *Molecules*, 2014, **19**, 19880–19891.
77. G. O. Occdownload, The Global Biodiversity Information Facility, 2022.
78. G. Senthilarasu and S. K. Singh, *Mycotaxon*, 2013, **121**, 69–74.
79. P. K. Chaudhary, M. Shrestha, B. H. Poudel and M. K. Adhikari, *Nepal J. Biotechnol.*, 2017, **5**, 27–31.
80. P. K. Chaudhary, M. Shrestha, B. H. Poudel and M. K. Adhikari, *Nepal J. Biotechnol.*, 2017, **5**, 27–31.
81. M. Kaur and Y. Singh, Proceedings of 8th international conference on mushroom biology and mushroom products (ICMBMP8), 2014.
82. T. Lakhanpal, A. Kumar and K. Kaisth, *Indian J. Mushrooms*, 1986, **12**, 1–4.
83. C. Pradeep, K. Vrinda, S. Mathew and T. J. M. R. Abraham, *Mushroom Res.*, 1998, **7**, 53–62.
84. M. Kirchmair, P. Carrilho, R. Pfab, B. Haberl, J. Felgueiras, F. Carvalho, J. Cardoso, I. Melo, J. Vinhas and S. Neuhauser, *Nephrol., Dial., Transplant.*, 2012, **27**, 1380–1386.
85. R. K. Verma and V. Pandro, *Int. J. Curr. Microbiol. App. Sci.*, 2018, **7**, 1694–1713.
86. J. Kalichman, P. M. Kirk and P. B. Matheny, *Taxon*, 2020, **69**, 425–447.
87. C. Simmons, T. Henkel and C. Bas, *Persoonia Mol. Phylog. Evol. Fungi*, 2002, **17**, 563–582.
88. P. Kirk, P. Cannon, D. Minter and J. Stalpers, *Dictionary of the Fungi*, 10th edn, CABI Publishing, Wallingford, UK, 2008.
89. P. E. Mortimer, S. C. Karunarathna, Q. Li, H. Gui, X. Yang, X. Yang, J. He, L. Ye, J. Guo and H. Li, *Fungal Divers.*, 2012, **56**, 31–47.
90. W. Fangfuk, R. Petchang, C. To-anun, M. Fukuda and A. Yamada, *Mycoscience*, 2010, **51**, 291–299.

91. M. Pavithra, A. A. Greeshma, N. C. Karun and K. R. Sridhar, *Mycosphere*, 2015, **6**, 421–432.
92. N. C. Karun and K. R. Sridhar, *Curr. Res. Environm. Appl. Mycol.*, 2014, **4**, 173–187.
93. A. Cabrera-Rodríguez, J. Pérez-Moreno, M. Torres-Aquino, G. Olmos-Oropeza, J. F. Martínez-Montoya, J. Palacio-Nuñez and J. A. Flores-Cano, *Rev. Chapingo Ser. Cienc. For. Ambiente*, 2022, **28**, 319–329.
94. C. Phosri, M. P. Martín and R. Watling, *IMA Fungus*, 2013, **4**, 347–356.
95. D. Arora and S. M. Dunham, *Econ. Bot.*, 2008, **62**, 376–391.
96. B. Buyck, F. Kauff, G. Eyssartier, A. Couloux and V. Hofstetter, *Fungal Divers.*, 2014, **64**, 101–121.
97. S.-C. Shao, P.-G. Liu, T.-Z. Wei and M. Herrera, *Mycologie*, 2021, **42**, 25–37.
98. S. Dhancholia, J. Bhatt and S. K. Pant, *Acta Botanica Indica*, 1991, **19**, 104–109.
99. K. Deepika, M. S. Reddy and R. C. Upadhyay, *Mycology*, 2013, **4**, 205–220.
100. D. Kumari, M. S. Reddy and R. C. Upadhyay, *Mycoscience*, 2011, **52**, 147–151.
101. S. Tedder, D. A. Mitchell and A. Hillyer, *Property rights in the sustainable management of non-timber forest products*, Citeseer, 2002.
102. R. Flores, M. d. C. Bran and M. Honrubia, Edible mycorrhizal mushrooms and their cultivation. Proceedings of the Second International Conference on Edible Mycorrhizal Mushrooms, Christchurch, New Zealand, 3–6 July, 2001, 2002.
103. I. Olariaga, G. Moreno, J. L. Manjón, I. Salcedo, V. Hofstetter, D. Rodríguez and B. Buyck, *Fungal Divers.*, 2017, **83**, 263–292.
104. C. Bijeesh, A. M. Kumar, K. B. Vrinda and C. K. Pradeep, *Phytotaxa*, 2018, **346**, 157–168.
105. M. Sevindik, *Indian J. Exp. Biol.*, 2021, **59**, 310–315.
106. L.-L. Niu, Y.-R. Wu, H.-P. Liu, Q. Wang, M.-Y. Li and Q. Jia, *J. Food Meas. Charact.*, 2021, **15**, 2842–2853.
107. I. Adamska and G. Tokarczyk, *Int. J. Food Sci.*, 2022, **2022**, 5773275.
108. N. Atri, Y. Sharma and S. Kumar, in *Microbial Diversity in Ecosystem Sustainability and Biotechnological Applications*, Springer, 2019, pp. 533–563.
109. S. Sharma, N. Atri, M. Kaur and B. Verma, *Kavaka*, 2017, **48**, 41–46.
110. D. Pegler and M. Vanhaecke, *Kew Bull.*, 1994, 717–736.
111. D. D. Tibuhwa, *J. Biol. Life Sci.*, 2012, **3**, 140–159.
112. N. C. Karun and K. R. Sridhar, *Czech Mycol.*, 2013, **65**, 233–254.
113. M. Kuo, D. R. Dewsbury, K. O'Donnell, M. C. Carter, S. A. Rehner, J. D. Moore, J.-M. Moncalvo, S. A. Canfield, S. L. Stephenson and A. S. Methven, *Mycologia*, 2012, **104**, 1159–1177.
114. K. C. Semwal and V. K. Bhatt, *Biodiversity Res. Conserv.*, 2019, **53**, 7–32.

115. H. K. Kanwal, K. Acharya, G. Ramesh and M. S. Reddy, *Curr. Microbiol.*, 2011, **62**, 1245–1252.
116. N. S. Atri, Y. P. Sharma, S. Kumar and Mridu, *Wild Edible Mushrooms of North West Himalaya: Their Nutritional, Nutraceutical, and Socio-biological Aspects*, 2019.
117. M. Kaliyaperumal, K. Kezo and S. Gunaseelan, in *Biology of Macrofungi*, ed. B. P. Singh, Lallawmsanga and A. K. Passari, Springer Cham, Switzerland AG, 2018, pp. 15–56.
118. K. C. Semwal, S. L. Stephenson, V. K. Bhatt and R. P. Bhatt, *Mycosphere*, 2014, **5**, 440–461.
119. O. J. Olatunji, J. Tang, A. Tola, F. Auberon, O. Oluwaniyi and Z. Ouyang, *Fitoterapia*, 2018, **129**, 293–316.
120. M. J. Winkelman, J. W. Allen, P. Y. Lamrood, M. Maillart-Garg, B. L. Sinha and N. Shah, *Indian Journal of Traditional Knowledge*, 2022, **21**, 341–352.
121. B. Patwardhan, D. Warude, P. Pushpangadan and N. Bhatt, *Evidence-Based Complementary Altern. Med.*, 2005, **2**, 465–473.
122. S. Debnath, B. Debnath, P. Das and A. K. Saha, *J. Appl. Pharm. Sci.*, 2019, **9**, 144–156.
123. N. Verma, A. Bhalla, S. Kumar, R. K. Dhimany and Y. K. Chawlay, *J. Clin. Exp. Hepatol.*, 2014, **4**, 361–365.
124. D. Moore and S. W. Chiu, in *Bio-exploitation of filamentous fungi*, ed. S. B. Pointing and K. D. Hyde, Fungal Diversity Press, Hong Kong, 2001, pp. 223–251.
125. S. W. Chiu, S. C. Law, M. L. Ching, K. W. Cheung and M. J. Chen, *J. Gen. Appl. Microbiol.*, 2000, **46**, 269–282.
126. E. F. Lambin and P. Meyfroidt, *Proc. Natl. Acad. Sci. U. S. A.*, 2011, **108**, 3465–3472.
127. M. K. Yadav, R. Chandra and P. K. Dhakad, *Indian J. Agric. Sci.*, 2016, **86**, 1070–1075.
128. T. Bhaben, G. Lisha and S. G. Chandra, *Indian J. Tradit. Knowl.*, 2011, **10**, 512–515.

CHAPTER 2

Genus Agaricus *Species Found in the Himalayas:* Agaricus campestris *L. and* Agaricus augustus *Fr.*

SWATI ALLEN,*[a] GARIMA BHARADWAJ[b] AND JOY K. CHAKMA[a]

[a] Central Procurement Cell, Indian Council of Medical Research, V. Ramalingaswami Bhawan, Ansari Nagar East, New Delhi, 110029, India; [b] Department of Chemistry, Sant Longowal Institute of Engineering and Technology, Sangrur 148106, India
*Email: allenswati@gmail.com

2.1 Introduction

In today's world, the consumption of mushrooms has been increasing day by day and they are therefore cultivated in several regions of the world. Almost 200 edible mushroom species have been identified in various regions of the world.[1] The majority of significant and essential edible mushrooms belong to the genus Agaricus and have been evaluated due to their nutritional value and pharmaceutical properties. The genus Agaricus belongs to the Agaricaceae family and includes nearly 300 different mushrooms, which include edible as well as poisonous species.[2] The fruiting part of Agaricus species has a pileus or fleshy cap along with either a dry, bare or scaly surface. They also have a number of radiating gills, called hymenophore. Agaricus species are valued and appreciated mainly due to their high nutritional value along with their

Phytochemistry and Nutritional Composition of Significant Wild Medicinal and Edible Mushrooms: Traditional Uses and Pharmacology
Edited by Ajay Sharma, Garima Bhardwaj and Gulzar Ahmad Nayik

Published by the Royal Society of Chemistry, www.rsc.org

bioactive compounds and antioxidant activity, including free radical scavenging and biological effects. These properties make them a valuable as well as significant reservoir with nutritional and pharmacological value.[3–5] Apart from being ingredients in the human diet, some species have also been used for centuries in traditional medicine. The button mushroom, *Agaricus bisporus* (*A. bisporus*), tops the list as the most frequently cultivated species in the world. It is also one of the most highly consumed mushrooms globally. Other Agaricus species include the medicinal mushroom *A. brasiliensis*, *A. arvensis* (a wild growing species, the fruiting bodies of which are commercially produced), *A. bitorquis, A. campestris*, and *A. silvaticus.*[6]

A. bisporus, an edible Basidiomycete, is also known as white mushroom, common mushroom, portobello mushroom, button mushroom, Swiss-brown, cremini, Roman brown mushroom, Italian, or cultivated mushroom.[7] It is mainly found in the grasslands of India, Europe, and North America.[8] *A. bisporus* represents 90% of all of the mushrooms consumed in the USA.[9] The earliest wild form of *A. bisporus* was noted to have a brownish cap with dark brown gills, but current mushroom variants have a white cap and a stalk with flesh and gills that are brown in colour.[10]

A. campestris is a white fleshy mushroom that is very similar to *A. bisporus*.[1] Due to loss of habitat and various chemical treatments, *A. campestris* is now less commonly seen. It is also popularly recognized as the field mushroom or meadow mushroom in North America. *A. augustus*, also known as "the Prince" is common in the state of California and the Pacific Northwest. It prefers to grow under conifers that are near to roads, gardens, parks, *etc.* This fungus is easy to identify due to its strong odor along with large size and scaly brown cap.[11]

2.2 Taxonomy, Botanical Names, Local Names, and Characteristics of Mushrooms

More than 400 Agaricus species are found worldwide. Bas in 1991,[12] and Zhao *et al.* in 2011[13] identified 386 species in the Agaricus genus. Presently, the total number of species is 434, which comprises 200 tropical species.[14] Although *Agaricus* is very well known worldwide and it is also recognized as a valuable genus of principally edible mushrooms, the taxonomy as well as delimitation of the species is very complex within the Agaricus genus.[15]

Most of the taxonomic monographs of *Agaricus* species are from Europe, Australia, the Americas, India, and New Zealand.[14] The most commonly cited monographs on this species are those of Heinemann.[16,17] Presently, researchers are busy developing new and efficient molecular technology-based tools for the recognition and classification of fungal reservoirs, such as DNA sequencing.[18] In recent years, phylogenic analyses have been widely used to identify and track the evolutionary relationship among taxa using ribosomal DNA sequences.[19–22] Due to tropical diversity, the systematics and phylogeny of agaricus species have failed to include the species that are originating from Asia. However, European and North American temperate Agaricus taxa have further evolved in recent years.[23]

Distinctive features of the family Agaricaceae are:

1. The fruiting body is well-differentiated into a stipe and a pileus.
2. The annulus is typically present on the stipe.
3. The lamellae are free.

Traditional ways of identifying Agaricus species have mainly been based on morphology characteristics and odor. The use of chemical-based reactions using Schäffer's reagent and potassium hydroxide have also been used for identification purposes.[23] However, due to great variations in species phenotypes resulting from various environmental factors, *e.g.*, different geographic conditions, along with separate developmental stages of Basidiomata, can also affect and therefore pose challenges in the morphology identification process of Agaricus. This could make identification difficult, particularly in differentiating *A. bisporus* species from other related Agaricus species.[24] Therefore, molecular tool-based applications that are primarily based on DNA study have proven to be vital and essential in identifying this species.[25]

Based on multigene molecular phylogenetic analysis such as the internal transcribed spacer (ITS), large subunit (LSU) of the rRNA gene, and translation elongation factor (TEF 1), morphological characteristics, and divergence time, in 2016, Zhao *et al.*[26] classified this species mainly into five subgenera and 20 sections. Chen *et al.* in 2017[27] and Parra *et al.* in 2018[28] revised the classification of Agaricus into six subgenera with 23 sections. He *et al.* in 2018[29] introduced a new section known as *Agaricus subg. Pseudochitonia.* At present, this species now contains six subgenera and 24 sections.

2.2.1 Representative Species

2.2.1.1 A. bisporus

The common mushroom or button mushroom is highest produced mushroom in the world due to its dietary, nutritional, pharmaceutical, and medicinal properties. However, the cultivation of this species is carried out traditionally as well as commercially. *A. bisporus* var hortensis is generally referred to as the common mushroom in the market, while *A. bisporus* var brunnescens, is known as cremini or portobello, depending upon its size as well as level of development. *A. bisporus* can attain a hat diameter of up to 18 cm; however, the size does not normally exceed 13 cm. Scales and spots can appear with age on the surface of this fungus, which is covered by a powdery cuticle.

2.2.1.2 A. augustus

This species can be identified due to its various characteristics, such as large size along with its scaly and brown cap. It is also very well known for its strong and sharp almond odor. It is found either under coniferous trees as well as along roadsides. This fungus can also grow in lawns and gardens during summer and fall. The fungal cap is approximately 10–35 cm in size. It is usually block-like in appearance and has an almost cylindrical look at first, later on changing from convex to broad convex or sometimes nearly flat.

2.3 Resource Availability of Wild or Cultivated Mushrooms: Sustainability and Distribution

Due to favorable environmental conditions, most mushrooms are found to naturally occur in forests. In developed as well as developing countries, mushroom trading and harvesting are important sources of income. The picking of mushrooms is considered to be a customary as well as recreational activity in rural communities. However, excessive and unrestricted collection of fungal reservoir for various medicinal, pharmaceutical, dietary and immunological properties have posed a threat to their conservation and future accessibility. Also, changes in microniches has been observed due to change in climatic conditions as well as improper management of natural ecosystems. This has led to deleterious effects on fungal bioresources. Therefore, mushroom cultivation plays a key role for economic benefits as well as serving as a conservation approach for society. This requires proper land management skills, supervised exploitation of fungal reservoir along with a strategic approach for sustainable use, which could also lead to future economic benefit. Governments and several NGOs have launched a few programs for the identification of fungal cultivation sites that can support rare as well as threatened mushroom species for their conservation. The urgent need of the hour is to identify and assess important fungal areas that need to be safeguarded.[12]

Mushrooms and their environmental responsive technologies

- Mycorrhizal fungi are primarily used to rehabilitate, restore and reclaim marginal and derelict land
- Fungi are also used as potential bioregulators of weeds
- Fungi are also used in the ecological management of natural plantations

Market-based diversification and novel fungal yields

- Fungi farming and truffle fruiting body of ascomycetes fungi as a sustainable alternative
- Collection of myco ecotypes and socioeconomic characteristics
- Unique and innovative bioproducts of commercial importance from fungi

2.3.1 Distribution of Mushrooms

The Indian subcontinent has various agro-climatic zones that support fungal diversity. In 2004, Deshmukh *et al.*[30] recorded approximately 850 species of mushrooms in India, which constitutes nearly one third of the myco-biodiversity on the globe.[31] In India, the collection as well as methodical study of mushrooms started during the 19th century.[32] Mushrooms have been widely researched in the west. However, tropical countries, such as India, have been less explored. As per recent studies in tropical forests, it has been suggested that compared to temperate areas, there is more fungal diversity in tropical regions.[33] In tropical forests, fungal species are particularly vital components of biodiversity as they contribute to the conservation of the biogeochemical cycle and biosphere as well as the earth's ecosystem.

Fungi are known for their immense beneficial roles, such as in nutrient cycling, as antibiotics, in the food industry and agriculture, as well as biofertilizers. In India, research has been carried out in a few specific areas, such as Sikkim,[34] Jammu and Kashmir,[35] the Amarkantak Biosphere Reserve,[36] Bangalore in Karnataka,[37] the Western Ghats,[38] Nagaland[39] and Meghalaya,[40] *etc.* Due to the presence of mushroom reservoirs on the Indian subcontinent, it is suggested that there is a direct correlation between mushroom variety and forest health. It has also been noted that in thick forests, mushrooms are more abundant compared to in areas of thin forest.

2.4 Medicinal Uses

In recent years, *A. bisporus* is considered as one of the best example of edible mushrooms that play a vital role in diet as well as being known for its excellent medicinal and nutritional properties; it has long been used in traditional medicines and therapies, *etc.* Various research studies have shown that bioactive compounds that are present in *A. bisporus* extracts, such as in powder form, can be used to heal different human ailments. *A. bisporus* has potential anti-obesity, antioxidant, anticancer and anti-inflammation properties. The active compounds present in *A. bisporus* can be used to treat immune disorders, cardiological diseases, uncontrolled growth of tumor cells, and diabetes mellitus, as well as viral, fungal, and bacterial infections.[41,42] Hence, the consumption of *A. bisporus* boosts the immune functioning of the body and provides more resistance to various diseases.[42] Numerous studies have confirmed that *A. bisporus* exhibits anticancer,[43–47] antioxidant,[48–50,58] and anti-diabetic[51–56] properties (Table 2.1).

2.4.1 Antioxidant Properties

Compared to important edible mushrooms, such as *Grifola frondosa* Dicks. Grey, *P. eryngii DC. Quel.*, *L. edodes*, and *P. ostreatus*, *A. bisporus* has comparatively higher antioxidant potential.[14] In a study conducted by Oms-Oliu *et al.*,[57] a phenolic content of around 100.32–100.78 mg per 100 g of fresh weight of button mushrooms was documented.

Muszynska *et al.*[41] determined the serotonin content of 5.21 mg per 100 g dry weight of *A. bisporus* fruiting bodies. Tocopherol, an antioxidant, is a fat-soluble vitamin that is also involved in various metabolic functions. The content of β-tocopherol in white *A. bisporus* was found to be 0.85 μg per 100 g fresh weight.[59]

Table 2.1 Chief phenolic compounds contributing to the medicinal and biochemical properties of *A. bisporus.*

	Phenolic compounds	Ref.
Antioxidant properties	Myricetin, catechin, protocatechuic acid, ferulic acid and gallic acid	92
	Serotonin	99
	Tocopherol	
Anticancer properties	Lectins and phenolic compounds	100
	Arginine in the basidiocarp slows tumor growth. ABP-1 and ABP2 polysaccharides suppress the increase in breast cancer cells	87 and 101

2.4.2 Anticancer Properties

In 2018, Kothari *et al.* showed anticancer activity in polysaccharides extracted from some edible mushrooms.[60] In *A. bisporus,* polysaccharides were shown to exhibit a significant mode of action against uncontrolled growth of cancerous cells by increasing cellular immunity. *A. bisporus* contains a considerable amount of several vital polysaccharides. As confirmed by *in vivo* and *in vitro* assays, these polysaccharides exhibit anti-tumor and intense immunostimulatory activity.[60]

Smiderle *et al.*[61] reported that in *A. bisporus* there are three principal polysaccharides, namely α- (alpha, 20.4%) and β- (beta, 23.7%) glucans, and galactomannan (55.8%). In another study carried out by McCleary and Draga,[62] low β-glucan content in *A. bisporus* was beneficial in increasing mucosal based immunity.

Jagadish *et al.*[63] reported that in *A. bisporus,* phenolic compounds and lectins suppress the growth of tumor cells. They also showed that the bioactive compounds, isolated from *A. bisporus*, also affected cancer cell lines. Their results concluded that multiplication of human leukemia-60 cells was repressed by button mushroom extract. This led to programmed cell death.[48] Similarly, in another study conducted on mice, button mushroom extract stopped the progression of prostate cancer cells, namely PC3 and DU145.[49] The extract of *A. bisporus* contains lectins that are extremely effective against lung cancer cells. Their active mechanism helps augment the cellular mechanism involved in the antioxidant-based defense system and thus hinders the growth of malignant cancerous cells.[63]

One of the amino acids extracted from *A. bisporus* that delays the growth of tumor cells is Arginine; it is found in the basidiocarp of *A. bisporus.* A study recommended that patients suffering from cancer should use the button mushroom as a dietary supplement owing to its anti-tumor properties. Selenium is one of the essential trace elements for humans and animals. According to Clark *et al.*,[64] compared to fruits and vegetables, a higher Se content is found in mushrooms. Therefore, *A. bisporus* also serves as a valuable and vital reservoir of trace minerals and therefore promotes increased health benefits. Selenium plays a key role in preventing successful chemotherapy in the treatment of cancer. In another anticancer study by Spolar *et al.*,[65] it was noted that selenium uptake plays a key role in building a strong immune system in order to fight against several diseases, such as cancer. However, in a contradictory study by Rzymski *et al.*,[66] it was reported that higher doses of selenium significantly led to an increase in cancer.

2.4.3 Anti-diabetic Activity

Various studies, literature surveys and research articles have confirmed the anti-diabetic properties of bioactive compounds present in *A. bisporus*.[67] Antioxidants, such as vitamins B12, C, and D; polyphenols; folates; and dietary fiber; which are found in *A. bisporus*, exhibit inhibitory effects against cardiovascular disorders and diabetes.[46] Calvo *et al.*[55] isolated various bioactive compounds that have the ability to prevent type 2 diabetes. In another study by Ekowati *et al.*,[68] it was found that *A. bisporus* extract increased the production of insulin as well as Glucose 6 phosphate dehydrogenase activity in rats, which resulted in a significant lowering of their glucose levels. Similarly, the intake of *A. bisporus* also caused an increase in cholesterol, triglyceride and low-density lipoprotein LDL levels. The glucose lowering effects of *A. bisporus* may be due to its elevated amounts of dietary-based fiber of around 19% as well as its other carbohydrates. Dietary fiber isolated from *A. bisporus* has a similar effect to that observed for highly-rated edible mushrooms.[67] The lectin-like molecules present in *A. bisporus* are also accountable for the decrease in glucose and simultaneous increase in insulin. Intake of *A. bisporus* therefore contributes to decreasing blood glucose levels due to its elevated fiber content. This is because it acts as an obstacle against the activities of digestive enzymes, ultimately resulting in lowering of blood glucose levels.[52,53] Propionate, a known ester of propionic acid, which is isolated from *A. bisporus*, has antidiabetic properties.[69,70] Propionate is connected to gluconeogenesis, as well as regulating the levels of serum lipids and thus decreasing blood glucose levels.

Wong *et al.*[71] used *A. bisporus* extracts and observed molecular mechanisms that are involved in decreasing blood glucose levels. In another study carried out by Volman *et al.*,[72] α-glucan was shown to lower the production of lipopolysaccharides. This ultimately resulted in a significant lowering of glucose concentration. However, in the case of rats, Ekowati *et al.*[68] reported that secondary compounds such as terpenoids and flavonoids increased superoxide dismutase (SOD) activity. In another study conducted by Huang *et al.*,[73] an increase in the expression levels of endogenous antioxidant genes was observed due to flavonoids, which consequently led to an increase in SOD activity. Terpenes or terpenoids are also involved in the alteration of SOD activity.[74] In the basidiocarp of *A. bisporus*, the presence of a high concentration of zinc (7.5–15 mg per 100 dry weight) significantly helps to improve SOD activity.[41]

2.4.4 Anti-obesity Activity

As a known dominant risk factor, hyperlipidemia is associated with increased levels of cholesterol or triglycerides. As a result of which, this illness leads to serious heart conditions and atherosclerosis in human beings.[75] As per a literature survey, edible mushrooms provide a high nutrition content as a dietary supplement as they are high in fiber and also contain bioactive compounds, such as polyphenols, alkaloids, flavonoids, polysaccharides, and sterols, as well as terpenes, which have high antioxidant potential. All these compounds have positive effects on various cardiac biomarkers that are helpful in treating weight gain problems such as obesity and its associated cardiovascular diseases.[76] Lin *et al.*[77] reported that use of plant sterols or phytosterols led to a reduction in cholesterol absorption. Extracts of *A. bisporus* also contain phytosterols, which help to reduce LDL cholesterol as well as plasma cholesterol. In another study conducted by Xu *et al.*[78] a substantial amount of lovastatin was detected in *A. bisporus*, which lowers the cholesterol levels in the body and decreases the chances of cardiac disorders. In another study by Jeong *et al.*,[46] it was shown that the basidiocarp of *A. bisporus* when eaten by rats helped regulate anti-cholesterolemic and anti-glycemic responses. These rats were fed a diet rich in cholesterol and were suffering from T2D induced by streptozotocin injection, which was injected at around 50 mg per kg body weight. The results showed that the button mushroom has anti-hypercholesterolemic as well as anti-glycemic effects in rats. Apart from these effects, it also showed a positive effect on lipid metabolism and liver functioning.

2.4.5 Antimicrobial Activity

A. bisporus contains several bioactive compounds that have immense potential in terms of their antimicrobial activities. As per recent studies, the use of traditional and synthetic medicines has proven to be ineffective in providing relief against animal, human, and plant diseases.[79] Various research surveys have examined and documented that the growth and development of Gram-positive bacteria is widely inhibited using a methanol extract from wild *A. bisporus*. Although, compared to Gram-positive bacteria, the methanol-based extract had less impact on the development and growth of Gram-negative bacteria.[80] Shang *et al.*[81] reported similar findings where they showed an extract from a Chinese strain of the button mushroom had no impact on Gram-negative bacteria. Soltanian *et al.*[82] also observed and studied the impact of a crude extract from *A. bisporus* on the growth and development of both Gram-negative and Gram-positive bacteria. They showed that crude extracts from wild as well as domesticated mushrooms had inhibitory effects on Gram-positive bacteria. Secondary metabolites of *A. bisporus* extract contain compounds that are considered to exhibit antimicrobial activity. To date, very few studies have been carried out to determine the specific compounds that are responsible for the antimicrobial activity of *A. bisporus*. Although, during the past few years, some insightful research findings have been published.[83] Significant antibacterial activities against *Staphylococcus aureus* have been observed in an aqueous protein extract from cultivated button mushrooms.[84] In another study, the growth of *Escherichia coli* and *S. aureus* was considerably stopped by ethanol cold-water extracts of *A. bisporus.*[85]

2.4.6 Anti-inflammatory Properties

A. bisporus exhibits anti-inflammatory properties.[86,87] Polysaccharides isolated from *A. bisporus,* such as mannogalactan, fucomannogalactan, and fucogalactan exhibit anti-inflammatory as well as analgesic properties.[87] Heterogalactan is also an important polysaccharide that is extracted from *A. bisporus* and can be used to fight sepsis in mice.[86] Golak-Siwulska *et al.*[88] documented that heterogalactan isolated from *A. bisporus* can be helpful in treating sepsis, which is a very serious medical issue and is one of the main causes of mortality in intensive care units around the world. A study carried out by Ruthes *et al.*[86] showed that extracts isolated from the button mushroom exhibit activity against sepsis as they showed strong anti-inflammatory effects.

2.5 Phytochemistry

In today's world, mushrooms are known for their immense valuable and crucial bioactive compounds that have immense nutritional value and pharmacological properties. Due to wide benefits that mushroom-based compounds show toward several clinical conditions, they have gained the interest of the scientific research community worldwide. Since the last decade, efforts have been made to understand the molecular mechanisms responsible for the biochemical actions that mushroom-based compounds have on the human body.[89] Several polysaccharides, as well as other myco-based compounds, such as lypopolysaccharides, proteins, glycoproteins, and secondary metabolites, have been shown to have powerful effects on the immunological process.[89] In the identification of such biomolecules, many polysaccharides have been isolated from basidiomycetes. These polysaccharides are either linear or branched glucans as well as heterogalactans, and contain either *O*-methyl groups or a wide range of side chains.[90,91] Heterogalactans have a central chain composed of $1 \rightarrow 6$-linked-D-Galp units with either mannose or fucose as a substituent,[92–94] although there are less research-based studies related to their comprehensive structure as well as biological activity. Few polysaccharides have also been evaluated for their antitumor activity.[95] Some reports have shown that mushroom extracts can have other effects, such as anti-inflammatory and anti-nociceptive activity.[96] However, as per detailed research studies, it is not possible to associate a relationship between compound structure and activity, as many investigations are done using crude polysaccharide extracts.[97]

Button mushrooms contains good quality protein at 28–42.5%, fiber at 8.3–16.3, and a moisture content of 90–93%. They also contain lysine, as well as tryptophan, which are usually absent in cereals. Carbohydrate content constitutes almost 59.4%, while ash content varies from 9.4 to 14.5%. Button mushrooms also contain chitin, glycogen, and hemicellulose, as well as linoleic acid. However, they do not contain cholesterol; this is replaced by ergo-sterol. Nearly 100 g dry weight of the button mushroom contains 8.8 mg of iron, 912 mg of phosphorous, 71.0 mg of calcium, 2850 mg of potassium, and 106 mg of sodium. Being a decent reservoir of vitamin C and the vitamin B complex (biotin, riboflavin, niacin, thiamine, and pantothenic acid) mushrooms also contain vitamin B12, as well as folic acid, which are lacking in most vegetables.[98] *A. bisporus* extract has also been used in Alzheimer's disease therapy. The extract exhibits both acetylcholinesterase and butyrylcholinesterase inhibition activity, which are effective against Alzheimer's disease.[99] The bioactive compounds present in *A. bisporus* extract are also used as anticancer, antioxidant, and anti-inflammatory agents against various heart diseases, diabetes, bacterial and fungal infections, and human immune system disorders.[100] *A. bisporus* extract contains three major polysaccharides, namely α-glucan, β-glucan, and galactomannan.[101] Fatty acids, such as linoleic, palmitic, and stearic acids, are also present in the extract.[102] The primary phytochemicals present in *A. bisporus* are presented in Table 2.2.

Secondary phytochemicals are tabulated in Table 2.3. The gas chromatography–mass spectrometry analysis of *A. bisporus* also shows the presence of 14 compounds.

2.6 Nutritional Value

The nutritional significance of *A. bisporus* can be attributed to its chemical configuration and composition. It has been observed that *A. bisporus* extract contains sugars, fat, 28–42.5% protein, 90–93% moisture, 8.3–16.2% fiber, 59.4% carbohydrates, and 9.4–14.5% ash. Lysine and tryptophan are not found in cereals. Glycogen, chitin, and hemicellulose are present *A. bisporus* instead of starch, and it has promising carbohydrate, water, protein, fat, dietary fiber, pantothenic acid, niacin, vitamin C, riboflavin, ash, amino acid and iron content[103] (Table 2.4). In *Agaricus* species, significant levels of phosphorus and low sodium are observed. Although, dry *Agaricus* matter comprises 85–90% water. Nevertheless, during growth and storage, the moisture content of *Agaricus* species is greatly affected by temperature and relative humidity. The protein content of *Agaricus* species is its most important component and also contributes to its immense nutritional value. In *A. bisporus,* the protein content is 34–44% of the total dry weight.[104]

The crude fat content in *A. bisporus* is 1–20% of total dry weight. Apart from protein content, free as well as combined fatty acids are also found in *A. bisporus.* Increased concentrations of oleic, palmic and stearic acids are also present.[105] However, one of the main active compounds present in the

Table 2.2 Chief phytochemicals present in *A. bisporus* extract.[a]

	Solvents used	
Phytochemicals	Petroleum ether	Methanol
Saponins	−	+++
Flavonoids	−	+
Tannins	+++	−
Quinones	++	−
Polyphenols	++	++
Terpenoids	+	++
Cardiac glycosides	+	++
Alkaloids	+++	−

[a] + poorly present; ++ moderately present; +++ luxuriantly present; – absent.

ethyl acetate fraction of *A. bisporus* is unsaturated fatty acids such as conjugated linoleic, linolenic, and linoleic acids.

Fresh mushrooms contain large quantities of carbohydrates, trehalose, pentoses, disaccharides, *etc.* which account for almost 3–28% of the total dry weight.[106] They are a good source of different vitamins, such as the vitamins A, B, C, D, thiamine, riboflavin, biotin, and niacin; minerals such as iron, sodium, calcium, and potassium; and the essential amino acids ornithine, methionine, and citralline.[107] One of the diverse group of carbohydrate-binding proteins commonly present in microorganisms, animals, and plants is lectins; they act as cellular mediators and also exhibit molecular recognition in a variety of biological systems.[108]

2.6.1 Amino Acid and Protein Content

Despite having relatively higher amino acid content compared to animal products, mushrooms are generally ranked lower then animal meats in terms of their nutrition.[109] The nutritional and chemical compositions of mushrooms are affected by pre- and post-harvest conditions. The different developmental stages of mushroom exhibit different amounts of amino acid and protein contents.[110,111] As per a literature survey, mushrooms display variable protein content if grown on different substrates. A study conducted by Braaksma and Schaap[112] showed that in *A. bisporus*, the major amino acids are valine, cysteine, aspartic acid, histidine, glutamine, serine, threonine, glycine, tyrosine, isoleucine, leucine, alanine, lysine, proline, arginine, and norleucine. However, the crude protein extract isolated from *A. bisporus* varies from 19 to 38% on a dry weight base. Sadiq *et al.*[113] reported 11.01% protein content in *A. bisporus*, whereas Mohiuddin *et al.*[114] determined a 17.7–24.7% protein content in the same species. In another study carried out by Ahlavat *et al.*,[115] a 29.1% protein content was reported for *A. bisporus*. All these results indicate that the protein content may vary in *A. bisporus* depending upon different growth substrates used.

Recently, Citores *et al.* in 2019 and Fogarasi *et al.* in 2020 discovered a novel and unique class of ribonucleases, popularly known as "ribotoxin-like

Table 2.3 Key secondary metabolites found in *A. bisporus*.

Name of compound	Chemical structure	Biological properties
Myristic acid	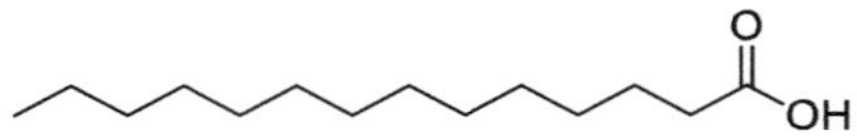	Larvicidal repellent activity, nematicide, antioxidant, hypercholesterolemic, anticancer[154,155]
Methyl-2-diazo-3-oxo-4-propylhept-6-enoate	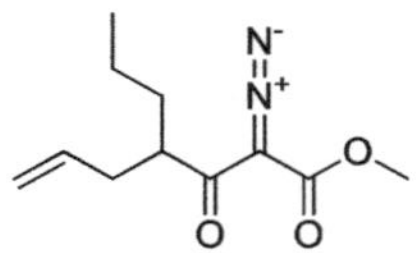	—
Lauric acid	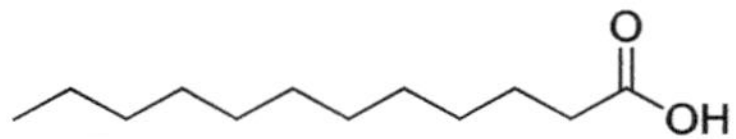	Pesticide, anti-microbial, nematicide[152,153]
6,10,14-Trimethyl-2-pentadecanone	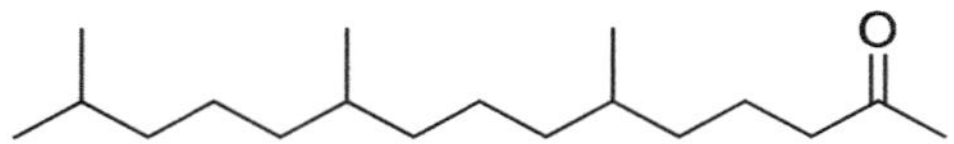	Lotions, cosmetic products, skin creams[156]
9,12,15-Octadecatrienoic acid, ethyl ester	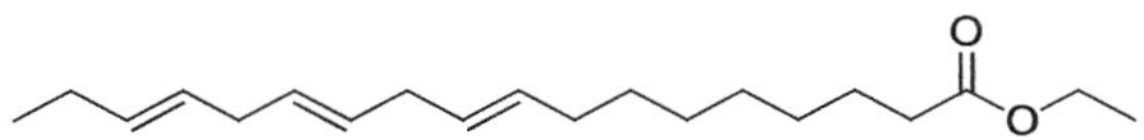	Anti-cancer[161]

Palmitic acid, methyl ester	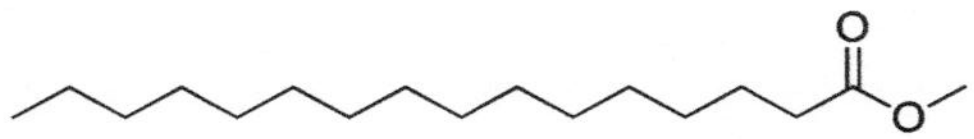	Antibacterial and antifungal[158]
Neophytadiene	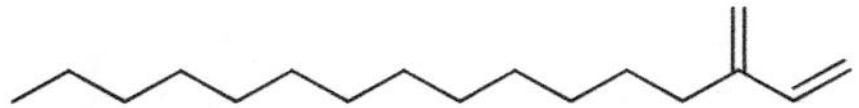	Anti-inflammatory, antimicrobial, antioxidant, analgesic, antipyretic.[159]
2-Bromolauric acid	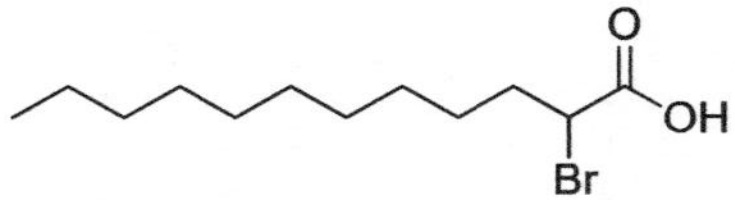	—
Palmitic acid	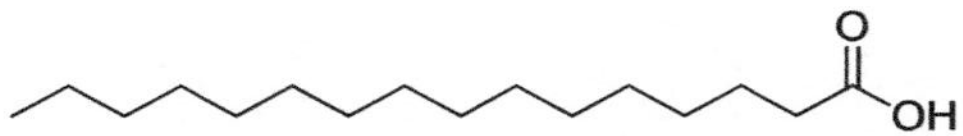	Antioxidant, nematicide, pesticide, lubricant, antiandrogenic, flavoring, hemolytic, 5-alpha reductase inhibitor, hypocholesterolemic.[160]
4-6,6-Dimethyl-2-methylenecyclohex-3-enylidene pentan-2-ol	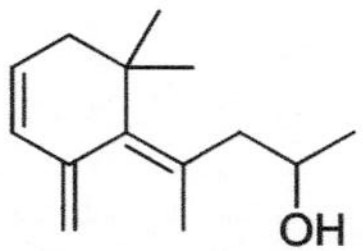	Melamine, dyes.[157]

Table 2.4 Nutritional value of *A. bisporus*.[a]

Raw nutritional value per 100 g	
	A. bisporus
Energy	94 kJ, 22 kcal
Carbohydrate	3.28 g
Vitamin C	2.1 mg, 4%
Protein	3.09 g
Dietary fiber	1.0 g
Niacin (vit. B3)	3.607 mg, 24%
Fat	0.34 g
Water	92.43 g
Riboflavin (vit. B12)	0.402 mg, 27%
Sugars	1.65 g
Pantothenic acid (vit. B5)	1.497 mg, 30%
Iron	0.50 mg, 4%

[a]Source: USDA nutrient database.

proteins,"[116,117] extensively distributed between different species of mushrooms. In related study, Citores *et al.*[116] also isolated ribotoxin, known as ageritin from the mushroom *Agrocybe aegerita*. Their study showed that ribotoxin has nuclease, antibacterial, endonuclease, and antiviral as well as cytotoxic activities and can be employed in plants with the help of transgenic techniques. This gene transfer in plants can help to enhance their resistance against viruses, fungi and bacteria.

2.6.2 Lipids

In *A. bisporus*, significant amounts of essential fatty acids are present, such as linoleic acid, compared to crude fats, which are present in small amounts. In a study conducted by Barros *et al.* in 2008,[48] wild *Agaricus* sp. were shown to contain moderately greater quantities of polyunsaturated fatty acids and less monounsaturated fatty acid content compared to cultivated strains. Their study showed that in *A. bisporus* the total fatty acid content varies from 180–5818 mg per kg dry weight. Almost 90% of the fat content in *A. bisporus* was shown to be represented by linoleic acid. In 2008, Sadiq *et al.*[113] reported that in *A. bisporus*, the fatty acid content comprises erucic, stearic acid, caprylic, oleic, linoleic, palmitic, and eicosanoic acids. Together they total an extracted fatty acid content of around 44.19%. However linoleic acid (61.82–67.29%) and palmitic acid 12.67–14.71% were the main fatty acids among the 13 fatty acids detected in *A. bisporus*.[80] In another study carried out by Shao *et al.*,[102] linoleic, palmitic, and stearic acids were reported to be the major fatty acids in *A. bisporus*.

2.6.3 Carbohydrate and Fiber

Mushrooms are a remarkable reservoir of both digestible and non-digestible carbohydrates. In *A. bisporus*, glucose and mannitol are the major valuable digestible carbohydrates. However, they exist in small quantities of <1% of the dry weight. Another important digestive-friendly carbohydrate in *A. bisporus* is glycogen, the concentration of which varies from 5 to 10% of the dry weight. A major portion of the carbohydrate in *A. bisporus* comprises mannans, trehalose, and β-glucan.[109] Reis *et al.*[118] reported that among the carbohydrate content, trehalose and mannitol were the most abundant sugars in *A. bisporus*. In 2016, Baars *et al.* reported that mannitol was the most abundant sugar found in *A. bisporus.*[119] Tsai *et al.* in 2007 and Owaid *et al.* in 2017 reported that in fresh fruiting bodies of *A. bisporus*, mannitol is one of the main forms of soluble sugar, followed by glucose, which varied from 17.6–28.1 mg per g depending on the stage of maturity of the fungus.[120,121]

Chitin is predominately found in the fungal cell wall and is present in dietary fiber.[122] Hemicelluloses, mannans, and glucans have also been found in the fungal cell wall, and contribute to the medicinal properties of mushrooms. In a study conducted by Atila *et al.*[43] the chitin content of *A. bisporus*, was found to be 9.60 g per 100 g dry matter.

2.6.4 Minerals

A. bisporus is considered to be a crucial and valuable source of essential minerals as it is mainly rich in iron (Fe), copper (Cu), potassium (K), manganese (Mn), cobalt (Co) and selenium (Se).[121] Phosphorus (P) and potassium (K) are the main mineral constituents in mushrooms, followed by calcium (Ca), magnesium (Mg), zinc (Zn), iron (Fe), and sodium (Na). Due to the presence of these minerals in *A. bisporus,* it is a healthy dietary supplement.[123] Mohiuddin *et al.*[114] also studied the mineral content of *A. bisporus* grown under various locations. However, the mineral contents varied in the samples in the ranges of 56.2–91.1 mg per kg for magnesium (Mg), 37.2–61.9 mg per kg for sodium (Na), 143.6–396 mg per kg for iron (Fe), 54.6–163.4 mg per kg for copper (Cu), and 36.6–58.0 mg per kg for zinc (Zn). Another study carried out by Caglarırmak[124] showed the mineral contents in *A. bisporus* as presented in Table 2.5.

Similarly, as per work done by Ahlavat *et al.*,[115] the mineral contents, especially in the fruiting bodies, of sodium and selenium in *A. bisporus* were shown to be 500.8 mg per kg and 1.34 mg per kg, respectively.

Table 2.5 Mineral content in *A. bisporus*.

Mineral	Range (mg kg^{-1})
Magnesium	88.0–76.3
Potassium	213.3–238.8
Zinc	8.1–8.7
Phosphorus	7.4–7.9
Sodium	2652–2500
Iron	7.4–7.9
Calcium	534.2–554.8

2.6.5 Vitamins

Due to their high phytochemical constituents and pharmaceutical value, it is now widely known that mushrooms are a valuable and treasured source of vitamins. A study carried out by Bernas and Jaworska[125] reported that in the button mushroom, niacin and riboflavin are the most abundant vitamins. Other important vitamins in the button mushroom include ascorbic acid, vitamin B1, α-tocopherol, and vitamin B3. In another study, Caglarırmak[124] reported that the button mushroom is a good source of the vitamins thiamine, folic acid, riboflavin and niacin, but also has a low amount of vitamin C. In another study carried out by Furlani and Godoy,[106] the average values of riboflavin and thiamine were determined from a fresh sample of *A. bisporus.* They found that the values of vitamin B1 and vitamin B2 were around 0.03 and 0.25 mg per 100 g. However, except for cultivated mushrooms, the riboflavin contents in the button mushroom, *Pleurotus* spp., and *L. edodes*, were higher than those of different vegetables.

Ahlavat *et al.*[115] showed that vitamin D is present in good quantity in the button mushroom, at 984 IU g per kg. However, Simon *et al.*[126] showed that the content of vitamin D varies between wild and cultivated *A. bisporus*. Further, Reis *et al.*[118] showed that the reason for poor or low vitamin D content might be due to the farming of mushroom species in the absence of suitable light availability. Another study reported by Roberts *et al.*[127] noticed that the exposure of skin to UV radiation helps to increase vitamin D levels. Ergo-sterol, a precursor of vitamin D, is also found in the myco-cell wall. However, Shao *et al.*[102] stated that the content of ergo-sterol is usually associated with the antioxidant activity of the mushroom. The nutritional components present in the basiocarp of *A. bisporus* are shown in Table 2.6.

2.6.6 Cosmetic Value

In recent years, mushrooms have also been researched for their medicinal as well as nutritional value.[128] However, inconsequential work has been carried out so far in order to identify and explore the hidden biotechnological potential of edible fungal extracts to formulate cosmetic products.[129–131] As per a literature survey, it is observed that skin whitening in humans is achieved by inhibiting the biochemical activity of the tyrosinase enzyme.[87,92] In the past few years, the cosmetic industry has isolated beneficial ingredients from mushrooms for their aesthetic value. This industry has also used them in cosmetics and skin-whitening products, such as skin lotions and creams.[132]

Several natural or intrinsic mechanisms lead to slow skin aging and have considerable effects on skin and other body organs through hormonal changes that occur with age and also due to exposure to UV rays. This exposure results in the creation of reactive oxygen species (ROS), which finally result in oxidative stress.[133] One of the important transcription factors that is accelerated by ROS is activator factor-1 (AP-1). AP-1 increases elastin protein and also carries out collagen breakdown by up-regulating matric metalloproteinase.[134] Due to

Table 2.6 Nutritional components in basiocarp of *A. bisporus*.

Nutritional group	Components	Amount	Unit	Ref.
Protein amino acids	Alanine	5.8	g per 100 g total protein fresh weight	82, 164 and 165
	Tyrosine	4.2		
	Proline	6.1		
	Cysteine	1.1		
	Asparagine acid	3.4	mg per g dry weight	
	Arginine	2.2		
	Glycine	2.0		
	Isoleucine	1.0		
	Serine	3.1		
	Methionine	0.8		
	Lysine	3.5		
	Phenylalanine	2.1		
	Leucine	0.8		
	Threonine	1.3		
	Histidine	14.1		
	Valine	2.3		
	Total amount of amino acids	44.2		
Lipid fatty acids	Palmitic acid	13.35	mg per 100 g dry weight	146
	Arachidic acid	0.92		
	Caprinic acid	0.85		
	Oleic acid	6.07		
	Linoleic acid	67.29		
	Myristic acid	0.94		
	Stearic acid	3.72		
	Caprylic acid	1.08		
	Pentadecanoic acid	0.23		
	Palmitoleic acid	4.84		
	Lauric acid	0.11		
	Total saturated fatty acids	20.28		
	Linolenic acid	1.52		
	Total unsaturated fatty acids	79.72		
Carbohydrates	Mannitol	23.62	g per 100 g fresh weight	82 and 166
	Total sugars	4.50		

Table 2.6 (*Continued*)

Nutritional group	Components	Amount	Unit	Ref.
	Trehalose	1–3		
	Fructose	2.62		
Indol compounds	L-Tryptophan	0.39	mg per 100 g dry weight	162 and 167
	Melatonin	0.11		
	Kynurenic acid	6.21		
	Serotonin	5.21		
	Tryptamine	0.06		
	Indoloacetic acid	0.19		
Vitamins	Niacin	42.0	mg per 100 g dry weight	164, 168 and 169
	α-Tocopherol	1–4		
	Vitamin B1	0.6		
	β-Tocopherol	1.0		
	Vitamin B2	5.1		
	Tocopherol	2–3		
	Vitamin B3	43.0		
	Vitamin C	17.0		
	Folic acid	450	μg per 100 g	
	Vitamin D	3.0		
	Vitamin B12	0.8		
Phenolic compounds	Caffeic acid	392.51	mg per kg dry weight	82, 163, 170 and 171
	Gallic acid	280.45		
	Catechins	56.74		
	p-Coumaric acid	2.31		
	Protocatechuic acid	83.26		
	Ferulic acid	42.83		
	Cinnamic acid	0.38		
	Myricetin	2729.46		
	Free phenols	176–487		
	Total phenols	277–687		
Sterols	Ergosta-5,7-deiniol	6.05	mg 100 per g dry weight	172 and 173
	Ergosterol	186.1		
	Ergosta-7,22-deiniol	2.45		
	Ergosta-7-enol	1.73		

increased interest in using natural or plant-based compounds for skincare regimes, there has been an increase in demand for natural compounds. This is mainly due to their protective and defensive roles in response to free radicals as well as their decrease in oxidative enzyme production. In one study, tyrosinase (isolated from *A. bisporus*) was found to be extremely efficient in protecting against inflammatory diseases, as well as inflammation.[135]

Ethanolic extract of the button mushroom exhibits strong antioxidant properties and is widely used in treating skin aging. *A. bisporus* extracts are known to show antibacterial activity against harmful microorganisms, such as *Staphylococcus aureus* and its methicillin-resistant form. These fatal bacteria are widely known as they colonize the skin pores during inflammation or injury. A cosmetic cream base prepared from mushroom extracts has been shown to display anti-inflammatory as well as antioxidant activity in response to the production of melanin and nitric oxide by inhibiting tyrosinase activity.[136]

Various bioactive compounds that are present in cream bases, are extracted from mushrooms. They are known to enrich beauty and also play crucial roles in treating hyperpigmentation, inflammation, and skin aging[86] A cream base based on *A. bisporus* extract has been shown to exhibit significant antioxidant as well as antibacterial action against bacterial strains.[133] The extract contains bioactive compounds, such as 2-amino-3*H*-phenoxazin-3-one, which when used in precise concentrations such as 0.5, 1, and 2 M can restrain melanin formation by 80, 54.1, and 39.4%, respectively.[136] However, the phenolic compounds found in mushroom extracts also exhibit antibacterial activity, disturbing bacterial membranes and therefore causing significant damage to bacterial DNA.[137]

In another study carried out on *A. bisporus* extract by Alves *et al.*,[138] it was seen that there was significant resistance against *P. aeruginosa* and *E. coli*. As *A. bisporus* accumulates selenium in its fruiting bodies,[139,140] its extract is also used in hair shampoos in order to treat hair problems, such as hair loss, dandruff, oily hair, *etc.* Moreover, it is also used for better growth of hair by cleansing the hair and scalp.[141] In another study, it was shown that in addition to selenium, *A. bisporus* extract also contains vitamin D and several minerals, such as iron and copper. Together this provides richness and also plays a key role in building healthy hair.[128]

2.7 Safety, Toxicity and Regulation

Due to the severe impact that environmental pollution has on human health, there has been an increase in the concerns of food safety worldwide. As is known, food products not only provide essential nutrients but should also be of appropriate quality for human health. Consumers or end users should be self-assured that the products that are available are completely safe for consumption and will definitely not cause any adverse health effects. As is known, mushrooms uptake heavy metals with the help of their spacious mycelia. Although, due to short life span of fruiting bodies, the proportion of contents that comes from atmospheric depositions seem to be of less importance.[142] However, several heavy metals, for example, lead and cadmium, are deleterious environmental pollutants. As per a few research studies, it has been reported that the contents of heavy metals in food products are higher than the allowed permissible levels.[143] The consumption of such heavy-metal-loaded food can cause severe health issues that lead to either acute or chronic poisoning.

In today's global acceptance of mushrooms as a key dietary food, the occurrence of mushroom poisoning is a matter of huge concern. However, mushroom poisoning differs according to lifestyle as well as economic factors.[144] According to the North American Mycological Association, in 2014 there were 78 reported mushroom poisoning cases in humans,[145] while in the years 1994–2012, there were 183 reported deaths in China[146] and around 102 cases at Iraq in the period of 2006–2012[147] accredited to mushroom poisoning. Although in Thailand, mushroom poisoning is seen to be closely related to lifestyle; around 38.73% of the victims of mushroom poisoning are agricultural workers.[148] It has been quite noticeable in the past few years that mushroom poisoning is emerging as a global issue.[149] However, in the Indian subcontinent, several cases of poisonings have been seen in specific regions where people are heavily dependent on wild resources for their food consumption.

In 2013, George and Hegde[148] recorded a case of muscarinic poisoning, and several local regional newspapers, such as The Kaimudi Online and The Shillong Times have reported mushroom toxicity in the past. Therefore it is essential to distinguish and differentiate between toxicological profiles of wild mushrooms before they can be considered safe for human consumption.[150] In 2017, Bijeesh *et al.*[150] observed that in India there has been an increase in mushroom poisoning cases in Kerala, mainly due to the misidentification of mushrooms.

In global guidelines formulated by the European Union, the regulations that are currently in force outline the maximum permissible limits of cadmium and mercury only in the case of fresh mushrooms, while for fresh cultivated mushrooms only the maximum permissible limit of lead is documented. However, in the case of arsenic, there is no limit that has been specified in mushrooms.[151] In the case of wild grown mushrooms, there are no legal rules or regulations that demarcate the maximum amount of lethal elements allowed in a dried sample. Therefore, it is the need of the hour to question the safety of mushroom consumption, especially of dried wild fungi.

2.8 Trade, Storage Stability and Transport

India has an agriculture-based economy, with more than 60% of its population, directly as well as indirectly, dependent on agro products. These agriculture-based products contribute nearly 18% to the nation's GDP. However, about 30–35% of yearly crop production is wasted due to poor pest management techniques. Due to unsuitable and inadequate postharvest management, annual losses in horticultural crops are around 35–40%, which is equivalent to Rs 40 000 crores per year. This includes millions of Rupees worth of produce, along with labor wastage, energy input, and loss of efforts which are needed in horticultural production. Moreover, the situation in India is alarming. As per sources, in India, annual wastage and discarded horticultural crops are almost equal to the annual consumption of the United Kingdom. Due to insufficient storage and processing facilities, 10–15% of horticultural crops perish every year that could be used to feed thousands of poor people. Therefore, the developing of efficient and effective postharvest technology (PHT) requires an amalgamation of different scientific disciplines and approaches that can lead to proficient storage and processing facilities. This includes scientific skills and unique and sustainable ideas, alongside commercial business partnerships that can work in collaboration in order to meet the growing food requirements due to an increase in population size.[152]

Edible mushrooms have the capacity to convert industrial, household, and agricultural waste, as well as forestry waste, into nutrient-rich food. Advanced PHT can help to meet the increasing food demands as it can help to minimize losses. It can also enable the production of nutritional dietary items from low-level raw product with the help of processing as well as advanced fortification techniques. PHT also has the potential to enhance or maintain food quality, which can help make it easily marketable. Some of the important stages of PHT are harvesting, transportation, handling, and proper storage, along with proper processing or preserving, and finally, packaging. All these stages help minimize food loss both after harvesting as well as during storage. Therefore, this adds to value to the products produced and can eventually help to generate income in rural and agro-dependent areas. Since mushrooms are known for their perishable nature and short life span, very effective and efficient, as well as advanced, post-harvest management strategies are required.

Due to high water content, insect pests, and their active metabolism, fresh mushrooms have higher chances of spoilage and quality deterioration postharvest. PHT processes involve proper harvesting, efficient mushroom handling, correct storage facilities, and suitable processing, packaging, and conveyance, in addition to marketing.[153] Instead of selling their products at local market places without first processing the raw material, farmers could go to primary centers for processing and to add commercial value to their products. This will generate more employment opportunities and will increase income in rural areas. With advanced PHT, the processed products are more stable, are known to have improved digestibility, and increase diet diversity. This will lead to varied products choices for customers along with a better range of minerals and vitamins. Proper packing and suitable storage are the most important postharvest practices which are needed for the availability of fresh mushrooms. Instant mushroom products as well as ready-to-prepare mushroom food products are also being considered by research experts of mushroom-based industries.

2.9 Conclusions

In recent years, studies on mushrooms have gained much attention due to their wide and comprehensive attributes, including antidiabetic, antioxidant, anti-cancer, and anti-microbial properties. However, the use of *Agaricus* species as a popular dietary supplement has not been limited to specific regions of the country and has entered the commercial market owing to their immense food value. As a known reservoir of several bioactive compounds, suitable postharvest strategies are required for wide commercial profiting from this fungus. This chapter detailed the various aspects of Agaricus species, including their taxonomy, medicinal, phytochemical, cosmetic, and pharmaceutical properties, alongside information on their trade and storage.

Abbreviations

A. bisporus	*Agaricus bisporus*
A. campestris	*Agaricus campestris*
A. brasiliensis	*Agaricus brasiliensis*
A. arvensis	*Agaricus arvensis*
A. bitorquis	*Agaricus bitorquis*
A. silvaticus	*Agaricus silvaticus*
DW	Dry weight
NGO	Non-government organization
SOD	Superoxide dismutase
T2D	Type 2 diabetes

References

1. J. Glamočlija, D. Stojković, M. Nikolić, A. Ćirić, F. S. Reis, L. Barros, I. C. Ferreira and M. Soković, *Food Funct.*, 2015, **6**(6), 1900.
2. R. Walleyn and J. Rammeloo, *Scr. Bot. Belg.*, 1994, **10**, 1.
3. W. Guan, X. Fan and R. Yan, *Food Control*, 2013, **34**, 554.
4. S. P. Therkelsen, G. Hetland, T. Lyberg, I. Lygren and E. Johnson, *PLoS One*, 2016, **11**(7), e0150191.
5. X. Wu, W. Guan, R. Yan, J. Lei, L. Xu and Z. Wang, *Postharvest Biol. Technol.*, **118**, 51–58.
6. M. Gąsecka, Z. Magdziak, M. Siwulski and M. Mleczek, *Eur. Food Res. Technol.*, 2018, **244**, 259–268.
7. R. T. Have, H. Wijngaard and N. A. E. Aries-Kronenburg, *J. Agric. Food Chem.*, 2003, **51**, 2242.
8. P. N. Hood, M. O. Beets and J. F. Kimberley, *For. Ecol. Manage.*, 2004, **196**, 311.
9. J. F. Grove, *Phytochemistry*, 1981, **20**, 2021.

10. R. Parslew, K. T. Jones, J. M. Rhodes and G. R. Sharpe, *Br. J. Dermatol.*, 1999, **140**, 56.
11. M. Kuo, *Agaricus augustus*, 2014, (http://www.mushroomexpert.com/agaricus_augustus.html).
12. C. Bas, Proceedings of the First International Seminar on Mushroom Science, Pudoc Wageningen, 1991.
13. R. L. Zhao, S. C. Karunarathna, O. Raspé, L. A. Parra, J. Guinberteau, M. Moinard, A. De Kesel, G. Barroso, R. Courtecuisse, K. D. Hyde, A. K. Gelly, E. D. Desjardin and P. Callac, *Fungal Diversity*, 2011, **51**, 279.
14. S. C. Karunarathna, J. Chen, P. E. Mortimer, J. C. Xu, R. L. Zhao, P. Callac and K. D. Hyde, *Mycosphere*, 2016, **7**(4), 417.
15. A. Cappelli, *Libreria editrice Biella Giovanna*, 1984, p. 558.
16. P. Heinemann, *Bulletin du Jardin Botanique National de Belgique*, 1956, **26**(1), 1.
17. P. Heinemann, *Darwiniana*, 1987, **28**, 283.
18. J. Geml, D. M. Geiser and D. J. Royse, *Mycol. Prog.*, 2004, **3**(2), 157.
19. J. Geml, D. M. Geiser and D. J. Royse, *Mycol. Prog.*, 2004, **3**(2), 157.
20. M. P. Challen, R. W. Kerrigan and P. Callac, *Mycologia*, 2003, **95**, 61.
21. R. W. Kerrigan, P. Callac, J. Guinberteau, M. P. Challen and L. A. Parra, *Mycologia*, 2006, **97**, 1292.
22. R. Kerrigan, P. Callac and L. Parra, *Mycologia*, 2008, **100**(6), 876.
23. L. A. Parra, *Fungi Europaei, Volume 1: Agaricus: Allopsalliota, Nauta and Bas*, Candusso Edizioni, Alassio, 2008, p. 824.
24. R. W. Kerrigan, *The Agaricales of California*, Mad River Press, Elureka, 1986, vol. 6.
25. N. Thongklang, J. Chen, A. R. Bandara, K. D. Hyde, O. Raspé and L. A. Parra, *Mycoscience*, 2016, **57**, 239.
26. R. L. Zhao, J. L. Zhou, J. Chen, S. Margaritescu, S. Sánchez-Ramírez and K. D. Hyde, *Fungal Divers*, 2016, **78**, 239.
27. J. Chen, P. Callac, L. A. Parra, S. C. Karunarathna, M. Q. He and M. Moinard, *Persoonia*, 2017, **38**, 170.
28. L. A. Parra, C. Angelini, B. Ortiz-Santana, G. Mata, C. Billette and C. Rojo, *Phytotaxa*, 2018, **345**, 219.
29. M. Q. He, B. Chuankid, K. D. Hyde, R. Cheewangkoon and R. L. Zhao, *MycoKeys*, 2018, **40**, 53.
30. S. K. Deshmukh, *Microbiology and Biotechnology for Sustainable Development*, CBS Publishers and Distributors, New Delhi, India, 2004, p. 121.
31. C. Manoharachary, K. Sridhar, R. Singh, A. Adholeya, T. S. Suryanarayanan, S. Rawat and B. N. Johri, *Curr. Sci.*, 2005, **89**(1), 58.
32. T. N. Kaul, *Trop. Mycol.*, 2002, **1**, 131.
33. T. S. Suryanarayanan, G. Venkatesan and T. S. Murali, *Curr. Sci.*, 2003, **85**, 489.
34. K. Das, *NeBIO Int. J. Environ. Biodiversity*, 2010, **1**(2), 69.

35. S. Kumar and Y. P. Sharma, Proceedings of the 7th International Conference on Mushroom Biology and mushroom Products (ICMBMP7), 2011, p. 568.
36. S. Dwivedi, M. K. Tiwari, U. K. Chauhan and A. K. Pandey, *Int. J. Pharm. Life Sci.*, 2012, **3**, 1363.
37. H. Pushpa and K. B. Purushothama, *Eurasian J. Agric. Environm. Sci.*, 2012, **12**, 750.
38. G. Thiribhuvanamala, V. Prakasam, G. Chandrasekar, K. Sakthivel, S. Veeralakshmi, R. Velazhahan and G. Kalaislvi, Proceedings of the 7th International Conference on Mushroom Biology and Mushroom Products (ICMBMP7), 2014, p. 155.
39. R. Kumar, A. Tawal, S. Pandey, R. K. Borah, D. Borah and J. Borgohain, *Nusantara Biosci.*, 2013, **5**, 1.
40. P. Khaund and S. R. Joshi, *Indian J. Nat. Prod. Resour.*, 2013, **4**, 197–204.
41. B. Muszynska, K. Kala, J. Rojowski, A. Grzywacz, W. Opoka and J. Pol, *Food Nutr. Sci.*, 2017, **67**, 173.
42. B. Das, B. De, R. Chetree and S. C. Mandal, in *Herbal Medicine in India*, Springer, Berlin/Heidelberg, Germany, 2020, vol. 5, p. 509; G. Dhamodharan and S. Mirunalini, *Pharmacol. Online*, 2010, **2**, 456.
43. F. Atila, M. N. Owaid and M. A. Shariati, *J. Microbiol. Biotechnol. Food Sci.*, 2017, **7**, 281.
44. L. S. Adams, S. P. Phung, X. Wu, L. Ki and S. Chen, *Nutr. Cancer*, 2008, **60**, 744.
45. M. R. C. G. Novaes, V. M. C. R. Fabiana, D. R. Goncalves and M. C. Menezes, *Clinics*, 2011, **66**, 2133.
46. S. C. Jeong, S. R. Koyyalamudi, Y. T. Jeong, C. H. Song and G. J. Pang, *Med. Food*, 2012, **15**, 58.
47. Z. U. Khan, G. Aisikaer, R. U. Khan, J. Bu, Z. Jiang, Z. Ni and T. Ying, *Postharvest Biol. Technol.*, 2014, **95**, 36.
48. L. Barros, T. Cruz, P. Baptista, L. M. Estevinho and I. C. F. R. Ferreira, *Food Chem. Toxicol.*, 2008, **46**, 2743.
49. G. Oms-Oliu, I. Aguilo-Aguayo, O. Martin-Belloso and R. Soliva-Fortuny, *Postharvest Biol. Technol.*, 2010, **56**, 216.
50. J. Liu, L. Jia, J. Kan and C. H. Jin, *Food Chem. Toxicol.*, 2013, **51**, 310.
51. L. Yu, D. G. Fernig, J. A. Smith, J. D. Milton and J. M. Rhodes, *Cancer Res.*, 1993, **53**, 4627.
52. N. A. Talpur, B. W. Echard, A. Y. Fan, O. Jaffari, D. Bagchi and H. G. Preuss, *Mol. Cell. Biochem.*, 2002, **237**, 129.
53. R. Estruch, M. A. Martinez-Gonzalez, D. Corella, J. Basora-Gallisa, V. Ruiz-Gutierrez, M. I. Covas, M. Fiol, E. Gómez-Gracia, M. C. López-Sabater and R. Escoda, *J. Epidemiol. Community Health*, 2009, **63**, 582.
54. S. C. Jeong, Y. T. Jeong, B. K. Yang, R. Islam, S. R. Koyyalamudi, G. Pang, K. Y. Cho and C. H. Song, *Nutr. Res.*, 2010, **30**, 49.
55. M. S. Calvo, A. Mehrotra, R. B. Beelman, G. Nadkarni, L. Wang, W. Cai, B. C. Goh, M. D. Kalaras and J. Uribarri, *Plant Foods Human Nutr.*, 2016, **71**, 245.

56. O. A. A. Zaid, S. E. Sonbaty and M. A. Neama, *Int. J. Pharma Sci.*, 2017, **7**, 1740.
57. G. Oms-Oliu, I. Aguilo-Aguayo, O. Martin-Belloso and R. Soliva-Fortuny, *Postharvest Biol. Technol.*, 2010, **56**, 216.
58. B. Muszyńska, A. Krakowska, K. Sułkowska-Ziaja, W. Opoka, W. Reczyński and B. Baś, *J. Food Sci. Technol.*, 2015, **52**, 7337–7344.
59. G. Bohus, *Annales Historico Naturales Musei Nationalis Hungarici*, 1990, vol. 81, p. 37.
60. D. Kothari, S. Patel and S. K. Kim, *Biomed. Pharmacother.*, 2018, **105**, 377.
61. F. R. Smiderle, G. Alquini, M. Z. Tadra-Sfeir, M. Iacomini, H. J. Wichers and L. J. Van Griensven, *Carbohydr. Polym.*, 2013, **94**, 91.
62. B. V. McCleary and A. Draga, *J. AOAC Int.*, 2016, **99**, 364.
63. L. K. Jagadish, V. V. Krishnan, R. Shenbhagaraman and V. Kaviyarasan, *Afr. J. Biotechnol.*, 2009, **8**, 654.
64. L. C. Clark, G. F. Combs, B. W. Turnbull, E. H. Slate, D. K. Chalker, J. Chow and A. Krongrad, *JAMA*, 1996, **276**, 19.
65. M. R. Spolar, E. M. Schaffer, R. B. Beelman and J. A. Milner, *Cancer Lett.*, 1999, **138**, 145.
66. P. Rzymski, M. Mleczek, P. Niedzielski, M. Siwulski and M. Gąsecka, *J. Sci. Food Agric.*, 2017, **97**, 923.
67. D. H. Kim, B. K. Yang, S. C. Jeong, J. B. Park, S. P. Cho, S. Das, J. W. Yun and C. H. Song, *Biotechnol. Lett.*, 2001, **23**, 513.
68. N. Ekowati, N. I. Yuniati, H. Hernayanti and N. I. Ratnaningtyas, *Biosaintifika J. Biol. Biol. Edu.*, 2018, **10**, 655.
69. K. Yamashita, K. Kawai and M. Itakura, *Nutr. Res.*, 1984, **4**, 961.
70. T. Todesco, A. V. Rao and O. Bosello, *Am. J. Clin. Nutr.*, 1991, **54**, 860.
71. J. M. Wong, R. de Souza, C. W. Kendall, A. Emam and D. J. Jenkins, *J. Clin. Gastroenterol.*, 2006, **40**, 235.
72. J. J. Volman, R. P. Mensink, L. J. L. D. Van Griensven and J. Plat, *Eur. J. Clin. Nutr.*, 2010, **64**, 720.
73. C. Huang, C. Lii, A. Lin, Y. Yeh, H. Yao, C. Li and T. Wang, *Arch. Toxicol.*, 2013, **87**, 167.
74. N. Leyva-Lopez, E. P. Gutierrez-Grijalva, O. G. Vazquez and J. B. Heredia, *Molecules*, 2017, **22**, 1.
75. A. Esmaillzadeh and L. Azatbakth, *J. Nutr.*, 2008, **138**, 1469.
76. M. Gasecka, Z. Magdziak, M. Siwulski and M. Mleczek, *Eur. Food Res. Technol.*, 2018, **244**, 259.
77. X. Lin, L. Ma, S. B. Racette, C. L. A. Spearie and R. E. Ostlund, *Am. J. Physiol. Gastrointest. Liver Physiol.*, 2009, **296**, 931.
78. H. Xu, Y. J. Yang, T. Yang and H. Y. Qian, *Ageing Res. Rev.*, 2013, **12**, 1.
79. A. Zepeda–Bastida, D. Ojeda-Ramirez, S. Soto-Simental, N. Rivero-Perez and M. Ayala-Martínez, *J. Agric. Sci.*, 2016, **8**, 43.
80. M. Ozturk, M. E. Duru, S. Kivrak, N. Mercan-Dogan, A. Turkoglu and M. A. Ozler, *Food Chem. Toxicol.*, 2011, **49**, 1353.

81. X. Shang, Q. Tan, R. Liu, K. Yu, P. Li and G. P. Zhao, *Int. J. Med. Mushrooms*, 2013, **15**, 165.
82. H. Soltanian, S. H. Rezaeian, A. Shakeri, J. Janpoor and H. R. Pourianfar, *Biomed. Res.*, 2016, **27**, 56.
83. G. Gebreyohannes, A. Nyerere, C. Bii and D. Berhe Sbhatu, *J. Evidence-Based Complementary Altern. Med.*, 2019, **2019**, 1–7.
84. M. H. H. Tehrani, E. Fakhrehoseini, M. K. Nejad, H. Mehregan, M. Hakemi-Vala and J. Iran, *Pharmaceut. Res.*, 2012, **11**, 145.
85. M. H. Risan, S. H. Taemor, A. H. Muhsin and S. Hussan, *Eur. J. Biomed.*, 2017, **4**, 9.
86. A. C. Ruthes, Y. D. Rattmann, E. R. Carbonero, P. A. J. Gorin and M. Iacomini, *Carbohydr. Polym.*, 2012, **87**, 1620.
87. D. L. Komura, E. R. Carbonero, A. H. P. Gracher, C. H. Baggio, C. S. Freitas, R. Marcon, A. R. Santos, P. A. Gorin and M. Iacomini, *Bioresour. Technol.*, 2010, **101**, 6192.
88. I. Golak-Siwulska, A. Kałuzewicz, S. Wdowienko, L. Dawidowicz and K. Sobieralski, *Herba Pol.*, 2018, **64**, 71.
89. S. P. Wasser, *Appl. Microbiol. Biotechnol.*, 2010, **89**, 1321.
90. S. P. Wasser, *Appl. Microbiol. Biotechnol.*, 2002, **60**, 274.
91. M. Zhang, S. W. Cui, P. C. K. Cheung and Q. Wang, *Trends Food Sci. Technol.*, 2007, **18**, 4.
92. D. L. Komura, E. R. Carbonero, A. H. Gracher, C. H. Baggio, C. S. Freitas and R. Marcon, *Bioresour. Technol.*, 2010, **101**, 6192.
93. E. R. Carbonero, A. H. Gracher, M. C. Rosa, G. Torri, G. L. Sassaki and P. A. J. Gorin, *Phytochemistry*, 2008, **69**, 252.
94. F. R. Rosado, E. R. Carbonero, R. F. Claudino, C. A. Tischer, C. Kemmelmeier and M. Iacomini, *FEMS Microbiol. Lett.*, 2003, **221**, 119.
95. M. F. Moradali, H. Mostafavi, S. Ghods and G. A. Hedjaroude, *Int. Immunopharmacol.*, 2007, **7**, 701.
96. C. M. P. G. Dore, T. C. G. Azevedo, M. C. R. Souza, L. A. Rego, J. C. M. Dantas and F. R. F. Silva, *Int. Immunopharmacol.*, 2007, **7**, 1160.
97. U. Lindequist, T. H. Niedermeyer and W. D. Julich, *J. Evidence-Based Complementary Altern. Med.*, 2005, **2**, 285.
98. M. Shrivastava, *Studies on Mushroom Dehydration (Pleurotus florida)*, PhD thesis, IIT KGP, W.B., 1998.
99. B. K. Mehta, S. K. Jain, G. P. Sharma, A. Doshi and H. K. Jain, *Int. J. Adv. Biotechnol. Res.*, 2011, **2**, 201.
100. E. P. Winer, C. Hudis and H. J. Burstein, *J. Clin. Oncol.*, 2002, **20**, 3317.
101. F. R. Smiderle, A. C. Ruthes, J. Van Arkel, W. Chanput and M. Lacomin, *BMC Complementary Altern. Med.*, 2011, **11**, 58.
102. S. Shao, M. Hernandez, J. K. G. Kramer, D. L. Rinke and R. Tsao, *J. Agric. Food Chem.*, 2010, **58**(22), 11616.
103. J. Ying, X. Mao, Q. Ma, Y. Zong and H. Wen, *Icons of medicinal fungi from China*, Science Press, 1987, p. 307.
104. B. J. Grube, E. T. Eng, Y. C. Kao, A. Kwon and S. Chen, *J. Nutr.*, 2001, **131**, 3288.

105. M. Sadler, *Nutr. Bull.*, 2003, **28**, 305.
106. R. P. Z. Furlani and H. T. Godoy, *Food Chem.*, 2008, **106**, 816.
107. F. J. Irazoqui, F. E. Zalazar, G. A. Nores and M. A. Vides, *Glycoconjugate J.*, 1997, **14**, 313.
108. C. A. Presant and S. Kornfeld, *J. Biol. Chem.*, 1972, **247**, 6937.
109. D. A. Shbeeb, M. F. Farahat and H. M. Ismail, *Prog. Nutr.*, 2019, **21**, 203.
110. R. Gothwal, A. Gupta, A. Kumar, S. Sharma and B. Alappat, *J. Biotechnol.*, 2012, **2**, 249.
111. R. V. C. Cardoso, A. Fernandes, J. C. M. Barreira, S. C. Verde, A. L. Antonio, A. M. Gonzalez-Paramas, L. Barros and I. C. F. R. Ferreira, *Food Chem.*, 2019, **278**, 760.
112. A. Braaksma and D. J. Schaap, *Postharvest Biol. Technol.*, 1996, **7**, 119.
113. S. Sadiq, H. N. Bhatti, M. A. Hanif and J. Iran, *Chem. Chem. Eng.*, 2008, **27**, 151.
114. K. M. Mohiuddin, M. Alam, T. Arefin and I. Ahmed, *Asian J. Med. Biol. Res.*, 2015, **1**, 495.
115. O. P. Ahlavat, K. Manikandan and M. Singh, *Mushroom Res.*, 2016, **25**, 1.
116. L. Citores, S. Ragucci, J. M. Ferreras, A. Di Maro and R. Iglesias, *ACS Chem. Biol.*, 2019, **14**, 1319.
117. M. Fogarasi, Z. M. Diaconeasa, C. R. Pop, S. Fogarasi, C. A. Semeniuc, A. C. Farcaş, D. Ţibulca, C.-D. Salagean, M. Tofana and S. A. Socaci, *Agronomy*, 2020, **10**, 1972.
118. F. S. Reis, A. Martins, L. Barros and I. C. Ferreira, *Food Chem. Toxicol.*, 2012, **50**, 1201.
119. J. J. P. Baars, A. S. M. Sonnenberg, R. Mumm, I. Stijger and R. Wehrens, *Metabolites contributing to taste in Agaricus bisporus*, Wageningen, The Netherlands, 2016.
120. S. Y. Tsai, T. P. Wu, S. J. Huang and J. L. Mau, *Food Chem.*, 2007, **103**, 1457.
121. M. N. Owaid, A. Barish and M. A. Shariati, *Open Agric.*, 2017, **2**, 537.
122. P. C. K. Cheung, *Nutr. Bull.*, 2010, **35**, 292.
123. J. Falandysz and J. Borovicka, *Appl. Microbiol. Biotechnol.*, 2013, **97**, 477.
124. N. Caglarırmak, *J. Sci. Food Agric.*, 2009, **89**, 634.
125. E. Bernas and G. Jaworska, *J. Food Compos. Anal.*, 2016, **49**, 1.
126. R. R. Simon, K. M. Phillips, R. L. Horst and I. C. Munro, *Agric. Food Chem.*, 2011, **59**, 8724.
127. J. S. Roberts, A. Teichert and T. H. McHugh, *J. Agric. Food Chem.*, 2008, **56**, 4541.
128. Y. Wu, M. H. Choi, J. Li, H. Yang and H. J. Shin, *Cosmetics*, 2016, **3**, 22.
129. D. Kim, J. Park, J. Kim, C. Han, J. Yoon, N. Kim, J. Seo and C. Lee, *J. Agric. Food Chem.*, 2006, **54**, 935.
130. T. S. Chang, *Int. J. Mol. Sci.*, 2009, **10**, 2440.
131. J. S. Lee, D. B. Shin, S. M. Lee, S. H. Kim, T. S. Lee and D. C. Jung, *Korean J. Mycol.*, 2013, **41**, 104.

132. O. Taofiq, A. M. Gonzalez-Paramas, A. Martins, M. F. Barreiro and I. C. F. R Ferreira, *Ind. Crops Prod.*, 2016, **90**, 38.
133. E. Papakonstantinou, M. Roth and G. Karakiulakis, *Dermatoendocrinology*, 2012, **4**, 253.
134. K. H. Leem, *Int. J. Biosci. Biotechnol*, 2015, **7**, 127.
135. T. S. Thring, P. Hili and D. P. Naughton, *BMC Complemen. Altern. Med.*, 2009, **9**, 27.
136. M. Miyake, S. Yamamoto, O. Sano, M. Fujii, K. Kohno, S. Ushio, K. Iwaki and S. Fukuda, *Biosci., Biotechnol., Biochem.*, 2010, **74**, 753.
137. N. Martins, L. Barros, M. Henriques, S. Silva and I. C. F. R. Ferreira, *BioMed. Res. Int.*, 2015, 1.
138. M. J. Alves, I. C. Ferreira, A. Martins and M. Pintado, *J. Appl. Microbiol.*, 2012, **113**, 466.
139. L. Racz, A. Bumbalova, M. Harangozo, J. Tolgyessy and O. J. Tomecek, *Radioanal. Nucl. Chem.*, 2000, **245**, 611.
140. Y. Ogra, K. Ishiwata, J. R. Encinar, R. Łobinski and K. T. Suzuki, *Anal. Bioanal. Chem.*, 2004, **379**, 861.
141. Y. T. Jeong, B. K. Yang, S. C. Jeong, S. M. Kim and C. H. Song, *Phytother. Res.*, 2008, **22**, 614.
142. P. Ziarati, S. Arbabi, S. Arbabi Bidgoli and M. Qomi, *IJFAS J.*, 2013, **2**, 268.
143. H. Persson, *Medicine*, 2012, **40**, 135.
144. M. W. Beug, NAMA Toxicology Committee Report for 2014 North American Mushroom Poisonings, North American Mycological Association, 2014.
145. Z. Chen, P. Zhang and Z. Zhang, *Fungal Divers.*, 2014, **64**, 123.
146. A. Badsar, M. Rahbar Taramsari, A. Amir Maafi, M. Rouhi Rad, G. Chatrnour and S. Khajeh Jahromi, *Iran. J. Toxicol.*, 2013, **7**, 798.
147. J. Saoraya and P. C. Inboriboon, *Emerg. Med.*, 2013, **3**, 1.
148. P. George and N. Hegde, *Toxicol. Int.*, 2013, **20**, 113.
149. A. D. Lima, R. C. Fortes, M. G. Novaes and S. Percário, *Nutr. Hosp.*, 2012, **27**, 402.
150. C. Bijeesh, K. B. Vrinda and C. K. Pradeep, *J. Mycopathol. Res.*, 2017, **54**, 477.
151. Commission Regulation (EU) No. 488/2014 of 12 May 2014 amending Regulation (EC) No. 1881/2006 as regards maximum levels of cadmium in foodstuffs Text with EEA relevance, 2014.
152. M. Singh, Mushroom production: an agribusiness activity, in *Mushroom: cultivation, marketing and consumption*, ed. M. Singh, B. Vijay, S. Kamal and G. C. Wakchaure, Directorate of Mushroom Research (Indian Council of Agricultural Research), Solan, 2013, p. 1.
153. C. G. Mrema and S. R. Rolle, Status of the postharvest sector and its contribution to agricultural development and economic growth, 9th JIRCAS international symposium—value addition to agricultural product, 2002, p. 13.

154. M. V. K. Sivakumar, R. Motha, D. Wilhite and J. Qu, Towards a compendium on national drought policy: Proceedings of an expert meeting, *Am. Meteorol. Soc.*, 2011.
155. E. P. Priya and S. Subhashini, Thin layer and deep bed drying basic theories and modeling, *Int. J. Agric. Eng.*, 2016, **18**, 314–325.
156. M. M. Hassan, A. O. Oyewale, J. O. Amupitan, M. S. Abduallahi and E. M. Okonkwo, Preliminary phytochemical and antibacterial investigation of crude extracts of the root bark of Detarium microcarpum, *J. Chem. Soc. Niger.*, 2004, **29**, 26–29.
157. A. Kumar, M. Mazzanti, M. Mistrik, M. Kosar and G. V. Beznoussenko, *et al.*, ATR mediates a checkpoint at the nuclear envelope in response to mechanical stress, *Cell*, 2014, **158**, 633–646.
158. S. Chandrasekaran, G. Gaurav, S. Serino and Miranda, Ringing and springing response of triangular TLPs, *Int. Shipbuild. Prog.*, 2011, **58**, 141–163.
159. B. Venkata Raman, L. A. Samuel, M. Pardha Saradhi, B. Narashimha Rao and A. N. Vamsi Krishna, *et al.*, Antibacterial, Antioxidant Activity and GC-MS Analysis of Eupatorium odoratum, *Asian J. Pharm. Clin. Res.*, 2012, **5**, 99–106.
160. T. Sudha, S. Chidambarampillai and V. R. Mohan, GC-MS Analysis of bioactive components of aerial parts of Fluggea leucopyrus Willd (Euphorbiaceae), *J. Appl. Pharm. Sci.*, 2013, **3**, 126–130.
161. Y. Yu, C. Lee, J. Kim and S. Hwang, Group-specific primer and probe sets to detect methanogenic communities using quantitative real-time polymerase chain reaction, *Biotechnol. Bioeng.*, 2005, **89**, 670–679.
162. D. A. Shbeeb, M. F. Farahat and H. M. Ismail, Macronutrients analysis of fresh and canned Agaricus bisporus and Pleurotus ostreatus mushroom species sold in Alexandria markets, Egypt, *Int. J. Nutr.*, 2019, **21**, 203–209.
163. F. S. Reis, A. Martins, L. Barros and I. C. Ferreira, Antioxidant properties and phenolic profile of the most widely appreciated cultivated mushrooms: A comparative study between *in vivo* and *in vitro*, *Food Chem. Toxicol.*, 2012, **50**, 1201–1207.
164. E. Bernas, G. Jaworska and Z. Lisiewska, Edible mushrooms as a source of valuable nutritive constituents, *Acta Sci. Pol., Technol. Aliment.*, 2006, **5**, 5–20.
165. F. Pei, Y. Shi, X. Gao, F. Wu, A. M. Mariga, W. Yang, L. Zhao, X. An, Z. Xin and F. Yang, *et al.*, Changes in non-volatile taste components of button mushroom (Agaricus bisporus) during different stages of freeze drying and freeze drying combined with microwave vacuum drying, *Food Chem.*, 2014, **165**, 547–554.
166. N. Gheibi, A. A. Saboury, K. Haghbeen and A. A. Moosavi-Movahedi, The effect of some osmolytes on the activity and stability of mushroom tyrosinase, *J. Biosci.*, 2006, **31**, 355–362.

167. B. Muszynska, K. Sułkowska-Ziaja and H. Ekiert, Indole compounds in fruiting bodies of some edible Basidiomycota species, *Food Chem.*, 2011, **125**, 1306–1308.
168. S. Y. Tsai, T. P. Wu, S. J. Huang and J. L. Mau, Antioxidant properties of ethanolic extracts from culinary-medicinal button mushroom Agaricus bisporus (J. Lange) Imbach (Agaricomycetes) harvested at different stages of maturity. *Int. J. Med. Mushrooms* 2008, **10**, 127–137.
169. J. Kalbarczyk and W. Radzki, Cultivated mushrooms as a valuable diet constituent and a source of biologically active substances, *Herba Pol.*, 2009, **55**, 224–232.
170. J. Czapski, Antioxidant activity and phenolic content in some strains of mushrooms (Agaricus bisporus), *Veg. Crops Res. Bull*, 2004, **62**, 165–173.
171. K. Labus, A. Turek, J. Liesiene and J. Bryjak, Efficient Agaricus bisporus tyrosinase immobilization on cellulose-based carriers, *Biochem. Eng. J.*, 2011, **56**, 232–240.
172. S. A. Heleno, P. Diz, M. A. Prieto, L. Barros, A. Rodrigues, M. F. Barreiro and I. C. Ferreira, Optimization of ultrasound-assisted extraction to obtain mycosterols from Agaricus bisporus L. by response surface methodology and comparison with conventional Soxhlet extraction, *Food Chem.*, 2016, **197**, 1054–1063.
173. B. Muszynska, K. Sułkowska-Ziaja and A. Wojcik, Levels of physiologically active indole derivatives in the fruiting bodies of some edible mushrooms (Basidiomycota) before and after thermal processing, *Mycosciene*, 2013, **54**, 321–326.

CHAPTER 3

Armillaria mellea: *Honey Fungus*

MANJU NEHRA,*[a] VIKASH NAIN,[a] RAHUL THORY[b] AND AMANJYOTI[b]

[a] Chaudhary Devi Lal University, Sirsa, India; [b] Guru Nanak Dev University, Amritsar, India
*Email: manju.vnehra@gmail.com

3.1 Introduction

One of the most harmful pathogenic fungi to impact trees, both in Britain and internationally, is the honey fungus, which can afflict trees in both urban and rural settings. It is essential to realize that a Honey Fungus attack on a roadside tree or garden tree could have more detrimental effects on surrounding organisms or road users than this fungus. Previously, *Armillaria mellea* (Vahl ex Fr.) Kummer was blamed for all tree infestations; however, one of the first signs that Honey Fungus was more diverse than widely believed was in Romagnesi's description of three more species.[1,2] An experiment in Finland showed that *Armillaria* species could be recognized in culture by genetic tests, which was a response to a pilot project that identified British species based on the fruiting bodies.[3,4] This research was of practical significance because the formation of fruiting bodies is frequently inconsistent and, when visible, they are not always readily identifiable. The results of the same experiment showed that Finland is home to three separate species with various physical and ecological traits. The word *Armillaria*, from the Latin name for bracelet, bangle, or epaulette. This word is relevant since the majority of *Armillaria* species have a broad, persistent ring that resembles skin that is linked to the upper half of the stipe.[5]

Phytochemistry and Nutritional Composition of Significant Wild Medicinal and Edible Mushrooms: Traditional Uses and Pharmacology
Edited by Ajay Sharma, Garima Bhardwaj and Gulzar Ahmad Nayik

Published by the Royal Society of Chemistry, www.rsc.org

3.2 Taxonomy

The honey mushroom *Armillaria mellea*, also known as Fungi, Basidiomycota, Dikarya, Agaricomycotina, Agaricales, Agaricomycetidae, Tricholomataceae, Agaricomycetes, *Armillaria*, is a member of this diverse biological group.[6] Although the word "melea" means "honey," it specifically relates to the way caps look, not to the flavor or aroma of the fungus, which might occasionally deceive. *A. mellea* is a white-rot fungus, which is a large group of saprophytic, parasitic, and mycorrhizal fungi. It belongs to Basidiomycota phylum, Agaricomycetes class and falls in the Agaricales order, in the family Physalacriaceae.

Armillaria mellea is a species that may be found throughout mainland Europe, except in Scandinavia, where it is only found rarely or on occasion but is becoming more ubiquitous farther south. It is very abundant and widespread throughout the entirety of Britain and Ireland. This particular species can also be discovered in many other regions, such as North America.

Armillaria fungi use mycelial threads to propagate all around a tree and, more importantly, from that tree to another tree, and the black bootlace-like rhizomorphs (which means 'root forms') are made up of parallel bundles of fungal hyphae as depicted above. The dark rhizomorphs in this image evolve beneath the bark of a hardwood tree. Finally, the bark peels off and falls away, exposing the mycelial threads hiding underneath. Individual *Armillaria mellea* rhizomorphs normally measure 2 mm wide; but, when they gather together, they can produce far larger threads, measuring up to 5 cm in diameter. These rhizomorphs can link the mycelium of Honey Fungus in an infected tree to the mycelium of a nearby host tree located up to a few meters away when they grow through soil.

The Danish mycologist Martin Vahl (1749–1804) was the one who first described and identified this species in 1790, giving it the scientific name *Agaricus mellea*. (Most gilled fungus were previously classified as belonging to the massive Agaricus genus, but they have now been moved to a variety of other genera.) The well-known German mycologist Paul Kummer was the one responsible for moving honey fungus into its current genus in 1871; as a result, its scientific name was changed to *Armillaria mellea* at that time. *Armillaria mellea* is the species that serves as the model for the *Armillaria* genus. Various other synonyms of *Armillaria mellea* are used, such as *Agaricus sulphureus* Weinm., *Armillaria mellea* var. sulphurea (Weinm.) P. Karst., *Armillariella mellea* (Vahl) P. Karst., *Armillaria mellea* var. maxima Barla, *Armillaria mellea* var. minor Barla, and *Clitocybe mellea* (Vahl) Ricken.

3.3 Resource Availability (Habitat, Characteristics and Species) of Honey Fungus

Honey fungus are parasitic on or close to broad-leaf trees, predominantly on fruit trees in orchards; some can even be saprobic on stumps and dead roots, and occasionally on fallen branches. Honey fungus may spread to an extensive range of plants and trees, such as alder (*alnus*), amelanchier, apple, peach, birch, ceanothus (*california lilac*), cedar, cotoneaster, fir, hawthorn, holly, horse chestnut (*aesculus*), katsura tree (*cercidiphyllum*), pear, privet (*ligustrum*), rhododendron and sorbus (*rowan*), *etc.* To date, there has been no spray or control measure developed for the elimination of honey fungus. Therefore, when a plant is infected by honey fungus there remains no other alternative but to destroy it. Proper removal of all parts of the plant is necessary as once the host plant is removed the bootlace rhizomorphs spreading out from it can no longer survive.

Honey fungus has been divided into multiple species by some researchers, while others group all of them together. For instance, certain types of honey fungus, such as *Armillaria mellea,* have highly-scaled caps and a stem with relatively small rings, whereas other types have almost no scales but large, white rings instead. Even if other *Armillaria* species have a similar overall look, it is still possible to differentiate between them in the wild by carefully examining their macroscopic characteristics. *Armillaria mellea* has a strongly acidic flavor with a hint of a faintly sour odor (if you plan on undertaking a taste test, you should be aware that some individuals believe that honey fungus is edible after appropriate processing while some people have found it to be indigestible, and it may even be dangerous to a small percentage of people).

Armillaria ostoyae: the Dark Honey Fungus, as it is more generally known, can be identified by the unique dark brown or black scales that cover the underside of its persistent stem ring.

Armillaria tabescens: the Ringless Honey Fungus is exceptionally similar to the Ringed Honey Fungus, with the exception that it does not have a stem ring and its gills mature to a pinkish-brown color.

Armillaria gallica has a bulbous stem and a temporary cobweb-like ring that matures into a yellowish ring zone. When the plant is fully developed, the ring zone is all that is left. It maintains an inrolled border, its gills turn evenly rusty-brown, and it has a fragrance and flavor similar to radish.

Pholiota squarrosa is generally the same color all over and is covered in scales.

3.4 Medicinal Uses

A common species found in deciduous, coniferous, and mixed forests on many continents is the honey mushroom, *Armillaria mellea* (Vahl.) P. Kumm. (*Armillariella mellea* (Vahl) P. Karst., upper *Basidiomycetes, Agaricomycetideae, Tricholomataceae*). Although this species is frequently used for its nutritional quality and therefore is commonly available in Europe, it is typically classified as a conditionally edible fungus (young fruiting bodies that are properly prepared are edible).

Research on the chemical makeup of fungal species is interesting and beneficial for expanding our knowledge as well as determining the medicinal and culinary benefits of mushrooms in addition to the toxicological elements of their actions. Polysaccharides are among the most well-known mushroom metabolite components and are frequently studied because of their importance in the treatment of cancer. Terpene molecules, a class of metabolites, are extensively researched in large quantities; it has been demonstrated that several of them have anticancer properties.

Numerous proteins- and non-proteinogenic peptides, enzymes, amino acids, and heterocyclic compounds were found in higher fungus species when nitrogen compounds were studied. Research on indole compounds has mainly concentrated on the hallucinogens and indole alkaloids found in mushrooms.[7] Tryptophan was the research focus on non-hallucinogenic indole alkaloids in many upper Basidiomycete species. Serotonin, melatonin, 5-hydroxy-tryptophan, indole, indoleacetic acid, and skatole, as well as tryptamine and its metabolites, have also been investigated.[8] The vast majority of species also serve as important sources of sterols, polyunsaturated, proteins, certain vitamins (especially those in the B complex), and essential minerals. Recent investigations into the chemical make-up of *A. mellea* reported the existence of a variety of chemicals that are advantageous to its biological characteristics. Carbohydrates, sterols, sphingolipids, fatty acids, sesquiterpenoids, indole compounds, peptides, enzymes, and mineral components are among the chemical classes to which these substances belong. These compounds protect brain[9] and bone marrow cells and have potential as antibacterial, fibrinolytic, anticancer, anti-atherogenic, and immunostimulant treatments.

3.4.1 Phytochemistry of Honey Fungus

3.4.1.1 Protein

Most of the essential amino acids are known to be present in the protein found in mushrooms. Calculating biological value, nutritional value, and protein score has been based on the quantitative spectrum of essential amino acids.[10] Protein has been observed to be present in *A.* mellea in significant amounts (23.0 g per 100 g of dry matter).[11] In higher fungi, the quantity of protein compounds ranges from 19.0 to 39.0 g per 100 g of dry matter and accounts for more than half of the total nitrogen.[12] The majority of the proteins in the fruiting bodies of *A. mellea* are albumins (50%). Prolamines and gluteins make up around 10% and 5% of the total nitrogen, respectively, whereas globulins make up about 25% of the nitrogen.[13]

3.4.1.2 Carbohydrates

Due to their role in anticancer therapy, polysaccharides are among the fungal metabolites that are the best understood among carbohydrates. Japanese researchers developed the first anticancer medications containing polysaccharide structures derived from fungi (lentinan, isolated from *Lentinus edodes* fruiting bodies; krestin from *Trametes versicolor* fruiting bodies; and schizophyllan isolated from the culture broth of *Schizophyllum commune*). Lentinan, krestine, and schizophyllan typically have the structures of a-mannans and -D-glucans. Molecular weight, water solubility, branching pattern, and tertiary structure all affect how they behave. Glucans with β-(1–6) linkages having β-(1–6) and β-(1–3) branches exhibit strong immunological activity as compared to the glucan with an α-(1–4) linkage. These polysaccharides have no negative impact on humans. Polysaccharides exhibit a variety of anticancer effects. By scavenging free radicals, boosting the immune system, or inducing apoptosis, they can decrease DNA damage, lower carcinogen concentrations, and limit the activation and growth of cancer cells. They can help treat infectious disorders as well as cancer because of their capacity to stimulate the immune system. Lentinan was first used in 2007 to treat advanced or recurring gastric cancer by the Japanese Foundation for Multidisciplinary Treatment of Cancer.[14] One important benefit of polysaccharides is they exhibit no harmful effects on humans.[10]

The Health Protection Agency Centre for Infection in London conducted the initial tests on polysaccharides from *A. mellea* fruiting bodies. Using a peptide component, it was shown that *A. mellea* contains -glucan. The anticancer activity of this component was reported by the authors. Studies on the peptide-sugar glucan component structure revealed that glucose was present and was linked to other molecules *via* -(1,3) and -(1,6) connections.[15]

A. mellea fruiting bodies contain polysaccharides with diverse chemical structures, according to more recent research conducted at Madrid's Centro de Investigaciones Biológicas. We extracted two polysaccharide components with β-glucan structures. While the second fraction contained -(1,3)-glucan, the main component was made up of linear chains of -(1,3)- and -(1,4)-glucan coupled to a protein.[16] Glycogen and indigestible substances, such as cellulose, other glucans, mannans, and chitin, are also present in the polysaccharides found in *A. mellea* fruiting bodies. These substances reduce the levels of LDL cholesterol in the blood and liver as well as the plasma triglyceride content, which lowers cardiovascular risk. Additionally, they affect how much cholesterol is absorbed from the diet, reducing both overall absorption and the LDL concentration of the blood's while having no impact on HDL. Eating edible mushrooms increases the elimination of bile acids and neutral steroids as they contain significant levels of dietary fiber, particularly glucans and chitin (which thicken intestinal contents). In the gastric environment, chitosan amine groups acquire a positive charge and interact with highly reactive bile acids. Low pH causes the chitosan and bile acid complexes to dissolve, and the organism then excretes them.[17]

In addition to having an impact on the immune system and blood pressure, chitins and glucans also have hypoglycemic, antibacterial, antiviral, and anti-inflammatory properties.[17] As such, they are used by the pharmaceutical sector as drug carriers and slimming aids. *A. mellea* were found to have 7.6 g of these polysaccharides for every 100 g of the edible components of the fruiting body. Trehalose is the only disaccharide found in significant amounts in the mushroom of higher fungi; mannitol is the most common sugar alcohol. Additionally, free-form monosaccharides such as glucose, galactose, mannose, fructose, and sedoheptulose are present relatively frequently.[18]

3.4.1.3 Fatty Acids

The amount of unsaturated fatty acids in the fruiting bodies of higher Basidiomycetes was calculated to be >70% of the overall amount of fatty acids, despite the fact that they are known for having a low fat content.[19] Sclerotia, mycelia, and fruiting bodies are all rich in unsaturated fatty acids as they serve as storage resources. The same fatty acids (unsaturated fatty acids) that are found in flower plants are the most plentiful. Numerous fatty acids have been found in the fruiting bodies of *A. mellea*, according to research by academics from the University of Georgia, Athens, USA. Four of the[20] fatty acids found were unsaturated, three of which were C18 compounds 9-*cis*, 12-*cis*-octadecadenoic acid (9,12-C18:2), 9-*cis*,10-cisexpoxy-octadecadenoic acid (9,12-Cexpoxy-18:2), and one was a C_{16} compound (5-*cis*-hexadecanoic acid) (5-C16:1). Due to their roles as bile precursors, building blocks for the production of prostacyclin and prostaglandins, and their antiatherogenic properties, unsaturated fatty acids are essential for human health.[19]

3.4.1.4 Sesquiterpenoids

In higher Basidiomycetes mushrooms, the group of metabolites known as sesquiterpenoids is highly prevalent. Fungi frequently exhibit antibacterial and anticancer properties. The most well-known are the derivatives of the winter mushroom *Flammulina velutipes*, iludan and kuparan. The category of protoilludan derivatives, which is made up of a number of chemicals, includes the sesquiterpenoids found in *A. mellea.* They appear in the form of aryl esters, such as orselinates.[20] The Chinese Academy of Sciences in Beijing conducted the most extensive research on them, finding the following substances in the *A. mellea* mycelium: 4′-methylmelledonal, judeol, armillol, armillarin, armillaricin, armillaripin, armillarridin, armillarigin, armillarikin, armillatin, armillaritin, and armillariwin.[21–23] The amounts of armillarin and melleolid in the mycelium, fruiting bodies, and rhizomorphs have been assessed by other authors, with the armillarin content calculated to be 0.0085%, 0.002%, and 0.0008%, and the melleolid content to be 0.0002%, 0.0011%, and 0.0011%, respectively.[24] It has been demonstrated that jedeol and 4′-methylmelleolid have antibacterial effects on Gram-positive bacteria. This suggests the possibility of using these compounds as antibiotics. Melledonal A, B, and C are other substances found in the honey mushroom.[25]

3.4.1.5 Sesquiterpene Aryl Esters

The main components of the *A. mellea* and its mycelium are reported in the literature as sesquiterpene aryl esters (sesquiterpenoids), with most of these compounds exhibiting antibiotic and antifungal activity.[26] The number of carbon atoms and the substituent type affect the potency of antibacterial compounds.[25] Melleolide was the first identified sesquiterpene aryl ester from *A. mellea* to exhibit antibacterial action.[27] Following the discovery of armillarin and armillaridin, more than 50 other sesquiterpenoids, including armillaricin, armillaribin, armillarigin, and judeol, were isolated.[21–23]

3.4.1.6 Indole Compounds

Among the chemicals derived from mushrooms, indoles, particularly indole alkaloids and hallucinogenic compounds, have drawn the most attention. Tryptophan has been the focus of research on non-hallucinogenic indole molecules, among other chemicals. Higher Basidiomycetes species, such as *Lactarius deterrimus*, *Xerocomus badius*, and *Tricholoma equstre*, as well as their *in vitro* mycelial cultures, have been found to contain tryptophan.[8,28]

At the Department of Pharmaceutical Botany, Collegium Medicum, Jagiellonian University in Kraków, the amount of non-hallucinogenic indole chemicals in *Armillaria mellea* fruiting bodies was examined (Poland). Melatonin, serotonin, tryptamine, tryptophan, indole, indoleacetamide, indoleacetonitrile, indoleacetic acid, kynureinic acid, and kynurenine sulphate were the ten indole compounds that were examined.[8,28,29] Tryptamine (2.74 mg per 100 g of dry weight), serotonin (2.21 g per 100 g dry weight), and tryptophan were discovered to be present in *A. mellea* fruiting bodies, according to Muszyska,[26] of the ten chemicals examined (4.47 g per 100 g dry weight).[29] The remaining substances were not found. However, the fruiting bodies of *A. mellea* contain a sizable amount of serotonin. Sleep, body temperature, mood, maturation, cell regeneration, and growth are all significantly regulated by serotonin, which also affects psychological inclination. Serotonin slows down cell ageing, enhances sleep quality, and elevates mood, all of which help the immune system function more effectively overall. Serotonin is a neurotransmitter that influences hunger and pain threshold in nervous system cells. Serotonin is the hormone that affects happiness. Depression must be treated with drugs that affect serotonin reuptake in synaptic clefts. According to recent reports, Alzheimer's disease might be brought on by low serotonin levels. Antioxidant properties are produced by indole compounds. Studies conducted *in vitro* have shown that melatonin and other indole derivatives (*N*-acetylserotonin, 6-methoxytryptamine) reduce lipid peroxidation in a dose-dependent manner.[30]

It is also important to note that *A. mellea* fruiting bodies contain significant amounts of tryptamine and tryptophan. Tryptamine can interact with monoamine oxidase (MAO) inhibitors and even result in fatal poisoning in those taking these medications.[31] Tryptamine is a psychotropic substance that binds to and activates the 5-HT2 receptor. The neurological system is harmed when tryptophan is consumed in excess.

3.4.1.7 Enzymes

The osmotic feeding of mushrooms is related to the high enzyme concentration in the mycelia. The majority of higher Basidiomycetes species generate hemicellulases, cellulases, and linginases, which participate in the breakdown of plant tissue cell walls. Different enzymes are produced by numerous species. Tyrosinase, for example, is produced by *Agaricus bisporus*, but proteases are common to many higher Basidiomycetes species. A nuclease found in *A. mellea* is likely involved in DNA recombination and nucleic acid degradation during the autolysis of the fruiting body. The enzyme exhibits endonuclease and exonuclease activity.[31,32] *A. mellea* also contains several proteases, one of which was discovered in fruiting bodies taken from their native environments and confirmed to be lysine-specific metalloendopeptidase. As it is utilized for protein sequencing, protease is a useful tool in molecular biology and genetic engineering.[33]

Protease also exhibits fibrinolytic characteristics, making it a viable treatment for thrombocytosis.[34] Similar outcomes were demonstrated for a chymotrypsin-like protease that was isolated from *A. mellea* mycelial cells. Cellulase, hemicellulase, and peroxidase, which are involved in the breakdown of cell walls, are also produced by *A. mellea* fruiting bodies, allowing them to invade tree tissues.[35]

3.4.1.8 Minerals

Higher fungi produce fruiting bodies with a high concentration of easily assimilated mineral components that are unevenly distributed, with concentrations typically higher in the cap than the stalk. Unfortunately, mushrooms also gather radioactive and heavy metals. These substances are found in particularly high concentrations in mushrooms that grow adjacent to highways and in heavily industrialized areas, especially near smelting factories.[36,37]

Heavy metal content in *A. mellea* fruiting bodies gathered in northern Poland was examined by a study team from the University of Gdansk (Poland), led by Falandysz. K, Na, Mg, Ca, Fe, Zn, Mn, Cu, Pb, Cd, Hg, and Ag were discovered to be present in the fruiting bodies. Zn, Cu, and Se were discovered by the Institute of Technology in Kraków, Poland. One of the primary antioxidants in the body, glutathione oxidase, contains a significant amount of selenium in its active core.[38] *A. mellea* fruiting bodies were found to contain Cr, Ni, and Cd by a Turkish research group at Gaziosmanpasa University.[38]

3.4.1.9 Vitamins

Since vitamins are vital for both human and animal organisms, vitamin content is incredibly valuable. According to Furlani and Godoy,[39] fungi are one of the finest sources of vitamins, particularly vitamin B.[40] Niacin and vitamin B1 are found in high concentrations in *A. mellea* when compared to other Basidiomycetes; however, its vitamin B2 content is lower than those of other fungi.[41,42]

3.4.1.10 Sterols and Sphingolipids

Mushrooms typically contain 5–8% dry weight total lipid content. *A. mellea* has a low total lipid content, estimated to be 1.8%. Mushrooms frequently contain sterols. Sterols are mainly present in mushroom fruiting bodies.

Typical mushroom sterols are extremely unsaturated. The most prevalent sterol found in all members of higher Basidiomycetes is ergosterol, also known as vitamin D. The primary components of this group found in *A. mellea* are ergosterol and ergosterol peroxide, the latter of which has been demonstrated to have anticancer properties.[12] Researchers from the Chinese Academy of Sciences in Kunming tested the discovery of a sphingolipid molecule called armillaramide, the chemical formula of which is (2*S*,3*S*,4*R*)-2-hexadecanoyl-aminooctadecane-1,3,4-triol. Scientists are very interested in this class of lipids due to their intriguing biological characteristics. There is evidence that several of the chemicals in this class exhibit anticancer, anti-hepatotoxic, and immunostimulatory properties.[43,44]

3.5 Safety, Toxicity, and Regulation

A. mellea is thought to be one of the most prevalent and harmful species of fungi that cause fungal root infections.[45,46] It has long been acknowledged as a significant plant parasite that primarily affects trees, but it can also harm fruit and nut crops, causing root rot. It also infests a number of orchids, although the orchid is viewed as the parasite and the mushroom as the host.

Attacks by *A. mellea* occur on trees of various ages and classes, including coniferous and deciduous types.[47,48] It is thought to have the largest host range of the *Armillaria* genus, even though its capacity for rhizomorphic spread is less developed than that of other *Armillaria* species (such as *A. bulbosa*).[49] Nevertheless, there is no cure for *Armillaria* root sickness. Although scientists from all over the world are working to find a way to control it, it presents a challenge as the mushroom feeds on living trees and kills other tree stumps that have naturally perished or were killed by other circumstances. Another problem is that, even after infected tree remains have been removed from the soil, mycelium can persist saprophytically for years inside woody residual roots, serving as an inoculum for the subsequent crop's infection.[45,50] As mentioned, *A. mellea* can function as a parasite, saprophyte, or in a mycorrhizal association.

3.5.1 Pathogen

A. mellea has a dual pathogenic mechanism of action: either it colonizes the cambium of living roots as a parasite or feeds on the dead tissue of decimated trees as a saprophyte.[44,51] According to reports, most *Armillaria* species are facultative necrotrophs, which implies that they switch between the two mechanisms at some point in their life cycle. *A. mellea* has been characterized as going through two phases in the literature: the parasitic phase, which involves invading living roots. Once the fungus destroys the cambium, this phase is over, and shortly after that the saprophytic phase begins, in which the fungus feeds on the dead tissue.[52] *A. mellea* attacks existing weakened and injured trees by penetrating the wound, but it can also enter through the roots without causing any damage. There are three ways in which *Armillaria* infections can occur. Due to their frequent occurrence, infections transmitted *via* mycelium and rhizomorphs, short-range dispersal routes, are the most often researched illnesses. Rhizomorphs from *A. mellea* are used to penetrate the corky layer protecting the root and adhere to the root.

First, a rhizomorph's tip applies mechanical force to conduct the entrance, softly compressing and pushing the host cells without causing damage.[53] Only later does the tip of the branch that is penetrating release enzymes that operate to break down the root bark, and once the penetration is complete, it only takes a short period of time for the fungus to penetrate and take control of the host.[54] Mycelium grows *via* immediate communication with uninfected trees, whereas rhizomorphs grow through the soil and produce wide and lengthy branches that can transmit the infection between infected and uninfected trees without any immediate communication between them.[55] The mycelium of *Armillaria* grows not only in tree roots but also along the tree trunk, sometimes reaching heights of a few meters.[46] A basidiospore-based infection is the final type of infection, which has received the least amount of attention in the literature. This occurs as the basidiospores are transported by wind to fresh, uninfected areas after being freed from the basidiocarp. Basidiospores appear to be a perfect tool for spreading the illness across a broader area, but they are ineffective since it is difficult for stumps to colonize them, and this type of infection thus only rarely occurs.[52]

3.6 Cultivation and Harvesting

In the phylum Basidiomycota, *Armillaria* is a genus of fungi that is found all over the world. Numerous species are root phytopathogens that typically infect herbaceous and woody hosts, including trees, shrubs, and vines. The species in this genus collectively have a substantial economic impact and can infect various hosts. Primary or secondary *Armillaria* infections are linked to abiotic stressors, such as root rot, flooding, and drought. Under these circumstances, *Armillaria* growth is aided by modifications in the status of secondary metabolites and root nutrients. Britain has long understood that *Armillaria* is primarily a problem for amenity/ornamental trees. Between 2001 and 2018, it was detected on 18% of the samples that members of the Royal Horticultural Society provided to the organization, making it the most often diagnosed disease. As previously mentioned, cultivated trees frequently experience stress due to unfavorable environmental factors and mishandled maintenance. In particular, deep burial and root abnormalities have been linked to heightened *Armillaria* susceptibility. Biotic stress also makes trees more vulnerable, as seen in acute oak decline, ash dieback, and defoliation. The quantity and viability of inoculum, or colonized material, present in the soil at a site is the primary determinant driving infection risk in addition to host stress. Ecologically speaking, *Armillaria* spp. are generalists with excellent plasticity, and together they have multiple ways of obtaining sustenance from a variety of sources. In regard to plants, various tactics occur on a continuum ranging from saprophytic to pathogenic, *via* parasitism. In the illness cycle, all three of these tactics are used. As saprobes, *Armillaria* spp. contribute significantly to ecology by degrading cellulose and lignin deadwood in the environment. Any infected substance, whether a saprophytic or pathogenic colonization, can serve as an infection reservoir as it can continue to exist for years after colonization and the demise of the host. As a result, entire sites are frequently referred to as being infested.

Many species of the genus *Armillaria* are phytopathogens, causing significant economic damage to a wide range of cultivated hosts. *In vitro* culture and experiments involving *Armillaria*-host interactions are required for disease management and ecology research. Inoculation relies on placing colonized woody substrates into root systems. Methodologies vary significantly between groups of researchers, with little standardization. Even for factors regarded as being controlled, *e.g.*, substrate stem diameter, there is potential for significant effects on, and variation in, the performance of the applied inoculum.

3.6.1 Pharmacological Potential

The orchid *Gastrodia elata* and the honey mushroom coexist harmoniously (Orchidaceae). Russia, Korea, China, Japan, Indonesia, India, Taiwan, Thailand, Nepal, Tibet (where it is used in folk medicine), Australia, and New Zealand are the native countries of this orchid species (where it is used as a food product). The orchid has been used for thousands of years in traditional Chinese medicine, and according to the People's Republic of China Pharmacopeia, it has anticonvulsant, analgesic, and sedative effects. The Chinese Pharmacopeia states that it is utilized in galenic medicines to treat tetanus, vertigo, migraines, paralysis, and epilepsy. There is no chlorophyll in orchids. Except during fluorescence, the plant spends its whole development cycle underground. *A. mellea* infection is necessary to grow a secondary tuber that bears flowers. However, the natural resources of Gastrodia elata have sadly declined, and today, collecting it from its natural habitats is illegal. Attempts to grow this orchid have been unsuccessful due to cost. However, some extremely encouraging results suggest that honey mushrooms may have a comparable impact on *Gastrodia elata*.

Various studies conducted (*in vitro* and *in vivo*) have shown that *A. mellea* is therapeutically effective in reducing the symptoms of epilepsy, vasomotor headaches, tinnitus, insomnia, and paraparesis. These effects are brought about by the anticonvulsant, sedative, hypotensive, and vasodilatory properties that increase cerebral blood flow.[44] Studies have revealed discrepancies in the chemical compositions of *A. mellea* and *Gastrodia elata*, despite pharmacological parallels between their actions. No reports have indicated that extracts from *A. mellea* that were discovered in *Gastrodia elata* included any active ingredients. However, it is believed that the active ingredients in orchids are related to the metabolites of the compounds found in honey mushrooms.

3.6.2 Nutritional Aspects

A. mellea is a species of higher fungi that is extremely intriguing. Much contradictory information about this species can be found in fungi indexes. It is usually classified as a conditionally edible fungus, and eating cooked young fruiting bodies is advised; the best specimens are found growing on coniferous trees. The fact that *A. mellea* caps contain dietary fiber (carbohydrates), unsaturated fatty acids, and important microelements should be stressed. Important potentially therapeutic substances in *A. mellea* include sesquiterpenes and polysaccharides (glucans), both of which exhibit anticancer effects. Studies have shown that the active ingredients in *A. mellea* and other higher Basidiomycete species effectively treat conditions such as hypertension, atherosclerosis, diabetes, and obesity. Therefore, adding mushrooms to the human diet is advised. It is important to note that *A. mellea* is frequently categorized as a parasitic fungus in monographs. It is a deadly parasite of trees, but it also aids in the decay of dead substances and has the potential to form a strong symbiotic relationship. More research should be conducted on the chemical makeup of this exceptionally interesting species, as well as observations on its physiology and biology.

3.7 Conclusions

Although *A. mellea* is a widely used and edible species of fungus, its range of characteristics and uses may come as a surprise. There has been much research on its species description, as evidenced by the literature, including the morphology, environment, life cycle, and elements of the fungus body. When the components of *A. mellea* were discovered, their medicinal and biological uses were also described and applied to a variety of fields or diseases with beneficial results, demonstrating that the world of fungi offers not only culinary delights or purely scientific results but also treatment solutions. *A. mellea*, a member of the Basidiomycota fungal family, is used in bioremediation to address environmental contamination problems. Although this subject has not been well studied, *A. mellea* holds great potential to be employed globally in the ecology field to serve as a primary tool for environmental conservation.

Acknowledgements

The authors acknowledge the Department of Food Science and Technology and the authorities of Chaudhary Devilal University, Sirsa, for providing all the resources for completing the work. The authors thank Prof. Suresh Gahlawat and Prof. Neeraj Dilbagi for their valuable advice. The authors declare no conflict of interest.

References

1. H. Romagnesi, Observations sur les Armillariel/a (1), *Bull. Soc. Mycol. Fr.*, 1970, **86**, 257–268.
2. H. Romagnesi, Observations sur les Armillariella (2), *Bull. Soc. Mycol. Fr.*, 1973, **89**, 195–206.
3. R. Watling, A pilot scheme, *Bull. Soc. Mycol. Fr.*, 1976, **10**, 43–44.
4. K. Korhonen, Interfertility and clonal size in the Armillariella mel/ea complex, *Karstenia*, 1978, **18**, 31–42.
5. A. K. Gębska, *Wpływ selenu na własności antyoksydacyjne grzybni Armillaria mellea*, Praca dyplomowa, Warszawski Uniwersytet Medyczny, 2009.
6. A. P. Kabulska, *Biosynteza, izolacja i badanie struktury analogu lentinanu - polisacharydowego leku z grupy modyfikatorów odpowiedzi biologicznej*, Praca dyplomowa, Warszawski Uniwersytet Medyczny, 2015.
7. M. Somei and F. Yamada, Simple indole alkaloids and those with a nonrearranged monoterpenoid unit, *Nat. Prod. Rep.*, 2003, **20**, 216–242.
8. B. Muszyńska, A. Maślanka, K. Sułkowska-Ziaja and J. Krzek, Analysis of indole compounds and nitric bases in fruiting bodies in Lactarius deterrimus by TLC-UV, *J. Planar Chromatogr.-Mod. TLC*, 2007, **20**, 50–58.
9. N. Watanabe, T. Obuchi, M. Tamai, H. Arak, S. Omura, Y. S. Yang, D. Q. Yu, X. T. Liang and J. H. Huan, A novel N6-substituted adenosine isolated from mi huan jun (*Armillaria mellea*) as a cerebral-protecting compound, *Planta Med.*, 1990, **56**, 48–52.
10. K. Sułkowska-Ziaja, B. Muszyńska and G. Końska, Biologically active compounds of fungal origin displaying antitumor activity, *Acta Pol. Pharm. Drug Res.*, 2005, **62**, 153–160.
11. A. A. Stasiak, *Badanie właściwości antyoksydacyjnych i stopnia zmiatania wolnych rodników wyciągów metanolowych i wodnych otrzymanych z grzybni nieselenowanej i selenowanej Armillaria mellea*, Praca dyplomowa, Warszawski Uniwersytet Medyczny, 2008.
12. J. Florczak and W. Lasota, Wchłaniane i wiązanie kadmu przez boczniaka ostrogowatego w warunkach uprawy, *Bromatol. Chem. Toksykol.*, 1995, **28**, 17–23.
13. I. Karkocha, Badania wartości odżywczej gąski zielonej (Tricholoma equestre) i opieńki miodowej (*Armillaria mellea*), *Rocz. PZH.*, 1964, **15**, 311–315.
14. S. Yoshino and M. Oka, Clinical trial of non-specific immunotherapy using Lentinan in advanced or recurrent gastric cancer, *Gan to Kagaku Ryoho*, 2008, **35**, 2239–2243.
15. C. Amar, J. M. Delaumèny and E. Vilkas, Chemical and biological properties of a peptido-glucan fraction from *Armillaria mellea* (Basidiomycetes), *Biochim. Biophys. Acta.*, 1976, **421**, 263–271.
16. M. E. Sánchez Hernández, C. Garcia Mendoza and M. Novaes-Ledieu, Two different alkali-soluble alpha-glucans in hyphal walls of the basidiomycete *Armillaria mellea*, *Microbiologia*, 1993, **9**, 34–42.
17. P. Pochanavanich and W. Suntornsuk, Fungal chitosan production and its characterization, *Lett. Appl. Microbiol.*, 2002, **35**, 17–21.

18. J. Florczak, A. Karmańska and A. Wędzisz, Comparison of the chemical contents of selected wild growing mushrooms, *Bromatol Toxykol Chem.*, 2004, **37**, 365–371.
19. E. Bernaś, G. Jaworska and Z. Lisiewska, Edible mushrooms as a source of valuable nutritive constituents, *Acta Sci. Pol. Technol. Aliment.*, 2006, **5**, 5–20.
20. A. Arnone, R. Cardillo, N. Gianluca and S. Valdo Meille, Secondary mould metabolites. Part 19. Structure elucidation and absolute United State using fatty acid methyl ester profiling, *Mycol. Res.*, 2006, **110**, 414–422.
21. J. S. Yang, Y. W. Chen, X. Z. Feng, D. Q. Yu and X. T. Liang, Chemical constituents of *Armillaria mellea* mycelium. I. Isolation and characterization of armillarin and armillaridin, *Planta Med.*, 1984, **50**, 288–290.
22. J. S. Yang, Y. W. Chen, Y. L. Su, L. Wang, X. Z. Feng, D. Q. Yu and X. T. Liang, Isolation and structure elucidation of armillaricin, *Planta Med.*, 1989, **55**, 564–565.
23. J. S. Yang, Y. L. Su, L. Wang, X. Z. Feng, D. Q. Yu and X. T. Liang, Chemical constituents of *Armillaria mellea* mycelium. VI. Isolation and structure of armillaripin, *Yao Hsueh Hsueh Pao*, 1990, **25**, 353–356.
24. C. Wang, S. Guo, X. Chen, W. Cao, J. Xu and P. Xiao, Contents of armillarin and melleolid at different stages of development of *Armillaria mellea* (Vahl ex Fr.), *Chung Kuo Chung Yao Tsa Chin.*, 1996, **21**, 274–276.
25. D. M. X. Donnelly, F. Abe, D. Coveney, N. Fukuda and D. O'Reilly, Antibacterial sesquiterpene aryl esters from *Armillaria mellea*, *J. Nat. Prod.*, 1985, **48**, 10–16.
26. L. W. Gao, W. Y. Li, Y. L. Zhao and J. W. Wang, The cultivation, bioactive components and pharmacological effects of *Armillaria mellea*, *Afr. J. Biotechnol.*, 2009, **8**, 7383–7390.
27. S. L. Midland, R. R. Izac, R. M. Wing, A. I. Zaki, D. E. Munnecke and J. J. Sims, Melleolide, a new antibiotic from *Armillaria mellea*, *Tetrahedron Lett.*, 1982, **23**, 2515–2518.
28. B. Muszyńska, K. Sułkowska-Ziaja and H. Ekiert, Indole compounds in fruiting bodies of some selected Macromycetes species and in their mycelia cultured in vitro, *Pharmazie*, 2007, **64**, 479–480.
29. B. Muszyńska, K. Sułkowska-Ziaja and H. Ekiert, Analysis of indole copounds in fruiting bodies of *Armillaria mellea*, *Acta Pol. Pharm. Drug Res.*, 2011, **68**(1), 93–97.
30. E. Sewerynek, M. Stuss, K. Oszczygieł, J. Kułak and A. Lewiński, Protective effect of indole compounds on lipopolysaccharide induced lipid peroxidation in in vitro conditions, *Pol. Endocrinol.*, 2005, **4**, 508–511.
31. G. K. Isbister, S. J. Bove, A. Dawson and I. M. Whyte, Relative toxicity of selective serotonin reuptake inhibitors (SSRIs) in overdose, *J. Toxicol., Clin. Toxicol.*, 2004, **42**, 277–285.
32. V. Healy, S. Doonan and T. McCarthy, Purification, characterization and cDNA cloning of an endo-exonuclease from the basidiomycete fungus *Armillaria mellea*, *Biochem. J.*, 1999, **339**, 713–720.

33. V. Healy, J. O'Connell, T. V. McCarthy and S. Doonan, The lysine-specific proteinase from *Armillaria mellea* is a member of a novel class of metalloendopeptidases located in Basidiomycota, *Biochem. Biophys. Res. Commun.*, 1999, **262**, 60–63.
34. W. G. Lewis, J. M. Basford and P. L. Walton, Specificity and inhibition studies of *Armillaria mellea* protease, *Biochim. Biophys. Acta.*, 1978, **522**, 551–560.
35. S. Lee, J. S. Kim, J. E. Kim and K. Sapkota, Purification and characterisation of fibrinolytic enzyme from cultured mycelia of *Armillaria mellea*, *Protein Expression Purif.*, 2005, **43**, 10–17.
36. F. P. Barry, S. Doonan and C. A. Ross, Cleavage by tripsin and by the proteinase from *Armillaria mellea* at ε-Nformyl- lysine residues, *Biochem. J.*, 1981, **193**, 737–742.
37. P. Kalač, L. Svoboda and B. Havlićkova, Contents of cadmium and mercury in edible mushrooms, *J. Appl. Biomed.*, 2004, **2**, 15–20.
38. P. Kalač, L. Svoboda and B. Havlićkova, Contents of detrimental metals mercury, cadmium and lead in wild growing edible mushrooms: a review, *Energy, Educ., Sci. Technol.*, 2004, **13**, 31–38.
39. J. Rajewska and B. Bałasińska, Biological active compounds in edible mushrooms and their profitable influence on health, *Postępy Hig Med. Doświad.*, 2004, **58**, 352–357.
40. R. P. Furlani and H. T. Godoy, Vitamins B_1 and B_2 contents in cultivated mushrooms, *Food Chem.*, 2008, **106**, 816.
41. M. Majdańska, *Badanie właściwości redukcyjnych i antyoksydacyjnych wyciągów wodnych i metanolowych otrzymanych z grzybni Armillaria mellea nieselenowanej i selenowanej*, Praca dyplomowa, Warszawski Uniwersytet Medyczny, 2007.
42. A. Podlewska, *Porównanie właściwości biobójczych wyciągów alkoholowych i wodnych otrzymanych z grzybni Armillaria mellea*, Praca dyplomowa, Uniwesytet Medyczny w Warszawie, 2006.
43. J. M. Gao, Z. J. Dong and J. K. Liu, A new ceramide from the Basidiomycetes *Armillaria mellea*, *Chin. Chem. Lett.*, 2001, **12**, 139–140.
44. J. M. Gao, X. Yang, C. Wang and J. Liu, Armillaramide, a new sphingolipid from the fungus *Armillaria mellea*, *Fitoterapia*, 2001, **72**, 858–864.
45. K. Baumgartner, M. Coetzee and D. Hoffmeister, Secrets of the subterranean pathosystem of *Armillaria*, *Mol. Plant Pathol.*, 2011, **12**, 515–534.
46. K. Mańka, Badania terenowe i laboratoryjne nad *opieńką miodową-Armillaria mellea* (Vahl) *Quel*, *Pr. Inst Bad Śn.*, 1953, **94**, 1–96.
47. Z. Sterota, *Choroby lasu, Centrum Informacyjne Lasow Państwowych*, Warszawa, 2001.
48. P. Lech and Z. Sterota, Poziom zagrożenia lasow grzybami patogenicznymi, in *Stan uszkodzenia lasow w Polsce w 1999 roku na podstawie badań monitoringowych*, Biblioteka Monitoringu Środowiska, Warszawa, 2000, pp. 59–65.
49. J. Rishbeth, Species of *Armillaria* in southern England, *Plant Pathol.*, 1982, **31**, 9–17.

50. D. B. Redfem and G. M. Filip, Inoculum and Infection, in *Agricultural Handbook*, USDA Forest Service, Washington D.C., 1991, pp. 48–61.
51. E. Hasegawa, Y. Ota, T. Hattori, N. Sahashi and T. Kikuchi, Ecology of Armillaria species on conifers in Japan, *For. Pathol.*, 2010, **41**, 429–437.
52. J. Rishbeth, Infection cycle of *Armillaria* and host response, *Eur. J. For. Pathol.*, 1985, **15**, 332–341.
53. S. M. Zelter, Observations on infections of apple and prune roots by *Armillaria mellea* Vahl, *Phytopathology*, 1926, **16**, 479–484.
54. T. E. Harold, Studies on *Armillaria melea* (Vahl) Quel, Infection, Parasitism and Host Resistance, *J. Agric. Res.*, 1934, **48**, 187–218.
55. D. B. Redfem and G. M. Filip, Inoculum and Infection, in *Agricultural Handbook*, USDA Forest Service, Washington D.C., 1991, pp. 48–61.
56. K. Mańka, *Fitopatologia leśna*, PWRiL, Warszawa, 1998.

CHAPTER 4

Amanita Species

A. BHARTI MITTU,*[a] B. ABIDA BHAT,[b] C. SANDALDEEP KAUR,[c] D. ZARINA BEGUM,[d] E. MAHALDEEP KAUR[e] AND E. NEHA[f]

[a] National Institute of Pharmaceutical Education and Research (N.I.P.E.R.), S.A.S. Nagar, Mohali, India; [b] Department of Immunology and Molecular Medicine, Sher-e-Kashmir Insititute of Medical Sciences, Kashmir, India; [c] Post Graduate Government College for Girls, Sector-11, Chandigarh, India; [d] Shaheed Udham Singh College of Engineering and Technology, India; [e] Department of Microbial Biotechnology, Panjab University, Chandigarh, India; [f] Center of Innovative and Applied Bioprocessing (CIAB), Mohali, India
*Email: bharti9mittu@yahoo.com

4.1 Introduction

The Amanita genus includes almost 500 species, comprising both edible and poisonous mushroom varieties. Amanita species (mushrooms) are native to the temperate and cold ecosystems of the northern hemisphere, such as those of Canada, USA, Asia, Ireland, and Europe. Belonging to the kingdom Fungus and family Amanitaceae, it exhibits an ectomycorrhizal symbiotic relationship with trees, such as pinus, pice, betula, and eucalyptus. The word muscaria comes from the Latin word "Musca", which means fly; hence, the common name of these species is fly agaric. Linnaeus formally classified these species into *Agaricus muscarius*, which later on were transferred to the genus Amanita by Hendrik. The global population of amanita is 900–1000 species, among which 100 are poisonous (*Phalloideae*), 50 are edible (*Caesareae*), and the function of the remaining species is unknown.[1] *Amanita muscaria* (*A. muscaria*) and *A. pantherina* (*Amanitaceae*) mushroom species are responsible for pantherina poisoning, a dysfunction disorder of the

Phytochemistry and Nutritional Composition of Significant Wild Medicinal and Edible Mushrooms: Traditional Uses and Pharmacology
Edited by Ajay Sharma, Garima Bhardwaj and Gulzar Ahmad Nayik

Published by the Royal Society of Chemistry, www.rsc.org

central nervous system.[2] The physical characteristic features of Amanita species are the presence of a cap (orange, red or yellow), gills (free to narrow or crowded), and stipe (with a skirt-like annulus and bulb). A literature review presented the bioactive compounds present in amanita species in terms of their pharmacological and toxicological benefits.[3,4] This chapter will emphasize the important bioactive compounds of these species regarding their nutritional and dietary benefits.

4.2 Taxonomy and Classification

A. muscaria is one of the species belonging to the genus Amanita, and is considered to be a poisonous mushroom. Fly agaric, fly mushroom, and fly amanita are the common names of this species group.

Kingdom: Fungi
Division: Basidiomycota
Class: Homobasidiomycetes
Subclass: Hymenomycetes
Order: Agaricales
Family: Amanitaceae
Genus: *Amanita*

These species often grow in groups with basidiocarps during their different stages of development. *A. muscaria* exhibits morphological variation within a number of known sub-species, such as *A. muscaria* spp. muscaria, *A. muscaria* spp. flavivolvata, *A. muscaria* spp. Formosa and *A. muscaria* spp. Americana.[5,6]

4.3 Resource Availability

Mushrooms not only help in diversification but also addresses issues related to food, health, and the environment. The recycling of agro-waste helps in the conservation and enhanced production of natural resources. Income as well as high sustainability can be achieved using the waste from growing mushrooms. Cultivation and production through commercial sources helps in bioconverting industrial, agricultural, household and forestry waste into nutritious food for mushrooms. However, the production of this food requires vertical space and it is considered to be the highest protein producer per unit area. Advanced technology has been provided with a promising scope to meet the shortage of food, without having undue pressure on land.[5]

4.3.1 Sustainability

Sustainability experts around the world have stated that mushrooms are the most sustainable vegetable – but why is that? The growth process of mushrooms is one of the reasons why fungus agriculture is highly sustainable compared with other types of agriculture. According to the American Mushroom Institute, farmers employ upcycled materials from other agricultural divisions for substrate, such as the manure of cows, litter from poultry, cotton hulls, almond husks, wheat straw, sawdust, and maize cobs. The environmental impact of mushrooms is then negligible throughout their growing process. According to South Mill Mushrooms, one pound of button mushrooms uses less than two gallons of water, which is remarkably low when compared to the tens of gallons of water necessary to cultivate other vegetables and fruits. Mushroom cultivation uses less energy than other crops, consuming about 1 kW h per pound, which is about the equivalent of one person boiling a coffee pot.[5,6]

4.3.2 Distribution

Amanita is a cosmopolitan mushroom that is distributed worldwide, the species of which exhibit mycorrhizal relationships with trees of different species. There are an estimated 1000 species of *Amanita* in the world, only half of which have been described. There are reports of the origin of *Amanita* in the Siberian region, from where it was then introduced to Asia, Europe, and North America.[6,7]

Amanita sp. is native to the subarctic and temperate forests of the northern hemisphere (including North America), and is also found in coniferous and deciduous woodlands. It has also moved across the southern highlands of warmer latitudes, such as Central America, the Mediterranean, and Hindu Kush. Mushrooms prefer to grow in soil rich in organic matter in New Zealand, Australia, South America, and South Africa.[8]

In India, various species and sub species can be found in the Hindu Kush belt and also in the forests of Central India. *A. muscaria* (L.) Lam has been found to grow on soil in Idukki, Kerala, Narkanda, Himachal Pradesh and Sonamarg, Jammu and Kashmir.[9,10]

A. muscaria is a species of the genus *Amanita* and is distributed worldwide, from Europe[10] to North America,[12] and eastern Asia.[13] This species has at least three different groups: Eurasian, Eurasian subalpine and North American.[14] *Amanita gemmata* is a species that was first discovered in Europe and has spread to North America[10,15] and eastern Asia.[16] The European variety of this species is found to be phylogenetically different from the one found in Asia and North America.[17] On the basis of such reports, it could be decisively revealed that *Amanita* species are endemic. *Amanita exitialis* has been found to be restricted to Southern China,[17] *Amanita fuliginea* to East Asia, and *Amanita virosa* to parts of Eurasia. However, there are a few species that are widely dispersed.

Another widespread species is *Amanita pantherina*, which have been reported from the Asian,[11] European,[18] African and American subcontinents. Two different groups of this species are prevalent; the North American and Eurasian groups[19] (Table 4.1).

Table 4.1 List of *Amanita* species with their characteristic features present in different habitats.[6,20–22]

Scientific name	Common name	Location	Sub-genus	Features
Amanita xenokommosis		Northeastern Brazil	Basidiomycota	Brown pileus with bulb covered by large saccate veil
Amanita manginiana	Mangins false death cap	Vietnam and China	Amanitina (Phalloideae)	Darker brown pileus with white gills and fleshy bulb
Amanita modesta	Modest death cap	Malaysia and China	Basidiomycota	Bluish pileus with small or medium sized annulus (membranous) and saccate volva
Amanita zambiana	Christmas mushroom	Africa	Caesareae	Cap brown in color with crowded white gills and base of stem with a volva
Amanita cyanochlorinosma	Chlorine lipedella	Guyana	Lepidella	Greyish blue pileus with white hymenophore, veil and stipe. They have a strong odor of chlorine
Amanita frostiana	Frosts amanita	Canada and US	Amanita	Orange to yellow cap with free gills and a yellow veil
Amanitafarinosa	American floury amanita	North America	Amanita	Gray-white cap with white gills and stipe with a small bulb
Amanita aprica	Sunshine amanita	North America	Amanita	Orange to yellow cap with gills and a wider stipe at the base
Amanita caesarea	Caesars mushroom	Southern Europe, northern Africa and east Asia	Amanita (Caesareae)	Orange cap with yellow gills and stipe
Amanita altipes	Yellow long stem amanita	China	Amanita	Yellow cap with a yellow stipe that is long and points upward
Amanita albocreata	Ringless panther	North America, Canada	Amanita	Yellow to white cap with gills (free) and thin flesh under the cap
Amanita gemmate	Jonquil amanita	Europe	Amanita	Yellow cap with white gills and white volva at the base of the stem

Amanita vaginata	Grisette	North America, Australia and Scotland	Amanita (Vaginatae)	Gray-brown cap with white gills and reddish brown volva
Amanita fulva	Tawny grisette	Europe and North America	Amanita (Vaginatae)	Orange-brown cap with gills close broad free and white stem with sack-like volva
Amanita cokeri	Cokers amanita	Southeastern North America	Amanitina (Roanokenses)	White cap with gills free from the stem and white stem with a large basal bulb
Amanita ocreata	Death angle	California in North America and the Pacific northwest	Amanitina (Phalloideae)	White cap with a brownish center with gills, stipe and volva that are white in color
Amanita muscaria	Fly agaric	Mediterranean, Central America, Siberia and Asia	Amanita	Bright red cap with free white gills
Amanita phalloides	Death cap	Europe, Ireland, Poland, western Russia, Italy, Spain, Portugal and Asia	Amanitina (Phalloideae)	Green cap with white stipe and gills
Amanita virosa	Destroying angel	Europe, India	Amanitina (Phalloideae)	Mushroom cap, gills, and stipe that are in white in color with a universal veil
Amanita verna	Fools mushroom	Europe	Amanitina (Phalloideae)	Mushroom cap, gills and stipe are white in color
Amanita pseudoporphyria	Hongos false death	Japan, north India, Nepal, south China and Thailand	Amanitina (Phalloideae)	White cap, gills and abundant volva cells
Amanita aestivalis	White American star footed amanita	Florida, US and Canada	Amanitina (Validae)	White cap with crowded white gills
Amanita volvata	Volvate amanita	United States	Amanitina (Amidella)	Saccate volva with elliptic spores

4.4 Medicinal Value

Edible fungus have been treasured for their tremendous health advantages and widely used in folk medicine for thousands of years. In a variety of ways, mushroom biochemical components are responsible for boosting human health. Polysaccharides, tri-terpenpoids, low molecular weight glycoproteins, and immunomodulatory chemicals are among the bioactive substances. As a result, mushrooms have been found to improve immune function; boost health; reduce inflammation; lessen cancer risk; limit tumour growth; assist in balancing blood sugar; fight against viruses, bacteria, and fungi; and support the body's detoxification mechanism. Mushrooms are becoming increasingly popular as a supplement to traditional medicine for treating a variety of ailments.[3,4,23]

4.4.1 Good for the Heart

Edible mushrooms are an excellent choice for heart patients because they have a low fat content, a high proportion of unsaturated fatty acids, and no cholesterol. They are also used for the treatment of cardiovascular disorders. The low sodium content combined with high potassium content of mushrooms improves salt balance and blood circulation in humans. As a result, mushrooms are very helpful for treating high blood pressure. Consumption of mushrooms, such as Lentinula and Pleurotus spp., on a regular basis lowers cholesterol levels. The lovastatin found in Pleurotus ostreatus and the eritadenine found in shiitake mushrooms have the ability to lower blood cholesterol levels.[4,24]

4.4.2 Low Calorie Food

Diabetic patients believe that mushrooms are an ideal food because of their low calorific value, lack of starch, and low fat and sugar content. Mushrooms consist of lean proteins that help in the burning of cholesterol in the body. As a result, they are the best food for people who are trying to lose weight.[4]

4.4.3 Prevents Cancer

Studies have reported that mushrooms contain constituents that are associated with anti-tumour activity. All edible mushrooms, but mainly button mushrooms, can help in preventing breast and prostate cancer. Fresh mushrooms have the ability to halt the action of 5-alpha-reductase and aromatase, two enzymes involved in the creation of malignant tumors. Polysaccharide-K (Kresin) is derived from *Trametes versicolor* (*Coriolus versicolor*), a plant that is utilized as a treatment for cancer.[25] Some polysaccharides derived from mushrooms have also been shown to lessen the negative effects of radiotherapy and chemotherapy. Such effects have been clinically proven in mushrooms such as *Lentinula edodes*, *Trametes versicolor*, and *Agaricus bisporus*.[26–30] Selenium, found in mushrooms in the form of selenoproteins, has anticancer effects.[31]

4.4.4 Anti-aging Properties

Mushroom polysaccharides are effective scavengers of superoxide free radicals.[32] These antioxidants prevent free radicals from acting in the body, slowing down the ageing process. Ergothioneine, an antioxidant found in *Flammulina velutipes* and *Agaricus bisporus*, is important for eye, kidney, bone marrow, liver, and skin health.[33,34]

4.4.5 Regulates the Digestive System

In the intestine, fermentable fibre and oligosaccharides from mushrooms work as prebiotics, anchoring useful bacteria in the colon. This dietary fiber helps in digestion and the suitable working of the digestive system.[35,36]

4.4.6 Strengthens Immunity

Mushrooms have the potential to boost the immune system. The immune system is up-regulated by polysaccharides (β-glucans) and minerals extracted from mushrooms.[37] These substances enhance the ability of the innate (non-specific) response acquired (specific) immune responses of the host and activate all immune cells. Mushrooms, like plants, have a lot of potential when it comes to producing high-quality food. These are a rich source of bioactive metabolites and a potent pharmacological source.[38,39] Advances in biochemistry, biotechnology, and molecular biology have boosted the use of mushrooms in medicinal research. In addition to their pharmacological effects, edible mushrooms and their by-products may provide very tasty, nutritious, and healthy meals. There are still plenty of obstacles to overcome. Until now, it has remained a mystery as to how these items work, and a great number of potential wild mushrooms have been unexplored. There is also a dearth of knowledge about their required dose, route and administration timings, the mechanism of action, and the site of activity. Various laboratories across the world are working to validate these medical qualities and isolate novel substances. These industries will play a leading role in the pharmaceutical and nutraceutical industries if these obstacles are overcome in the coming years. Mushrooms will become a popular food item as people become more aware of their nutrient content and medicinal properties. In some cases, mushrooms may even become a substitute for non-vegetarian cuisine. Mushroom cultivation is both commercially and environmentally beneficial. Mushroom consumption is good in every way. As a result, mushrooms are a true health food, as well as a promising nutraceutical.

4.5 Phytochemistry

Medicinal mushroom extracts and essences have been utilized as an alternative medicine in Korea, China, Japan, and Eastern Russia for decades. Approximately 200 wild mushroom species are used for therapeutic purposes, with 35 edible mushroom species being commercially produced. Edible mushrooms are eaten for their nutritional and therapeutic benefits. Many fungal species are picked indiscriminately or are farmed. *Lentinus*, *Auricularia*, *Hericium*, *Grifola*, *Flammulina*, *Pleurotus*, and *Tremella* are examples of edible mushrooms with medicinal and functional characteristics. *Ganoderma* and *Trametes* are two more species known solely for their therapeutic benefits. The *Amanita* genus, consisting both edible and poisonous species, has been screened to consist of six groups of chemical compounds: amino-acids, amavadin, peptide, sterols, isoxazoles, and ceramides.[40] *A. muscaria* mushroom species store vanadium in the form of amavadin complex in their dry mass weight, $\leq$100 mg kg^{-1}. Mushrooms such as Amanita absorb metals and non-metals *via* an active or passive mode directly from soil through their root-like structured mycelium. Amanita species caps during their stages of development contain varying concentrations of amino acids and lipids compared to their stipe, such as asparagine, alanine, valine, leucine, threonine, glycerol, glycerophosphocholine, and choline. Studies conducted to characterize the metals present in these species have shown that caps contain mercury (Hg) and selenium (Se) in higher concentrations compared to the stipe, which show an increase during its late stages of development. The developmental stage, soil properties and the nature of the species effect the concentration variability of elements such as Rb, Cu, Zn, V, Mg, Sb, As, and Se, which in *A. muscaria* are in high concentrations in the cap, whereas Al, Li, Pb, Sr, Ti, Mg, Co are in high concentrations in the stipe[41] *A. muscaria* is known as a poisonous mushroom due to its psychoactive compounds, such as muscimol and ibotenic acid, which have neurological effects on the brain. Bio-accumulation of metallic elements occurs in *A. muscaria*, and different geogenic areas show homeostasis of Ag, Ca, Co, Fe, Mg, Ba, Al, Cu, Mn, Zn and Sr in the fruiting parts of the mushroom.[31] The efficient geogenic accumulation of metallic compounds from soil makes *A. muscaria* a natural bio-collector in the forest ecosystem. In a recent study, a biochemical method (HPLC-ICPMS) was used to extract vanadium in the form of amavadin from the fruiting bodies of *A. muscaria* efficiently, with a concentration of $\leq$0.17 μg V L^{-1}.[42] Edible mushrooms such as *A. masasiensis*, *A. subiscosa*, and *A. Loosii* are wild varieties consumed as food or to treat slit hands and ulcers by the local population.[43] The *Lentinula edodes* mushroom, often known as Shiitake and believed to be the second most regularly produced edible mushroom in the world, is found in several Asian nations with temperate climates, such as China and Japan. Anti-microbial, anti-viral, cardioprotective, anti-fungal, anti-diabetic, hemagglutinating, anti-oxidative, hepatoprotective, hypolipidemic, immunomodulatory, and anti-neoplastic activities were observed for several well-studied preparations, extracts, and active components from *L. edodes*. Lentinan (-D-glucans) (1), lentinamicin, eritadenine (2), lentin, lectin, lentysine, and other physiologically active components have been isolated and identified from this fungus. Because of its anticancer and immunomodulatory properties, Lentinan (-1,3-glucan) is the most significant polysaccharide extracted from *L. edodes*. Carbohydrates (58–60%), proteins (20–23%), fibre (9–10%), and lipids are all present in dried L. edodes fruiting bodies. The oyster mushroom, *Pleurotus ostreatus*, is one of the most widely farmed edible mushrooms. The characteristics of *P. ostreatus* include anticancer, hypocholesterolemic, anti-atherogenic, and anti-oxidative properties. In early research, using a fermentation technique in a bioreactor, edible *P. ostreatus* was prepared. The bioactive metabolites, such as fatty acids, polyphenols, nucleotides, and alkaloids, were isolated on a large scale from fruiting bodies. Fruiting bodies were also rich in various bioactive compounds like indolo-3-carboxylic acid, *trans*-3,4-dihydro-3,4,8,-trihydroxynapthalen-1(2*H*) (3), 3-formylpyrrole (3), uracil (4), and 4-hydroxybenzaldehyde. *P. florida* is a protein-rich edible mushroom with a pleasant flavor. Several bioactive compounds, including ergosterol and ergosterol peroxide, were isolated from the *P. florida* fruit body; cerevisterol and a combination of palmitic and stearic acids, as well as linoleic and oleic acids, are all found in olive oil in a 1.5 : 1 : 0.5 ratio. The aqueous extract of *P. florida* have been found to exhibit anti-hepatotoxic properties. *Pleurotus sajor-caju* (*P. sajor-caju*) (grey oyster mushroom) is a commercially farmed edible fungus found all over the world. This mushroom has an acceptable amount of dietary fibre, glucans, B vitamins, and vitamin D, and has considerable immunomodulatory capabilities and good anti-oxidant properties. Protein, amino acids, and vitamins such as B1, B2, C, niacin, and biotin are abundant in *P. sajor-caju*. Protein, carbohydrate, crude fat, ash, and crude fiber are among the nutritional components found in *P. sajor-caju*.[44]

4.6 Secondary Bio-active Compounds

Bioactive compound composition in mushrooms is subject to the soil nature, environmental conditions, and developmental stage of the fruiting body. *A. muscaria* bio-accumulate many metals, metalloids, and other toxic metals that exhibit pharmacological activity, which could contribute to them being used to generate new drugs. Toxins present in *A. muscaria* are muscazone, muscimol, and ibotenic acid. Other bioactive compounds present in *Amanita* species are antioxidants (flavonoids, ascorbic acid, ergothioneine, tocopherols), pigments (muscarufin, muscaflavin), polysaccharides (fucomannogalactan, 1,3-1,6-D-glucan), and metals (amavadine, tricholomicacid, stizolobinic acid).[45]

Fly agaric compounds exhibit diverse biological activities, primarily muscarine, ibotenic acid, and muscimol. Depending on how these compounds interact, they can cause a wide range of symptoms. Based on the amount consumed, the relative amounts of compounds in the mushroom impact the overall physical condition of the patient, inducing conditions such as inebriation, hallucinations, agitation, increased psychomotor drive, central nervous system dysfunction, depression, and gastrointestinal problems. The typical clinical symptoms usually begin with nausea, vomiting, diarrhea, cramps, tremors, and other symptoms 30 minutes after the ingestion of a fly agaric. Further symptoms include changes to mental status, deterioration, agitation and unconsciousness. other common symptoms include hallucinations, auditory and visual distortions. Lethargy and deep sleep are the later effects exhibited after ingestion. The toxic effects are cholinergic and anticholinergic. Furthermore, after repeated consumption, mushrooms harvested from contaminated areas cause chronic toxicity due to their exposure to heavy metals. Intake of the cap is unlikely to cause death. The mushroom is considered to be edible after detoxification processes.

4.6.1 Isoxazole Compounds

In addition to muscarine toxin, *A. muscaria* has been found to contain isoxazole compounds, namely, ibotenic acid and muscimol, which have been associated with few pharmacological activities compared to other agents. These compounds act as neurotransmitter agonists, readily crossing the blood–brain barrier and influencing the central nervous system. Ibotenic acid is an agonist of *N*-methyl-D-aspartate (NMDA) glutamate receptors whereas muscimol is a γ-aminobutyric acid (GABA) agonist. These toxins have been associated with psychotropic effects upon the ingestion of mushrooms. Studies have reported that a 50–70 g fruiting body of *A. muscaria* contains 6 mg of muscimol and 70 mg of ibotenic acid. Moreover, ibotenic acid was found to be present in the spores of the mushroom. Upon consumption, these compounds are metabolized partially to muscimol *via* decarboxylation in the gastrointestinal tract. After ingestion, these compounds are excreted by the kidney and can be detected in urine samples. Similar to mucarine, it is also not degraded by the enzyme cholinesterase.[46]

4.6.2 Muscarine

The pharmacologists Schmiedeberg and Buchheim isolated muscarine for the first time in in *A. muscaria* mushrooms.

Muscarine is an alkaloid derivative, a non-selective acetylcholine agonist. This alkaloid acts on mucarine acetylcholine receptors in the parasympathetic nervous system. It has a long duration of action on neurons as once it becomes attached to a receptor it is not degraded by the enzyme cholinesterase. Thereby, it has a direct effect on various organs in the body. Symptoms related to this toxin include digestion issues, defecation, encompassing sexual arousal, salivation, lacrimation, and urination. The compound is unable to cross to the central nervous system *via* the blood–brain and blood–cerebrospinal fluid barriers. A study reported the presence of 0.02% dry weight of muscarine compound in basiodiocarps, which can cause health issues, such as hypersalivation, bradycardia, diarrhea, lachrymation, perspiration, and fatigue. The amount of this compound varies in the genera *Inocybe*, *Inosperma* and *Pseudosperma*.

A. muscaria by weight contains 1–3% muscarine compared to another member in this family, *A. pantherina*, which contains a higher concentration of this compound.[47] Muscarine cause "muscarinic" syndrome when ingested, which effects the human central nervous system. It is thermostable and does not dissolve in water, and thus it is not effected by cooking. Known symptoms associated with its poisoning are diarrhea, low blood pressure, decreased heart rate, bronchospasms, and blurred vision. The chemistry of this compound shows that it mimics acetylcholine (neurotransmitter) by activating its receptors (acetylcholine receptor).[48] Muscarine when digested is resistant to the plasma cholinesterase enzyme, and its effect persists for several hours. The dose concentration of this compound that is lethal to humans is in the range of 40–495 mg. Compounds used as antagonists against muscarine poisoning are atropine, pirenzepine, and scopolamine. Five types of muscarine receptors (M1, M2, M3, M4, and M5) localized in the central nervous system are encoded by different genes.[49]

4.6.3 Ibotenic Acid

The ibotenic acid (psychoactive) toxin present in *A. muscaria* is encoded by "ibo" genes present in the fungus.[50]

This compound shows structural similarity to GABA and glutamate (a neurotransmitter). The toxicity of ibotenic acid causes gastrointestinal symptoms such as vomiting, nausea, and abdominal pain, as well as central nervous system symptoms such as hallucinations and tremors. Ibotenic acid forms muscimol *via* a decarboxylation process.[51] Ibotenic acid is soluble in water and is metabolized to muscimol in the stomach, liver, and brain upon digestion. Ibotenic acid behaves as an excitatory neurotoxin.[52]

4.6.4 Muscimol

Muscimol, a heterocyclic alkaloid compound synthesized from ibotenic acid that is a psychoactive toxin, has structural similarity to GABA and is an agonist to all GABA-A receptors. In aqueous solution, muscimol forms neutral zwitterionic states.[53]

Muscimol is present in *A. muscaria* and *A. pantherina* species alongside ibotenic acid and muscarine. Consumption of this compound causes symptoms such as lucid state of mind, synesthesia, tremors, and euphoria.[54]

4.7 Pharmacological Potential

Amanita spp. exhibit proven health benefits as they are a rich source of biologically active compounds, such as phenolic compounds (flavonoids, lignans, oxidized polyphenols, phenolic acids, stilbenes, and tannins), proteins and peptides (glutathione and ergothioneine), vitamins and derivatives (ergosterol, biotin, ascorbic acid, carotenoids, and tocopherols), and minerals (zinc, iron, potassium, and selenium). *A. muscaria* has also been reported to accumulate significant amounts of metals (such as cadmium, copper, mercury and manganese, present in the mushroom caps) from the soil or substrate on which it is growing.[55] Reports have shown the uptake of some selected heavy metals (cadmium, zinc, chromium, copper, lead, *etc.*) by the wild fungus *A. princeps*, which thereby removes the metal pollutants from the soil.[56] Therefore, there has been a rise in the pharmacological and toxicological interest towards the nutritional and pharmaceutical properties of these bioactive compounds. Secondary metabolites of some species of the genus Amanita have been exploited for their pharmacological activities towards various human ailments and their potential has been proven as anti inflammatory, anti-oxidant, anti-diabetic, anti-viral, anti-malarial, anti-fungal, anti-bacterial, anti-cancer, and immunomodulatory agents.[57–59] High levels of tocopherols (γ-tocopherol and δ-tocopherol) and ascorbic acid in fly agaric extracts exhibit powerful antioxidant properties *in vivo*[60] as well as *in vitro*.[61]

Phenolic compounds are broad spectrum antioxidants which could prove to be therapeutically valuable with the medicinal properties to prevent oxidative stress associated pathologies, such as cancer, inflammation, microbial infections, and the effects of diabetes. Alkaloids from mushrooms exhibit therapeutic properties against pathogenic infections and various types of cancer. Saponosides exhibit a wide range of beneficial pharmacological effects in inflammatory conditions and diabetes and also act as immune biomodulators.[62] Flavonoids also exhibit pharmacological protection against inflammation, many forms of cancer, diabetes, and pathogenic infections (bacterial, viral, fungal, and protozoal).[63] These secondary metabolites also show beneficial therapeutic effects to treat skin problems (*A. crassiconusis*). The edible species *A. masasiensis* and *A. subvicosaare* have been reported to treat ulcers, hard skin on feet, and cuts on hands.[64]

A number of researchers have reviewed the role of extracts of *A. muscaria* against neurodegenerative diseases such as Parkinson's and Alzheimer's, and also in the treatment of cerebral ischaemia. The results suggest that Amanita have neuro-protective, hepato-protective, cardiovascular protective, anti-inflammatory, and anti-oxidant abilities. *A. muscaria* offers a great range of health benefits, especially in terms of cell protection, and hence may contribute to the development of new drugs.[44] The phytochemical muscimol in fly agaric extract has been demonstrated to exhibit neuroprotective properties on rat brain synaptosomes exposed to neurotoxin 6-hydroxydopamine. The results were significant as preserving synaptosomal viability led to reduced formation of reactive oxygen species and reactive quinones in the cells (a process that occurs in Parkinson's disease). This mushroom species is an established natural insecticide and its insecticidal properties (due to a high concentration of ibotenic acid) have been well documented in the housefly, *Musca domestica*.[65]

Wang *et al.* studied the antioxidant ability of *Amanita* in rodents with hepatic injury. The animal models were induced using either carbon tetrachloride, streptozotocin, or D-galactosamine (D-GalN). The results demonstrated free radical scavenging activity by activation of the $GABA_A$ receptor, which protects hepatic cells from injury. Moreover, treatment with muscimol (a $GABA_A$ receptor agonist) revealed that a complex GABA signaling system may exist in the rodent liver. This bioactive compound was found to protect the liver against toxic injury by decreasing the formation of pseudo bile ductiles and the enlargement of hepatocytes canaliculi in D-galactosamine treated rats.[66] The nutraceutical properties of Amanita sp. can be correlated to the presence of its wide array of phytochemicals or bioactive metabolites. These compounds hold great potential from the viewpoint of their application for improving nutrition and as therapeutic agents to better human health.[67–71]

Previous findings have shown that gastric cancer and hypertension related diseases share common etiological factors. This was evident from the findings by Ma *et al.*, who showed that patients with cerebrovascular ailments are also prone to gastric cancer. Studies have also demonstrated that intraperitoneal administration of muscimol (0.5 mg per kg body weight) over a long time period in spontaneously hypertensive rats resulted in significant reduction in gastric cancer in the animals. A similar protective role of muscimol was also displayed in rodents with oral squamous cell carcinoma[72] (Table 4.2).

Hence, such wild edible mushrooms could be important candidates for the food and pharmaceutical industries.[42]

4.8 Nutrition and Dietary Composition

Mushrooms are considered to be a valuable source of vitamins, minerals, and dietary fibre. The dry weight of the fruiting body of mushrooms contain 25% protein, 56.8% carbohydrate, 5.7% fat and 12.5% ash. These are cultivated and consumed in different parts of countries such as India, China, and Japan. They have been used as medicine and food since ancient times. Mushrooms are either edible, such as *Auricularia*, *Lentinus*, and *Agaricus* species, or poisonous, such as *Amanita* and *Psilocybe* species. The majority of *Amanita* species, particularly those belonging to *Amanita*, *Validae*, *Phalloides*, and *Lepidella* are poisonous and produce toxic metabolites.[73] The Himalayan regions (Jammu and Kashmir, Himachal, Sikkim, Uttarakhand, West Bengal and Assam) are associated with huge biodiversity and macrofungi distribution. The consumption of species such as *Amanita, Astraeus, Hericium,* and *Pleurotus*, as food and for medicinal (antioxidant and antimicrobial) purposes, has been documented from the regions of the Himalayas.[74] Edible mushrooms are a good source of nutrition as they contain vitamins (B1, B12, B2, C, E, and D) and antioxidants (carotenoids, phenolics, flavonoids, and glycosides). The nutritional compositions of wild mushrooms are incomplete due to poor information on the antioxidants and toxins present in the fungi, which could have scope in the field of medicine.[75,76]

Table 4.2 Biological properties of bioactive metabolites of *Amanita* sp.

Metabolite	Applications	Mechanism
Ibotenic acid, muscazone and muscimol	Antitumor, pesticidal, cytotoxic, antioxidant, anticancer, antibacterial, acetylcholinesterase, esterolytic, antiviral, anti-larvicidal, antifungal, and anti-inflammatory properties	Induces apoptosis cascade for cell death
Polysaccharides	Anti-inflammatory, antitumor, and immunomodulatory activities	Mannitol, mannose, glucose, trehalose, rhamnose, xylose, galactose, arabinose, xylose, fucose, fructose, *etc.*
Protein and peptides	Antiviral, antitumor, antifungal, antibacterial, and immunomodulatory properties, *etc.*	Inhibits enzyme activities, increases theabsorption and digestion of exogenous nutritional ingredients and modulates the immune system
Terpenes	Antiviral, anticholinesterase, anticancer, antioxidant, antimalaria, and anti-inflammatory activities	
Phenolic compounds	Antioxidant capacity as well as antiviral, anti-inflammatory, antiatherogenic, cardioprotective, anticancer, antimicrobial, antithrombotic, antiallergenic, and vasodilator effects	Tannins, phenolic acids, oxidized polyphenols, hydroxybenzoic acids, lignans, flavonoids, stilbenes, and hydroxycinnamic acids
Polyunsaturated fatty acids	Antioxidant properties; protection against microbial and cardiovascular diseases and degenerative malfunctions	Ergosterol

4.9 Mushroom Cultivation and Harvesting

Cultivation of mushrooms plays significant roles in the bioconversion of industrial and agricultural solid wastes into edible biomass, which can also be used as a functional food or a source of medications and pharmaceuticals. Scientists must study the basic biology of mushrooms and design cultivation equipment that is environmentally benign and can be used to cultivate mushrooms in an environmentally friendly manner. This is especially true of *Lentinula edodes*, *Volvariella volvacea*, and *Pleurotus* spp., which are consumed in Asian countries but are gaining favour as specialty mushrooms around the world. Biodiversity is a critical part of genetic resources for breeding programs that meet a variety of consumer demands.

Owing to human inhabitation, deforestation, and competition from a cultivar from wild germplasm, the protection of these mushrooms has become critical and urgent. However, the solid compost of spent mushrooms produced by the mushroom business can be used as a soil fertilizer and a potential bio remediating agent. Xenobiotics, from oyster mushrooms, are used to eliminate heavy metals from the environment.

4.10 Conservation of Mushrooms

The cultivation of mushrooms is highly profitable in terms of food, nutrition, and medicinal value. The scarcity of edible mushrooms in terms of quantity and quality is mainly due to high levels of deforestation. Other such factors, such as habitat disruption, timber exploitation, and forest fragmentation, have led to the extinction of around 28 species of wild edible fungi. In contrast, artificial cultivation represents a sustainable and safe process that enhances the yield and quality of edible fungi throughout the year.[77]

4.11 Safety

The safety of mushrooms is evaluated using different approaches that rely on five principles. First is the identification and confirmation of raw fungi. There are many challenges that are associated with fungal identification. Differences in morphological features of the same species complicate their taxonomy, which leads to genetically identical species of fungi being assigned numerous names (termed 'pleomorphy').[78] However, DNA identification resolves the issues of identification due to advances in the speed and cost of genetic analysis.[79]

Studies have reported the identification of fungi by internal transcription.[80] Second, the screening of raw fungus extract against a fungal metabolites database for quick identification of a known toxin can be used. Third, these collected extracts can be compared with those of culinary mushrooms from a store. Mushrooms that are consumed widely as food in a diverse range of populations can provide an innocuous baseline with which to compare common components. Fourth, review of available chemical and toxicological data sets for safety aspects. Fifth, the establishment of the presence of the product using marketplace data sets.

4.11.1 Toxicity

The toxicity of *A. muscaria* is generally associated with hallucination in young children or people. It is generally found to be toxic when a dose of 1 g of this mushroom or 50–100 mg of ibotenic acid is ingested, which acts as a neurotoxin and brain-lesioning agent. However, the side effects of fly agaricus are unpredictable, ranging from nausea to drowsiness to cholinergic effects, as well as a change in mood, such as euphoria. In addition, it can cause delirium, confusion, irritability, hallucination, and depression.

Upon severe poisoning it may also cause seizures and lead to a coma. Death associated with ingestion is rare. However, one study reported two fatalities in North America associated with this fungus. Prognosis is generally good with modern medical treatment.[81]

The elemental composition of *A. muscaria* differs from region to region, season to season. The ibotenic acid content is 10-fold higher in spring and summer than during fall fruiting. A vast number of fatalities are associated with greenish/yellowish to brownish mottled death caps (*A. phaloides*). Ibotenic acid is a toxic compound that is soluble in water and is sensitive to heat. *A. muscaria* can be detoxified partially using parboiling or leaching approaches, as ibotenic acid is converted to muscimol upon heating.[82]

In comparison to the hallucinogenic mushroom Psilocybe, Amanita mushrooms are rarely consumed. US sales related to this mushroom are monitored by the FDA. Many countries do not have any issues against the sale of this mushroom, as it is presently legal. However, following the outlawing of mushrooms containing psilocybin in the UK, there have been increases in the sale and consumption Amanita mushrooms. Upon consuming mushrooms, the active ingredients are excreted in the urine. It has been noted that it is customary for the shaman of a tribe to eat mushrooms and the rest of the tribe to drink the urine, which in effect partially detoxifies the ingredients. Moreover, in Siberia, the poor would consume the urine of wealthy people who had consumed mushrooms. However, if fly agaric is consumed, it must be dried or taken in consumed form, as ibotenic acid is converted to the more stable compound muscimol.[83]

4.11.2 Regulation

The taxonomy of mushrooms and their products as novel foods is challenging. Mushroom products are monitored by the E.U. 2015/2283 regulation on novel foods. Novel food regulations state the regulatory requirements for the authorization of the sue of food. The authorization of the biomass fermentation of mushroom and mycelium products (MMP) is particularly relevant. The new sector of MMP relies on starts ups with little capitalization. The stringent regulatory framework set by novel food regulation hampers the innovation sector badly. Moreover, the use of food and agri-byproducts as substrates for fermentation may increase the sustainability of the industry, but this is restricted by food safety regulations.[84]

4.12 The Mushroom Trade

The numerous health benefits of mushrooms, rising per capita consumption, cost-effective production, rising demand for vegan and natural foods in the diet, and an increasing number of health-conscious people around the world are all expected to contribute to the significant growth in the global mushroom cultivation market. With a cultivation area of 4.37% and a range of agroclimatic conditions, India is primarily an agricultural nation and produces 620 million tons of agricultural waste every year.[78] Growing mushrooms not only aids in the recycling of agricultural waste, it also helps to satisfy the nutritional deficit of the enormous population of India. Between 2010 and 2017, India produced over 0.13 million tons of mushrooms, representing an average increase in production of 4.3%. In 2019, the button mushroom market segment held the biggest market share. Almost all of the major mushroom-producing nations, including China, Japan, the US, the UK, Germany, and Poland, commercially farm this particular variety of mushroom. The market for mushrooms is expected to increase significantly, from 15.25 million tons in 2021 to 24.05 million tons by 2028. Fresh and canned mushroom exports from the United States grossed up to 7212 mt and 709 mt, in 2007/08 and 8119 mt and 1281 mt in 2008/09, respectively. These amounts are tiny in comparison to imports. Since 2003/04, Canada has been the primary destination for US exports of fresh mushrooms, accounting for almost 90% of its annual export volume. Other countries exported to include Singapore, Hong Kong, the Philippines, and India. Concentration and durations of 30 or 60 min proved to be factors that led to a decrease in total phenolic content (TPC) and total antioxidant activity (TAA).[85]

4.12.1 Storage of Mushrooms

Mushrooms must be refrigerated immediately as their respiratory activity is 3.5 times higher at 10 °C than it is at 0 °C. Therefore, the heat should be turned off as soon as the mushrooms are harvested, and the temperature should be reduced to 4–5 °C as rapidly as possible. Low temperatures prevent moisture loss, slow down post-harvest metabolic activity in the mushroom tissues, and slow down microbial development. Vacuum-cooling is a different way of storing mushrooms in which the mushrooms are exposed to very low pressure and the water evaporates, releasing the latent heat of vaporization. Compared to conventionally chilled mushrooms, those that have been vacuum cooled have a better colour.[86] A "cold sterilization" technique called radiation preservation allows for the preservation of mushrooms without significantly altering their natural characteristics. The shelf-life of mushrooms could be increased and microbial contamination decreased using low doses of radiation.[87] After harvest, it has been effectively demonstrated that modified atmosphere packing (MAP) of mushrooms can postpone senescence and retain quality.[88] In modified atmospheric packaging, in addition to changing the atmosphere inside the box, various chemicals can be included to further extend the shelf life of the product. Activated carbon, sorbital, chitosan, and potassium permanganate are a few other packing materials that can be utilized in MAP to preserve the quality of mushrooms at room temperature. Another way to change the oxygen and carbon dioxide contents inside the package and the respiration rate is by controlled atmospheric storage. The shelf life is increased and brown discoloration (enzymatic browning) is reduced with controlled atmospheric packaging. Ozone has been utilized to lengthen storage times and slow deterioration. The TPC and total antioxidant activity TAA of mushrooms decreased after being stored at 2 °C for 14 days following ozonization at 0.5, 1, or 2 $mg L^{-1}$ concentrations for periods of 30 or 60 min.[84]

4.12.2 Transport of Mushrooms

For transportation to market, mushrooms are firstly packed. In India, 200 g or 400 g of mushrooms are packaged in polythene packs with a thickness of 100 gauge. Pulp-board punnets or large polythenes are used for packaging a greater quantity of mushrooms, which can withstand long-distance transport. The impact of packing and pre-cooling can be negated if the stored mushrooms are later transported to a warmer environment. Moreover, during short-distance transport in a cool environment, mushrooms are kept in polypacks containing ice and are stacked in small boxes or wooden cases. However, long-distance transportation of mushrooms is costly as refrigerated trucks are required.

4.13 Conclusions

Mushrooms play an important role in the forest ecosystem and bio-accumulation of essential nutrients that have tremendous medicinal value. The dispersal of Amanita species across the continents is likely due to human activities. These species include both poisonous and edible mushrooms. The genus *Amanita* includes 1550 names that have been presented from around the world in the fungorum index (http:/www.indexfungorum.org). Many varieties of Amanita species are still unknown due to the little knowledge related to their location and toxicity. Many toxic secondary metabolites present in these species have biological and therapeutic applications.

References

1. P. Zhang, L. P. Tang, Q. Cai and J. P. Xu, *Mycology*, 2015, **6**(2), 86–93.
2. D. Michelot and L. M. Melendez-Howell, *Mycolo. Res.*, 2003, **107**(2), 131–146.
3. A. Bhambri, M. Srivastava, V. G. Mahale, S. Mahale and S. Kumar Karn, *Front. Microbiol.*, 2022, **13**, 837266.
4. G. Venturella, V. Ferraro, F. Cirlincione and M. LetiziaGargano, *Int. J. Mol. Sci.*, 2021, **22**(2), 634.
5. Q. Carboué and M. Lopez, Amanita muscaria: Ecology, Chemistry, Myths, *Encyclopedia*, 2021, **1**, 905–914.
6. K. S. Mighell, T. W. Henkel, R. A. Koch, A. Goss and M. C. Aime, *Fungal Syst. Evol.*, 2019, **3**(1), 1–2.
7. J. Geml, G. A. Laursen, K. O'Neill, H. C. Nusbaum and D. L. Taylor, *Mol. Ecol.*, 2006, **15**, 225–239.
8. M. Moor-Smith, R. Li and O. Ahmad, *BCMJ*, 2019, **61**(1), 20–24.
9. R. K. Verma and V. Pandro, *Indian J. Trop. Biodiversity*, 2018, **26**(1), 42–54.

10. K. Shukla, B. Giri and R. V. Shukla, *Developments in Fungal Biology and Applied Mycology*, 2017, pp. 501–523.
11. M. Moser and R. Phillips, *Keys to Agarics and Boleti (Polyporales, Boletales, Agaricales, Russulales)*, London, 1983.
12. D. T. Jenkins, *Amanita of North America*, Mad River Press, Eureca (CA), 1986.
13. R. Imazeki and T. Hongo, *Colored illustrations of mushrooms of Japan*, Hoikusha, Osaka, 1987, vol. 1.
14. T. Oda, C. Tanaka and M. Tsuda, *Mycol. Res.*, 2004, **108**, 885–896.
15. W. C. Coker, The Amanitas of the eastern United States, *J. Elisha Mitchell Sci. Soc.*, 1917, **33**, 1–88.
16. E. Nagasawa and T. Hongo, *Some agarics from the San-in district, Japan*, Memoirs of National Science Museum, Tokyo, 1985, vol. 18, pp. 73–88.
17. L. F. Zhang, J. B. Yang and Z. L. Yang, *Fungal Divers.*, 2004, **17**, 219–238.
18. Q. Cai, R. E. Tulloss, L. P. Tang, B. Tolgor, P. Zhang, Z. H. Chen and Z. L. Yang, *BMC Evol. Biol.*, 2014, **14**, 143–158.
19. D. A. Reid and A. Eicker, South African fungi: the genus Amanita, *Mycol. Res.*, 1991, **95**, 80–95.
20. E. Wartchow, *Plant Biosyst. Int. J. Dealing Aspects Plant Biol.*, 2020, **154**(5), 700–704.
21. E. Davison and D. Giustiniano, *Muelleria*, 2020, **32**, 59–73.
22. E. Wartchow, R. E. Tulloss and M. A. Cavalcanti, *Mycologia*, 2009, **101**(6), 864–870.
23. A. Assemie and G. Abaya, *Int. J. Microbiol.*, 2022, **2022**, 8744788.
24. L. Dicks and S. Ellinger, *Nutrients*, 2020, **12**(4), 1134.
25. S. Habtemariam, *Biomedicines*, 2020, **8**(5), 135.
26. Y. Zhang, M. Zhang, Y. Jiang, X. Li, Y. He, P. Zeng, Z. Guo, Y. Chang, H. Luo and Y. Liu, *J. Cancer Res. Clin. Oncol.*, 2018, **144**, 2177–2186.
27. H. Ahn, E. Jeon, J. Kim, S. Kang, S. Yoon, H. Ko, P. Kim and G. Lee, *Sci. Rep.*, 2017, **7**, 1314.
28. C. Li and N. H. Oberlies, *Life Sci.*, 2005, **78**, 532–538.
29. L. Blumfield, K. Abbott, E. Duve, T. Cassettari, S. Marshall and F. Fayet-Moore, *J. Nutr. Biochem.*, 2020, **84**, 108453.
30. L. S. Adams, S. Phung, X. Wu, L. Ki and S. Chen, *Nutr. Cancer*, 2008, **60**, 44–56.
31. J. Falandysz, M. Mędyk and R. Treu, *Environ. Sci. Pollut. Res.*, 2018, **25**, 25190–25206.
32. A. Soares, A. de Sá-Nakanishi, A. Bracht, S. da Costa, E. Koehnlein, C. de Souza and R. M. Peralta, *Molecules*, 2013, **18**, 7609–7630.
33. A. J. Weigand-Heller, P. M. Kris-Etherton and R. B. Beelman, *Prev. Med.*, 2012, **54**, 75–78.
34. G. A. Martinez-Medina, M. L. Chávez-González, D. K. Verma, L. A. Prado-Barragán, J. L. Martínez-Hernández, A. C. Flores-Gallegos, M. Thakur, P. P. Srivastav and C. N. Aguilar, *J. Funct. Foods*, 2021, **77**, 104326.
35. T. I. S. Nguepi and H. Song, *Int. J. Med. Mushrooms*, 2020, **22**(6), 509–519.

36. T. Sawangwan, W. Wansanit, L. Pattani and C. Noysang, *Agric. Nat. Res.*, 2018, **52**, 519–524.
37. H. A. El Enshasy and R. Hatti-Kaul, *Trends Biotechnol.*, 2013, **31**, 668–677.
38. J. Martel, Y. F. Ko and D. M. Ojcius, *et al.*, *Trends Pharmacol. Sci.*, 2017, **38**(11), 967–981.
39. D. K. Patel, S. D. Dutta, K. Ganguly, S.-J. Cho and K.-T. Lim, *Molecules*, 2021, **26**(5), 1359.
40. C. Li and N. H. Oberlies, *Life Sci.*, 2005, **78**(5), 532–538.
41. J. Falandysz, A. Hanć, D. Barałkiewicz, J. Zhang and R. Treu, *Fungal Biol.*, 2020, **124**(3–4), 174–182.
42. S. Braeuer, M. Walenta, L. Steiner and W. Goessler, *J. Anal. At. Spectrom.*, 2021, **36**(5), 954–967.
43. O. G. Fadeyi, F. M. Assogba, D. D. Chabi, N. S. Yorou and J. D. Gbenou, *J. Pharmacogn. Phytochem.*, 2019, **8**(3), 335–341.
44. D. Sharma, V. P. Singh and N. K. Singh, *Mini-Rev. Med. Chem.*, 2018, **18**(13), 1095–1109.
45. M. Voynova, A. Shkondrov, M. Kondeva-Burdina and I. Krasteva, *Pharmacia*, 2020, **67**, 317.
46. A. Poliwoda, K. Zielińska, M. Halama and P. P. Wieczorek, *Electrophoresis*, 2014, **35**, 18.
47. M. R. Lee, E. Dukan and I. Milne, *J. R. Coll. Physicians Edinburgh*, 2018, **48**(1), 85–91.
48. F. Xu, Y. Z. Zhang, Y. H. Zhang, G. Y. Guan, K. P. Zhang, H. J. Li and J. J. Wang, *Toxicon*, 2020, **179**, 72–75.
49. F. Nascimento, L. R. Spindler and G. B. Miles, *Sci. Rep.*, 2019, **9**(1), 1–6.
50. S. Obermaier and M. Müller, *Angew. Chem.*, 2020, **132**(30), 12532–12535.
51. C. N. Filer, *J. Radioanal. Nucl. Chem.*, 2018, **318**(3), 2033–2038.
52. C. E. Housecroft, *Chimia*, 2019, **73**(1/2), 96–97.
53. T. Kupka, M. A. Broda and P. P. Wieczorek, *Magn. Reson. Chem.*, 2020, **58**(6), 584–593.
54. J. Patocka and B. Kocandrlova, *Mil. Med. Sci.*, 2017, **86**(3), 122–134.
55. N. Z. Zahari, A. Calder, P. M. Tuah, F. N. Cleophas, F. A. F. Adnan, K. Budin, S. A. M. Ali and S. A. Rahim, *Earth Environ. Sci.*, 2021, **765**, 012085.
56. A. Das, C. Chen, S. Mu, S. Yang, Y. Ju and S. Li, *Pharmaceutics*, 2022, **14**(2), 436.
57. U. Lindequist, T. H. J. Niedermeyer and W. D. Jülich, *Evidence-Based Complementary Altern. Med.*, 2005, **2**, 285–299.
58. S. T. Sushri, R. Ashutosh and G. Nibha, *Int. J. Innovative Drug Discovery*, 2014, **4**, 124–129.
59. R. M. Chugh, P. Mittal, M. P. Namratha, T. Arora, T. Bhattacharya, H. Chopra, S. Cavalu and R. K. Gautam, *Front. Pharmacol.*, 2022, **13**, 925387.
60. F. S. Reis, S. A. Heleno, L. Barros, M. J. Sousa, A. Martins and C. Santos-Buelga, *J. Food Sci.*, 2011, **76**, C824–C830.
61. S. Wei, J. P. Helsper and L. J. Van Griensven, *Int. J. Med. Mushrooms*, 2008, **10**(1), 1–13.

62. O. G. Fadeyi, F. M. Assogba, D. D. C. B. Chabi, N. S. Yorou and J. D. Gbeno, *J. Pharmacogn. Phytochem.*, 2019, **8**(3), 335–341.
63. S. Morel, PhD, University of Angers, French, 2011, p. 266.
64. M. Lingarao and N. Savithramma, *Der Pharm. Lett.*, 2011, **3**, 51–55.
65. M. Kondeva-Burdina, M. Voynova, A. Shkondrov, D. Aluani, V. Tzankova and I. Krasteva, *Food Chem. Toxicol.*, 2019, **132**, 110687.
66. S. Wang, Y. Xiang, J. Zhu, F. Yi, J. Li, C. Lui and W. Lu, *Am. J. Physiol.: Gastrointest. Liver Physiol.*, 2017, **312**(3), 2018–2218.
67. I. C. F. R Ferreira, P. Baptista, M. V. Boas and L. Barros, *Food Chem.*, 2007, **100**, 1511–1516.
68. H. Rathore, S. Prasad and S. Sharma, *PharmaNutrition*, 2017, **5**, 35–46.
69. J. Ishara, A. Buzera, G. N. Mushagalusa, A. R. A. Hammam, J. Munga, P. Karanja and J. Kinyuru, *J. Food Biochem.*, 2022, **46**(1), e14025.
70. J. L. Cadet, K. Bolla, E. Byrne, S. DiMauro, R. S. J. Frackowiak, R. T. Johnson, and Y. Mizuno, *et al.*, Neurology and Clinical Neuroscience, *Mosby, Philadelphia, PA, 2007, pp. 1477–1488.*
71. A. Pišlar, J. Sabotič, J. Šlenc, J. Brzin and J. Kos, *Cell Death Discovery*, 2016, **2**, 16021.
72. J. Ma, Y. Zhang, J. Wang, T. Zhao, P. Ji, J. Song and W. Luo, *Int. J. Mol. Med.*, 2016, **38**(1), 305–311.
73. B. A. Bunyard, *Fungi*, 2018, **10**(4), 40–48.
74. S. Kaul, M. Choudhary, S. Gupta, D. C. Agrawal and M. K. Dhar, in *Medicinal Mushrooms*, Springer, Singapore, 2019, pp. 371–389.
75. M. Borthakur and S. R. Joshi, in *New and Future Developments in Microbial Biotechnology and Bioengineering*, Elsevier, 2019, pp. 1–12.
76. A. A. Greeshma, K. R. Sridhar and M. Pavithra, *J. Fungal Biol.*, 2018, **8**(1), 54–68.
77. V. G. Thiribhuvanamala, G. Prakasam, K. Chandrasekar, S. Sakthivel, S. Veeralakshmi, R. Velazhahan and G. Kalaiselvi, *Proceedings of the 7th International Conference on Mushroom Biology and Mushroom Products (ICMBMP7)*, 2011.
78. J. Sugiyama, K. Nagai and K. Komagata, *J. Gen. Appl. Microbiol.*, 1987, **33**, 197–204.
79. H. A. Raja, A. N. Miller, C. J. Pearce and N. H. Oberlie, *J. Nat. Prod.*, 2017, **80**(3), 756–770.
80. F. I. Rampolli, P. Kamler, C. Carnevale Carlino and F. Bedussi, *Eur. J. Case Rep. Int. Med.*, 2021, **8**(3), 002212.
81. C. Quentin and M. Lopez, *Encyclopedia*, 2021, **1**, 905–914.
82. J. Hillebrand, D. Olszewski and R. Sedefov, *European Monitoring Centre for Drugs and Drug Addiction (EMCDDA)*, Lisbon, 2006.
83. A. Molitorisová, A. Monaco and K. P. Purnhagen, 2021. Available at SSRN 3955899.
84. J. Raman, S. K. Lee, J. H. Im, M. J. Oh, Y. L. Oh and K. Y. Jang, *J. Mushrooms*, 2018, **4**, 239–249.
85. M. Zalewska, E. Górska-Horczyczak and M. Marcinkowska-Lesiak, *Agriculture*, 2021, **11**, 748.

86. A. Koorapati, D. Foley, R. Pilling and A. Prakash, *J. Food Sci.*, 2004, **69**(1), 25–29.
87. M. K. Roy, S. R. Chatterjee, D. Bahukhandi, R. Sharma and A. S. Philips, *J. Food Sci. Technol.*, 2000, **37**(1), 83–86.
88. M. Zalewska, M. Marcinkowska-Lesiak, A. Onopiuk, A. Stelmasiak and A. Półtorak, *J. Food Process. Preserv.*, 2018, **42**(12), e13839.

CHAPTER 5

Astraeus hygrometricus *(Pers.) Morgan*

RITEE BASU,[†a] SUKANYA DASGUPTA,[†,a] IRENE ROSE ANTONY,[a] JESSICA PANDOHEE (iD 0000-0003-4343-7192)*[b] AND AYESHA NOOR[a]

[a] Centre for Bio Separation Technology, Vellore Institute of Technology (VIT), Vellore-632 007, India; [b] Telethon Kids Institute, Nedlands, Western Australia 6009, Australia
*Email: jessica.pandohee@gmail.com

5.1 Introduction

Mushrooms have been consumed for centuries and are popularly known for their culinary values and therapeutic properties.[1] For example, the ancient Greeks thought that mushrooms offered vitality for soldiers in battle. As for the Romans, they regarded mushrooms as a "Food of the Gods." For hundreds of years, the Chinese have also considered mushrooms to be the "elixir of life." Mushrooms are now known to be a functional food because they are rich in bioactive compounds and are low in carbohydrates, sodium, and fat; therefore, they are regarded as a cholesterol-free food.[2] Different species of mushroom are cultivated abundantly and are divided into edible and medicinal mushrooms.[3] Mushroom production is steadily increasing, with China being the world's largest producer. Moreover, wild mushrooms are becoming increasingly significant due to their sensory, nutritional, and, most importantly, their pharmacological properties.[4]

[†] These authors contributed equally.

Phytochemistry and Nutritional Composition of Significant Wild Medicinal and Edible Mushrooms: Traditional Uses and Pharmacology
Edited by Ajay Sharma, Garima Bhardwaj and Gulzar Ahmad Nayik

Published by the Royal Society of Chemistry, www.rsc.org

Native people on the west coast of India and in the Western Ghats commonly consume wild mushrooms that are well-known for their nutritional and medicinal values, such as *Agaricus* spp., *Astreaus* spp., *Amanita* spp., *Boletus* spp., *Auriculria* spp., *Lentinus* spp., *Termitomyces* spp., and *Pleurotus* spp.[5] Moreover, mushrooms are ranked as the second most popular food product in Australia, after potatoes. The growing public knowledge of the high nutritional content of mushrooms, as well as the possibility of health-promoting and therapeutic characteristics, has resulted in a rise in their consumption, notably in Australia.[6] The pharmacological potential of different kinds of mushrooms is noteworthy and there is a necessity to know more about the mechanistic view of the therapeutic actions to confirm and commercialize the different findings on mushrooms.[7]

Astraeus hygrometricus (Pers.) Morg, also known as false earthstar *Astraeus hygrometricus*, is a wild-type edible mushroom that has not been explored much.[8] It belongs to the class Gastromycetes, which belongs to the lineage of Basidiomycota. Gastromycetes includes many groups of fungi – puffballs, earthstars, false truffles, and stinkhorns, which are well known for the diverse nature of their fruiting bodies. The spore mass is enclosed within the basidiocarp or the gasterocarp, which makes them different from other fungi.[9] *A. hygrometricus* belongs to the genus *Astraeus*, which includes 13 different species. The genus *Astraeus* is a well-known for its therapeutic and nutritional properties.[10] The *Astraeus* species are ectomycorrhizal with various tree species, such as *Ericaceae*, *Pinaceae*, and *Betulaceae*. Among the 13 species, four species are very important and are being studied by scientists around the world. The four most common species are *A. hygrometricus, A. thailandicus, A. odoratus,* and *A. asiaticus. A. hygrometricus* is considered in many countries to be invaluable mushroom due to its various nutritional and pharmaceutical properties.

A. hygrometricus has been considered beneficial for human health and it can be incorporated into the human daily diet as it is rich in proteins, mycochemicals, terpenoids, tannins, flavonoids, phenolic compounds, heterocyclic compounds, crude fibers, essential amino acids, vitamins, and minerals.[11] In this chapter, an attempt is made to understand the impact that *A. hygrometricus* has in terms of health benefits, its traditional uses, phytochemistry, and pharmacological potential. Further, safety, toxicity, and the regulation of *A. hygrometricus* are been discussed.

5.2 Botany, Taxonomy, and Geographical Distribution

5.2.1 Taxonomy

Astraeus hygrometricus (Pers.) Morg, also known as barometer earthstar, false earthstar or hygroscopic earthstar, is a gasteroid fungus that belongs to the family Diplocystidiaceae or Astraeaceae. The different colloquial names attributed to the fungus are mainly because of its hygroscopic nature.[9] Among the tribal people in India, it is commonly called *puttu* or *rugra*.[12]

In some parts of India, it is also known as the kurkuri mushroom. *A. hygrometricus* has often been confused with *Geastrum* species as they are superficially similar. However, it has been noted that *A. hygrometricus* lacks the presence of columella in its younger fruiting bodies and its spore sac is divided into locules, unlike in the *Geastrum* species.[13]

5.2.2 Physical Characteristics

The microscopic features of *A. hygrometricus* were first described by Morgan in 1885.[14] Its fruiting bodies are 1–8 cm in length and covered with a thick exoperidium. The exoperidium consists of four different layers. First, the mycelial layer, which contains the branched hyphae; second, the fibrous layer of the branched hyphae; third, the collenchyma type layer with the branched hyphae, and fourth, the soft layer with the hyphae.[15] On the top of the fruiting body, a sessile spore case is present which can be unwrapped by an irregular tear or cut. The internal part of the spore case is called the gleba, which is divided into ovular locules. This characteristic feature helps to distinguish the *Astraeus* species from the *Geastrum* species[13] (Table 5.1). From the base of the fruiting body thin hair-like threads called rhizomorphs arise, which break off once the fruiting bodies reach maturity. The spores are 7–11 micrometers long and are spherical in shape, covered with warts and spines.[5] Among the spores, masses of thread-like fibers arising from the peridium are distributed, called the capillitia.[5] The basidia or the spore-bearing cells are arranged in clusters of long strings and there are usually 4–8 spores.[5]

The young and mature fruiting bodies can be differentiated by various characteristics. The young fruiting bodies are almost spherical in shape and are covered with a whitish mycelial layer. As the fruiting body matures, it emerges into a stellate pattern with around 4–20 rays (dark grey to black in color) from

Table 5.1 Characteristic features that distinguish the *Astraeus* species from the *Geastrum* species.[13]

Characteristic feature	Younger fruiting bodies	Mature fruiting bodies
Shape	Roughly spherical	Star-shaped
Development	Partially embedded in the substrate	Emerges into a stellate shape with 4–20 rays from the mycelial layer and exoperidium
Mycelial layer	Smooth whitish layer that partially or fully covers the fruiting body	It gets torn as the fruiting body matures
Endoperidium	Gray-brown in color and sparsely warty	Brownish gray, thin and papery
Spore sac	Unexposed	Exposed
Gleba	White and solid	Brown and powdery
Rhizomorphs	Present during the younger stages	Breaks down as the fruiting body matures

the mycelial layer and the outer exoperidium. This can concurrently help the spore case enclosed in a thin endoperidium to be exposed above the ground. The rays of the fungus are visually appealing, but they also play an important role in helping the species to survive under harsh atmospheric conditions.[14] The inner-most fibrous layer of the exoperidium is highly sensitive to the moisture content present in the atmosphere. The rays tend to curl outwards during high humidity and curl inwards when the air is dry. This can help the fungus to reduce the rate of evaporation during dry periods. Under optimum humidity, the fruiting bodies have the habit of releasing spores. As fruiting bodies dry up, spores are easily transported by the wind and dispersed to different places.[16]

5.2.3 Geographical Distribution

The genus *Astraeus* contains around 10 different mushroom species, found all around the world. They are commonly found in temperate, tropical and sub-tropical ecosystems in Africa, North America, South America, Europe, Australia and Asia. It is found on all the continents, except for Antarctica. The wide distribution indicates the process of translocation as well as the minor differences in basidiospore morphology exhibited by the species in different regions.[10] In India, *A. hygrometricus* was first identified in the Western Ghats of south-west India and is considered to be a local delicacy by various tribes and is used by the local herbal healers for its medicinal values.[13] In several parts of Asia, in Japan, Thailand, Lao and India, *Astraeus* are considered to be a highly invaluable mushroom. In some areas of India and China, it is used as a herbal medicine by tribal people.[17] On the Indian subcontinent, it is seen widely distributed in Himachal Pradesh, West Bengal, Uttar Pradesh, Uttarakhand, Orissa, Jharkhand, Kerala and Karnataka.[18] Several locations for *A. hygrometricus* have been identified and it is important to protect its ecosystem as most of the tribal groups are solely dependent on the mushrooms for their stable livelihood. In other countries they are mostly found in the sandy soils of American, Africa, Europe, and Australia.[18]

5.2.4 Resource Availability and Sustainability

A. hygrometricus is mostly found in scattered groups in natural habitats and open spaces. It is mostly confined to laterite soils and sometimes found in abandoned paddy fields and mixed Arecanut plantations where lateritic soils are present.[19] *A. hygrometricus* is ectomycorrhizal with several species of trees – *Phyllanthus emblica, Syzigium cumini, Artocarpus hirsutus*, and *Hopea parviflora.*[20] It is also ectomycorrhizal with exotic tree species, such as *Anacardium occidentale* and *Acacia auriculiformis.* Forest fires can reduce the diversity of macro-fungi or mushrooms, but ectomycorrhizal fungi are able to survive in the sub-soil region supported by the roots of the host tree species.[12] Being a fungus that it ectomycorrhizal with different tree species, there are no means of cultivating it.[21] Locally, people collect mushrooms during early to late monsoon, with the maximum yields being witnessed in the months of July and August. When collected in bulk quantity, the mushrooms can be preserved by storing them underneath soil for 3–4 days. Nowadays, cleaned mushrooms are kept at room temperature wrapped in a wet cloth or they can be preserved for a few days in the refrigerator. In India, the highest yields of mushrooms are mostly obtained in the eastern lateritic regions.[22] Both the cooked and uncooked fruiting bodies of *A. hygrometricus* have many advantages; they can be used as a raw material for producing pharmaceuticals and functional foods.[23]

5.3 Traditional Uses

Mushrooms are often considered to be the most important food source in the diet of rural communities.[24] Apart from a being source of high protein, mushrooms also contribute to the curing of human ailments and they are often used to control pathogenic diseases in cultivated crops and domesticated animals.[25] Collecting wild mushrooms can also provide a source of additional income to rural communities. In rural or tribal communities, mushrooms play a vital role in social, economic, and ecological balance.[26] In Thailand, the mushroom is considered as a traditional delicacy and is sold at local markets for high prices. In North America, unlike other wild-type mushrooms, *A. hygrometricus* is not considered to be desirable for eating because of its hardness and tough texture.[10]

5.3.1 Folklore and Ancient Uses

A. hygrometricus has several functions, such as its edibility, medicinal uses, as a plant growth promoter, and it is also used as a raw material for making food products. In different countries, the traditional use of *A. hygrometricus* varies according to the social and cultural aspects of that country. According to Chinese folk medicine, *A. hygrometricus* is used in powdered form to treat burns and wounds.[21] The ethnic tribes residing in India often use the mushroom for medicinal purposes.[27] Local people also consume these mushrooms with the belief that it can help in curing age-related disorders.[28]

5.3.2 Modern Uses

A. hygrometricus, in its cooked or uncooked form, is balanced in nutrients as it contains a moderate amount of protein, high carbohydrate and fiber content, and low lipid content. After the cooking process, an oily and herbal flavor is imparted on the mushrooms, which may be due to the presence of volatile compounds. In addition to the balanced nutritional value, the bioactive constituents and phytochemicals present in the mushroom make it an excellent natural nutraceutical.[29] Some of the physical features of the tender and young fruiting bodies of the mushroom also make them suitable for added-value. The young and tender fruiting bodies sink when immersed in water because of their high wet mass content compared to matured fruiting bodies, which tend to float on the surface when immersed in water. This makes them appropriate for post-harvest handling.[23]

The ethanolic extract of *A. hygrometricus* exhibits many therapeutic activities, such as hypoglycemic, chemo-protective, hepatoprotective, cardioprotective, anti-inflammatory, anti-cancer, immune-enhancing, anti-candidal, leishmanicidal, and free radical scavenging properties.[30] It is therefore used as a raw material for producing pharmaceuticals, nutraceuticals, confectionary, and baking products.[31] Even though the research studies on *A. hygrometricus* are limited, it is evident that the mushroom has great potential to prevent and cure human ailments and is also a source of high nutritive value. Some of the bioactive constituents and their biological activities are listed in Table 5.2. The presence of high amounts of phenols also makes it an excellent candidate for developing drugs.[11]

5.4 Phytochemistry

5.4.1 Nutritional and Dietary Composition

A. hygrometricus contains many compounds with nutritional value. The fungus contains a high polysaccharide/carbohydrate, protein, and mineral content, which is beneficial to human health. The mushroom has a moisture content of 83.87%. With a moderate protein content (14–14.7%), high carbohydrate content (44.9–54.4%), high crude fiber content (10.8–12.3%), and low fat content (2.7–4.4%), the fruits of *A. hygrometricus* in northern Thailand are an excellent source of nutrients. A study of the exterior and interior parts of *A. hygrometricus* fruits from India revealed a low amount of fiber, fat, and protein and a moderate amount of carbohydrates.[32] The exterior and interior sections of *A. hygromatricus* contain significant amounts of calcium, sulfur, potassium, magnesium, iron, phosphorus, copper, as well as zinc, making it a good source of nutrients in the diet. The concentrations of trace minerals, such as iron and zinc, are quite substantial. Although trace elements account for less than 0.0001% of the total body proportion, they play essential roles in hemoglobin, myoglobin, and oxygen transfer. Zinc, for instance, is also a co-factor for a number of enzymes involved in energy metabolism and acts as an immunological factor.[33] Potassium is the most abundant element, followed by phosphorus and sulfur in *A. hygrometricus*. Therefore, the fungus is a good source of minerals. Five saturated and one unsaturated fatty acid were identified in *A. hygrometricus*, with uncooked samples containing substantially more palmitic acid than cooked samples.[23] Oleic acid is the second most abundant fatty acid, which is present in considerably higher amounts in raw samples.[23] The exterior and interior parts of fruit bodies exhibit a fat content of 1.04% and 0.24%, respectively, which is low compared to in other edible mushrooms.[33]

Table 5.2 Some bioactive constituents and their biological activities found in mushrooms.

Bioactive constituents	Structure	Biological properties
Astrakurkurone	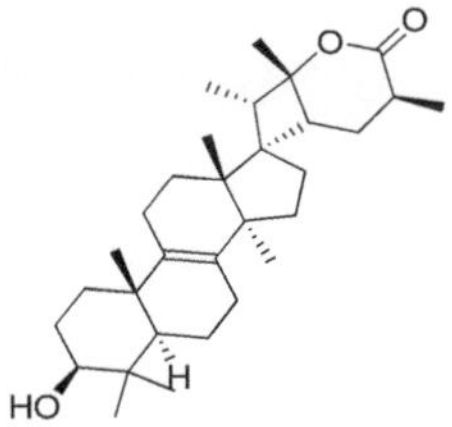	Anticandidal effects and promastigocidal activity[45]
Ascorbic acid	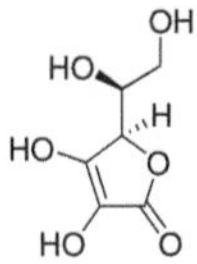	Antioxidant activities[5]
Oleic acid	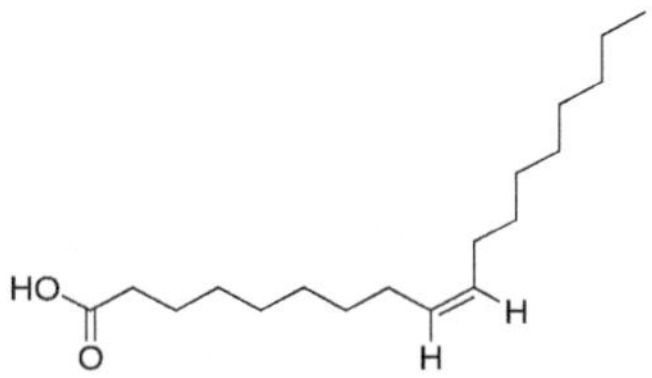	Delivers nutritional value[23]
Ferulic acid	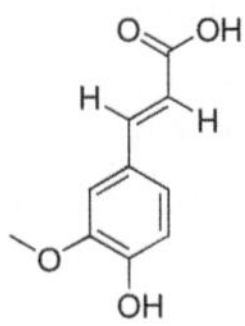	Antioxidant and anti-inflammatory activities[23]
Palmitic acid	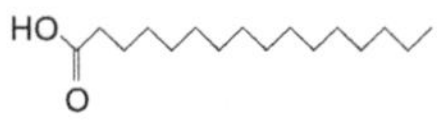	Increases the nutritional value of mushrooms
β-Carotene	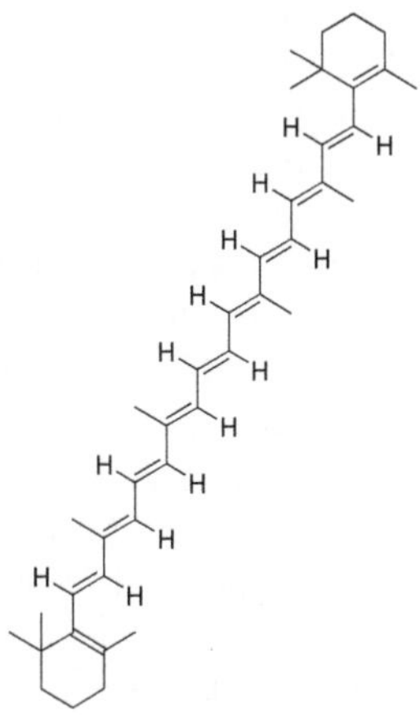	Antioxidant activities[33]

Table 5.2 (*Continued*)

Bioactive constituents	Structure	Biological properties
Syringic acid	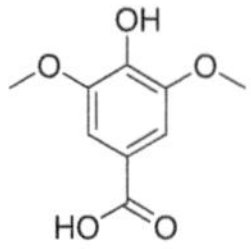	Anti-inflammatory and antioxidant properties[28]
Astrakurkurol	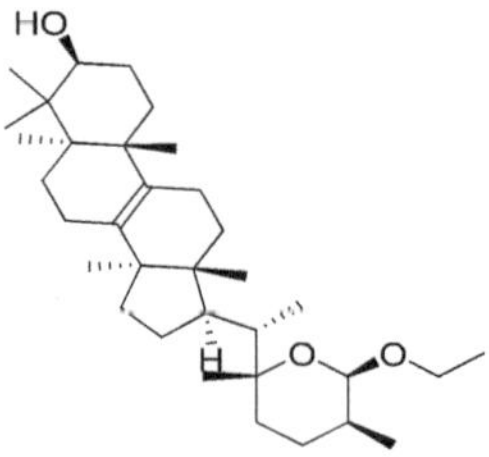	Anticandidal effects[45]

The extensive use of *Astraeus* species polysaccharides in pharmacological applications can be attributed to their non-toxicity, biodegradability, biocompatibility, and availability as a natural source. Polysaccharides from *Astraeus* species have been reported to exhibit anticancer, immunomodulatory, antioxidative, anti-inflammatory, antiviral, and anti-diabetic properties, amongst others.[34] Studies on *A. hygrometricus* have revealed the presence of polysaccharides, protein and lectins, among other bioactive compounds in this fungus, which exhibit biological attributes, such as hepatoprotective, anti-inflammatory, anticancer, antitumor, and antioxidant properties.[11] The first polysaccharides were derived from an aqueous extract of *A. hygrometricus* in a 1.9 : 0.9 : 1 ratio of D-glucose, D-galactose, and D-galacturonic acid. D-Glucose, D-mannose, and L-fucose have been found in a 1 : 2 : 1 molar ratio in an alkaline extract of *A. hygrometricus* fruits. Upon fractionation by chromatography, the water-soluble fraction comprises two homogenous fractions. An analysis of the two fractions revealed (AQSI) as a α-(1→4), β-(1→6) linked glucan and AQS-II as a heteropolysaccharide with a 2 : 1 : 1 molar ratio of glucose, galactose, and fucose in the polysaccharide portion.[11] According to a study carried out by Chakraborty *et al.*, it was reported that the glucan (AQS-I) induced splenocyte activation.[35] Polysaccharides, such as glucans, are effective anticancer compounds obtained from mushrooms. Polysaccharides have shown anticancer properties and are used to treat cancers, such as ovarian, liver, and lymphatic cancers.[36] They function by suppressing cell growth and promote cell apoptosis. *A. hygrometricus* contains D-mannose, D-glucose, D-galactose, and D-galacturonic acid, which are responsible for the treatment of colon cancer, melanoma, and Dalton's lymphoma, as well as inhibiting cancer through cell cycle dysregulation and apoptosis.[36]

The polysaccharide AE2 is derived from *A. hygrometricus* extracts and is composed of mannose, galactose, glucose, and fucose and a putative β-glucan. In addition, AE2 has been demonstrated to stimulate macrophage potential *in vitro*. The presence of lectins in *A. hygrometricus* was first reported by Yagi *et al.*[11] Lectins are carbohydrate-binding proteins with anticancer properties that bind to glycosylated cells.[11] The transition towards plant polysaccharides and lectins is owing to their low toxicity and widespread availability.[5]

Astrakurkurone (a novel triterpene) has been derived from *A. hygrometricus*. In *Leishmania donovani* promastigotes, astrakurkurone induced selective reactive oxygen species (ROS)-triggered, mitochondria dependent, as well as caspase-independent apoptosis. In murine macrophages, ROS generated by astrakurkurone was shown to be effective against intracellular amastigotes. *In vitro*, astrakurkurone showed considerable antipromastigote efficacy, reducing total protein while considerably increasing neutral lipid droplets in 8.81% of promastigotes. As a result, changes in lipid and protein content may disrupt plasma membrane integrity and induce apoptosis.[37] The structural elucidation of new triterpenes, namely astrahygrol, 3-*epi*-astrahygrol, and astrahygrone (3-oxo-25*S*-lanost-8-eno-26,22-lactone), present in *A. hygrometricus*, has been reported In *A. odoratus,* astraeucins A–L, lanostane triterpenoids, were isolated from fruiting bodies. Studies have confirmed that some of these lanostanoids from the *A. odoratus* species exhibit antibacterial properties.[38] Steroids are found in various mushrooms that have been used to treat a variety of ailments, including rheumatoid arthritis, autoimmune diseases, inflammation, and lung complications.[38] Two known steroids, namely ergosta-7,22-diene-3-ol and ergosta-4,6,8-(14),22-tetraene-3-one, have been found in *A. hygrometricus*. Matsutakone, a sterol with an unprecedented polycyclic ring structure, and matsutoic acid, a novel norsteroid, have been reported to be present in the fruiting bodies of edible mushrooms.[38]

5.4.2 Vitamins

Mushroom fruiting bodies have been identified to be a rich source of vitamins. The major vitamins present in *A. hygrometricus* are the water-soluble vitamins ascorbic acid and thiamine, which have been reported to be found in both the internal as well as external parts of the fungus, with 3.26 mg per 100 g and 5.23 mg per 100 g of ascorbic acid and thiamine were found in the outer layer and 0.26 mg per 100 g and 3.54 mg per 100 g in the inner layer respectively.[11] In another study reported by Pavithra *et al.* that the antioxidant potential of *A. hygrometricus* was attributed to the significant amount of vitamin C present in the mushroom, which is diminished after cooking due to thermal or heat treatment.[5]

5.4.3 Phytochemicals

Polyphenols are abundantly present in *A. hygrometricus.* Singh *et al.* reported the presence of a high amount of ferulic acid, salicylic acid, protocatechuic acid, and syringic acid, as well as anthranilic acid in this fungus. The total phenolic content of the mushroom was reported to be 1.4% in the interior as well as exterior part of the fruit, respectively.[33] The *Astraeus* genus species have a number of therapeutic properties. In addition, the antioxidant activity of *A. hygrometricus* was also reported in 2010 by Singh *et al.*, attributed to the presence of phytoconstituents, such as phytic acid, phenolics, β-carotene, tannins, and flavonoids, present in the fungus.[33]

5.4.4 Volatile Organic Compounds

Characterization of the volatile organic compounds present in *Astraeus* sp. was conducted by Kakumyan *et al.*[39] The volatiles derived from raw *Astraeus* species contained high levels of C8 complexes such as (*E*)-2-octen-1-ol, 3-octane, 1-octen-3-one, 1-octen-3-ol, and 1-octanol. The most abundant of these was 1-octen-3-ol. An oxygenase enzyme was found to catalyze the formation of the compound 10-hydroperoxy-(*E*,*Z*)-8,12-octadecadienoic acid (10-HPOD) from linoleic acid. (*E*)-2-Octen-1-ol, 1-octanol, and (*E*)-2-octenal, have been reported to have been derived from a nonenzymatic reaction with 10-HPOD10-hydroperoxy-(*E*,*Z*)-8,12-octadecadienoic acid by transition metals. Apart from these compounds, reducing agents were also found to be present in raw *Astraeus* spp.[39]

5.5 Pharmacological Potential

5.5.1 Immunomodulatory Properties

Several bioactive components have been identified and characterized from mushrooms, amongst which immunomodulators have attained great interest due to their increasing development in the immunotherapeutic sector.[40] Mushrooms, which are a natural source of immunomodulatory agents, are safer alternatives to synthetic drugs in medicine.[41] At a dose of 10 ng mL^{-1}, fraction I (water soluble), derived from an aqueous extract of *A. hygrometricus* fruit bodies, demonstrated the strong activation of splenocyte and the stimulation of B cells, T cells, and macrophages, while also promoting immune responses.[35] Another polysaccharide, referred to as AE2, that was tested for macrophage-stimulating activity produced more nitric oxide and cytokines. *In vitro*, macrophages treated with AE2 demonstrated increased phagocytic potential increased cytokine production, and activated immune cells, from splenocytes.[42] Cibacron blue affinity eluted protein (CBAEP), a protein fraction derived from *A. hygrometricus*, stimulated splenocytes, bone marrow cells and thymocytes in an animal model.[42] It also increased the cytotoxicity of mouse natural killer cells and produced NO by activating macrophages. These results support considering *A. hygrometricus* as a potential immunomodulator source. Polysaccharides derived from mushrooms have long been studied for their medicinal potential, but the mechanism of action of their pharmacological activities is still unknown. Macrophages and monocytes recognize heteroglucans or glucans by interacting with receptors such as scavenger receptors that are present on the cell surface, Toll-like receptors, and Dectin1, which activate the MAPK (mitogen-activated protein kinases) and Akt pathways. Mallick *et al.* in their study demonstrated the involvement of p38 MAPK pathways, PKC, ERK1/2, MEK (MAPK/ERK kinase), and JNK in RAW cell activation using AE2 in *A. hygrometricus*, although it failed to promote phosphorylation of the JAK2/STAT1 pathway.[42]

5.5.2 Antioxidant, Anti-inflammatory, Antimicrobial and Antifungal Activities

ROS are produced by a variety of endogenous and external sources. Although nearly all organisms have antioxidant defense mechanisms, they are frequently insufficient to totally prevent oxidative stress-induced disease. As a result, antioxidant supplements or natural items containing antioxidants may be employed to protect the human body from oxidative damage. For hundreds of years, mushrooms have been a staple part of the human diet, and consumption has increased dramatically in recent years, encompassing a diverse range of species. Phenolic compounds (such as phenolic acids and flavonoids), tocopherols, ascorbic acid, and carotenoids are the primary bioactive components of mushroom. Additionally, polysaccharides are significant active substances.[43] A study by Khatua *et al.* reported that ethanolic extract of *A. hygrometricus* showed inhibition of hydroxyl and superoxide radicals, lipid peroxidation and DPPH scavenging activity.[43] Another study compared the antioxidant capacity of raw and cooked *A. hygrometricus* fruit bodies, and it was discovered that cooking considerably reduced antioxidant activity.[5] Similar studies have also been conducted in south-western India, where *A. hygrometricus* is considered to be a delicacy.

Additionally, it has been shown that cooked mushrooms lose up to one-third of their overall antioxidant activity, as heating reduces antioxidant activity; therefore, it has been suggested that boiled mushroom tissue, as well as its broth, can be consumed to take advantage of antioxidant activity and to retain the nutritional value of fungi.[5] *A. hygrometricus* has been extensively investigated due to its free radical scavenging activity, and its IC_{50} values suggest that it might be a promising defender towards oxidative damage induced by free radicals. Another study has shown that a methanolic extract of *A. hygrometricus* is rich in phenolic compounds and has significant overall antioxidant activity (IC_{50} value of $9.3 \pm 0.3\ \mu g\ mL^{-1}$).[11] Phenolic fractions are frequently associated with antioxidant and antibacterial activities. Numerous phenolic constituents analyzed from *A. hygrometricus* extracts were found to be rich in protocatechuic, ferulic, salicylic, anthranilic, and syringic acids. Due to the presence of phenolic constituents in *A. hygrometricus*, they contribute to different significant pharmacological effects.[11]

Biswas *et al.* showed that food containing high levels of polyphenols can reduce inflammation.[28] The presence of flavonoids and polyphenols in an ethanolic extract of *A. hygrometricus* was confirmed by preliminary chemical analysis. The antioxidant and anti-inflammatory properties of the fungus may be attributed to these compounds. Carrageenan, dextran, or formalin-induced paw oedema in mice was significantly reduced by an ethanolic extract of *A. hygrometricus*. The impact of the extract is equivalent to diclofenac, at a concentration of 125 mg per kg body weight.[28] *A. hygrometricus* mushrooms have been shown to have potent antibacterial and antifungal effects against resistant human diseases.[44] *A. hygrometricus* has a limited spectrum of antimicrobial properties, with *Bacillus cereus*, *Escherichia coli*, and *Staphylococcus aureus* being the most sensitive bacteria among the strains examined.

A methanolic extract of *A. hygrometricus* was shown to exhibit high inhibitory action toward *Proteus vulgaris, Escherichia coli, Pseudomonas aeruginosa,* and *Candida albicans.*[44] Two pure lanostane type triterpenoid were isolated that demonstrated significant anticandidal activity against the pathogenic *Candida albicans* strain, which is entirely resistant to antifungal medications such as nystatin, fluconazole, and amphotericin. Biswas *et al.*, reported that various extracts of *A. hygrometricus* exhibited anti-leishmanial activity.[11] It was reported that the terpenoid fraction (AHFa) inhibits the proliferation of *L. donovani* promastigotes *in vitro* by apoptosis, while the water-soluble polysaccharide fraction (AHFb) inhibits intracellular amastigotes *in vitro* by significantly increasing the release of the vital anti-leishmanial compound, pro inflammatory cytokine IL-12, in addition to NO.[28] Another study reported that astrakurkurone and astrakurkurol (Table 5.2) both exhibited anticandidal effects, whereas astrakurkurone, an active molecule, exhibited considerable promastigocidal activity against *Leishmania donovani* promastigotes.[45]

5.5.3 Antitumor Properties

Mushroom polysaccharides have been shown to exhibit anticancer properties by either activating the immune system of the host (immuno-enhancing activity) or by having an antiproliferative effect on tumour cells by modifying the expression of signals within tumour cells. Tumour regression mediated by AE2 was achieved by reversing tumour-induced immunosuppression. Macrophage activation has been elevated in AE2-treated tumour-bearing mice, as observed by the increased surface expression of co-stimulatory molecules and MHC-II on the macrophages of these mice. In addition to increasing NK cell activity, AE2 therapy was also shown to increase tumour microenvironment Th1 cytokine levels while decreasing tumour microenvironment Th2 cytokine levels. Following immuno-activation, Dalton's lymphoma (DL) cells experienced apoptosis, reducing tumour development and increasing tumour-bearing host survival. Furthermore, it was shown that AE2 administration at 10 and 20 mg per kg of body weight restored the tumour-mediated reduction of proliferation and was comparable to that of normal mice. As a result, the reversal was shown to be optimal at 10 mg per kg in AE2 therapy. In addition, dose-dependent reversal of the tumour-induced inhibition of bone marrow cells in mice that had been treated with AE2 has also been reported.[30] In another investigation, CBAEP, the protein fraction from *A. hygrometricus*, was found to have antiproliferative activities in a mouse model by the stimulation of apoptosis in tumour cell lines, such as melanoma (B16-F0), colon cancer (HT 29), HeLa, Dalton's lymphoma (DL), and sarcoma-180.[11]

5.5.4 Hepatoprotective Properties

The liver is known for its important functions in fat, protein, and carbohydrate metabolism. Hepatotoxicity is described as liver injury associated with altered liver activity caused by the administration of a medication or another non-infectious substance. Mushroom extracts have been shown to protect the liver from injury caused by hazardous substances.[46] Therefore, mushrooms may be a potential therapeutic option for treating liver ailments or their symptoms, and future research should explore this possibility. The hepatoprotective properties of mushroom extracts appear to be attributed to phenolics, triterpenes, polysaccharides, and peptides.[46]

In another study reported by Biswas *et al.*, an ethanolic extract of *A. hygrometricus* was shown to protect the liver against chronic hepatotoxicity induced by CCl_4.[47] The results revealed that the antioxidants in the extract were able to suppress the chain reaction cascade of CCl_4 toxicity through scavenging the intermediate free radicals. Serum indicators such as glutamate pyruvate transaminase, glutamate oxaloacetate transaminase, bilirubin, and alkaline phosphatase were reduced to normal levels after treatment with the *A. hygrometricus* extract, while antioxidant defense enzymes in the liver were increased. Histopathological investigations also revealed that varying levels of fatty infiltration and necrosis were caused by radical-mediated cytotoxicity. The direct radical-scavenging activity of the extract might be implicated in its hepatoprotective properties against chronic CCl_4 exposure. Based on these findings, it was suggested that the antioxidant defense mechanism is involved in the hepatoprotective activity of the ethanolic extract of *A. hygrometricus.*[47]

5.5.5 Anti-diabetic Properties

Mushrooms are known to contain specific constituents that aid in the appropriate functioning of the liver, pancreas, and other endocrine glands, boosting the secretion of insulin and associated hormones that enable optimal metabolic activity.[48] These include naturally occurring insulin-like components and enzymes that aid in the breakdown of starch or sugar and alleviate insulin resistance. A study reported that in alloxan-induced diabetic mice, an ethanolic extract of *A. hygrometricus* exhibited exceptional hypoglycaemic activity. In both acute and subacute studies, oral administration of the ethanolic extract at a concentration of 500 mg per kg of body weight lowered blood glucose levels. The dosage appeared to enhance glucose tolerance, implying increased peripheral glucose consumption in the diabetic mice during oral glucose tolerance tests. Additionally, it was discovered that mushrooms containing a high concentration of mannitol are regarded as a beneficial diet for diabetes patients. *A. hygrometricus* has been reported to contain a high concentration of mannitol. As a result, when ingested in sufficient amounts, *A. hygrometricus* can be used as a medical food for diabetes patients.[11]

5.5.6 Cardioprotective Properties

Arterial hypertension is a type of cardiovascular disease (CVD) that affects people and may be worsened by atherosclerosis, stroke, or myocardial infarction. Mushrooms are thought to be a potent source of naturally generated cardioprotective biological molecules that have hypocholesterolemia, hypoglycaemic, hypotensive, anti-inflammatory, antioxidative, fibrinolytic, and thrombolytic properties.[49] Ethanolic extract of *A. hygrometricus* has been shown to suppress platelet aggregation, prostaglandin production, and increase NO generation in human blood platelets, consequently decreasing the risk of myocardial infraction, cardiac hypertrophy, and vascular mortality in patients with cardiovascular disease. Evidence indicates that the ethanolic extract of *A. hygrometricus* may be beneficial in reducing the formation and progression of coronary artery disease and serve as a novel treatment for cardiac hypertrophy.[11]

5.6 Safety, Toxicity and Regulation

Mushrooms are a part of the human diet. Some mushrooms, such as *Agaricus bisporus*, *A. campestris*, and *Tricholoma matsutake*, are well-known across the world for their nutritional and medicinal benefits.[50] However, some species, such as *A. phalloides, A. verna,* and *A. suballiacea*, are harmful because they induce toxicity.[50] Amatoxins (cyclopeptides), orellanus (*Cortinarius* species), gyromitrin (monomethylhydrazine), muscarine, ibotenic acid, psilocybin, and coprine are the seven primary types of mushroom toxins that have been reported.[50] Several reports have shown the therapeutic and toxic properties of mushrooms. There have been reports regarding *A. hygrometricus* and its constituents, where extracts of this fungus have been shown to be safe up to a dosage of 5000 mg per kg of body weight in acute toxicity studies for diabetes.[51] No evidence of lethality or toxic effects was observed during the study. Based on the literature available, it can be inferred that the different constituents of *A. hygrometricus*, such as polyphenols, flavonoids, vitamins, polysaccharides, minerals, and steroids, are responsible for its therapeutic properties.

5.7 Conclusions

Despite the fact that there are limited studies on the pharmacological and nutritional properties of *A. hygrometricus*, it is evident from the literature that this fungus holds enormous potential for the prevention of many human illnesses. *A. hygrometricus* has many nutrients, making it a good food source. Such prospects could be improved by promoting and implementing contemporary food technology standards to ensure that mushrooms, such as *A. hygrometricus,* are safe and appealing for human consumption. Although few studies have been carried out on the therapeutic and nutritional properties of *A. hygrometricus*, numerous efforts are required to further investigate this mushroom, as it has been identified to contain various bioactive compounds, but the biological properties of only a few them have been identified and reported. Furthermore, in depth research on the possible mechanisms of action is required as the existing evidence is inadequate. Further work has to be carried out, considering the indigenous evidence, for commercializing the pharmaceutical and nutraceutical products of this species.

References

1. K. Kumar, *South Asian J. Food Technol. Environ.*, 2015, **5168**, 211–218.
2. S. Rathee, D. Rathee, D. Rathee, V. Kumar and P. Rathee, *Rev. Bras. Farmacogn.*, 2012, **22**, 459–474.
3. G. Ma, W. Yang, L. Zhao, F. Pei, D. Fang and Q. Hu, *Food Sci. Hum. Wellness*, 2018, **7**, 125–133.
4. D. Jasso de Rodríguez, E. de J. Salas-Méndez, R. Rodríguez-García, F. D. Hernández-Castillo, M. L. V. Díaz-Jiménez, A. Sáenz-Galindo, S. González-Morales, M. L. Flores-López, J. A. Villarreal-Quintanilla, F. M. Peña-Ramos and D. A. Carrillo-Lomelí, *Ind. Crops Prod.*, 2017, **107**, 499–508.
5. M. Pavithra, K. R. Sridhar, A. A. Greeshma and K. Tomita-Yokotani, *Mycology*, 2016, **7**, 191–202.
6. X. Zeng, J. Suwandi, J. Fuller, A. Doronila and K. Ng, *Food Sci. Technol. Int.*, 2012, **18**, 367–379.
7. S. Patel and A. Goyal, *3 Biotech*, 2012, **2**, 1–15.
8. W. Fangfuk, R. Petchang, C. To-anun, M. Fukuda and A. Yamada, *Mycoscience*, 2010, **51**, 291–299.
9. V. Vishal, S. S. Munda, G. Singh and S. Lal, *Indian J. Appl. Pure Bio.*, 2021, **36**, 567–579.
10. C. Phosri, M. P. Martín and R. Watling, *IMA Fungus*, 2013, **4**, 347–356.
11. G. Biswas, S. Nandi, D. Kuila and K. Acharya, *Pharmacogn. J.*, 2017, **9**, 799–806.
12. R. K. Verma, V. Pandro, S. N. Mishra, D. Raj and A. J. K. Asaiya, *Int. J. Curr. Microbiol. Appl. Sci.*, 2019, **8**, 563–575.
13. M. Pavithra, A. A. Greeshma, N. C. Karun and K. R. Sridhar, *Mycosphere*, 2015, **6**, 421–432.
14. Y. Hayat, *SSRN Electron. J.*, 2014, 11.
15. I. G. Baseia and T. C. de. O. Galvão, *Acta botánica Bras.*, 2002, **16**, 8.
16. D. Maiti, K. Chandra, S. Mondal, A. K. Ojha, D. Das, S. K. Roy, K. Ghosh, I. Chakraborty and S. S. Islam, *Carbohydr. Res.*, 2008, **343**, 817–824.

17. S. K. Mallick, S. Maiti, S. K. Bhutia and T. K. Maiti, *J. Med. Food*, 2010, **13**, 665–672.
18. P. Pradhan, A. K. Dutta, A. Roy, S. K. Basu and K. Acharya, *Biodiversity*, 2012, **13**, 88–99.
19. N. Karun, *Curr. Res. Environ. Appl. Mycol.*, 2014, **4**, 173–187.
20. A. W. Wilson, M. Binder and D. S. Hibbett, *New Phytol.*, 2012, **194**, 1079–1095.
21. P. E. Mortimer, S. C. Karunarathna, Q. Li, H. Gui, X. Yang, X. Yang, J. He, L. Ye, J. Guo, H. Li, P. Sysouphanthong, D. Zhou, J. Xu and K. D. Hyde, *Fungal Divers.*, 2012, **56**, 31–47.
22. S. Manna, D. Ray and A. Roy, *Ethnobot. Res. Appl.*, 2014, **12**, 15–24.
23. M. Pavithra, K. R. Sridhar and A. A. Greeshma, in *Fungi and their Role in Sustainable Development: Current Perspectives*, Springer Singapore, Singapore, 2018, pp. 367–382.
24. W. A. Elkhateeb, *J. Chem. Res.*, 2020, **5**, 106–118.
25. H. Rathore, S. Prasad and S. Sharma, *PharmaNutrition*, 2017, **5**, 35–46.
26. K. Yongabi, M. A. Gho and D. Martínez-Carrera, *Micol. Apl. Int.*, 2004, **16**, 34–36.
27. M. K. Panda and K. Tayung, *Asian J. Pharm. Clin. Res.*, 2015, **8**, 139–143.
28. G. Biswas, S. Sarkar and K. Acharya, *Lat. Am. J. Pharm.*, 2010, **29**, 549–553.
29. M. Pavithra, K. R. Sridhar and A. A. Greeshma, *Kavaka*, 2017, **49**, 22–27.
30. S. K. Mallick, S. Maiti, S. K. Bhutia and T. K. Maiti, *Food Chem. Toxicol.*, 2010, **48**, 2115–2121.
31. A. A. Greeshma, K. R. Sridhar and M. Pavithra, *Microb. Biosyst.*, 2018, **3**, 34–44.
32. K. R. Sridhar and M. Pavithra, *Bioprospect potential of gasteroid mushrooms of the genus Astraeus*, Elsevier B.V., 2021.
33. N. Singh, *Nat. Environ. Pollut. Technol.*, 2010, **9**, 597–600.
34. P. Maity, I. K. Sen, I. Chakraborty, S. Mondal, H. Bar, S. K. Bhanja, S. Mandal and G. N. Maity, *Int. J. Biol. Macromol.*, 2021, **172**, 408–417.
35. I. Chakraborty, S. Mondal, M. Pramanik, D. Rout and S. S. Islam, *Carbohydr. Res.*, 2004, **339**, 2249–2254.
36. J. Gupta, A. Ahuja and R. Gupta, *Anticancer Agents Med. Chem.*, 2021, **22**, 101–114.
37. S. Mallick, S. Dey, S. Mandal, A. Dutta, D. Mukherjee, G. Biswas, S. Chatterjee, S. Mallick, T. K. Lai, K. Acharya and C. Pal, *Future Microbiol.*, 2015, **10**, 763–789.
38. M. Isaka, S. Palasarn, S. Sommai, S. Veeranondha, K. Srichomthong, P. Kongsaeree and S. Prabpai, *Tetrahedron*, 2017, **73**, 1561–1567.
39. P. Kakumyan and K. Matsui, *Biosci., Biotechnol., Biochem.*, 2009, **73**, 2742–2745.
40. D. K. Patel, S. D. Dutta, K. Ganguly, S.-J. Cho and K.-T. Lim, *Molecules*, 2021, **26**, 1359.
41. H. A. El Enshasy and R. Hatti-Kaul, *Trends Biotechnol.*, 2013, **31**, 668–677.
42. S. K. Mallick, S. Maiti, S. K. Bhutia and T. K. Maiti, *Cell Biol. Int.*, 2011, **35**, 617–621.

43. S. Khatua, S. Paul and K. Acharya, *Res. J. Pharm. Technol.*, 2013, **6**, 496–505.
44. A. Mukul Jana, S. Sharma and S. Singh Gill, *Artic. World J. Pharm. Pharm. Sci.*, 2014, **3**, 1009–1023.
45. T. K. Lai, G. Biswas, S. Chatterjee, A. Dutta, C. Pal, J. Banerji, N. Bhuvanesh, J. H. Reibenspies and K. Acharya, *Chem. Biodivers.*, 2012, **9**, 1517–1524.
46. A. A. Soares, A. B. De Sá-Nakanishi, A. Bracht, S. M. G. Da Costa, E. A. Koehnlein, C. G. M. De Souza and R. M. Peralta, *Molecules*, 2013, **18**, 7609–7630.
47. G. Biswas, S. Sarkar and K. Acharya, *Dig. J. Nanomater. Biostructures*, 2011, **6**, 637–641.
48. D. D. De Silva, S. Rapior, K. D. Hyde and A. H. Bahkali, *Fungal Divers.*, 2012, **56**, 1–29.
49. S. M. Badalyan, A. Barkhudaryan and S. Rapior, *Int. J. Med. Mushrooms*, 2021, **23**, 21–31.
50. B. A. Wani, R. H. Bodha and A. H. Wani, *J. Med. Plants Res.*, 2010, **4**, 2598–2604.
51. G. Biswas and K. Acharya, *Int. J. Pharm. Pharm. Sci.*, 2013, **5**, 391–394.

CHAPTER 6

Cantharellus cibarius *Fr.*/ Cantharellus subalbidius/ Cantharellus lateritius *Singer (Chanterelle)*

ROSANA DE CARVALHO CRUZ*[a,b] AND MOHAMED SHEASHEA[c]

[a] College of Dentistry, Faculdade Padre Arnaldo Janssen, Minas Gerais, Belo Horizonte, MG, Brazil; [b] Laboratório de Bioengenharia, Departamento de Engenharia Mecânica, Universidade Federal de Minas Gerais, Belo Horizonte, MG, Brazil; [c] Medicinal and Aromatic Plants Department, Desert Research Center, Cairo, Egypt
*Email: rosanabio@uol.com.br

6.1 Introduction

Edible mushrooms are a natural resource in developed and developing countries, providing high nutritional value to the human diet at low economic cost, and collecting them is usually considered to be a hobby in different regions around the world.[1] The purpose of edible mushroom cultivation varies between dietary intake and medicinal purposes, with 1200 species being utilized in 85 different countries. Wild edible mushrooms are revered as sacred objects; however, some tend to be poisonous due to them containing high concentrations of heavy metals. Unlike other plants, mushrooms are identified based on their visible spore-bearing structure; this is a limiting factor as some species only produce spores at certain times of the year (every ten years) and other species do not produce spores at all.[2]

Phytochemistry and Nutritional Composition of Significant Wild Medicinal and Edible Mushrooms: Traditional Uses and Pharmacology
Edited by Ajay Sharma, Garima Bhardwaj and Gulzar Ahmad Nayik

Published by the Royal Society of Chemistry, www.rsc.org

Therefore, fewer mushroom species have been identified, *i.e.*, only 148 000 species worldwide, compared with plant identification rates (350 000 species). In 2019, the continents that made the largest contributions to the identification of new mushrooms were Asia, Europe, and South America, with shares of 41%, 23% and 9.5%, respectively, according to KEW reports.

Mushrooms are considered to be a reliable source of digestible proteins, fiber, and mineral intake that are low in fat and calories. In general, edible mushrooms contain 280–399 g per kg dry weight of carbohydrates. Mushroom intake provides the following major minerals, Cu, Fe, K, Mg, Se, P and Zn, with very low levels of Na.[3] Mushrooms are also considered to be a good source of fiber and unsaturated fatty acids (particularly oleic and linoleic acids).[4] Several studies have shown that mushrooms contribute to daily vitamin intake by providing vitamins B, D, and K. *Cantharellus cibarius* (*C. cibarius*) and *Cantharellus tubaeformis* (*C. tubaeformis*) contain high amounts of ergocalciferol (vitamin D2), with 128 and 298 mg per kg fresh weight, respectively. Therefore, *C. cibarius* and *C. tubaeformis* are good sources of vitamin D for vegetarians.[5] The nutritional value of proteins from mushrooms is considered to be equal to the quality of proteins derived from animal sources. The most abundant amino acid in mushrooms is leucine and the least abundant is phenylalanine. In addition, some mushrooms contain all the essential amino acids, such as *Termitomyces umkowaani,* and others lack one or more amino acids, *i.e.*, isoleucine, leucine, methionine and cytosine, which are the limiting amino acids in *Hericium erinaceus.*[6]

Moreover, dried mushrooms and concentrated extracts are used as nutraceuticals for medicinal purposes, with some mushroom species exerting strong antitumor and antioxidant effects by enhancing immune system functions.[7] In this context, the most common mushroom species used in medicine are *Claviceps purpurea*, *Cordyceps sinensis*, *Ganoderma lucidum*, *Laricifomes officinalis*, and *Trametes versicolor.*[8] The antitumor effects of mushrooms are manifested in β-1,3 glucan isolated from *Lentinula edodes* exerting its effects by increasing T-helper cells and macrophages.[9] Moreover, other medicinal properties of mushrooms have been explored, and the findings have shown the protective effects on the liver of *Trametes versicolor* and the lowering of cholesterol levels by *Flammulina velutipes.*[10,11]

The most appreciated species of wild edible mushrooms are *Boletus edulis* and *C. cibarius*, which are popular foods eaten throughout the world. *Boletus edulis* is very popular owing to its aroma and both nutritional and medicinal value.[12] The Cantharellus species, mainly *C. cibarius,* are considered to be the most prized edible mushrooms, with an estimated market share of over \$1.5 billion US.[13] In this regard, the fruiting bodies of *C. cibarius* contain large amounts of antioxidant metabolites, *i.e.*, anisole, BHT, tocopherols, vitamins, and elements such as manganese, zinc, and selenium, supporting their use in gastronomy.[14] *C. cibarius* is a desirable trade mushroom because of its taste, stability during transport, storage, and level of resistance to insect larvae, in addition to its nutritional value, as it contains high levels of potassium and vitamin D.[15] Within this chapter, light will be shed on the health benefits and the chemical constituents of three Cantharellus species, *C. cibarius*, *C. subalbidius*, and *C. lateritius*.

6.2 Taxonomy

The Cantharellus species belongs to the Cantharellaceae family, and it is most related to the Clavariaceae family, distinct from true gill fungi as indicated by the presence of basidia. Within the two groups, basidia bear two to eight spores and the spindle of the nuclei are present in parallel to the longitudinal axis of the basidium. Other distinguishing factors include the spore type, type of fruiting bodies produced, and configuration of the hymenium, where Cantharellus species are distinguished from Craterellus species by the presence of hymenium in the form of distinct diverged ridges or folds or obtuse forked lamellae rather than the smooth hymenium in the Craterellus species. The fruiting body of this genus is a fleshy to moderately tough structure and sometimes adopts a woody form upon drying. The shape of the pileate is tubular and the pileus and stipe are not always distinct. The hymenium is often located on the underside of the pileus and upper portion of the stipe, usually in the form of narrow gill-like folds in anastomosing or dichotomous branches. The basidia typically has flexuous pedicels with two to eight spores with a white to yellowish color. The hyphae of the gill and pileus are barely distinguishable in shape and form, typically with the absence of cystidia.[16] *Cantharellus cibarius* Fr., the most characterized of the species and where the most studies have been concentrated, is known as Chanterelle, and belongs to Basidiomycota; class: Agaricomycetes; order: Cantharellales; family: Cantharellaceae (see Figure 6.1).[17]

Many organisms have a so-called folk classification, also known as cultural significance, which refers to when a certain species plays a particularly important role in a certain culture. Different criteria are adopted to indicate whether an organism has cultural significance, among others, archaeological studies of human ecology, strategies for consumption and subsistence, perception, and knowledge of the specimens by a population, in addition to the use of names that indicate their function. However, many of these classifications are of a subjective nature. In 2007, Garibay-Orijel *et al.* proposed a randomized study by means of a questionnaire and established a pyramid for identifying edible mushrooms that were of cultural significance in Mexico. The results showed that the genus *Cantharellus*, especially *Cantharellus cibarius*, appeared to be one of the most well-known and used in popular culture and had a high degree of cultural significance.[18]

6.3 Geographic Distribution

When considering the geographic distribution of fungi, they are perceived to be cosmopolitan. *Cantharellus cibarius* and their related species are highly

Figure 6.1 Image of the fungus *Cantharellus cibarius* Fr. (chanterelle). Reproduced from ref. 17 with permission from Royal Society of Chemistry, Copyright 2015.

representative groups of wild fungi in the Northern hemisphere, distributed among different geographic regions, such as America, Asia, and Europe.

For example, Henkel *et al.* identified two new species of the genus *Cantherellus* sp. in South America and named them *C. guyanensis* and *C. excelsus* by means of comparative morphological analyses of the hyphae, basidia, and basidiospores, in addition to philogenetic studies. This same group also identified three species in Brazil in the Atlantic Forest (Mata Atlantica) region. These had a morphology like that of *C. guyanensis* but with differences in their structure, and more robust than those found in French Guinea.[19]

As far back as 2017, studies using *C. cibarius* as a reference species and robust molecular biology tools, such as analysis of the rDNA region transcribed spacer 2 (ITS-2), 28S DNA sequences, and transcription elongation factor 1-alpha (Tef-1α), revealed the existence of species of this genus in North America, which had been unknown until then. Molecular analyses in combination with morphological descriptions allowed the identification of a new yellow species of *Cantherellus* in the mid-western region of the United States, denominated *Cantharellus chicagoensis*, in addition to showing evidence of a diversity of this genus present in this region.[20]

In 2020, Buyck *et al.* also by using the region ITS-2, in addition to final alignment of the Tef-1 sequences α for philogenetic studies, identified two new species of *Cantharellus* in South Korea: *C. citrinus* (the name was chosen due to the lime green color of the basidia) and classified it in the sub-genus *Cinnabarini*, and *C. koreanus*, which was clearly shown to be a new species.[21]

In Mexico, Montoya *et al.* identified two new species of Cantharellus in the east of the country, also by means of alignment of (Tef-1α) studies and transcription of the larger sub-unit of the ribosome, in addition to proposing a new name, *C. furcatus*, to substitute for *C. confluens*. The discoveries were made by comparing microscopic descriptions of the morphology and taxonomic analysis based on 78 samples collected in the field and at different stages of basidioma.[22]

Zhang *et al.* conducted a geographic distribution study and compared the mitochondrial DNA sequence of 363 samples of the genus *Cantherelles* from 30 geographic locations in six different countries. The study revealed known species and five potentially new species. However, the major contribution appears to have been the suggestion of strong evidence of heteroplasmia (the presence of a mixture of more than one type of mitochondrial genome in one and the same species). This perception led to strong indications that mitochondrial recombination occurred repeatedly in the evolution of the genus *Cantharellus*.[23]

Fungi of the genus *Cantharellus* are known to be capable of guiding a culture or even changing the dietary habits of a population. Over the last 140 years, this has occurred in Sweden. Despite industrialization, the Swedish population still maintains forestry habits, such as hunting animals and harvesting wild fruits. Many people are returning to or rediscovering the habits of collecting and using wild mushrooms in the quest for a healthier diet. Edible fungi of the genus *Cantharellus* are becoming increasingly present on Swedish tables and the Swedes are seeking a better understanding of the species, taking courses, classes or even specializing in learning about the diversity of taxa, either as a hobby or to know how to distinguish edible from non-edible types.[24]

6.4 Cultivation

Wild edible mushrooms, especially *C. cibarius*, are difficult to cultivate under controlled hot-house conditions, and it is very hard to obtaining an isolate culture from this species. Research seeking to perform the ectomycorrhizae of *C. anzutake* "*in vitro*" and to induce the mushroom growth in acclimatized plants or growth under controlled environmental conditions has shown promising results. A percentage of 84% of seedlings retained and developed ectomycorrhizae in a controlled environment in a system of jars with a pine host. Repeated fruitifications were verified in these environments, which increased the expectations of achieving an *in vitro* cultivation of this genus.[25]

There are few studies about the nutritional requirements of *C. cibarius* that would allow it to be cultivated in the laboratory. A recent study determined the ideal conditions for the mycelial growth of this species *in vitro*. The conditions tested were pH, temperature, various sources of carbon (glucose, fructose, sucrose, galactose, and xylose) and nitrogen. Peptone and yeast extract supplemented with different concentrations of ammonium sulfate were used as sources of nitrogen. The conditions suggested to be optimal for *in vitro* mycelial growth were a pH of 6.0, temperature of 22.5 °C, 2% sucrose as a source of carbon and peptone supplemented with 0.5% ammonium sulfate and sodium nitrate. A change in the pigmentation of the colony from golden in color to dark gray was detected. Additional investigations are necessary for the purpose of verifying the relationship between this change in color and the nutritional alterations in the cultivation in a medium containing peptone.[26]

6.5 Chemical Profiling of *Cantharellus* Species (*i.e.*, *C. cibarius*, *C. subalbidus* and *C. lateritius*)

Edible mushrooms are considered to be a source of vitamins, minerals, and amino acids. Furthermore, the secondary metabolites present in these species play an important role in supporting their medicinal use. In this regard, species of mushrooms have shown chemical variability to a greater extent than that of plants because each fruit body of mushrooms could lead to crossbreeding of different hyphae. Moreover, substrate composition, soil, enzyme activity, and pH have been shown to be factors affecting the mineral content between species.[14] In the next subsection, the chemical profile of *Cantharellus* species will be discussed based upon their vitamins and minerals, and categories of primary (see Figure 6.2) and secondary metabolites (see Figure 6.3).

6.5.1 Vitamins and Minerals

6.5.1.1 Vitamin D

The golden chanterelle (*C. cibarius*) mushroom guarantees high amounts of vitamin D in the form of ergocalciferol in 128 μg per kg fresh weight. In the *Cantharellus* genus, the pileus and gills are often effectively exposed to light; therefore, this fungus frequently contains larger amounts of vitamin D than other fungal species. Moreover, studies on humans have found that ergocalciferol from lyophilized and homogenized *Cantharellus* was easily absorbed.[3] For *C. tubaeformis*, a clinical bioassay study conducted with 27 volunteers compared ergocalciferol supplementation (14 μg) with the intake of *C. tubaeformis* mushrooms to provide an equal amount of ergocalciferol. Serum 25-hydroxyvitamin D concentrations did not differ significantly between the groups at week 3.[27]

6.5.1.2 Vitamins B12 and B9

Six edible mushrooms, *i.e.*, *Boletus* spp., *Macrolepiota procera*, *Pleurotus ostreatus*, *Morchella conica*, *Craterellus cornucopioides* and *C. cibarius* (the mushrooms most frequently consumed by European vegetarians) were investigated using a microbiological method based on *L. delbrucekii* ATCC 7830 to characterize B12 levels. *Craterellus cornucopioides* and *C. cibarius* demonstrated

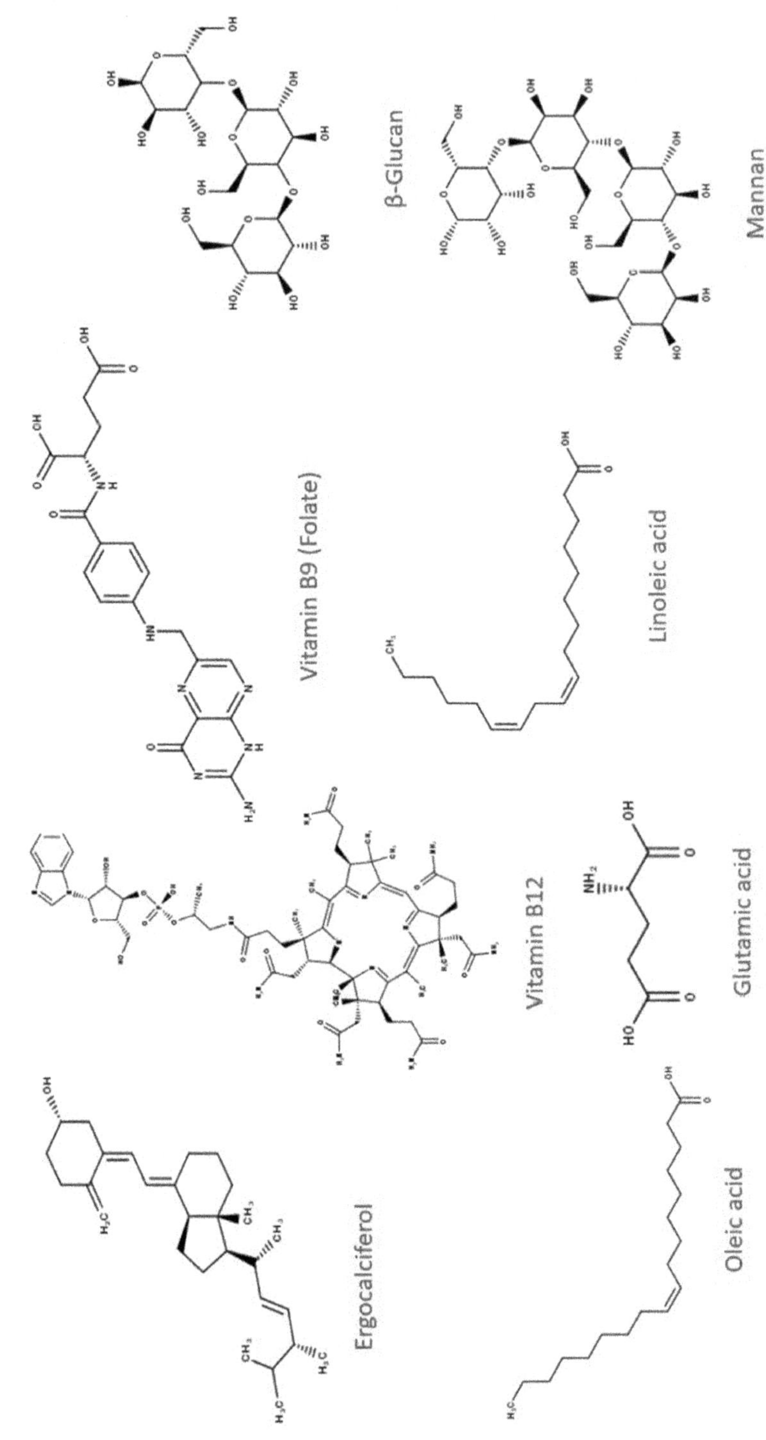

Figure 6.2 Chemical structures of major primary metabolites present in *Cantharellus cibarius*.

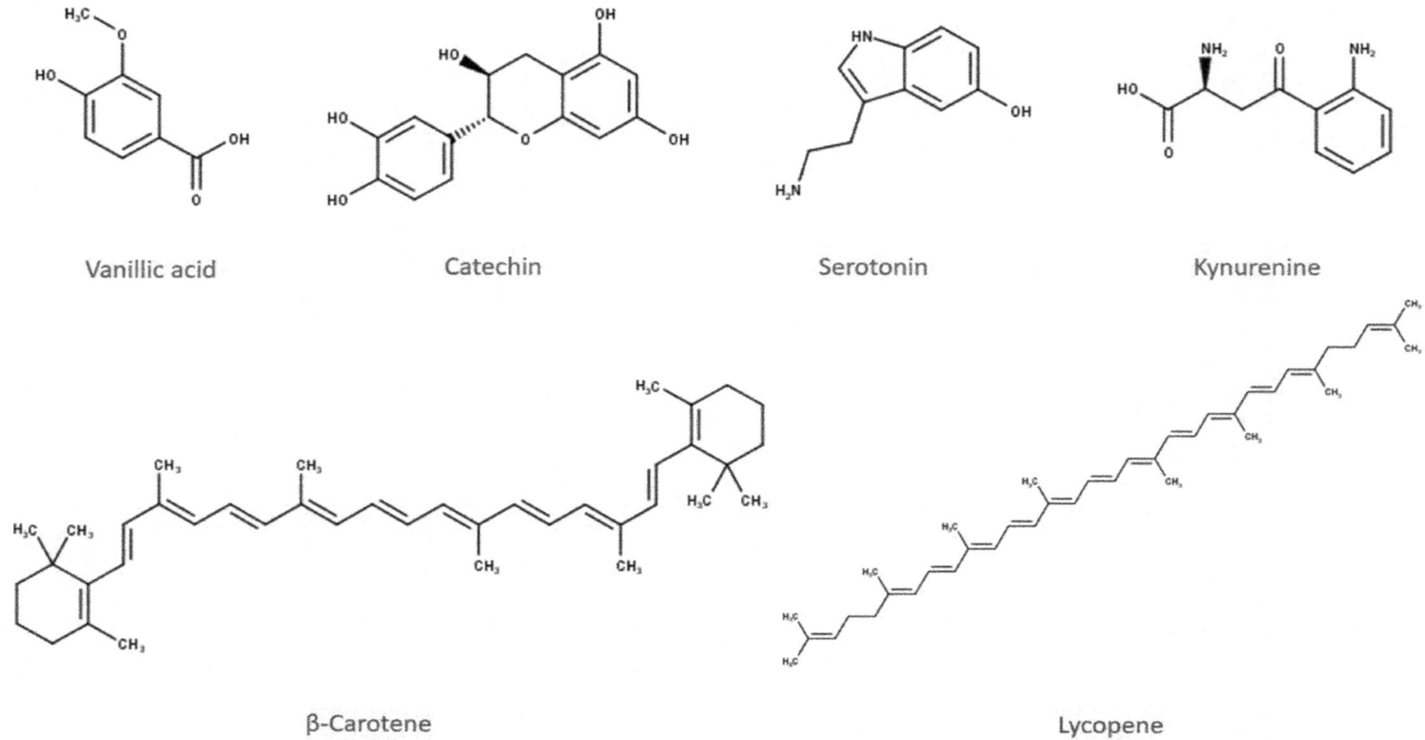

Figure 6.3 Chemical structures of major secondary metabolites that are present in *Cantharellus cibarius*.

the highest corrected levels of B12 in the range of 1.09–2.65 μg per 100 g dry weight. Moreover, liquid chromatography–electrospray ionization tandem mass spectrometry (LC/ESI–MS/MS) was applied to investigate the type of B12 present in both species, either in the authentic form or pseudo form (inactive form in humans). Compounds were purified from *C. cornucopioides* and *C. cibarius* using silica gel 60 thin-layer chromatography (TLC) and analyzed by LC–MS in comparison with authentic B12. Authentic B12 produces a divalent ion with an *m*/*z* of 678.2931 in positive mode, and the main peaks in the mass spectra of *C. cornucopioides* and *C. cibarius* were *m*/*z* 678.2892 and 678.2884, respectively, supporting the absence of pseudo vitamin B12 in the selected species.[28]

Vitamin B9 (folate) plays an important role in fetal neural development and reducing cardiovascular risk, thromboembolitic processes, colon cancer and neuropsychiatric disorders (Wein *et al.*, 2012). The most common sources of folate are fruits, vegetables and whole grains, with the main vitamers being methyltetrahydrofolate (5-CH_3–H_4 folate), 5-formyltetrahydrofolate (5-HCO–H_4 folate), 10-formylfolate (10-HCO-folate), 10-formyldihydrofolate (10-HCO–H_2 folate) and tetrahydrofolate (H_4 folate).[29] Folate vitamers were analyzed using LC–MS after tri-enzyme extraction in white button, cremini, shiitake, maitake, enoki, oyster, chanterelle (*C. cibarius*), morel, portabella, and UV-treated portabella mushrooms. In this regard, chanterelle and morel expressed the lowest total folate (2–6 μg per 100 g) content, and oysters had the highest levels, with 44.2 μg per 100 g.[30]

6.5.1.3 Minerals

Minerals are categorized into toxic elements (Al, Ba, Cd, As, Hg and Pb) and biologically active elements (Ca, Co, Cu, Fe, K, P, Mn, Na, Ni, Sr and Zn), in which the concentration of toxic minerals in mushrooms is a factor that indicates the degree of environmental pollution. The mineral content of mushrooms is dependent on the type of substrate, place of origin, environmental pollution, and mushroom species.[31] Edible mushrooms contain high levels of zinc, establishing mushrooms as a good alternative to minerals from plant intake. In a simulation of the human digestive tract, the release of zinc into digestive juices was found to be highest for *Boletus badius* and *C. cibarius*, with values of 4.13 and 3.29 mg per 100 g dry weight, respectively. These outcomes have paved the way for using these mushrooms as zinc supplements in nutraceuticals.[32]

6.5.2 Primary Metabolites

6.5.2.1 Amino Acids

Edible mushrooms have been shown to be a rich source of essential amino acids and a good alternative to animal protein for vegetarians. They are valued as highly nutritious throughout the world owing to their protein content, low total fat content and high proportion of polyunsaturated fatty acids, which make them suitable for healthy diet regimens.[33,34] In this regard, the total crude protein of four types of edible mushrooms was determined using AOAC parameters, revealing that *C. cibarius* and *Ramaria botrytis* had the highest values in term of protein, *i.e.*, 4.09 g per 100 g fresh weight and 4.08 g per 100 g fresh weight, respectively.[35] Moreover, the methanol extract of *C. cibarius* exhibited a protein content of 9.3 mg per g dry weight of extract using the Bradford method (BSA standard).[36] Amino acid analysis of 11 edible mushrooms from Turkey revealed the abundance of aspartic acid, glutamic acid, isoleucine, leucine, and arginine in the fungi, with 11.8, 17.2, 9.32, 11.3 and 7.57 μg mg dry weight of *C. cibarius*, using an acid hydrolysis method.[37] Indeed, glutamic acid was also manifested to be the most abundant amino acid present in *C. cibarius* (found widely in Poland), at 29.99 mg per g dry weight, in addition to threonine, lysine, alanine and histidine in 8.98, 5.74, 4.98 and 3.15 mg per g dry weight amounts, respectively.[5] In a comparative study of amino acid composition in 11 wild edible mushrooms from Poland, *C. cibarius* was shown to contain the highest amount of glutamic acid, with 888.2 mg per kg dry weight, making *C. cibarius* a good nutraceutical source of glutamic acid.[35]

The taste of edible mushrooms is due to the presence of both volatile and non-volatile compounds, *i.e.*, free amino acids and 5′-nucleotides, which make them a potential raw ingredient for use in the food industry. The taste of umami (a meaty or brothy taste) is mainly caused by glutamic acid and aspartic acid salts binding to T1R1 + T1R3 receptors.[38] Furthermore, the 5′-nucleotides 5′-ionosine monophosphate (5′IMP), 5′-guanosine monophosphate (5′GMP), 5′-xanthosine monophosphate (5′XMP) and 5′-adenosine monophosphate (5′AMP) also contribute to the taste of umami. Both nucleotides and amino acids work in synergy by intensifying the receptor binding affinity.[39] A Finnish study was conducted on four edible Nordic forest mushrooms (*Lactarius camphoratus*, *Boletus edulis*, *C. cibarius* and *Craterellus tubaeformis*) to analyze their free amino acid and nucleotide content using high-performance liquid chromatography analysis. *C. cibarius* exhibited the lowest content of nucleotides and arginine was its most abundant amino acid, which give the mushroom its bitter taste.[40]

6.5.2.2 Fatty Acids and Phospholipids

As previously mentioned, wild edible mushrooms are considered to be a healthy food because of their low fat content. In this regard, the fatty acid composition of *C. cibarius* was analyzed and it was revealed that the highest fat content comprised polyunsaturated fatty acids (PUFA) and monounsaturated fatty acids (MUFA), at 45.6 and 37.3 mg per g dry weight, respectively. In addition, essential linoleic and oleic acids were the most abundant fatty acids present in *C. cibarius*, at 45.6 and 28 mg per g dry weight, respectively. This composition of fatty acids paved the way for *C. cibarius* to be recommended for a healthy diet as a preventive strategy for hyperlipidemia.[37]

Confirming the latter results, the cyclohexane extract of *C. cibarius* was analyzed to detect the degree of MUFA and PUFA with respect to saturated fatty acid (SFA) and fatty acid contents. The extract showed the presence of high percentages of oleic and linoleic acids, of 31.54 and 31.80%, rationalizing the lower ratio of SFA (20.29%) to MUFA (45.43%) and PUFA (34.28%).[41] Moreover, the fatty acid compositions of seven Portuguese wild edible mushrooms (*C. cibarius*, *Hypholoma fasciculare*, *Lepista nuda*, *Lycoperdon molle*, *Lycoperdon perlatum*, *Ramaria botrytis*, and *Tricholoma acerbum*) were investigated for MUFA, PUFA and SFA content. In this regard, *C. cibarius* exhibited the highest percentage of MUFA of 37.54% and lowest percentage of SFA of 12.04% among the other mushrooms studied, in which eicosanoid and nervonic acid were the abundant fatty acids, at 27.98% and 1.23%, respectively.[35]

6.5.2.3 *Compositions of Polysaccharides and Oligosaccharides*

Polysaccharides and oligosaccharides are considered to be complex carbohydrates composed of repeating units of monosaccharides or their derivatives linked together with glycosidic bonds to form a polymeric structure.[42] Alcohol-insoluble and water-soluble fractions were extracted and purified from *C. cibarius* using anion-exchange chromatography that yielded three polysaccharide-rich fractions. One of these fractions (CC1) was not retained and yielded 36.9% sugar content. The other two fractions (CC2a and CC2b) with 69.2 and 53.2% sugar content were eluted. Fraction CC2a was mainly composed of O-2 and O-3 branched 1,6-linked mannan (54%) and linear 1,3-glucan (28%). The other fraction (CC2b) also contained linear 1,3-glucan (73%). The non-retained fraction (CC1) was composed of mannan (31%) and other sugars. The polysaccharide composition in the CC2a fraction was supported using Fourier-transform infrared (FTIR) spectroscopy to inspect the functional groups, where a broad band was observed in the region from 3567 to 3057 cm^{-1}, assigned to the stretching vibration of OH carbohydrate groups and hydrogen bonds. The structural information obtained from the carbohydrate region (1200–950 cm^{-1}) was attributed to the overlapping bands of C–O and C=C stretching vibrations in glycosidic and pyranose rings, and the band at 810 cm^{-1} was characteristic of mannans.[43] Furthermore, another mannan derivative was isolated in the study of naturally-occurring *N*-glycans from *C. cibarius* using matrix-assisted laser desorption–ionization time-of-flight mass spectrometry (MALDI-TOQ MS). The study revealed the presence of *N*-glycans containing deoxyhexose identical to fucose and these were sensitive to L-fucosidase activity.[44] In another structural characterization study, a new heteropolysaccharide was extracted and purified from the polysaccharide fraction of the fruiting bodies of *C. cibarius* that consisted of glucose and xylose in a ratio of 5 : 1.[45] In addition, a crude polysaccharide extract from the fruiting bodies of *C. cibarius* (obtained from a local Chinese market) was analyzed and applied to a DEAE-cellulose column, yielding a novel linear 3-*O*-methylated galactan (WCCP-N-b) that exhibits an antitumor immune response.[46]

6.5.3 Secondary Metabolites

6.5.3.1 Phenolic Compounds

Phenolic compounds are the most abundant natural antioxidants in the human diet and alleviate various disorders and diseases. The total phenolic and total flavonoids are directly correlated to the antioxidant activity of *C. cibarius*. In this regard, the total phenolic content was calculated as 23.9 mg g^{-1}, equivalent to gallic acid; and total flavonoid content was determined as 14.4 m g^{-1} equivalent to quercetin from *C. cibarius* widely collected in Iran.[47] Furthermore, two varieties of *C. cibarius* were analyzed to determine the total phenol and total flavonoid contents, where the contents of *C. cibarius* var. intermedius were 7.67 and 1.46 mg g^{-1} and those of *C. cibarius* var. *multiramis* were 9.4 and 0.81 mg g^{-1}, respectively.[48] Wild Romanian *C. cibarius* was investigated to determine the phenolic content in different fractions of the mushroom, in which the highest quantities of 2,4-dihydroxybenzoic acid (202.48 and 90.33 μg per g dry weight were found in aqueous and hydro-alcoholic extracts, respectively. Protocatechuic acid-4-*O*-glucoside was found to be abundant in three fractions, hydro-alcoholic, etheric and hexanic, at 253.54, 20.67 and 5.4 μg g^{-1}, respectively. Gallic acid, syringic acid, 1-*O*-galloyl-glucose, *P*-hydroxybenzoic acid-4-*O*-glucoside, catechin, quercetin-3-*O*-acetyl-rhamnoside, epicatechin, myricetin-3-*O*-glucoside, rutin, quercetin-3-*O*-malonyl-glucoside and quercetin-3-*O*-glucosylrhamnosyl-glucoside were the other phenolic compounds detected in reasonable quantities in *C. cibarius* fractions.[49] Confirming the latter results, vanillic acid was found to be the main compound in an acidic aqueous extract of *C. cibarius* powder.[50] Moreover, gallic acid was found to be the prime metabolite in an acidic hydro-acetonic extract of *C. cibarius* mushrooms. In addition, naringenin was demonstrated to be the most abundant metabolite in a hydro-methanolic extract of *C. cibarius*, in addition to other constituents detected, such as catechin, epicatechin, and quercetin-3-*O*-rutioside.[51] Furthermore, vanillic acid was found to be the key metabolite of *C. cibarius*, unlike in *B. edulis*, and its highest concentration accounted for 3.32 mg per kg dry weight.[5] Catechin was only found to be the key metabolite of *C. cibarius* and *A. bisporus*; however, the concentration of this compound in *C. cibarius* was 11.4 times higher than it was in *A. bisporus*.[52]

6.5.3.1.1 Indole Compounds. Indole is an aromatic heterocyclic organic compound (with the formula C_8H_7N) and can be divided into two groups: hallucinogens and non-hallucinogens. In this context, an analytical method using high-performance liquid chromatography–diode array detector–electrospray ionization–mass spectrometry (HPLC–DAD–ESI–MS) was developed to detect indole compounds in a methanolic extract of both the fruiting bodies and mycelium of *C. cibarius*. Serotonin and kynurenine sulfate were found to be the major indole compounds in the fruiting bodies of *C. cibarius* along with other indole compounds detected, *i.e.*, L-tryptophan, 5-hydroxytryptophan, tryptamine, melatonin, indole, and indoleacetonitrile sulfate.[5] However, the mycelium contained larger quantities of these latter compounds than the fruiting bodies. Moreover, most of the Basidiomycota species were found to contain higher levels of hydroxytryptophan when an *in vitro* culture technique was used.[53]

6.5.4 Miscellaneous

Carotenoids are pigments with a yellow, orange, or red color that are produced by plants and some fungi. The major carotenoids found in *C. cibarius* are β-carotene and lycopene, at 5.77 $\mu g\ g^{-1}$ and 1.95 $\mu g\ g^{-1}$, respectively. Fungal carotenoids fortify mushrooms against UV damage, and *C. cibarius* has been found to be a good source of these pigments.[35] Volatile compounds were assessed and separated using headspace in-tube extraction coupled with gas chromatography–mass spectrometry (HS-ITEX/GC–MS). Thus, 34 compounds were revealed, divided into different functional groups: alcohols, aldehydes, ketones, terpenoids, acids, and sulfur compounds. The most abundant metabolites were aldehydes and ketones, paving the way for them to be the metabolites responsible for the odor of *C. cibarius*. Furthermore, 2-methyl-pentanal, hexanal and benzaldehyde were the dominant metabolites in the volatile composition of the mushroom.[51]

6.6 Pharmacological and Immunological Potential of *Cantharellus* Species

Mushrooms are considered to be a source of highly nutritional and functional compounds that make them of high importance to the nutraceutical industry. There is huge interest in consuming *C. cibarius* because of its organoleptic properties, medicinal uses, and economic significance. *C. cibarius* is used traditionally in many different cultures for alleviating and preventing various health disorders.[54] In the last decade, the interest in the pharmaceutical applications of *C. cibarius* has rapidly increased, suggesting *C. cibarius* to a reservoir of new bioactive agents that can be used in the dietary supplement industry.

6.6.1 Antioxidant Activity

The antioxidant capacity of *C. cibarius* is performed through the synergistic effect of flavonoids and phenols present in this mushroom. In this context, the antioxidant activity of a *C. cibarius* methanol extract was achieved through chelating iron compared to hydrogen atom and/or electron transfer.[17] In a survey of 16 Portuguese edible mushrooms including *C. cibarius*, the phenol content was assayed using HPLC-UV and the antioxidant capacity was determined *via* the DPPH method. *C. cibarius* exhibited weak antioxidant activity, with an EC_{50} of 19.65 $mg\ mL^{-1}$ owing to the presence of cinnamic acid only (14.97 mg per kg dry weight). [35] Moreover, more confirmative results were obtained from comparing five edible mushrooms (including *C. cibarius*), in which Boletus edulis was found to be the most powerful antioxidant agent, with an EC_{50} value of >2.6 mg mL^{-1}. *C. cibarius* was the weakest antioxidant agent, with an EC_{50} value of 7.49 mg mL^{-1}, in agreement with it exhibiting the lowest phenolic content of 0.88 mg g^{-1}.[55] Moreover, *C. cibarius* displayed stronger antioxidant capacity, with an EC_{50} value of 8.02 mg mL^{-1} compared with 12.1 mg mL^{-1} for *Suillus luteus* ferrous ion chelating.[56] Furthermore, spectroscopic analysis revealed a novel heteropolysaccharide extracted from the fruiting bodies of *C. cibarius*. The novel compound was identified as CC-1 with the backbone of 1,4-linked-β-D-glucose with a 6-1-α-D-xylopyranose residue, which showed dose-dependent DPPH and ABTS scavenging activities of 35.23% and 40.7%, respectively.[45]

6.6.2 Immunological Activity

The appearance and development of an immune response almost always involves a complex interaction between pattern recognition receptors (PRRs) that are able to recognize pathogen-associated molecular patterns (PAMPs); different types of cells; chemical mediators, such as different types of cytokines, lymphokines, and chemokines; and complement proteins.[57] Many fungus sugars can connect to PRRs such as the toll-like 2 receptor (TLR2), toll-like 4 receptor (TLR4), and Dectin-1 protein and stimulate the innate immune system. Macrophages are agranulocyte cells that are capable of performing different functions in the innate response, in addition to bridging the gap between the innate and adaptive responses.[46] Among these polysaccharides, especially α- or β-glucans.

A purified extract of *C. cibarius* contained a 3-*O*-methylated galactan that was investigated for its immunostimulatory activity in a macrophage culture. When assessing the capacity for nitric oxide (NO) production, phagocytic capacity, and the production of cytokines that increase the inflammatory response, such as tumor necrosis factor (TNF-α), interleukin 6 (Il-6) and interleukin 1, beta (IL1-β), an increase in the production of all these immunostimulatory markers was verified. These results opened up new pathways of research for polysaccharides purified from the fruiting structures of this species that have clinical application.[58]

The versatility of the polysaccharides of *C. cibarius* has frequently been studied. Galactan, a polysaccharide present in this species was later tested "*in vivo*" in a female mouse strain (C57 BL/6) in an endeavor to reproduce the results obtained by Yang *et al.* in 2019. After sacrificing the animals and later processing their macrophages, the same immunological markers as those mentioned previously were evaluated and the results indicated that this polysaccharide had the capacity to polarize macrophages "*in vivo*". Macrophages are an abundant cell type in the tumor environment. Macrophages of M2 type are known to be immunosuppressors and facilitate angiogenesis, which leads to favoring maintenance of the cancer. Macrophages of the M1 type produce pro-inflammatory cytokines, activating the immune response, protecting against microbial infection, and increasing immunological resistance to tumors. Research has shown that the galactan of *C. cibarius* can convert macrophage M2 into M1, contributing to inhibiting cancer cells, thereby opening up the pathway to understanding the mechanism of immune response and their use in therapy against cancer. An example will be mentioned in another topic of this chapter.[46,59]

The polysaccharide β-glucan of *C. cibarius* has also been studied to understand immunomodulatory activity in macrophages. This polysaccharide was purified and its structure determined in the fruiting bodies of this species and assessed against immunological markers (NO, TNF-α, IL-6) by means of ELISA assay. Evidence of the pro-inflammatory activity by means of activation of these markers was clearly shown. Nevertheless, additional studies are necessary to corroborate the immunomodulatory potential of the β-glucan reported in this study.[60]

Inflammation is the natural response of the body to pathogens. Some mushrooms contain polysaccharides that exhibit anti-inflammatory activity. This is the case of trehalose, which inhibits pro-inflammatory proteins such as cyclooxygenase-2 and induces nitric oxide synthase (iNOS). Mushrooms also exhibit the capacity for assimilating bio elements directly from the soil during their growth. Metals such as zinc present in *C. cibarius* cultures may have antioxidant and anti-inflammatory characteristics. Zinc is a microelement that leads to reduction in immunity when present is smaller quantities. When present in larger quantities, it induces the production of pro-inflammatory cytokines, thereby increasing the inflammatory process. When this increase occurs in a disorderly manner, the immune response itself may cause damage to the host. The presence of different micronutrients and biomolecules isolated from *C. cibarius*, especially zinc, may help with understanding the inflammatory process and minimize damage to the host by inhibiting the synthesis of inflammatory mediators. Therefore, they may exhibit therapeutic effects, particularly when consumed by patients that have autoimmune diseases.[5,61]

Furthermore, the biomolecules of *C. cibarius* that exhibit anti-inflammatory properties can be used as healing agents. Extracts of *C. cibarius* obtained from forests in Iran were characterized to determine their phenolic and flavonoid compounds. These compounds were tested experimentally in rats using COX-2 as an anti-inflammatory marker. The results allowed us to suppose that the extract exhibited a high level of antioxidant and anti-inflammatory activity, allowing rapid healing of damaged tissues.[47,62]

6.6.3 Anticancer Activity

According to the WHO, colon cancer is considered to be one of the most abundant types of cancer globally, as its risk factors are related to lifestyle, diet, and genetic factors. The potential anticancer activity of the crude polysaccharides of *C. cibarius* was investigated in the human colon adenocarcinoma cell line LS180 by MTT assay and they revealed the ability to inhibit colon cancer cell proliferation in a dose-dependent manner. However, antiproliferative activity in tumor cells plays a crucial role without affecting normal cells. Thus, the study also revealed that a decrease in LDH released from normal CCD 841 CoTr cells exposed to polysaccharide fractions of *C. cibarius* suggested its use in possible colon cancer treatment protocols.[63]

Moreover, another study was conducted confirming the latter results on colon cancer, where retained fractions rich in polysaccharide content of *C. cibarius* (CC2a with 69.2% mainly mannan and CC2b with 53.2% mainly glucan) from an ion-exchange column were found to inhibit DNA synthesis in LS180 cells, unlike the non-retained fraction with a lower polysaccharide content (36.9%) that showed no effect. Additional studies on human colon epithelial CCD 841 CoN cells demonstrated both fractions rich in polysaccharides affected DNA synthesis. In examining their antiproliferative activity, the polysaccharide-rich fractions were found to more specific and sensitive to LDH assay.[64]

However, the potential of other mushroom polysaccharides has been evaluated for treating colon cancer using different cell lines. In this context, a polysaccharide fraction isolated from *Boletus edulis* was found to exhibit antiproliferative activity in colon cancer cells.[65] Furthermore, inhibition of colon cancer proliferation in the HCT-15 cell line was achieved in an *in vitro* study using the polysaccharide fraction of *Clitocybe alexandri*.[66] For *Pleurotus ostreatus*, the water-soluble polysaccharide fraction induced colon cancer cell (COLO-205) death by triggering the mitochondrial apoptosis pathway and G0/G1-phase cell cycle arrest.[67]

Mushrooms have also been evaluated for their potential in treating glioblastoma – the most common brain tumor in adults. In this context, the anti-glioma activity of the ethanol extract of *C. cibarius* was found to be better than that of the aqueous extract when examined using U87MG, LN-18 glioblastoma and SVGP12 normal human astroglia cell lines.[68] Based on molecular oncology, a novel linear 3-*O*-methylated galactan isolated from *C. cibarius* converted tumor-promoting M2-like macrophages into their tumor-inhibiting M1-Like phenotype. Tumor-associated macrophages that resemble M2-like macrophages are considered to be a novel strategy for the treatment of different types of cancer, as they play crucial roles in tumor progression, immune suppression, metastasis and establishing premalignant niches and angiogenesis.[69] The novel compound WCCP-N-b increased the phosphorylation of mitogen-activated protein kinases (MAPKs) and degradation of IKB-α through targeting the toll-like receptor-2 (TLR-2). Thus, the key factors in converting M2-like macrophages into M1-like macrophages were the activation of MAPKs and degradation of IKB-α, paving the way for the potential use of *C. cibarius* as a macrophage-based tumor immunotherapy.[46]

6.6.4 Consumption and Processing in the Food Industry

Although mushrooms have long been consumed in the human diet, few studies have included a sensory profile analysis. The aim of a study conducted in Finland was to analyze sensory differences among edible mushrooms by comparing wild species with cultivated mushrooms commercially available. Wild, bleached, and fresh *C. cibarius* and *C. tubaeformis* mushrooms were described by consumers as having high intensity and various descriptions of odor, with a low intensity of flavor as opposed to that reported for cultivated mushrooms, possibly because the latter had already lost many of their ancestral characteristics.[70]

Edible mushroom harvesting is widely practiced as, apart from being a leisure activity among many people, in general they are also delicious and are excellent ingredients for popular or gourmet dishes. Nevertheless, care needs to be taken about the stages of processing and their toxicity, as well as with the best way to process them without losing their nutritional characteristics. Moreover, the real quantification of the micronutrients present in the dish has also become interesting to certify that mushrooms can be a good source of these compounds. *C. cibarius* has a large quantity of indole compounds, especially tryptophan, important in processes such as translation and protein synthesis. The increase in temperature partially reduces the availability of the above-mentioned compounds. However, simple cooking allows a high level of availability of this biomolecule, showing that the type of mushroom preparation influences the nutrients that will be available.[71]

There has been growing concern about the accumulation of toxic substances in wild edible mushrooms and possible intoxication, as in general they are collected and consumed directly from the soil. Research has been conducted with the aim of assessing the availability of toxic substances after some type of processing or method of cooking. Using different approaches for preparation (dried, deep freezing, blanched or in a preservative composition), approximately 1565 fruiting bodies of *C. cibarius* were assessed regarding the presence of cadmium residues. The results provided evidence that the choice of type of preparation was capable of interfering with the heavy metal content of the fungus. Processes such as the bleaching of deep-frozen mushrooms promoted the loss of up to 40% of cadmium and *C. cibarius* scalded in a preservative composition lost up to 71% of this metal. Thus, these findings may help to guide consumers about the safest way to prepare these foods.[72]

Several species of fungus are also known to accumulate radioactive material, especially cesium. Therefore, many species may be used as bioindicators of soil quality. However, there is concern about the risk to health from the consumption of these mushrooms. When investigating simple methods for reducing radiocesium activity in *C. cibarius* and Chanterelles from Russia, Austria, and Romania, the furrows produced during the cooking of these mushrooms were observed to be disproportionately rich in cesium. When these furrows were removed by means of prior washing, followed by the removal of these same furrows after cooking, the presence of radiocesium was reduced by 30% in Chanterelles. These results demonstrated the need for some type of processing of these foods to guarantee lower risk to health.[73]

Edible mushrooms, in general, are good sources of carbohydrates, proteins, vitamins, and fiber. Therefore, there has been growing interest in their applicability in the food industry. *C. cibarius* (Chanterelle) are ranked among the five mushrooms most frequently consumed that are collected directly from nature. Production for commercial purposes generates around 150 000 to 200 000 tons per annum, representing a market value of up to US$1.7 billion.[26]

However, during industrial processing, many substances produced by fungi may be lost, which results in a real problem for industry. When comparing the antioxidant activity of *C. lutescens* after being submitted to industrial processes, such as freezing and preserves, with raw mushrooms (used as an experimental control), mushrooms in canned preservative compositions were perceived to show significant losses ($p > 0.05$) of antioxidant activity when compared with a frozen product. Therefore, the frozen form may be promising for industrial processing, as it allows consumption of the mushrooms throughout the entire year with minimal loss due to oxidation.[74]

An industrial application of *C. cibarius* is its use in the fermentation industry. On submitting this species for lactic fermentation using lactic bacteria as initiators, a detectable increase in phenolic acid production and antioxidant activities were perceived, with consequent reduction in the hydrogen ionic potential (pH) when compared with the production arising from bleaching processes. Phenolic compounds have been studied in healing processes.[75]

Another use for *C. cibarius* in industry is its applicability as a food preservative. In the sausage industry, there is growing concern about the contamination of this type of food as different types of processes and handling are necessary during the manufacture of sausages. The storage of different types of meat may cause changes that interfere in the consumption and useful life of this product on supermarket shelves. Synthetic antioxidants are normally used as preservatives; however, many are toxic and may affect the health of consumers. *C. cibarius* was tested as an adjuvant in *in vitro* decoction and in sausages to verify its antioxidant and antimicrobial properties. The results showed that its addition significantly reduced the proliferation of aerobic bacteria without compromising the texture and without promoting a change in the sensory quality of the product. *C. cibarius* has thus been shown to be a promising substitute for synthetic preservatives without the concern of ingesting a substance that is harmful to health.[76]

For a long time, gastronomy has reported the presence of a fifth flavor, umami. This is a characteristic flavor present in different foods, such as bacon, monosodium glutamate, and mushrooms. This flavor has increasingly been appreciated by people and has therefore caught the attention of the food industry. When analyzing the release of umami amino acid components in *C. cibarius* and in *C. tubaeformis*, a higher level of release of these amino acids could be perceived to occur in cooked food (90 °C). This could possibly be because fragmentation occurred during this process, which increased the availability of these amino acids. Nevertheless, when the temperature is excessively increased, a reduction occurs in the release of these compounds. Further studies are necessary, particularly on the ideal cooking time, in order to standardize an improved protocol for the adequate release of umami for consumption.[77]

6.7 Conservation

When foods that are consumed in a fresh state or *in natura* are introduced in the diet, a larger intake of nutrients is obtained; however, the digestive process is a complex set of biochemical activities that frequently varies according to the characteristics of individuals. The ingestion of processed or ultra-processed foods may affect the availability of the bio elements in the food. When the fruiting bodies of fresh or processed (by freezing or drying) of *C. cibarius* are consumed, it is possible to verify that methods of conservation or lyophilization or drying in the sun are more advantageous in terms of quantity and availability of bio elements that manage to remain preserved and are used in the process of human digestion, when compared with frozen material. Both naturally and commercially available dried mushrooms can deliver increased dietary and therapeutic potential.[78]

Some storage methods allow the achievement of greater stability in the conservation of biomolecules that measure the integrity of the cell membrane and, consequently, the integrity of the fungal structure. For example, storage at −20 °C (freezing) with UV radiation treatment allows the integrity of the ergosterol of the membrane to be maintained, with a reduction in the oxidation conditions, thereby maintaining the integrity of the cell. This is a promising discovery that contributes to the potential benefits to health and pharmacological applicability of storing mushrooms.[79]

6.8 Safety and Toxicity

As previously mentioned in this chapter, wild mushroom consumption has become increasingly popular. The activity of collecting them is constantly increasing in popularity. In addition to being excellent sources of biomolecules, they are also good providers of the micronutrients necessary for the different metabolic processes that occur in the human body. Moreover, they are considered to be good indicators for monitoring environmental pollution.[31] Therefore, growing numbers of research studies are being conducted with the aim of detecting, characterizing, and controlling the presence of toxic substances of different natures in edible mushrooms. Various techniques and methodologies are used for characterizing and quantifying these elements in fruiting bodies, examples among these are flame atomic absorption spectroscopy (FAAS), gamma spectrometry, optical emission spectrometry, gas chromatography, and atomic absorption spectrometry.[80–83] This research is based on the concern of the health of consumers who appreciate wild edible mushrooms or their use as bioindicators of soil quality. Many studies about safety and toxicity have concentrated on Poland; however, it is possible to find studies conducted in Turkey and China, nations that are large-scale consumers of this type of food.

When studying species of mushrooms isolated from the eastern region of the Black Sea in Turkey, in locations free of pollution and with a low population, high concentrations of cadmium, cobalt, and lead have been detected in samples of *C. cibarius*. This result indicates a high risk for contamination and consequent risk to human health.[84]

Mushrooms are considerate good source of riboflavin, vitamin B12, selenium, copper, potassium, and fiber, and have a low carbohydrate and fat content. Some of these elements are extremely important to human metabolism, such as zinc, magnesium, copper, and selenium, for example, as they may function as enzymatic cofactors. Approximately 22 species of mushrooms collected in Poland were evaluated in terms of the availability of these elements. The fungus species and its geographic origin influence the composition of the minerals that compose them. *C. cibarius* was one of the species that was shown to be capable of making available good quantities of calcium, sodium, potassium, and magnesium, and is therefore considered to be a good source of these micronutrients.[80]

Many pesticides used in the large-scale industrial production of cultivated mushrooms are based on organochlorines, which have a long-term permanence in the environment and may pose a danger to consumption. When analyzing the levels of chlorinated hydrocarbon residues in 45 samples of fungi in northeastern Poland, a strong presence of this residue was detected in samples of *C. cibarius*, meaning that this fungus could serve as a bioindicator of environmental contamination by organochlorines. Of the quantity found, 90% was not metabolized by the fungi, which indicates an accumulation of the pollutants and potential risk to health when these mushrooms are collected in crops that make use of pesticides containing this type of compound.[85]

The same authors mentioned above also conducted a study related to the capacity of *C. cibarius* for accumulating toxic metals, especially cadmium and lead. The results were promising because of the species tested in the study (*Boletus badius*, *Boletus edulis*, and *C. cibarium*), of which *C. cibarium* was the species that showed the lowest rates of the accumulation of these elements that do not exceed the values that are deemed to be admissible by the European Union (EU). This makes them apt for consumption relative to these elements.[86]

The contamination of fungi by heavy metals appears to be species dependent. This affirmation was supported in a study conducted in Galicia, Spain. The study focused on the detection of different types of radioisotopes, especially potassium (^{40}K), in samples of different species of wild or cultivated edible fungi, including *C. cibarius*. The values found were considered to be effectively low in all the samples studied, with rates well below those admissible by the EU. Therefore, they not only show no risk to health, but can also be considered to be a good dietary source of potassium.[82]

Poland and China are considered to be great consumers of mushrooms of the genus *Cantharellus*. In China, ingestion of these fungi exceeds 20 kg of fresh product per capita. In this context, constant information about the content of metal elements in these fungi is necessary for the purpose of knowledge and updating data. When analyzing the presence of different mineral constituents

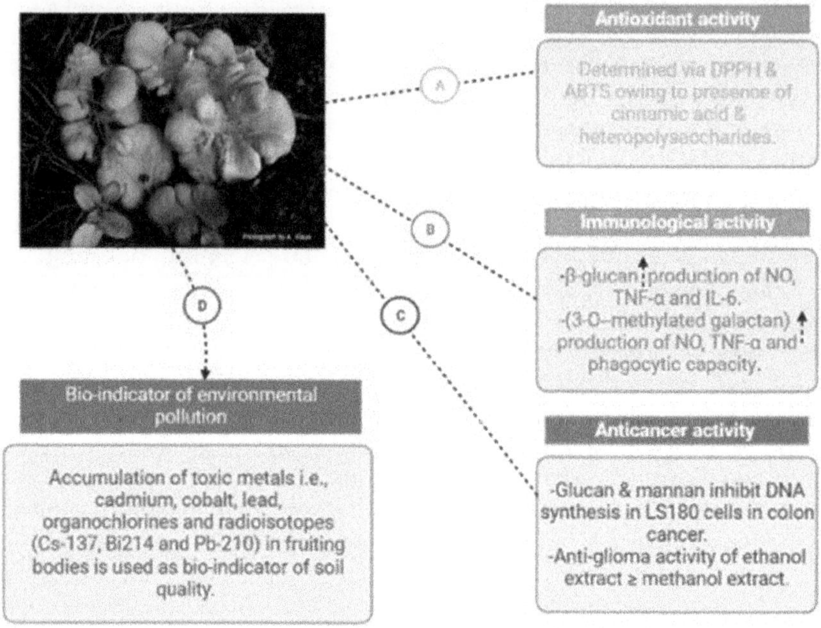

Figure 6.4 Summary of the main biological functions of *C. cibarius*.

in this genus by means of cell-mass spectroscopy, it is possible to detect and measure the levels of metals, such as lead, nickel, barium, and arsenic in the fruiting bodies of the fungi. The values found in the mushrooms isolated in China have been shown to be much lower than those recorded in Poland.[83]

A more recent studied warned about the consumption of species of fungus, such as *C. cibarius*, contaminated with the isotope Cs-137. When measuring the activity of different radioisotopes (Cs-137, K-40, Bi-214 and Pb-210) as well as the concentration of other types of heavy metals in Poland, it was noted that even 34 years after the incident at Chernobyl, mushroom species still had the ability to accumulate Cs-137. Species of *C. cibarius* showed great capacity for the accumulation of radioactive isotopes, thus drawing attention to the need for further research in this area for the purpose of more extensive monitoring of the region in which these mushrooms are frequently consumed.[87] All the studies mentioned in this section allow consideration of the capacity of *C. cibarius* as a model organism for studies on soil quality indicators. Figure 6.4 summarizes the main biological functions of *C. cibarius* (see Figure 6.4).

6.9 Conclusions and Perspectives

Fungi are primordial components of diverse natural ecosystems and play important in human health and popular culture. Species of the genus

Cantharellus sp. was the focus of this chapter, which have been shown to be extremely versatile for a wide range of purposes. In this study, it was possible to understand the relationship between the species of this genus and cultural significance, especially in the countries of Asia, Europe, and America, in which the species *C. cibarius* in particular was shown to be one of the types most known and used in popular culture.

These species of fungi in general are large providers of vitamins, minerals, and biomolecules in the form of special polysaccharides, amino acids, and fatty acids. Fragments of polysaccharides of the genus *Cantharellus* sp. have demonstrated enormous potential for applicability in treatments against cancer, thanks to studies that have aimed to understand their interaction with the cells of the immunological system, especially macrophages and their pro-inflammatory activity. The flavonoids and carotenoids of these fungi have also been shown to be potent antioxidant agents with a promising future in pharmacological applications.

Various species of this genus have been shown to be very closely related phylogenetically, which allows an understanding of their colonization in different types of ecosystems. Phylogenetic studies have also sought to establish genetic and morphological connections to gain a better understanding of their intrinsic relations, thereby also allowing suggestions for the reclassification of fungi based on molecular and morphological studies.

Cantharellus sp. still require many further phylogenetic, comparative, and morphological characterization studies of their structures. New species are constantly being discovered. Furthermore, research toward the improved characterization of the variety of biomolecules of this genus have become of interest. This is especially true for polysaccharides, such as glucans, which have an eventual relationship with our immunological response, with a view to their future applicability in the pharmacological industry, either as bioactive molecules or therapeutic adjuvants.

Considering its applicability in ecosystems, this genus has been shown to be highly promising as an indicator of soil quality, in which different research in this field is being conducted. However, on a planet where there is increasing concern about the use of renewable resources and sustainability, the continuity of research in this area has become imperative. The species of this genus have provided us with gifts of surprising discoveries.

Acknowledgements

The authors would like to thank to Bioengineering Laboratory, Department of Mechanical Engineering, Universidade Federal de Minas Gerais, Belo Horizonte, and Margery Galbraiht for by technical support. The authors also thank you to the Royal Society of Chemistry for permission granted to use Figure 6.1 in this work.

References

1. M. E. Valverde, T. Hernández-Pérez and O. Paredes-López, *Int. J. Microbiol.*, 2015, **2015**, 376387.
2. M. Dowlati, H. R. Sobhi, A. Esrafili, M. FarzadKia and M. Yeganeh, *Trends Food Sci. Technol.*, 2021, **109**, 527–535.
3. M. de Roman, E. Boa and S. Woodward, *Proc. Nutr. Soc.*, 2006, **65**, 190–197.
4. D. Sande, G. P. de Oliveira, M. A. F. e Moura, B. de Almeida Martins, M. T. N. S. Lima and J. A. Takahashi, *Food Res. Int.*, 2019, **125**, 108524.
5. B. Muszyńska, K. Kała, A. Firlej and K. Sułkowska-Ziaja, *Acta Pol. Pharm.*, 2016, **73**, 589–598.
6. W. Ma, Z. Yang, L. Liang, Q. Ma, G. Wang and T. Zhao, *Front. Recent Dev. Plant Sci.*, 2021, **12**, 34–43.
7. J. Ishara, A. Buzera, G. N. Mushagalusa, A. R. Hammam, J. Munga, P. Karanja and J. Kinyuru, *J. Food Biochem.*, 2022, **46**, e14025.
8. H. P. Molitoris, *Feddes Repertorium*, 2002, **113**, 165–182.
9. V. Vetvicka, T. V. Teplyakova, A. B. Shintyapina and T. A. Korolenko, *J. Fungi*, 2021, **7**, 250.
10. C. Pezzella, G. Macellaro, G. Sannia, F. Raganati, G. Olivieri, A. Marzocchella, D. Schlosser and A. Piscitelli, *PLoS One*, 2017, **12**, e0178758.
11. D. Kothari, S. Patel and S.-K. Kim, *Biomed. Pharmacother.*, 2018, **105**, 377–394.
12. S. Novaković, *IOP Conf. Ser.: Earth Environ. Sci.*, 2021, **854**, 012068.
13. R. G. Thorn, A. Banwell, T. H. Pham, N. P. Vidal, C. F. Manful, M. Nadeem, A. G. Ivanov, B. Szyszka Mroz, M. B. Bonneville, N. P. A. Huner, M. D. Piercey-Normore and R. Thomas, *Sci. Rep.*, 2021, **11**, 20590.
14. M. Galgowska and R. Pietrzak-Fiecko, *Molecules*, 2020, **25**(16), 3588.
15. O. Burhan, E. Havsut and K. Yildiz, *LWT–Food Sci. Technol.*, 2021, **138**, 110639.
16. A. H. Smith and E. E. Morse, *Mycologia*, 1947, **39**, 497–534.
17. W. Elkhateeb and G. Daba, *Open Access J. Mycol. Mycol. Sci.*, 2021, **4**, 1–8.
18. R. Garibay-Orijel, J. Caballero, A. Estrada-Torres and J. Cifuentes, *J. Ethnobiol. Ethnomed.*, 2007, **3**, 4.
19. T. W. Henkel, A. W. Wilson, M. C. Aime, J. Dierks, J. K. Uehling, M. Roy, H. Schimann, F. Wartchow and G. M. Mueller, *Mycologia*, 2014, **106**, 307–324.
20. P. R. Leacock, J. Riddell, A. W. Wilson, R. Zhang, C. Ning and G. M. Mueller, *Mycologia*, 2016, **108**, 765–772.
21. B. Buyck, V. Hofstetter, R. Ryoo, K. H. Ka and V. Antonin, *MycoKeys*, 2020, **76**, 31–47.
22. L. Montoya, M. Herrera, V. M. Bandala and A. Ramos, *MycoKeys*, 2021, **80**, 91–114.
23. Y. Zhang, S. Wang, H. Li, C. Liu, F. Mi, R. Wang, M. Mo and J. Xu, *Front. Microbiol.*, 2021, **12**, 699598.

24. I. Svanberg and H. Lindh, *J. Ethnobiol. Ethnomed.*, 2019, **15**, 42.
25. W. Ogawa, Y. Takeda, N. Endo, S. Yamashita, T. Takayama, M. Fukuda and A. Yamada, *Mycorrhiza*, 2019, **29**, 519–530.
26. S. Deshaware, S. J. Marathe, D. Bedade, J. Deska and S. Shamekh, *Arch. Microbiol.*, 2021, **203**, 1539–1545.
27. T. A. Outila, P. H. Mattila, V. I. Piironen and C. J. Lamberg-Allardt, *Am. J. Clin. Nutr.*, 1999, **69**, 95–98.
28. F. Watanabe, J. Schwarz, S. Takenaka, E. Miyamoto, N. Ohishi, E. Nelle, R. Hochstrasser and Y. Yabuta, *J. Nutr. Sci. Vitaminol.*, 2012, **58**, 438–441.
29. C. M. Pfeiffer, Z. Fazili, L. McCoy, M. Zhang and E. W. Gunter, *Clin. Chem.*, 2004, **50**, 423–432.
30. K. M. Phillips, D. M. Ruggio and D. B. Haytowitz, *Food Chem.*, 2011, **129**, 630–636.
31. M. V. Dimitrijevic, V. D. Mitic, J. S. Nikolic, A. S. Djordjevic, J. J. Mutic, V. P. Stankov Jovanovic and G. S. Stojanovic, *Chem. Biodiversity*, 2019, **16**, e1800492.
32. M. Zajac, B. Muszynska, K. Kala, A. Sikora and W. Opoka, *J. Physiol. Pharmacol.*, 2015, **66**, 763–769.
33. F. L. Oyetayo, A. A. Akindahunsi and V. Oyetayo, *Nutr. Health*, 2007, **18**, 383–389.
34. F. Bach, C. V. Helm, M. B. Bellettini, G. M. Maciel and C. W. I. Haminiuk, *Int. J. Food Sci. Technol.*, 2017, **52**, 2382–2392.
35. L. Barros, B. A. Venturini, P. Baptista, L. M. Estevinho and I. C. Ferreira, *J. Agric. Food Chem.*, 2008, **56**, 3856–3862.
36. M. Kozarski, A. Klaus, J. Vunduk, Z. Zizak, M. Niksic, D. Jakovljevic, M. M. Vrvic and L. J. Van Griensven, *Food Funct.*, 2015, **6**, 1875–1886.
37. F. A. Ayaz, L. T. Chuang, H. Torun, A. Colak, E. Sesli, J. Presley, B. R. Smith and R. H. Glew, *Int. J. Food Sci. Nutr.*, 2011, **62**, 328–335.
38. Y. Zhang, C. Venkitasamy, Z. Pan and W. Wang, *Trends Food Sci. Technol.*, 2013, **33**, 78–92.
39. O. G. Mouritsen and H. Khandelia, *FEBS J.*, 2012, **279**, 3112–3120.
40. H. Manninen, M. Rotola-Pukkila, H. Aisala, A. Hopia and T. Laaksonen, *Food Chem.*, 2018, **247**, 23–28.
41. M. Kolundzic, T. Stanojkovic, J. Radovic, A. Tacic, M. Dodevska, M. Milenkovic, F. Sisto, C. Masia, G. Farronato, V. Nikolic and T. Kundakovic, *J. Med. Food*, 2017, **20**, 790–796.
42. J. H. Prestegard, J. Liu and G. Widmalm, Oligosaccharides and Polysaccharides, in *Essentials of Glycobiology*, ed. A. Varki, R. D. Cummings, J. D. Esko, P. Stanley, G. W. Hart, M. Aebi, A. G. Darvill, T. Kinoshita, N. H. Packer, J. H. Prestegard, R. L. Schnaar and P. H. Seeberger, Cold Spring Harbor Laboratory Press, Cold Spring Harbor (NY), 3rd edn, 2015.
43. M. K. Lemieszek, F. M. Nunes and W. Rzeski, *Food Funct.*, 2019, **10**, 5816–5826.
44. J. Grass, M. Pabst, D. Kolarich, G. Poltl, R. Leonard, L. Brecker and F. Altmann, *J. Biol. Chem.*, 2011, **286**, 5977–5984.

45. D. Zhao, X. Ding, Y. Hou, W. Hou, L. Liu, T. Xu and D. Yang, *Int. J. Mol. Med.*, 2018, **41**, 2744–2754.
46. G. Yang, Y. Qu, Y. Meng, Y. Wang, C. Song, H. Cheng, X. Li, L. Sun and Y. Zhou, *Carbohydr. Polym.*, 2019, **214**, 34–43.
47. D. Nasiry, A. R. Khalatbary and M. A. Ebrahimzadeh, *Int. J. Med. Mushrooms*, 2017, **19**, 893–903.
48. D. Kumari, M. S. Reddy and R. C. Upadhyay, *Food Sci. Technol. Int.*, 2011, **17**, 557–567.
49. M. Fogarasi, M. I. Socaciu, C. D. Salagean, F. Ranga, A. C. Farcas, S. A. Socaci, C. Socaciu, D. Tibulca, S. Fogarasi and C. A. Semeniuc, *Molecules*, 2021, **26**(24), 7508.
50. B. Muszyńska, K. Kała, K. Sułkowska-Ziaja, A. Szewczyk, M. Łojewski and J. Rojowski, *Med. Int. Rev.*, 2015, 146–152.
51. M. Fogarasi, S. A. Socaci, F. V. Dulf, Z. M. Diaconeasa, A. C. Farcas, M. Tofana and C. A. Semeniuc, *Molecules*, 2018, **23**(12), 3272.
52. I. Palacios, M. Lozano, C. Moro, M. D'arrigo, M. Rostagno, J. Martínez, A. García-Lafuente, E. Guillamón and A. Villares, *Food Chem.*, 2011, **128**, 674–678.
53. B. Muszynska, K. Sulkowska-Ziaja and H. Ekiert, *J. Food Sci. Technol.*, 2013, **50**, 1233–1237.
54. S. Patel and A. Goyal, *3 Biotech*, 2012, **2**, 1–15.
55. B. Queiros, J. C. Barreira, A. C. Sarmento and I. C. Ferreira, *Int. J. Food Sci. Nutr.*, 2009, **60**(Suppl. 6), 160–172.
56. A. M. Witkowska, M. E. Zujko and I. Mirończuk-Chodakowska, *Int. J. Med. Mushrooms*, 2011, **13**, 335–341.
57. T. H. Mogensen, *Clin. Microbiol. Rev.*, 2009, **22**, 240–273.
58. G. Yang, Y. Qu, Y. Meng, Y. Wang, C. Song, H. Cheng, X. Li, L. Sun and Y. Zhou, *Carbohydr. Polym.*, 2019, **214**, 34–43.
59. M. Orecchioni, Y. Ghosheh, A. B. Pramod and K. Ley, *Front. Immunol.*, 2019, **10**, 1084.
60. Y. Qu, X. Zhao, H. Guo, Y. Meng, Y. Wang, Y. Zhou and L. Sun, *Int. J. Mol. Med.*, 2021, **47**, 50.
61. L. B. Nicholson, *Essays Biochem.*, 2016, **60**, 275–301.
62. K. Kala, A. Krakowska, A. Szewczyk, B. Ostachowicz, K. Szczurek, A. Fijalkowska and B. Muszynska, *Food Chem.*, 2021, **352**, 129456.
63. N. Nowacka-Jechalke, R. Nowak, M. Juda, A. Malm, M. Lemieszek, W. Rzeski and Z. Kaczynski, *Food Chem.*, 2018, **268**, 355–361.
64. M. K. Lemieszek, F. M. Nunes, G. Marques and W. Rzeski, *Int. J. Biol. Macromol.*, 2019, **134**, 770–780.
65. M. K. Lemieszek, C. Cardoso, F. H. Ferreira Milheiro Nunes, A. I. Ramos Novo Amorim de Barros, G. Marques, P. Pożarowski and W. Rzeski, *Food Funct.*, 2013, **4**, 575–585.
66. J. A. Vaz, S. A. Heleno, A. Martins, G. M. Almeida, M. H. Vasconcelos and I. C. Ferreira, *Food Chem. Toxicol.*, 2010, **48**, 2881–2884.
67. S. Arora and S. Tandon, *Arch. Iran. Med.*, 2015, **18**, 284–295.

68. P. Nowakowski, R. Markiewicz-Zukowska, K. Gromkowska-Kepka, S. K. Naliwajko, J. Moskwa, J. Bielecka, M. Grabia, M. Borawska and K. Socha, *Biomed. Pharmacother.*, 2021, **133**, 111090.
69. S. I. Grivennikov, K. Wang, D. Mucida, C. A. Stewart, B. Schnabl, D. Jauch, K. Taniguchi, G.-Y. Yu, C. H. Österreicher and K. E. Hung, *Nature*, 2012, **491**, 254–258.
70. H. Aisala, O. Laaksonen, H. Manninen, A. Raittola, A. Hopia and M. Sandell, *Food Res. Int.*, 2018, **109**, 526–536.
71. B. Muszynska and K. Sulkowska-Ziaja, *Food Chem.*, 2012, **132**, 455–459.
72. M. Drewnowska, A. Hanc, D. Baralkiewicz and J. Falandysz, *Environ. Sci. Pollut. Res. Int.*, 2017, **24**, 21733–21738.
73. G. Steinhauser and V. Steinhauser, *J. Food Prot.*, 2016, **79**, 1995–1999.
74. M. A. Murcia, M. Martinez-Tome, A. M. Jiménez, A. M. Vera, M. Honrubia and P. Parras, *J. Food Prot.*, 2002, **65**, 1614–1622.
75. E. Jablonska-Rys, A. Slawinska and D. Szwajgier, *Food Sci. Biotechnol.*, 2016, **25**, 439–444.
76. S. Novakovic, I. Djekic, A. Klaus, J. Vunduk, V. Djordjevic, V. Tomovic, B. Sojic, S. Kocic-Tanackov, J. M. Lorenzo, F. J. Barba and I. Tomasevic, *Foods*, 2019, **8**(12), 635.
77. M. Rotola-Pukkila, B. Yang and A. Hopia, *Food Chem.*, 2019, **278**, 56–66.
78. K. Kala, A. Krakowska, P. Zieba, W. Opoka and B. Muszynska, *J. Sci. Food Agric.*, 2021, **101**, 3481–3488.
79. A. Villares, L. Mateo-Vivaracho, A. Garcia-Lafuente and E. Guillamon, *Food Chem.*, 2014, **147**, 252–256.
80. J. Brzezicha-Cirocka, M. Grembecka, I. Grochowska, J. Falandysz and P. Szefer, *Ecotoxicol. Environ. Saf.*, 2019, **173**, 353–365.
81. M. Galgowska and R. Pietrzak-Fiecko, *Food Chem.*, 2021, **353**, 129441.
82. M. J. Melgar and M. A. Garcia, *Environ. Sci. Pollut. Res. Int.*, 2021, **28**, 52925–52935.
83. J. Falandysz, M. Chudzinska, D. Baralkiewicz, M. Drewnowska and A. Hanc, *Environ. Sci. Pollut. Res. Int.*, 2017, **24**, 11472–11482.
84. C. Sarikurkcu, I. Akata, G. Guven and B. Tepe, *Environ. Sci. Pollut. Res. Int.*, 2020, **27**, 26419–26441.
85. M. Galgowska and R. Pietrzak-Fiecko, *J. Environ. Sci. Health, Part B*, 2017, **52**, 214–217.
86. M. Galgowska and R. Pietrzak-Fiecko, *Molecules*, 2021, **26**, 7289.
87. O. Ronda, E. Grzadka, I. Ostolska, J. Orzel and B. M. Cieslik, *Food Chem.*, 2021, **353**, 129441.

CHAPTER 7

Craterellus cornucopioides *(L.)*/ Craterellus odoratus *(Schwein.) Fr. (Black Chanterelle)*

A. TRIDIP BORUAH,* B. PRASTUTI PARASHAR, C. DEBASISH DAS, D. HIMASHRI DEVI, E. ABHINOB KR NATH, F. APARNA BARMAN AND G. HANGSHA BARMAN

Plant Ecology Laboratory, P.G. Department of Botany, M.C. College, Barpeta, Assam-781301, India
*Email: tridip_boruah@gauhati.ac.in

7.1 Introduction

Mushrooms are usually macro-fungi with very unique features not similar to animals or plants and are classified in a different kingdom due to their unique characteristics. These mushrooms can be observed without a microscope, unlike many other fungi. To date, almost 14 000 different mushrooms have been found on earth, with China contributing >900 types of mushrooms to the already identified list of species. As expected, all mushrooms present in the wild are not edible; only a few of them are edible and have the necessary constituents to provide a pleasant sensation to our taste buds. Both wild and edible mushrooms can be either saprophytic or parasitic depending upon their taxonomic characteristics and ecological functioning.[1] Although mushrooms have been used in different parts of the world for generations, they only became popular in the late 1990s due to their hundreds of different types of medicinally beneficial attributes, such as antidiabetic, detoxification, anticancer, immunomodulatory, antiviral, antiallergic, cardiovascular protective, antibacterial, antifungal, and

Phytochemistry and Nutritional Composition of Significant Wild Medicinal and Edible Mushrooms: Traditional Uses and Pharmacology
Edited by Ajay Sharma, Garima Bhardwaj and Gulzar Ahmad Nayik

Published by the Royal Society of Chemistry, www.rsc.org

antiparasitic properties, which have been reported from all parts of the globe.[2] In the Indian sub-continent, the vernacular names of various mushrooms mainly depend on the taxonomy used by the local people in different areas, which is inherited generation after generation as their educational syllabus does not include detailed knowledge of mushrooms. However, in developed nations, this is purely based on basic educational knowledge. In many countries, mushroom knowledge is shared between local people in local languages on a generational basis and little of this is documented.[3] In ancient Egypt, mushrooms were used to prevent death. Roman people regarded them as a food consumed by the gods, and it was found that poisonous mushrooms were used by inside enemies to kill the emperor. The great philosopher Pliny described the use of mushrooms in human diseases thousands of years ago.[4] Therefore, mushrooms have been used for thousands of years in different parts of the world, which indicates their importance as a source of food.

Wild mushrooms are consumed directly or indirectly in many parts of the world as a major source of non-timber yielding food products, and are used directly as the main source of foodstuff in many parts of the world. Nowadays, mushrooms are the most valuable and important source of nutrients, not only due to their taste, but also due to their high nutritional content, which is important for human beings as well as animals. Mushrooms have a very high content of pharmacologically important compounds. For that reason, in the present-day context, mushrooms have a high market and economic value.[5] Mushrooms have a high content of easily digestible proteins, around 10% to 40% of their dry weight, which is not more than that contained in milk and meats, but more than in most vegetables. Data related to studies on amino acids have found that mushrooms contain all of the biologically important amino acids. Mushrooms, in a fresh condition, contain 3% to 21% carbohydrates, which is a good amount of carbohydrates for a healthy diet.[6] Mushrooms provide nutritional value as well as health benefits. There has been consistent evidence that mushrooms have a positive effect on human health, due to their immunological, anticancer, antioxidant, antihypertensive and cholesterol-lowering properties, and liver protection, as well as anti-inflammatory, antidiabetic, antiviral, and antimicrobial properties.[7]

Craterellus cornucopioides was first named and described by Linnaeus in 1753, with this classic species often referred to in English as the "black trumpet" or in French as the "trompette de la mort". Having a deep vase-shaped form, it has a grey or black surface, with a whitish or cream-colored sheen developing on the outer surface of slightly older mushrooms as the spores mature.[8] *Craterellus cornucopioides* contain substances like protein, fat, and carbohydrate, essential amino acids, and non-essential amino acids, but the presence of these chemicals in the final product is completely dependent on the method of cooking and storage.[9] In addition, *Craterellus cornucopioides* also contain various other dietary, chemical, and medicinally important substances. *Craterellus cornucopioides* mushrooms are effective against liver diseases and are often utilized as cardiovascular, anticancer, antifungal, antioxidant, antiallergic, antiparasitic, antidiabetic, immunomodulatory, protective, anticholesterolemic, antiviral, antibacterial, and detoxification agents in various modern medicinal prescriptions.[10] Therefore, this mushroom can be regarded as one of the most important food sources for humans both nutritionally and medicinally.

Craterellus cornucopioides has well documented medicinal properties, meaning that this mushroom is important for the pharmaceutical industry, and for that reason it needs to be conserved. For conservation, people need adequate knowledge. *Craterellus cornucopioides* are naturally found in wild forms but no proper cultivation process has been reported to date. The mushroom has very distinctive characteristics morphologically as well as biochemically. There is a prominent scope for detailed study regarding the biochemical nature of the mushroom. A confirmatory biochemical study could be very helpful for understanding the physiochemical nature of the chemicals found in the *Craterellus cornucopioides* mushroom and how these chemicals are responsible for the treatment of various diseases in humans, such as cancer, microbial infections, *etc.* The lack of documented cultivation techniques hinders the cultivation of the mushroom in parts of the world that could benefit from it both economically and ecologically. This chapter gives an overview of the proven scientific knowledge regarding the taxonomy, characteristics, resource availability, ancient and modern medicinal uses, phytochemistry, pharmacology, cultivation, conservation, safety regulations, toxicity parameters, trade value, storage potential, and transportation-related issues of *Craterellus cornucopioides* to provide initial direction for anyone who is fascinated by wild edible mushrooms to enable them to carry out further research in this relatively less explored field of study.

7.2 Taxonomy of *Craterellus cornucopioides*

Donatha D. Tibuhwa,[11] a prominent researcher in the field of wild edible mushrooms, published a research paper and described the systematic classification of *Craterellus cornucopioides* as follows:

Kingdom: Fungi
Division: Basidiomycota
Class: Basidiomycetes
Order: Cantharellales
Family: Cantharellaceae
Genus: *Craterellus*
Species: *Craterellus cornucopioides*

7.2.1 Common Name

Craterellus cornucopioides, commonly known as the horn of plenty. It is also known as the trumpet of the dead and the black trumpet.

7.2.2 Synonyms

Scientifically used synonyms of this species are *Merulius cornucopioides* (L.) Pers. and *Dendrosarcus cornucopioides* (L.) Kuntze.

7.2.3 Characteristics of Mushrooms

Craterellus cornucopioides grows on oak trees or any other broad leaved trees, under beech trees, mainly over mosses in moist places or on heavy calcareous soil. The mushroom is generally dark in color, is small in size, and attains a height of up to 10 cm and a width of up to 5–7 cm. The mushroom looks like a hollow cone with its point on the land; its shape is similar to a horn or funnel and it is thus commonly referred to as a "Horn of plenty". The fruiting body of *Craterellus cornucopioides* does not differentiate into a stalk and cap. The cap is 1.2–2.8 cm in length and hairy when immature and metamorphoses into a dendritic small outgrowth such as another small pileus on the peak of the cap. The length of the stipe of this mushroom is more or less 3.5 cm and it is flattened in nature. *Craterellus cornucopioides* are found to grow singly, sometimes dispersed or in a dense bunch. The mushroom consists of a prominent upper and lower surface and its upper surface is dark grey or black, and barely yellow in colour. The lower surface is fertile in nature and it is light grey in colour and more or less flat in shape. The spore-producing surface which is on the outside of the mushroom expands up to the margin at its base. At the base of the mushroom, white fibers are connected; these fibers connect the mushroom to the main body, *i.e.*, a subterranean network of fungal cells. In this mushroom, anonymous dark pigments are found that are membrane-bound and insoluble in the solvents used for pigment analysis. The spores of *Craterellus cornucopioides* are acyanophilic, spheroidal, inamyloid and flat. There are 13–14 spores that are 7–8 μm in size. Relatively bigger cells found on the sporocarp of Basidiomycetes, *i.e.*, the cystidium (plural cystidia) are absent in this mushroom. The trama hyphae lay out parallel towards the hymenium, are 3–12 μm wide, somewhat puffed up, thin-walled, and short celled. In trama hyphae, the pigments are yellow in colour and farinaceous in nature, while in the surface hyphae, the pigments are brown in colour. Trama hyphae are intracellular; the surface hyphae are membrane bound. In this mushroom, clamp connection has not yet been reported.[12]

7.3 Resource Availability, Sustainability and Distribution of *Craterellus cornucopioides*

Black Trumpet or *Craterellus cornucopioides*, a category of edible wild mushroom, have universal distribution. Due to its edible qualities, this mushroom is a favorite in California as well as in the whole of Europe. *Craterellus cornucopioides* is widely distributed in countries such as Japan, Korea, and North America, as well as the Sichuan and Yunnan provinces of China. *Craterellus cornucopioides* has been reported multiple times as being found in North America.[8] In North America, the distribution of *Craterellus cornucopioides* was not initially specified in any detail; later, in 2005, Phillips reported it to be a rare species on the continent, with a very scantily distributed pattern.[13] In some countries, such as the Netherlands, it seems to be rare, although this has not been specified in detail. *Craterellus cornucopioides* is not known to exhibit a very wide host range. The European plant Quercus is the only molecularly confirmed plant associated with *Craterellus cornucopioides*. The range of host unions in other *Craterellus* species is not well documented. In a description of the genera *Cymatoderma*, *Thelephora*, and *Craterellus*, the authors described that of the entire continent of Africa, *Craterellus cornucopioides* is found only in South Africa.[11] Generally, it appears from June to November in all countries, apart from the UK, where it appears from August to November. Among seven wildlife sanctuaries of the Indian state Sikkim, the Barsey Rhododendron Sanctuary is famous for various wild mushrooms, including *Craterellus cornucopioides*. A broad range of microclimatic conditions and elevation variation coupled with various groups of habitats leads to an intense amount of leaf litter, which is the reason why several mycorrhizal trees are present in this sanctuary. The decomposition of various parts of the trees and leaf litters nourishes the growth of *Craterellus cornucopioides*. Due to these reasons, *Craterellus cornucopioides* is widely found in Barsey Rhododendron Sanctuary.[14]

Although *Craterellus cornucopioides* is distributed in wild areas, it has very high commercial value. Due to this high commercial value, farmers from several countries are trying to cultivate this mushroom in their native countries. Farmers from Nepal have achieved a good amount of success as far as the cultivation of this mushroom is concerned. Recently in Portugal, the consumption of this mushroom has increased due to successful cultivation practices. This species accumulates selenium and in addition to its edible nature it is of broader interest to the people of Portugal.[15] Although the collection and the cultivation of this mushroom have drawn significant attention among researchers there is still some scope for creating awareness and interest among the general public for better comprehension and use of their economic perspective. During the last decade of the 20th century, Spain became a mycophilic country, due to their knowledge of various types of mushrooms. This interest has helped the nation to cultivate the more difficult mushrooms to grow, including *Craterellus cornucopioides*.[16] All the edible mushrooms, such as *Craterellus cornucopioides*, are a highly precious asset. Hence, they should be considered to be a key product for a green economy, and sustainable development needs to be at forefront of its cultivation and collection. *Craterellus cornucopioides* as a food source has great importance, so countries should focus on this type of food source in a proper scientific way to secure a better future for mankind.

7.4 Medicinal Uses of *Craterellus cornucopioides*

Craterellus cornucopioides, or the 'Black Trumpet' mushroom, is an edible mushroom that has many uses. Since ancient times, it has been used as food because of its many nutritional benefits for the human body. Edible mushrooms are beneficial to the human diet due to their curative and nourishing properties; they have been used by people for millennia.

Wild mushrooms are acquired from forests, mountains or undisturbed areas and are very rarely cultured. Various sources of information have outlined that wild mushrooms are a reliable source of inherent antimicrobial and antioxidant compounds. Since ancient times, mushrooms have been heavily used by human beings for food and medicinal purposes. Around 1069 edible mushroom species have been observed to be consumed by various communities, one of which is *Craterellus cornucopioides.*[17] Although, it is known that cultivated mushrooms have less fiber content than wild mushrooms, they contain better bioactive compounds.[18] The benefits of mushrooms are not only restricted to them being a tasty food; they also contain a wealth of beneficial minerals in a low-calorie food for a healthy human diet. Mushrooms contain amino acids and fatty acids, unsaturated and polyunsaturated, such as oleic, linoleic, and linolenic acids.[19] *Craterellus cornucopioides* contains highly beneficial and important nutritious constituents, including 88–94% water, 15–42% protein, 2–6% crude fat, 42–71% carbohydrates, and 6–13% vitamins. Since ancient times, in various countries, *Craterellus cornucopioides* has been used as a source of medicine for performing various traditional practices.[17] *Craterellus cornucopioides* has some unique medicinal properties; for which, its extracts can be found in many medicines designed to provide therapy for Parkinson's, Alzheimer's, hypertension, and people at high risk of stroke. Some of these medicines can also be used to reduce the probability of cancer cells and progression due to their anti-tumoral attributes. Mushrooms contain antibacterial cholesterol-lowering agents; they also additionally contain bioactive compounds, and *Craterellus cornucopioides* is no different in this regard.[2] We cannot easily differentiate between the edible and medicinal mushrooms of the family of these mushrooms as many of the edible species of these mushrooms can be used for the treatment of many diseases as well.[19] In excess of 100 medicinal practices are being connected to edible mushrooms and their medicinal antioxidant, anticancer, antidiabetic, antiallergic, immunomodulating, cardiovascular, anticholesterolemic, antiviral, antibacterial, antiparasitic, antifungal, detoxification, and liver protective properties, as well as their prevention of tumor enlargement and anti-inflammatory activity, which are also encouraging signs for future drug design.[10]

Various recent studies have revealed that the edible mushroom *Craterellus cornucopioides* is now being used more often compared to in ancient times. Recent studies have indicated that these mushrooms are now consumed daily as an important food in various parts of the world because of their high nutritional and medicinal value.[20] Mushrooms not only have nutritional value but are also useful constituents for the health of the human body. Their uses have continuously been exhibited to have a useful effect on the health of the human body as different types of immunological and anticancer substances are found in this type of edible mushroom. Antioxidant, antihypertensive, and cholesterol-lowering substances, as well as liver protective, anti-inflammatory, anti-diabetic, antiviral, and antimicrobial substances are also found in this type of mushroom. A recent study reported effective compounds extracted from *Craterellus cornucopioides*, including keto esters and sesquiterpenoids that exhibit captivating antitumor activities.[20] In ancient times, this mushroom was not well understood, so people used it mainly as a source of food; however, nowadays due to various research work on this edible mushroom, it is being used increasingly more in many countries. The best part of this development is the fact that the medicinal and nutritional properties of *Craterellus cornucopioides* have become fairly well known and are being utilized among everyday people belonging to many communities.[21]

7.5 Phytochemistry of *Craterellus cornucopioides*

Anything that is wild but edible always catches the attention of scientists and researchers working in the field of phytochemistry. The medicinal value associated with *Craterellus cornucopioides* makes it an interesting candidate for drug developers and molecular biologists.

7.5.1 Important Bioactive Compounds of *Craterellus cornucopioides*

Mushrooms were formerly only valued for their edible qualities, nutritional value, medicinal properties, and for their unique taste and flavor; however, now many mushrooms are widely believed to have therapeutic properties.[22] Mushrooms are considered to be a very good source of digestible protein, with a content of 10–40% on a dry weight basis, which is greater than the protein content of green vegetables but much less than that of meat and milk. Statistics from studies on amino acids in mushrooms include all the most important amino acids. Fresh mushrooms contain 3–21% carbohydrate and 3–35% fiber content on a dry weight basis; consequently, an extensive share of the carbohydrate content of mushrooms is nutritional fiber, meaning that the caloric cost of mushrooms is low. Wild mushrooms are rich in minerals and are a high-quality source of vitamins. *Craterellus cornucopioides* is a rare type of edible mushroom with good medicinal properties. Detailed studies have shown that *Craterellus cornucopioides* exhibits important biological activity, *viz.* antioxidation and bacteriostasis properties.[23] A recent study has revealed the extraction of a water-soluble polysaccharide, CC-M, from the fruiting bodies of *Craterellus cornucopioides*. Concentrations of CC-M of 5, 10, and 20 $\mu g\,mL^{-1}$ were found to be sufficient to trigger T cell proliferation whenever required. When the concentration level of CC-M reached 10 $\mu g\,mL^{-1}$, the proliferation efficiency of T cells reached a maximum. In terms of cell morphology, cells treated with CC-M formed big clusters, which increased in numbers.[24] B cells originated from the bone marrow's pluripotent stem cells which can produce a large number of plasma cells when it was stimulated by antigens. When the final concentration of CC-M was 5–20 $\mu g\,mL^{-1}$, it was recorded that the proliferation effect of B cells enhanced but when the concentration level of CC-M was

10 μg mL^{-1} the proliferation effect of B cells reached a maximum. RAW264.7 cells exhibit diverse immunological functions and have a phagocytic effect on antigens and the adhesion of antigens. RAW264.7 cells are involved in the immune system as well as in the regulation of the immune system.[25] When the concentration of the polysaccharide was 5–20 μg mL^{-1} the level of proliferation of RAW264.7 cells increased in line with the concentration. When the CC-M concentration was 10 μg mL^{-1} the activities of these cells reached a maximum. A number of studies have been found to describe the chemical composition and bioactivity of *Craterellus cornucopioides*. Craterellins D and E, two new sesquiterpenoids and 4-hydroxy-4-isopropenyl-cyclohexane, a menthane monoterpene, were extracted from *Craterellus cornucopioides* and evaluated for their cytotoxicity, with the results being found to be satisfactory.[26] In a similar study, Yang and co-workers isolated a polysaccharide fraction from *Craterellus cornucopioides* and investigated its antioxidant properties.[27] The analysis of phenolic compounds from this mushroom was performed using high-performance liquid chromatography with diode-array detection (HPLC-DAD). By comparing the retention time to reference compounds, as well as its characteristic UV spectrum, peak identity was established. The quantitative determination of phenolic compound was carried out based on calibration curves. The studies of Palacios and co-workers found that the phenolic compounds in *Craterellus cornucopioides* were gallic acid (118.8 μg g^{-1}) and ferulic acid (14.03 μg g^{-1}) and confirmed the presence of *p*-coumaric acid (8.6 μg g^{-1}), gallic acid (0.7 μg g^{-1}), caffeic acid (16.2 μg g^{-1}), and quercetin (39.6 μg g^{-1}) in *Craterellus cornucopioides*.[10]

Many edible mushrooms have antimicrobial properties. The antimicrobial effect of *Craterellus cornucopioides* varies with the concentration of the extract and the organism tested. In an experiment, antibacterial effect was observed against Gram-positive and Gram-negative bacteria, but the Gram-negative bacteria were found to be more resistant. According to general reports, Gram-negative bacteria are more resistant to antibiotics than Gram-positive bacteria. Due to the complex structure of their cell walls, fungi are resistant to bacteria.[20] *Craterellus cornucopioides* contains some important antimicrobial compounds, such as ferulic acid, *p*-coumaric acid, gallic acid, and quercetin.[28] An acetone concentrate of *Craterellus cornucopioides* in the tested amounts did not affect the concentration of manganese (Mn), suggesting that the mushroom extract exhibited no genotoxic potential *in vitro*. This recommends that the concentrate eased back the cell energy. In the consolidated treatment with mytomysin C (MMC), the concentrations of extract tested (100 and 200 μg mL^{-1}) showed antimutagenic effects. The antimutagenic properties of the acetone concentrate of *Craterellus cornucopioides* can be clarified using phytochemical substances. Phytochemical investigation showed that the tested concentrate was rich in polyphenolics, such as ferulic acid and quercetin. The *Craterellus cornucopioides* mushroom separate had a moderate cytotoxic impact on pre-existing malignant growth cells, among which HeLa cells were generally susceptible. There is almost no data available with regards to the anticancer capability of the *Craterellus cornucopioides* species. A few studies in the literature have reported *Craterellus cornucopioides* to be an exceptionally intriguing reservoir of bioactive mixtures, which gives limitless freedom to new cell reinforcement, antimicrobial, antigenotoxin, and anticancer specialists. Cell reinforcement movement was assessed by free-radical scavenging and superoxide radical scavenging. The free-radical scavenging activity was estimated utilizing 2,2-diphenyl-1-picrylhydrazyl (DPPH) assay, as indicated by Dorman and his co-workers.[29] In a globally accepted experiment, the antimicrobial activity of *Craterellus cornucopioides* was evaluated against ten microorganisms, including *Staphylococcus aureus* (ATCC 25923), *B. subtilis* (ATCC 6633), *B. cereus* (ATCC 10987), *Escherichia coli* (ATCC 25922), *Proteus mirabilis* (ATCC 12453), *Aspergillus niger* (ATCC 16888), *Candida albicans* (ATCC 10259), *Penicillium italicum* (ATCC 10454), *Mucor mucedo* (ATCC 20094), and *Trichoderma viride* (ATCC 13233) obtained from the American Type Culture Collection (ATCC). A 96-well microtiter assay using resazurin as an indicator of cell growth was employed for the determination of minimum inhibitory concentration (MIC) for the active extract.[20] *Craterellus cornucopioides* contains some very important and unique medicinally important chemicals. Therefore, the conservation, domestication, and culture of this mushroom are very important for both the food and pharmaceutical industries. The names of the compounds extracted from *C. cornucopioides* and their pharmacological importance are shown in Table 7.1.

7.5.2 Nutritional and Dietary Composition

Mushrooms have high nutritional value as they contain a high protein and carbohydrate content, vitamins, and minerals. Moreover, mushrooms also contain unsaturated fats, meaning that they are regarded as a low-calorie food. Recent studies have shown that mushroom exhibit anti-diabetic, antioxidant, immunoregulatory, and antitumor properties.[30] Different types of mushrooms contain different types anti-microbial, nutritional, and phytochemically important molecules. There are very few doubts that the physical and chemical reactions and nutritional composition vary across mushroom species.[31] In general, mushrooms are cooked and consumed or used directly on salads. There are different methods used to cook mushrooms, such as boiling, steaming, microwaving, and frying. Textural, sensory, and nutritional changes occur in mushrooms while they are cooking. The presence of bioactive compounds in mushrooms is directly related to the cooking methods of mushrooms. The effect of bioactive compounds on human health is dependent on their biochemistry. The process of digestion changes the biochemical state of the micro and macronutrients to determine their final bio-accessibility.[32] The process of digestion and absorption is very complex, involving enzymatic actions that lead to physiochemical changes in the nutrients. In the last 20 years, different types of *in vitro* digestion methods have been found to influence the physiological conditions

Table 7.1 Compounds extracted from *C. cornucopioides* and their pharmacological importance.

Type of extract	Treatment in diseases	Responsible compound	Ref.
Aqueous dry extract	Modifies the production ability of A431 tumor cells; also acts as a great antioxidant agent	Phenolic compounds	15
Alcoholic dry extracts	Modifies the production ability of A431 tumor cells	Free sterols	15
Dry extracts	Tumor cells	β-Glucans	42
Dried powder	Boosts the innate and cellular immune system and excites tumor denunciation	A novel polysaccharide fraction CCPP-1	27
Dry extracts	Antioxidant activity	Quercetin, ferulic, gallic and *p*-coumaric acid	20
Wet extract	Antimicrobial activity against numerous bacteria and fungi	Craterellins D and E and a new methane monoterpene 4-hydroxy-4-isopropenyl-cyclohexanemethanol acetate	26
Acetone extract	Cytotoxic activity and antimutagenic properties	Polyphenolics such as ferulic acid and quercetin	20
Dry extract	Anti-hepatitis B viral activity	Caffeic acid	43
Dry extract	Anti-inflammation activity	Polysaccharide fraction CCPP-1	44
Powdery extract	Hypocholesterolemic properties	Ergosta-5,7,22-trien-3β-ol	45
Powdery extract	Anti-hyperglycemic properties	Phenols, flavonoids, ascorbic acid, ergosterols, tocopherols	46
Aqueous extract	Anticancer properties	Cyclohexane, dichloromethane	47

(agitation, temperature, enzyme, chemical composition, and pH) of oral, gastric and small intestinal phases.[33]

Craterellus cornucopioides is a precious wild mushroom that is widely distributed. Detailed studies have revealed the content and structural characterization of 77 important phytonutrients, such as sesquiterpenoids, polysaccharides, phenolics, and flavonoids, extracted from this wild edible mushroom. There is no available information regarding the *in vitro* digestion of this mushroom, but there is a traditional way of eating *Craterellus cornucopioides* by soaking. *Craterellus cornucopioides* also contains various macronutrients, such as protein (16.97 g per 100 g dry weight), fat (5.31 g per 100 g dry weight), and total carbohydrate (6.81 g per 100 g dry weight), and ash (9.62 g per 100 g dry weight), as revealed by a detailed nutritional study.[34] It is interesting to note that different types of cooking methods result in different protein content, with methods such as boiling (14.50 g per 100 g dry weight), steaming (15.82 g per 100 g dry weight), frying (16.02 g per 100 g dry weight), and microwaving (9.98 g per 100 g dry weight) having a different impact on the final product.[32] Although higher lipid levels were present in the mushroom, cooking induced a proper dilution of both the protein and lipid content. The slight decrease in the protein content can be due to the hydrolysis of parts of the proteins, their solubility, and leaching out of nitrogenous substances. The fat content of the mushroom is reduced because of the cooking methods, except for frying, might promote conversion of lipids into fatty acids and glycerol. Most studies are related to the protein content of meat and its digestibility as well as the cooking methods of meat. Physicochemical changes in meat proteins during cooking are mainly caused by heating (temperature and time), as well as the consequences of protein denaturation and fast chemical reactivity.[35] Cooking temperature and time are the major factors in the denaturation of the protein content of meat during cooking; this statement is also true of edible fungal mushrooms, including *Craterellus cornucopioides*. Recent studies have confirmed that the amino acid content of *Craterellus cornucopioides* is much higher than that of other plant substances. Amino acids were found to be reduced by frying but retained in microwaving. The ratio of the essential amino acids to non-essential amino acids of CCs, CCs-s, CCs-b, CCs-f and CCs-m were 0.78, 0.75, 0.79, 0.82, and 0.77, respectively. For example, cooking treatments increased the content of isoleucine, leucine, and phenylalanine (except for in the case of frying). The release of amino acids could be enhanced by inducing favorable changes in protein structure using various cooking methods, but higher temperatures could result in unfavorable structural changes that reduced the susceptibility of proteins to enzymatic hydrolysis. A total of 17 kinds of amino acids were freed in the digestive juices. However, cooking and digestion resulted in the release of fewer essential amino acids than non-essential amino acids. Among the amino acids, aspartic acid, glutamic acid, glycine, alanine, and arginine are responsible for the palatable taste of the mushroom.[36] In *C. cornucopioides*, the amino acid content from high to low was glutamic acid, aspartic acid, glycine, and alanine. Branched-chain amino acids comprise leucine, isoleucine, and valine, which have unique properties with diverse physiological and metabolic roles and are essential in immunological and brain functions.[27] This simply means that *Craterellus cornucopioides* plays a key role in the immune defense mechanism.

There has been no significant difference observed in free fatty acid absorption between steaming, boiling, and microwaving. By contrast, steaming, boiling, and microwaving have been shown to change the composition of C18 fatty acids, particularly the ratio of 18 : 0 to 18 : 2. Although raw materials have been shown to have a high C18:2 content, it was clear that their bioaccessibility was low. Also, Shen and co-workers found that dietary stearic acid reduced metastasis tumor burden in an athymic (nude) mouse breast cancer metastasis model and that dietary stearic acid reduced visceral adipose tissue.[37] Long-chain polyunsaturated fatty acids release more slowly than polyunsaturated fatty acids. Other studies have also shown that lipids composed of long-chain triglycerides (*e.g.*, fish oils) are digested slower than fats, which are composed of short-chain or medium-chain triglycerides, such as coconut oil.[26] Besides, saturated fatty acids (C16:0 and C18:0) have also been reported to be released faster than unsaturated fatty acids (C18:1, C18:2n-6, and C18:3n-3) as noticed by Guo and co-workers.[35] The release of C18:2, and C18:3 during frying was significantly higher than in other treatments. As mentioned above, the fatty acid composition could dramatically influence lipid digestion efficiency. Compared with the other samples, the digestibility was improved significantly by frying (71.75%). Additionally, there was no significant difference in digestibility between steamed and boiled *Craterellus cornucopioides*. By combining these results, it could be concluded that although the lipid profiles of the different samples were different, the more abundant fatty acids could be more easily released during digestion. Apart from Mn, Pb, Na, Mg, and K, the mineral content was not significantly affected by microwaving; however, steaming, boiling, and frying significantly affected the mineral content. In general, during the boiling of foods, minerals were lost in the boiling water,[38] which resulted in an inevitable decrease in the mineral content. Cooking also affected the separation of minerals (enrichment), the destruction of inhibitors, and the formation of complexes between food components and metal ions. Manditsera and co-workers found that boiling did not significantly affect the Cu and Mn content of beetles and the Ca, Cu, and Mn content of chicks.[39] However, contrary to the available findings, Hassan and co-workers found that iron and zinc content remained the same after boiling.[40] The degree of bioaccessibility is based on how much of the original content is available for intestinal absorption.[41] Cu had the highest percentage of bioaccessibility (83.35%), and Fe had the lowest (3.96%). This study also revealed that the bioaccessibility of several elements (K, Ca, and P), was extremely low during digestion. Adding boiling water to *Craterellus cornucopioides* significantly reduced the mineral bioaccessibility of Cd, while frying significantly reduced the mineral bioaccessibility of Mg.

The extracts of *Craterellus cornucopioides* have already been shown to contain some pharmacologically important constituents, including phenolic compounds, organic acids, amino acids, long chain unsaturated fatty acids, stilbenes and β-glucans.[48] On a dry weight basis this mushroom contains important metals and nutrients, such as a phosphorus concentration of 2.1 mg per g dry weight, a potassium concentration of 26.1 mg g^{-1}, a sulfur concentration of 2.5 mg g^{-1}, a Mg concentration 0.5 mg g^{-1}, a Ca concentration of 0.2 mg g^{-1}, a zinc concentration of 37.8 mg kg^{-1}, an iron concentration of 623 mg kg^{-1}, a Mn concentration of 22.1 mg kg^{-1}, a Cu concentration of 48.7 mg kg^{-1}, and a boron concentration of 6.0 mg kg^{-1}.[49] This mushroom also contains a high amount of sugar content, composed of mainly D-glucose, D-fructose, D-xylose, D-mannose, trehalose, D-ribose, D-arabinose, D-fucose, L-rhamnose, and D-galactose. The composition of sugar alcohol content in terms of dry weight is glycerol at 0.11 mg g^{-1}, mannitol at 17.1 mg g^{-1}, xylitol at 0.01 mg g^{-1}, dulcitol at 0.11 mg g^{-1}, D-sorbitol at 0.02 mg g^{-1}, D-arabitol at 0.01 mg g^{-1}, myo-inositol at 0.05 mg g^{-1}, and *meso*-erythritol at 0.02 mg g^{-1}, *etc.*[49] Functionality studies have revealed the anticancer, anti-inflammatory or hypoglycemic, antimicrobial, antioxidant, immunomodulatory, hypocholesterolemic, and anti-hyperglycemic properties of this mushroom.[15] The aqueous dry extract, obtained by mixing the dried parts of the specimens with distilled water, functions as a great antioxidant agent and contains a high content of phenolic substances, and alcoholic dry extracts obtained by mixing dried mushroom parts with ethanol are a vital source of free sterols. These extracts possibly modify the production ability of A431 tumor cells. Selective cytotoxicity of the dry extract toward the A431 tumor cells can be mainly attributed to the polysaccharides in the fungus.[15] β-Glucans, which are considered to be biologically active modifiers that can boost the innate and cellular immune system and may excite tumor denunciation, can be extracted from this mushroom. These polysaccharides stimulate the dectin-1 receptor (a non-toll-like structure detection receptor that is articulated on myeloid cells, neutrophils, or macrophages), which promotes the development of dendritic cells both under *in vitro* and *in vivo* conditions and increases the responses of antigen-specific CD8T and CD4 cells.[50] According to IC_{50} values, the aqueous extracts (IC_{50}: 12.01±1.72 mg mL^{-1}) are less efficient in comparison to methanolic extracts (IC_{50}: 5.26±0.67 mg mL^{-1}) of this mushroom. There are so many bioactive compounds found in the extracts of *C. cornucopioides*; these bioactive compounds are mainly phenolic compounds, gallic acid, *p*-coumaric acid, lycopene, β-carotene, syringic acid, epicatechin, epigallocatechin gallate, catechin, quercetin, *t*-resveratrol, and rosmarinic acid. Out of these compounds, gallic and *p*-coumaric acids are present in the lowest amounts in the *C. cornucopioides* extracts.[51] The extract of this mushroom has great antimicrobial potential. Kol and co-workers experimentally tested six microorganisms, *Escherichia coli, Staphylococcus aureus, Bacillus licheniformis, Bacillus subtilis, Agrobacterium tumefaciens,* and *Enterococcus faecalis,* which showed antibacterial activities in both phenolic and aqueous dry extracts.[52]

Craterellus cornucopioides has high nutritional and economical value due to its high protein, fatty acid, carbohydrate, mineral, and secondary metabolite content. This mushroom is widely used in medicine, pharmacy, and industry as it exhibits cytological activity, excellent antibacterial properties, excellent antioxidant features, and inhibits unwanted cell growth. *C. cornucopioides* has also been used in therapeutics as its extract is a source of antibacterial, antioxidant, and cytotoxic natural compounds, which have the potential (antibacterial drugs) to work against bacterial as well as infectious diseases.[53] *C. cornucopioides* contains a high amount of polysaccharides and macromolecules that stimulate bioactivity. The extract of *C. cornucopioides* also contains sesquiterpenoids and keto esters. These molecules have proven anti-tumor activity in humans.[35] The polysaccharides act as antiradicalization, immunoregulatory, and anticoagulant agents, reducing blood lipid activity and scavenging reactive oxygen species. Immunomodulatory functions are also carried out by polysaccharides. The immune system of humans contains different immune defense mechanisms, which work against numerous pathogenic microorganisms. Phagocytic cells and macrophages are found more or less in all the human body cells; therefore, they play significant roles in triggering the innate and adaptive immune responses.[53] Recently, several researchers have recommended that polysaccharides might improve the phagocytic occupation of macrophages and control immunity functions. In dry mushroom extract, the highest and lowest bioaccessibility was Cu (83.35%, frying) and Fe (3.96%, frying), respectively, as determined recently. The dry extract exhibits a high quantity of reducing sugar and great antioxidant power; they are also the victims of protein bioaccessibility because of their lack of structural uniformity.[54] Chemically this mushroom has a very complex array of substances, so the process of digestion and absorption of food materials is very complex and requires lots of enzyme and physiochemical processes.[55] By using different types of *in vitro* methods, absorption abilities and quantities can be determined, which is vital for stimulating physiological activities in the small intestinal part, gastric, and oral or upper gastrointestinal tract. Due to its highly efficient, reproducible, less complex, stagnant models, *in vitro* digestion was preferred over anthropogenic and animal studies. The isozyme of 11Bita-hydroxysteroid dehydrogenase (11Bita-HSD1) and the cytotoxins of *Craterellus cornucopioides* have been shown to be potent against several (five) human cancer cell lines, such as hepatocellular carcinoma SMMC-7721, PANC-1 pancreatic cancer, A-549 lung cancer cells, breast cancer SK-BR-3, and human myeloid leukemia HL-60 cells.[56] Guo and co-workers reported that ten of the compounds extracted from this mushroom show considerable cytotoxicity against SMMC-7721 (IC value 0.69 μM), A-549 (IC_{50}, 0.64 μM), HL-60 (IC_{50}, 0.50 μM), SW-480 (IC_{50}, 0.54 μM) and MCF-7 (IC_{50}, 1.10 μM).[35]

7.6 Cultivation and Harvesting

Craterellus cornucopioides is widely distributed, mainly in parts of China, particularly in the southeast region. Black trumpets are mainly venerated for their cooking qualities in Europe and are much loved in places like California.[8] This mushroom is also reported to be found in Asia, South America, North America, Central America, and Europe, as well as occasionally in Australia *Craterellus cornucopioides* contains some very important nutritive substances or properties, so we cannot depend upon its wild counterparts alone to provide us with the essential nutritious and medicinally important compounds; it is important to cultivate this mushroom to fulfil the nutritional needs of society. It is found growing mostly under broad leaved trees or oak, under beech, and is also found in deciduous forests, under a range of deciduous trees, and in hardwoods or mixed woods.[57] The *Craterellus* species is a truly wild mushroom, this species thrives in a dead woody forest environment, where anthropogenic activities are restricted. In the presence of low temperature and high humidity, this mushroom grows abundantly. The mushroom prefers to grow during the rainy season, when the temperature is around 17–20 °C. This species grows in clusters, predominantly from July to September. *Craterellus cornucopioides* is naturally found in wild forms, but there are no so-called perfect cultivation processes that have been recorded to date. This mushroom has very distinctive characteristics, morphologically as well as biochemically. We know the physiochemical nature of the chemicals found in this mushroom and we are aware of how these chemicals are responsible for the treatment of various diseases in humans, such as cancer, microbial infections, *etc.* As the importance of mushrooms is known, they needs to be conserved; however, there has been no available flawless method of cultivation reported from anywhere in the world. Therefore, we must find a way to sustainably cultivate fungi. If a cultivation method was available then we could cultivate the mushroom in various parts of the world to gain economic and ecological benefits out of it. As the first step in the conservation of this mushroom, we have to popularize this mushroom among local communities by citing its high nutritional and medicinal values. Popular mushrooms, such as *Agaricus bisporus*, have gained popularity among the people of the world and are used as a good source of protein and minerals in various parts of the world. People are aware or have adequate knowledge on *Agaricus bisporus* and its nutritional composition but *Craterellus cornucopioides* does not have the same popularity as *Agaricus bisporus*. As people are not aware of the nutritional composition of *Craterellus cornucopioides,* this mushroom has much better nutritional content to offer than any other so-called popular mushroom. Wild edible fungi include several species of chanterelles fungi, such as *Craterellus cornucopioides,* boletus fungi (Boletus spp.), and matsutake fungi (Tricholoma spp.), which participate in an ectomycorrhizal symbiotic relationship with specific vascular plants species. *Craterellus cornucopioides* is used to produce a high number of commercial products as well as varieties of edible foods. So, this mushroom is predictably harvested in wild forests to fulfil social needs. There are no known cultivation processes to produce a commercially large quantity of *Craterellus cornucopioides*. As the *Craterellus cornucopioides* mushroom is only sourced from the wild and good cultivation processes are still unknown, this species is collected by people without any mechanical instruments. In the literature, the complexity of wild edible mushrooms, such as *Craterellus*, and their fungal ecology and habitat are recognized. This complexity has hindered the commercial productivity of the cultivation of this mushroom; financial success is yet to be seen in this area.[8] This mushroom is collected in the early morning to ensure its good quality. While harvesting this mushroom, equipment such as a basket, knife, brush, waxed paper, and a stick, is required. The basket should contain some holes drilled in it to promote air circulation as well as release the rainwater from the surface of the mushrooms. As a first step in harvesting, the base of the mushroom is removed with the help of a knife and dirt is quickly cleaned from the mushrooms with the help of a brush, whereafter they are collected in the basket. After collection, is the mushrooms are cleaned with water and dried in a bamboo tray. After that, they are packed and sold for consumption. One of the most common methods used to preserve methods is drying. However, the dried parts of this mushroom can sometimes be dangerous and dried mushrooms have a very strong flavor; therefore the use of dried mushrooms in cooking is generally avoided. Dried mushrooms are generally used in sauces, pickles, soups, and stew.

7.7 Processing in the Food Industry

In recent decades, humans have become more concerned about their food; what they are eating, as well as how the food is produced and marketed. While searching for varieties of food for our changing taste buds, mushrooms play a great role in fulfilling our needs. Wild mushrooms contain not only nutritional value but are also a good source of medicinal compounds. Wild mushrooms are not only used in food for humans, they are also used to feed animals. The sporocarps of wild mushrooms are a good source of food for wild animals. Sanmeea and co-workers reported that in Thai forests, wild mushrooms are a source of food for wild pigs, insects, snails, rats, *etc.*[58] As *Craterellus cornucopioides* is a wild edible mushroom, it contains many bioactive compounds, holding medicinal and high nutritional value for humans. After collecting mushrooms, they have to be cleaned, dried, frozen, canned, or preserved by other methods for processing in the food industry. *Craterellus cornucopioides* has not been used on a large scale in the food industry to date. However, this mushroom is eaten for its nutritional composition. For this reason, it is used by tribal communities around the world. However, there is no known appropriate processing technique that is known. The extract of *Craterellus cornucopioides* mainly comprises polysaccharides that exhibit several bioactive effects. Guo and co-workers reported that a polysaccharide extract increased the proliferation activity over a specific range of concentration and time, inducing the activation of RAW264.7 cells.[35] The polysaccharide extract of this mushroom contains phenolics as well as flavonoid compounds, which act as excellent antioxidants. Therefore, the naturally occurring antioxidant properties of this mushroom make it ideal for use in food applications.

7.8 Conservation

Craterellus cornucopioides is not only beneficial for humans but also helps in the balancing of the ecosystem by nutrient cycling. As it is saprotroph, it gets its nutrients by degrading the organic matter of trees by secreting various enzymes. Due to its various uses, the demand for this mushroom is increasing to meets the demands of a growing population. People are exploiting this mushroom to fulfil their nutritional and medicinal requirements. However, the overexploitation of this mushroom will have a massive consequence on its diversity and will imbalance the ecosystem. Sometimes various natural factors are also responsible for the loss of these mushrooms. To keep the ecosystem in balance and make it available for the future generation it is important to conserve this mushroom. Various possible methods to conserve this mushroom are discussed herein.

7.8.1 Conservation in Habitat

This is also known as *in situ* conservation. As *Craterellus cornucopioides* is a rare wild edible mushroom, it is very important to conserve this mushroom in a sustainable manner. By conserving its habitat and by regular monitoring we can conserve this mushroom for the long term. This mushroom can be conserved at the government, community, and individual levels. Implementation of some rules and regulations by the government in the harvesting, exporting and importing, *etc.*, of this mushroom species can aid its conservation. This mushroom generally grows in moist soil in forests. People who live near these forests directly benefit from all the forest's timber and non-timber products. These community people have used *Craterellus cornucopioides* as food and medicine since ancient times. Creating awareness among these people about the sustainable use of this species can help to conserve this species.

7.8.2 Awareness of Generations

It is very important to generate awareness among people to conserve this species. Including this species in the syllabi of subjects such as Microbiology, Biotechnology, Botany, Life Sciences, and Environmental Science for undergraduate and postgraduate students could also help to conserve this species.[59] Besides these methods, *Craterellus cornucopioides* can also be conserved by drying in herbaria and using different storage techniques. To conserve the mushroom by these methods, it is very important to have well-equipped laboratories and skilled personnel near to mushroom-rich areas.

7.8.3 Cryopreservation

In this method, a cell suspension is stored at a temperature that is equal to or below the freezing point. This may be ordinary freezing (−5 to −20 °C), ultra-cold freezing (−50 to −80 °C), or freezing with liquid nitrogen (−150 to −196 °C). This helps to conserve the mushroom for the long term without changing its viability, purity, and genetic stability.[60]

7.8.4 Freeze-drying or Lyophilization

By applying this technique the mushroom can be conserved for a long period of time for up to 10 or more years. This method stops the metabolism of the species and extracts water by sublimating the ice under vacuum conditions. To avoid intracellular crystallization during freezing, dehydration is necessary.[61]

7.8.5 Tissue Culture

This method is used to produce mother spawn. Freshly collected fruiting bodies are washed with sterile distilled water; after that, surface disinfection is achieved using ethanol. Then, a small part of the tissue is aseptically taken from the fruiting body and inoculated in potato dextrose agar medium. After inoculation, it is incubated at 25 °C in a dark room. Cultured mushrooms are isolated and purified in fresh medium. These pure cultures are stored in a refrigerator at 4 °C.[62] Sub-culturing is done periodically, where the mushroom is transferred to new petri dishes containing potato dextrose agar media. This is important to provide the required environment for optimal development of the mushroom.[60]

7.9 Safety, Toxicity and Regulation

Normally, edible mushrooms are eaten as food, either directly or indirectly. Edible mushrooms are quite useful for humans and other domesticated animals and wild animals grazing in the forests. However, sometimes edible mushrooms contains toxic elements, such as heavy metals, so consumption beyond permissible levels can be harmful sometimes. The toxic elements of the mushroom can be deposited in human cells, which can lead to cell apoptosis. As a safety or control measure, we can cook mushrooms before consumption and avoid consuming them in a raw state. The cooking of mushrooms changes the biochemical nature of their bioactive compounds. Cooking processes change the structure of complex compounds into simpler ones that are easily usable by the human body and are less harmful in cooked form than in their natural state; the complex carbohydrate and protein structure changes in the cooked form have very little or negligible impact on health. The mineral content of mushrooms is the same as natural when cooked. Toxic elements are eliminated in the process of cooking.

Craterellus cornucopioides have few recorded cytotoxic effects on humans. The cytotoxicity of the dry extracts of *Craterellus cornucopioides* can be determined by means of Daphnia magna, MTT (3-(4,5-dimethylthiazol-2-yl)-2,5-diphenyltetrazolium bromide), and live and dead bioassays. Dry extracts are prepared when dried black trumpets are heated in a solvent such as water, 80% ethanol, or 80% methanol in a reflux condenser for 30 min. After that, filtrates are evaporated at 40 °C under reduced pressure and dried in a freezer.[15] The Daphnia magna bioassay is one of the most prominent methods for studying the cytotoxicity of this mushroom. In this method, crustaceans are exposed to diluted dry extracts and after 24, 48, and 72 h of incubation survivors are counted. Another reliable assay for the testing of cytotoxicity resulting from mushrooms is the MTT assay; it is one of the most popular test methods. By applying this test one can assess the anticancer effects of different compounds of mushroom extracts. In this test, cultured cells are removed from the culture medium after 24 h and 72 h. Then the monolayers of cells are incubated for 4 h in MTT solution at 37 °C, which allows the formation of formazan in the metabolically active cells. These formazan crystals are then solubilized in isopropanol. After that, the concentration of the solution is spectrophotometrically quantified. The live and dead assay is a fluorescence-based method used to determine cell viability. This uses two ways to detect cell viability, by detecting intracellular esterase activity in live cells and compromised plasma membrane integrity in dead cells.[63] Selenium is one of the major toxic elements present in *Craterellus cornucopioides*. As is known, a high concentration of selenium is harmful to the human body, so there is a need to concentrate on the elimination of selenium from processed mushrooms. If we can eliminate the harmful elements from this mushroom then people around the world can consume it without any worry. This mushroom would also gain entry to the international market and could be harvested in different parts of the world without any worry about selenium toxicity; this could be beneficial to the global economy. The regulation standard for toxic levels of harmful elements are set differently in the different countries of the world; however, the heavy metal toxicity of this mushroom is fortunately under the permissible levels of the major countries of the world, including those of the USA, UK, France, Germany and Russia.[64]

7.10 Trade, Storage Stability and Transport

Craterellus cornucopioides is a very important and highly admirable source of food globally. It is a good source of protein and fiber and is consumed to for their minerals and vitamins. Moreover, it has also been used for medicinal purposes since ancient times. For these reasons, this mushroom has great commercial value. The commercial harvesting of this mushroom has increased greatly in many regions of the world over the past two decades. The market value of *Craterellus cornucopioides* in some forests may be as high as that of timber. The government in Finland tried to encourage the use of various wild mushrooms such as *Craterellus cornucopioides* by introducing rural policy measures. Since 1997, a company named Dalla Valle Oy purchased wild mushrooms from pickers and marketed most of its output in southern Europe. In each major town of southern Europe, various companies have placed permanent 'collection points' where pickers can sell their harvest of *Craterellus cornucopioides* throughout the season. The picker level prices for *Craterellus cornucopioides* are 4 to 7 Euros per kg in this region.[16] Among developing nations, Turkey is one of the most favored nations for the trade of wild mushrooms such as *Craterellus cornucopioides*. This mushroom was reported in a study to be an edible wild mushroom species that is exported from Turkey and a picking time was also suggested for better quality harvesting. This mushroom is collected from all over Turkey and exported by various companies through sellers using refrigerated automobiles. This mushroom is exported in fresh, dried, and frozen form from Turkey to other countries. As *Craterellus cornucopioides* is one of the most delicious mushrooms, it is largely used by the native people of Bulgaria for soup or fried with oil and salt. Therefore, this mushroom is commercially very popular in the local markets of Bulgaria; it is also very important for domestic and foreign trade.[65] Wild-growing mushrooms such as *Craterellus cornucopioides* have become a delicacy worldwide due to their specific smell and appearance.[22] The storage of this mushroom is very difficult because of its fragile nature. Drying is a common method used to store this mushroom. Mixing with salt is another method by which to preserve this mushroom. In a recent study, it was shown that if this mushroom is stored at 10 °C for 12 days it exhibits a maximum weight loss of around 11%. Hence, to preserve this mushroom for a long time, it needs to be stored at a temperature of between 10 and 15 °C.[66] Since the storage of the mushroom is very difficult, it is often direct consumed by local communities.

7.11 Conclusions and Future Perspectives

Craterellus cornucopioides is a wild edible mushroom that has nutritional, medicinal, and pharmacological value. It contains highly beneficial and important nutrients, such as water (88–94%), protein (15–42%), crude fat (2–6%), and carbohydrates (42–71%). Due to its highly beneficial impact on human health, its popularity has increased and it is exported and imported between countries. Increasingly more medicinal practices are being connected with these edible mushrooms and their medicinal uses as antioxidant, anticancer, antidiabetic, antiallergic, immunomodulatory, cardiovascular, anticholesterolemic, antiviral, antibacterial, antiparasitic, and antifungal agents is already being reported from many parts of the world. In spite of this, very little effort has been made towards their cultivation, conservation, and harvesting. This mushroom has great potential to become the "next big thing" for mushroom hunters due to its high medicinal and nutritional value. As very little scientific literature has been reported from the Indian subcontinent, it is highly recommended for researchers in this area to add this subject to their research interests to augment their research with instant novelty.

Acknowledgements

On behalf of all the authors, the corresponding author is thankful to the esteemed Editors of this book for granting the necessary consent and trusting us for writing this chapter. We are grateful to Dr Prakash Sarma, Principal, M.C. College, Barpeta for his guidance and words of encouragement while preparing this chapter which has given me the strength and courage to continue with this chapter even in the middle of the hectic I.Q.A.C. works of the college.

References

1. S. T. Chang, in *Mushrooms as Functional Foods*, ed. P. C. K. Cheung, Wiley Online Books, USA, 1st edn, 2008, p. 1.
2. M. E. Valverde, T. Hernández-Pérez and O. Paredes-López, *Int. J. Microbiol.*, 2015, **9**, 1.
3. D. D. Tibuhwa, *J. Ethnobiol. Ethnomed.*, 2013, **9**, 1.
4. M. A. Kotowski, *Microb. Biosyst.*, 2019, **4**, 1.
5. A. S. Sonnenberg, W. Gao, B. Lavrijssen, P. Hendrickx, N. Sedaghat-Tellgerd, M. Foulongne-Oriol, W. S. Kong, E. G. Schijlen, J. J. Baars and R. G. Visser, *Fungal Genet. Biol.*, 2016, **93**, 35.
6. P. Mattila, K. Könkö, M. Eurola, J. M. Pihlava, J. Astola, L. Vahteristo, V. Hietaniemi, J. Kumpulainen, M. Valtonen and V. Piironen, *J. Agric. Food Chem.*, 2001, **49**, 2343.
7. B. Das, B. De, R. Chetree and S. C. Mandal, in *Herbal Medicine in India*, ed. S. Sen and R. Chakraborty, Springer Nature, Singapore, 1st edn, 2020, p. 509.
8. D. Pilz, *Ecology and management of commercially harvested chanterelle mushrooms*, US Department of Agriculture, USA, 2003.
9. S. Chatterjee, M. K. Sarma, U. Deb, G. Steinhauser, C. Walther and D. K. Gupta, *Environ. Sci. Pollut. Res.*, 2017, **24**, 19480.
10. I. Palacios, M. Lozano, C. Moro, M. D'arrigo, M. A. Rostagno, J. A. Martínez, A. García-Lafuente, E. Guillamón and A. Villares, *Food Chem.*, 2011, **128**, 674.
11. D. Tibuhwa, *Tanz. J. Sci.*, 2018, **44**, 24.
12. D. Kumari, R. C. Upadhyay and M. S. Reddy, *Mycol. Prog.*, 2012, **11**, 769.
13. R. Phillips, in *Mushrooms And Other Fungi of North America*, Firefly Books, Richmond Hill, 2005.
14. K. Das, *NeBIO*, 2010, **1**, 1.
15. F. Costa-Silva, G. Marques, C. C. Matos, A. I. Barros and F. M. Nunes, *Food Chem.*, 2011, **126**, 91.
16. M. Cai, D. Pettenella and E. Vidale, *Forest Policy Econ.*, 2011, **13**, 221.
17. M. De Román and E. Boa, *Micol Aplicada Int.*, 2004, **16**, 25.
18. Z. M. Thu, K. K. Myo, H. T. Aung, M. Clericuzio, C. Armijos and G. Vidari, *Molecules*, 2020, **25**, 1972.
19. S. Beluhan and A. Ranogajec, *Food Chem.*, 2011, **124**, 1076.

20. M. M. Kosanic, D. S. Šeklic, M. M. Jovanovic, N. N. Petrovic and S. D. Markovic, *EXCLI J.*, 2020, **19**, 442.
21. N. Turfan, A. Y. Sezgin, A. Pekşen and Ş. S. Akin, *Uluslar. Tarım Yaban Hayatı Bilim. derg.*, 2020, **6**, 202.
22. P. Kalač, *J. Sci. Food Agric.*, 2013, **93**, 209.
23. G. Ma, W. Yang, L. Zhao, F. Pei, D. Fang and Q. Hu, *Food Sci. Hum. Wellness*, 2018, **7**, 125.
24. J. K. Yan, J. J. Pei, H. L. Ma and Z. B. Wang, *Carbohydr. Polym.*, 2015, **5**, 64.
25. S. J. Yeo, J. G. Yoon, S. C. Hong and A. K. Yi, *J. Immunol.*, 2003, **170**, 1052.
26. D. Zhu, R. Guo, W. Li, J. Song and F. Cheng, *Food Bioproc Tech.*, 2019, **12**, 587.
27. W. W. Yang, L. M. Wang, L. L. Gong, Y. M. Lu, W. J. Pan, Y. Wang, W. N. Zhang and Y. Chen, *Int. J. Biol. Macromol.*, 2018, **117**, 473.
28. A. Borges, C. Ferreira, M. J. Saavedra and M. Simões, *Microb. Drug Resist.*, 2013, **19**, 256.
29. H. J. Dorman and R. Hiltunen, *Food Chem.*, 2004, **88**, 193.
30. M. Oyaizu, *J. J. Nutr. Diet.*, 1986, **44**, 307.
31. Z. X. Ng and N. F. Rosman, *J. Food Sci. Technol.*, 2019, **56**, 865.
32. Y. Tian, Y. Zhao, J. Huang, H. Zeng and B. Zheng, *Food Chem.*, 2016, **15**, 714.
33. A. Brodkorb, L. Egger, M. Alminger, P. Alvito, R. Assunção and S. Balance, *Nat. Protoc.*, 2019, **14**, 991.
34. Y. Liu, X. Duan, S. Duan, C. Li, B. Hu, A. Liu, Y. Wu, H. Wu, H. Chen and W. Wu, *Food Funct.*, 2020, **11**, 6899.
35. M. Guo, M. Meng, J. Zhao, X. Wang and C. Wang, *Int. J. Biol. Macromol.*, 2020, **160**, 871.
36. B. Singh, B. Singh, A. Kishor, S. Singh, M. N. Bhat, O. Surmal and C. M. Musarella, *Sustainability*, 2020, **12**, 7526.
37. M. C. Shen, X. Zhao, G. P. Siegal, R. Desmond and R. W. Hardy, *PLoS One*, 2014, **9**, e104083.
38. J. Falandysz and J. Borovička, *Appl. Microbiol. Biotechnol.*, 2013, **97**, 477.
39. F. A. Manditsera, P. A. Luning, V. Fogliano and C. M. Lakemond, *Food Res. Int.*, 2019, **121**, 404.
40. L. G. Hassan, K. J. Umar, S. M. Dangoggo and A. S. Maigandi, *Pak. J. Nutr.*, 2011, **10**, 23.
41. K. Zhang, Y. Y. Pu and D. W. Sun, *Trends Food Sci. Technol.*, 2018, **78**, 72.
42. I. Mironczuk-Chodakowska, A. M. Witkowska, M. E. Zujko and K. M. Terlikowska, *Rocz. Panstw. Zakl. Hig.*, 2017, **68**, 3.
43. G. F. Wang, L. P. Shi, Y. D. Ren, Q. F. Liu, H. F. Liu, R. J. Zhang, Z. Li, F. H. Zhu, P. L. He, W. Tang and P. Z. Tao, *Antivir. Res.*, 2009, **83**, 186.
44. J. J. Xu, L. L. Gong, Y. Y. Li, Z. B. Zhou, W. W. Yang, C. X. Wan and W. N. Zhang, *J. Food Biochem.*, 2021, **45**, e13842.
45. A. Gil-Ramirez, C. Clavijo, M. Palanisamy, C. Soler-Rivas, A. Ruiz-Rodriguez, F. R. Marín, G. Reglero and M. Pérez, *ICMBMP7*, 2011, **2**, 110.

46. Y. T. Liu, J. Sun, Z. Y. Luo, S. Q. Rao, Y. J. Su, R. R. Xu and Y. J. Yang, *Food Chem. Toxicol.*, 2012, **50**, 1238.
47. J. Radović, A. Leković, A. Tačić, M. Dodevska, T. Stanojković, T. Marinković, Č. Jelić, T. Kundakovic-Vasović, A. Spórna-Kucab, A. Tekieli and K. Skalicka-Woźniak, *Polish J. Food Nutr. Sci.*, 2022, **72**, 171.
48. E. A. Bhat, N. Sajjad, J. S. Sabir, M. R. Kamli, K. R. Hakeem, I. A. Rather and A. Bahieldin, *Saudi J. Biol. Sci.*, 2020, **27**, 1562.
49. M. B. Bahadori, C. Sarikurkcu, O. U. Yalcin, M. Cengiz and H. Gungor, *Microchem. J.*, 2019, **150**, 104172.
50. Y. Masuda, H. Inoue, H. Ohta, A. Miyake and M. Konishi, *Int. J. Cancer Res.*, 2013, **133**, 108.
51. C. Moro, I. Palacios, M. Lozano, M. D'Arrigo, E. Guillamón, A. Villares, J. A. Martínez and A. García-Lafuente, *Food Chem.*, 2012, **130**, 350.
52. K. O. Sevim, A. Bostanci, A. Kocabaş, U. Z. Yasin and S. A. Gökhan, *Anatolian J. Bot.*, 2018, **2**, 60.
53. K. L. Rock, J. J. Lai and H. Kono, *Immunol. Rev.*, 2011, **243**, 191.
54. A. Mudhoo, S. K. Sharma, V. K. Garg and C. H. Tseng, *Crit. Rev. Environ. Sci. Technol.*, 2011, **41**, 435.
55. M. G. Weller, *Sensors*, 2012, **12**, 9181.
56. T. Feng, X. H. Cai, Y. P. Liu, Y. Li, Y. Y. Wang and X. D. Luo, *J. Nat. Prod.*, 2010, **73**, 22.
57. A. T. Kuiters and P. A. Slim, *Biol. Conserv.*, 2002, **105**, 65.
58. J. A. Matisoff, *Bull. Natl. Mus. Ethnol.*, 2011, **35**, 655.
59. J. M. Dunwell, S. Khuri and P. J. Gane, *Microbiol. Mol. Biol. Rev.*, 2000, **64**, 153.
60. K. Castro-Rios and L. P. Bermeo-Escobar, *J. Microbiol. Biotechnol. Food Sci.*, 2021, **10**, 620.
61. K. Kała, A. Krakowska, P. Zięba, W. Opoka and B. Muszyńska, *J. Sci. Food Agric.*, 2021, **101**, 3481.
62. E. M. Adongbede and A. F. Atoyebi, *J. Appl. Sci. Environ. Manag.*, 2021, **25**, 1497.
63. S. Singh, R. Thangjam, G. D. Harish, H. Singh, R. Kumar, D. P. Meena and A. Agrawal, *Plant Cell, Tissue Organ Cult.*, 2021, **144**, 195.
64. V. Mishra, S. Tomar, P. Yadav and M. P. Singh, *Int. J. Biol. Macromol.*, 2021, **182**, 1628.
65. B. Ivančević, M. Matavuly and M. Karaman, *Biol. Serb.*, 2012, **34**, 1.
66. K. Dhalsamant, S. K. Dash, L. M. Bal and M. K. Panda, *Sci. Hortic.*, 2015, **189**, 41.

CHAPTER 8

Enokitake (Flammulina velutipes)

ARI SATIA NUGRAHA,*[a] HENDRIS WONGSO,[b] DWI KOKO PRATOKO[a] AND PHURPA WANGCHUK[c]

[a] Drug Utilisation and Discovery Research Group, Faculty of Pharmacy, Universitas Jember, Jl. Kalimantan I No. 2, Sumbersari, Jember, Jawa Timur, 68121, Indonesia; [b] Research Center for Radioisotope, Radiopharmaceutical, and Biodosimetry Technology, Research Organization for Nuclear Energy, National Research and Innovation Agency, Puspiptek, Banten 15314, Indonesia; [c] Centre for Molecular Therapeutics, Australian Institute of Tropical Health and Medicine, James Cook University, Building E4, McGregor Rd, Smithfield, Cairns, QLD 4878, Australia
*Email: arisatia@unej.ac.id

8.1 Introduction

The historical evolution of mushrooms has been claimed to date back to 1430 Ma; however, archeological exploration has led to the successful excavation of a fossil of the first-known fruiting body mushroom, *Gondwanagaricites magnificus* gen. et sp. nov., from the Gondwana era in Brazil, dated to around 113–120 Ma.[1] Indeed, the early use of mushrooms in human culture was depicted in the psilocybin mushroom cult of the Maya people in around 1000 BC.[2] Mushrooms are considered to be macroscopic fungi with large fruiting bodies produced through the sexual reproduction cycle, predominantly by Ascomycete and Basidiomycete fungi. The fungus kingdom comprises macro and micro fungi, with a total of >2 million species, in which only 148 000 species have been fully described by taxonomists.[3] Despite their benefits, fungi can harm plants and humans. Around 14 000

Phytochemistry and Nutritional Composition of Significant Wild Medicinal and Edible Mushrooms: Traditional Uses and Pharmacology
Edited by Ajay Sharma, Garima Bhardwaj and Gulzar Ahmad Nayik

Published by the Royal Society of Chemistry, www.rsc.org

mushroom species have been well described globally, of which 2000 species are edible and 65 species are considered to possess medicinal benefits.[4] Mushroom species employed in the fungiculture industry include: *Agaricus bisporus* (button), *Lentinula edodes* (shiitake), *Pleurotus* species (oyster), *Auricularia* species (wood ear), *Volvariella volvacea* (straw), *Flammulina velutipes* (enokitake), *Tremella fuciformis* (jelly fungus), *Hypsizygus marmoreus* (bunashimeji), and *Grifola frondosa* (maitaki).[5] *Flammulina velutipes*, (golden needle or winter mushroom) is one of the most important mushrooms on the global market, which has different trivial names across the continents, such as: Enokitake (Japan), zimska panjevica (Serbia), fetuitui (Cameroon), Heand (India), Paengibeoseot (Korea), and Jin Zhen Gu (China).[6–8] This xylophagous mushroom is widely distributed under enoki, velvet stem, golden needle mushroom, winter mushroom, and velvet foot labels.[9]

8.2 Taxonomy and Characteristics of Mushrooms

Taxonomically, the macrofungus *Flammulina velutipes* belongs to the tricholomataceae family, falling in the basidiomycota division (Figure 8.1)[10,11] along with a further 15 *Flammulina* species, including: *F. callistosporioides* (Singer) Singer, *F. cephalariae* Pérez-Butrón & Fern.-Vic., *F. elastica* (Sacc.) Redhead & R.H. Petersen, *F. fennae* Bas, *F. ferrugineolutea* (Beeli) Singer, *F. glutinosa* G. Stev., *F. lupinicola* (Redhead & R.H. Petersen) C. Hahn, *F. Mexicana* Redhead *et al.*, *F. ononidis* Arnolds, *F. populicola* Redhead & R.H. Petersen, *F. rossica* Redhead & R.H. Petersen, *F. similis* E. Horak, *F. stratosa* Redhead *et al.*, *F. velutipes* (Curtis) Singer, and *F. yunnanensis* Z.W. Ge & Zhu L. Yang.[12] Among the Flammulina genus, *F. velutipes* and *F. rossica* are well distributed throughout the world, whereas edible *F. populonica* and *F. fennae* are found not as frequently in the wild in temperate regions, yet are still foraged in Russia. Other species, such as *F. mexicana* and *F. stratosa* seem to be geographically restricted.[12]

Flammulina mushrooms are well known as the winter mushroom, as its lipid and carbohydrate compositions and content enable the *Flammulina*

Figure 8.1 Cultivated (a) *F. velutipes* (b) wild *F. velutipes.*

species to withstand winter conditions. At hypothermic temperature, phospholipids with a high degree of nonsaturation dominate membrane lipids, while the antifreeze glycerol along with arabitol and trehalose are accumulated in the cell cytosol.[13]

The wild *F. velutipes* has an orange to yellow brown convex-shaped first cap (1.5–6.5 cm in diameter) with an in-rolled margin to flat hump and flaring margin when mature. The cap surface is shiny, moist, and sticky when fresh. The gills are broad, adnexed to sinuate, and subdistant with a yellowish-white color. The cap is supported by a flexible cylindric stipe with dimensions of 2–7 cm × 0.3–1.0 cm. This stalk has no ring and is yellowish-white to dark brown at the base during its mature stage. The spores have a smooth and ellipsoid shape with dimensions of 6.5–9.0 pm × 2.5–4.0 pm.[14] Cultivated *F. velutipes* has a distinct appearance compared to its wild-mushroom counterpart. The commercially farmed *F. velutipes* is recognized by a less sticky cap with a significantly smaller diameter (1–3 cm). The cap is supported by a finely clustered pubescent stipe with dimensions of 1.5–7 cm × 0.2–0.7 cm.[15] The cap and stipe are white in color due to the absence of light and the rich carbon dioxide exposure during the cultivation of the mushroom. This growth environment forms thin and soft stalks that are beneficial for culinary purposes (Figure 8.1).

8.3 Resources Availability, Sustainability and Distribution

Wild mushroom picking has been a popular practice across the globe, especially for culinary purposes, and has become a cultural activity in certain regions across the world.[16] Despite practical mushroom guides having been published in large numbers, misidentification still occurs, which sometimes leads to serious safety issues.[17] In addition, navigating skills and knowledge of the ecosystem are necessary for people who do not visit forests on a regular basis.[18] In the wild, the fruiting body of *F. velutipes* is commonly found on the dead stumps and trunks of deciduous trees, on diseased living trees, and in huge tiers and groups. *F. velutipes* are commonly foraged in temperate zones in the northern hemisphere (America, Europe, and Asia) where the mushroom is naturally grown over the winter. A long-term study on wild mushroom foraging in Switzerland revealed that this cultural practice neither effects the future yield of the fruiting body nor the species diversity of wild mushrooms.[19,20] In addition, wild mushrooms have become commercial products on the market on several continents.[19] Edible mushrooms are in high demand due to their nutritional and health benefits, so domestication and cultivation techniques have been developed as cultivated *F. velutipes* are preferred over wild mushrooms. This make *F. velutipes* the fourth most widely cultivated mushroom across China, Siberia, Asia Minor, Europe, Africa, North America, Australia, Taiwan, and Japan.[11,21]

F. velutipes has been cultivated in China since 800 AD, in which its artificial cultivation details are described in an ancient book written during the Tang Dynasty.[8] Nevertheless, Japanese growers have used sawdust and rice bran media since 1928, followed by a cultivation revolution in the 1960s that led Japan to be the largest *F. velutipes* producer in the global market until the 1980s, with total yield of 62 283 tons produced, the second most cultivated mushroom in Japan after *Lentinus edodes*.[22,23] Currently *F. velutipes* or enokitake has become the most cultivated mushroom in Japan, with total production of 127 914.2 tons in 2022.[24] In the 1990s, China made a remarkable leap in *F. velutipes* production, reaching 200 000 tons, meaning that China has had the largest market share globally ever since, with total mushroom production of nearly 40 million tons in 2020.[25,26]

Overall, the global consumption of edible macro fungi has reached an annual global production of 12.74 million tons and is estimated to be around 20.84 million tons by 2026, with the production exceeding 100 000 tons annually, in countries such as China, Japan, the United States, the Netherlands, South Korea, Poland, Vietnam, Spain, France, Thailand, and the United Kingdom.[27] The macro fungi business yields an economic value of a total of $63 billion in 2013.[27]

8.4 Medicinal Uses (Folklore and Ancient and Modern Uses)

Compared to medicinal mushrooms, *F. velutipes* is well known to be a good nutritional source, making this fungus one of the top mushrooms consumed globally as a delicacy.[9] Apart from its nutritional-related uses, *F. velutipes* has been used traditionally as a medicine for several ailments. To the people of Jammu and Kashmir (India), the fruiting body, referred to by them as the heand, is prepared as a tonic and immunomodulator.[6] Another survey in the Indian Himalayas discovered that the local herbalist used the wild fruiting body for diabetes.[28] In modern society, the scientific based uses of medicinal mushrooms are demanded, in which the bioprospecting protocol successfully revealed the *F. velutipes* fruiting body as a potential source of anticancer, antimicrobial, antioxidant, cholesterol-lowering, and immunomodulatory agents.[9,29]

8.5 Phytochemistry

8.5.1 Secondary Metabolites

F. velutipes is a cultivated mushroom that is rich in secondary metabolites, ranging from simple phenolic compounds to arrays of sesquiterpenes, including alkaloid norsesquiterpene. The mushroom also contains the sulfonated molecule 5-sulfosalicylic acid which might contribute to adaptation to bacterial biotic stress.

8.5.1.1 Phenolic Compounds

Phenolic compounds are typical organic molecules with at least one aromatic ring and one hydroxyl group, in which some molecular structures are in their fused form. In the fungi kingdom, especially in macro fungi, these compounds commonly exist in mushrooms as phenolic acids, flavonoids, tocopherols, tannins, and other form of phenols.[30] Phenolic acids constituents are commonly present in mushrooms as benzoic acid (C_1–C_6 backbone) and cinnamic acid (C_3–C_6 backbone) derivatives. Flavonoids are more developed phenolic compounds with a C_6–C_3–C_6 backbone that are distributed in several sub-groups, such as anthocyanidins, anthoxanthins, flavanones, flavanonols, flavans, and isoflavonoids.[31] The occurrences of flavonoids have been reported from several studies, both in edible and medicinal mushrooms.[32–34] Tannins are polymerised forms of simple phenolic compounds that are classified based on the scaffold unit as hydrolyzable and condensed tannins.[35] The occurrence of tannins is related to their ability to bind with alkaloids, amino acids, nucleic acids, and carbohydrates.[36]

At least 11 phenolic compounds have been successfully isolated and characterised from the fruiting body of *F. velutipes* (Figure 8.2). A study on a methanol–dichloromethane extract of *F. velutipes* using a liquid chromatography–tandem mass spectrometry protocol observed the occurrence of antioxidant phenolic compounds, including the benzoic derivatives protocatechuic acid **1**, ellagic acid **2** and cinnamic acid derivatives *p*-coumaric **3**.[37] Kim *et al.* conducted a study on an acidified acetonitrile crude extract of a fruiting body sourced from Korea and detected antioxidant active phenols, identifying the phenols as benzoic acid derivatives, protocatechuic acid **1** (21 μg per g dry weight), gallic acid **4** (21 μg per g dry weight), and 5-sulfosalicylic acid **5** (15 μg per g dry weight); cinnamic derivatives, caffeic acid **6** (17 μg per g dry weight), chlorogenic acid **7** (26 μg per g dry weight),

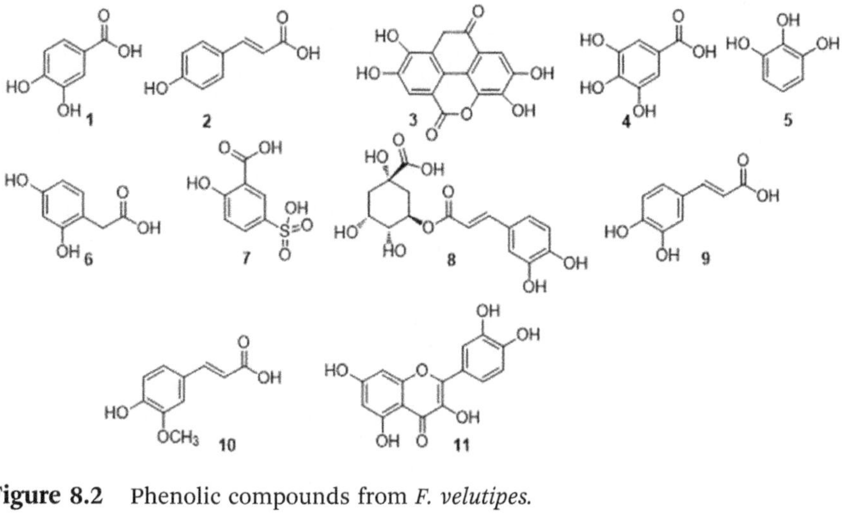

Figure 8.2 Phenolic compounds from *F. velutipes.*

ferulic acid **8** (9 μg per g dry weight); decarboxylated gallic acid, pyrogallol **9** (15 μg per g dry weight); intermediate components of amino acids, homogentisic acid **10** (15 μg per g dry weight), and common flavonoids, such as quercetin **11** (35 μg per g dry weight).[34] The study also reported protocatechuic acid to be the dominant phenolic compound, which is different from the plant kingdom, for example in rice, ferulic acid is present in high abundancy.[34]

8.5.1.2 Terpenes

The mevalonate pathway has evolved in terpene biosynthesis, and sesquiterpene biosynthesis in *F. velutipes* was studied through genomic analysis and found to produce arrays of sesquiterpene backbones, cuparane type sesquiterpenes, seco-cuparane type sesquiterpenes, eudesmane, cadinane, spiroaxane, isolactarane, stepurane and cucumane (Figure 8.3).[38] The study also revealed a distinct sesquiterpene skeleton divergence between Japan and China strains.

8.5.1.2.1 Cuparane and Seco-cuparane Type Sesquiterpenes. Cuparane and seco-cuparane type sesquiterpenes are the most abundant sesquiterpenes present in *F. velutipes*, with 19 derivatives having been reported from its

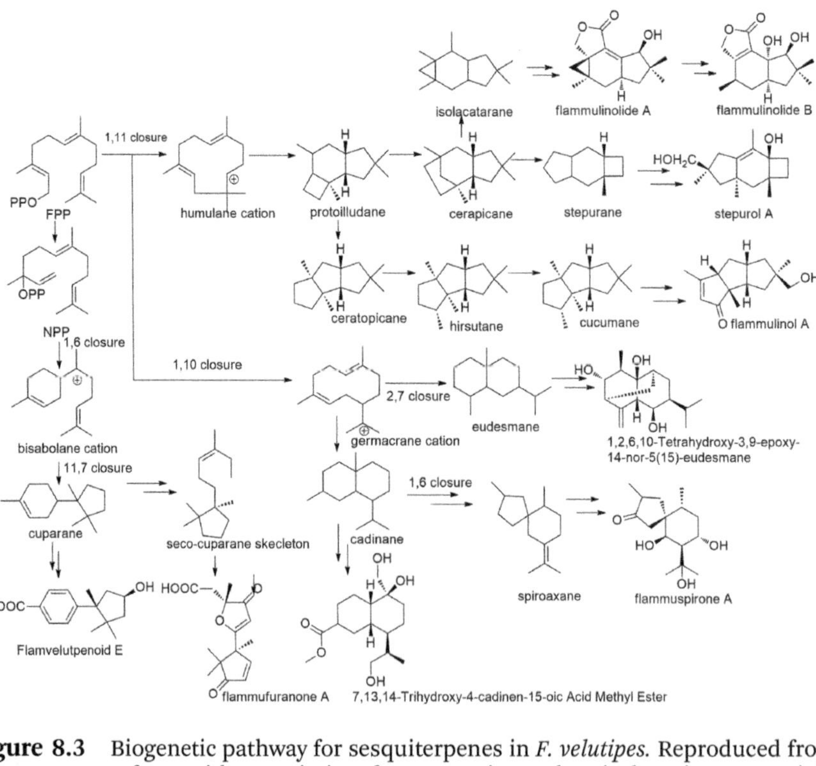

Figure 8.3 Biogenetic pathway for sesquiterpenes in *F. velutipes*. Reproduced from ref. 38 with permission from American Chemical Society, Copyright 2016.

mycelia-fermented culture (Figure 8.4). The highly oxidised cuparane-type sesquiterpene, enokipodin B **13**, was isolated from the malt agar plate mycelia fermentation of *F. velutipes* supplied by the Hokkaido Forest Products Research Institute, Asahikawa, Japan. Enokipodin B **13** was proposed as an autooxidised form of enokipodin A **12**.[39] Further study on the same strain by Ishikawa ended with the isolation of enokipodin D **15**, which is an autoxidised form of enokipodin C **14**.[40] A study on several *Flammulina* species confirmed not all enokipodins are produced, such as in some *F. velutipes* strains.[41] Wang *et al.* isolated C6–C5 skeleton cuparane sesquiterpenes, enokipodins E–J **16–21**, and 2,5-cuparadiene-1,4-dione **22**, from cooked rice seeded with *F. velutipes* sourced from collection at the Institute of Microbiology, Chinese Academy of Sciences, Beijing, China.[42] Other cuparane-type sesquiterpenes, flamvelutpenoids A–F **23–28**, were isolated from the Chinese strain *F. velutipes*.[38,43] Seco-cuparane type sesquiterpenes, flammufuranones A and B

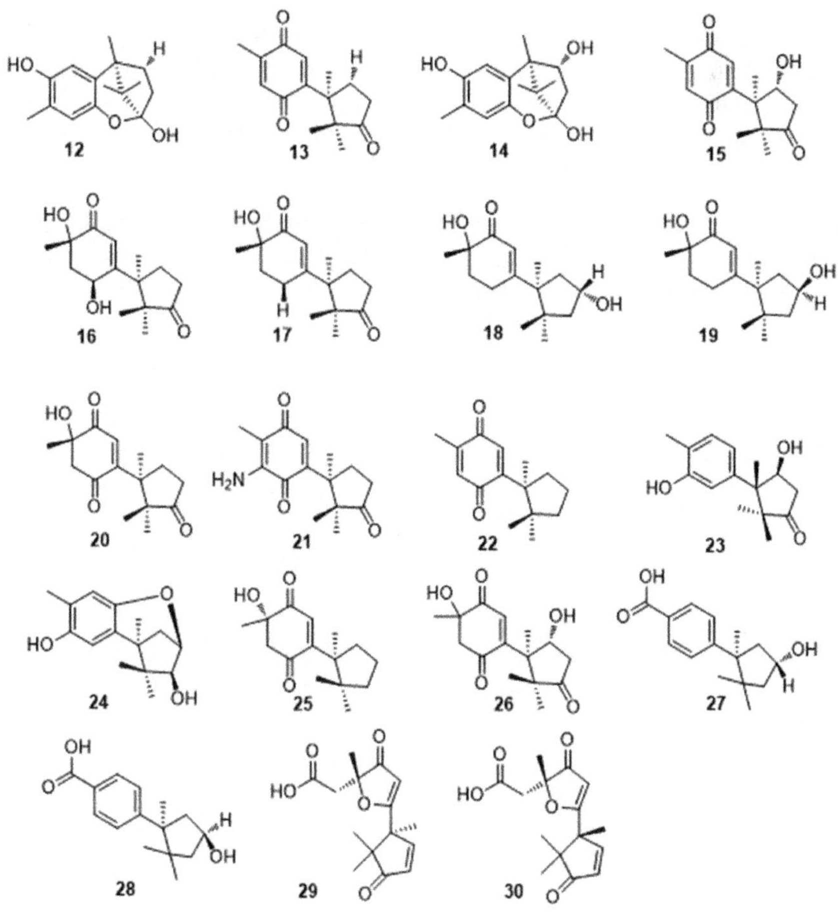

Figure 8.4 Molecular structures of cuparane and cuparane-type sesquiterpenes from *F. velutipes*.

29 and **30**, were obtained from the solid state fermentation of the fungal strain *F. velutipes* isolated from fruiting bodies collected from Yunnan–Guizhou. The seco-cuparane C5–C5 skeleton was formed through subsequent modification of the C6–C6 cuparane skeleton.[38]

8.5.1.2.2 Nor-eudesmane, Cadinane, and Spiroaxane-type Sesquiterpenes. A genome study conducted by Tao *et al.* confirmed that eudesmane, cadinene, and spiroaxane sesquiterpene backbones were initially biosynthesised from a germacrene cation intermediate through the 1,10 ring closure of farnesyl pyrophosphate.[38] Ten spiroaxane sesquiterpenes (Flammuspirones A–J, **31–40**), were obtained from a mycelial culture of *F. velutipes*, making these the second most common sesquiterpene skeletons present in *F. velutipes.*[38] Cadinane and nor-eudesmane type compounds, namely 7,13,14-trihydroxy-4-cadinen-15-oic acid methyl ester **42** and 1,2,6,10-tetrahydroxy-3,9-epoxy-14-nor-4(14)-eudesmane **41** were also obtained from the same culture[38] (Figure 8.5).

8.5.1.2.3 Isolactarane. One isolactarane (flammulinolide A, **47** (60)) and isolactarane-related norsesquiterpenes (flammulinolides B–G, **48–53**) were isolated from rice fermented by *F. velutipes* originating from fungal collection at the Institute of Microbiology, Chinese Academy of Sciences, Beijing, China (Figure 8.6).[44]

8.5.1.2.4 Sterpuranes. Three sterpurane sesquiterpenes (Figure 8.7), terpurols A **43** (0.16 mg per g dry extract), B **44** (0.8 mg per g dry extract), and sterpuric acid **45** (0.6 mg per g dry extract), were isolated from solid-

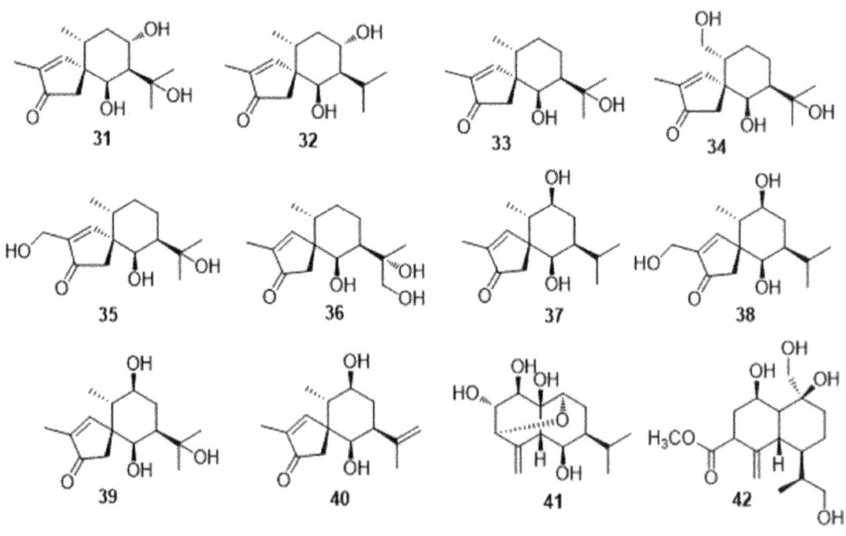

Figure 8.5 Nor-eudesmane, cadinane, and spiroaxane-type sesquiterpenes from *F. velutipes.*

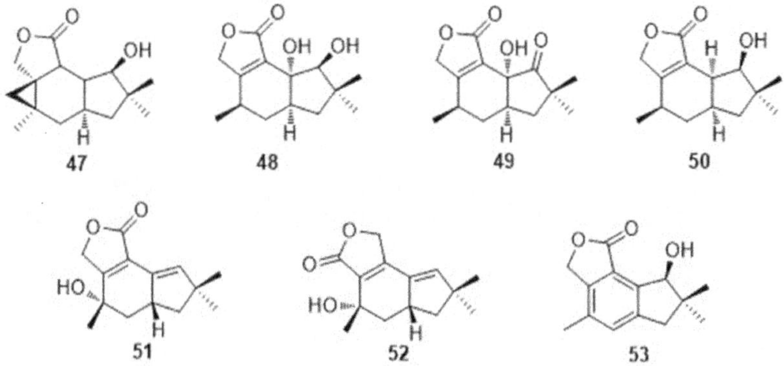

Figure 8.6 Molecular structures of isolactaranes isolated from *F. velutipes.*

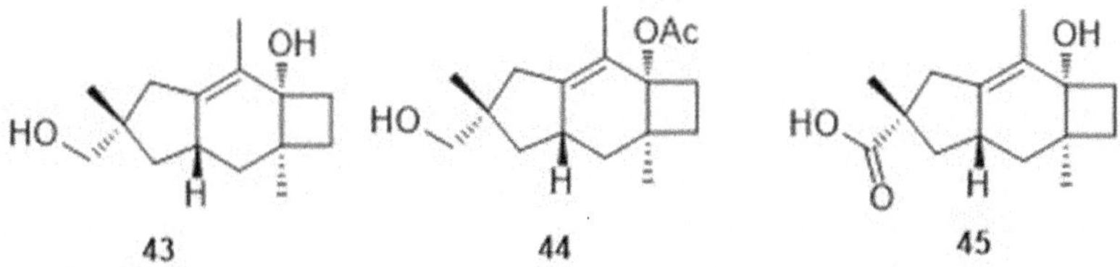

Figure 8.7 Molecular structures of sterpurane sesquiterpenoids from *F. velutipes.*

state grown *F. velutipes* supplied by the Institute of Microbiology, Chinese Academy of Sciences, Beijing, China.[42] These sterpurane sesquiterpenes were initially synthesized from the isolatactarane backbone, in which structural divergence was observed through cyclobutene formation.[38,42]

This distinct molecular feature, the cyclobutene ring, facilitated molecular elucidation.

8.5.1.2.5 Cucumanes. Yellow needles of flammulinol A **46** (Figure 8.8) were obtained at low yield (0.03 mg per g crude extract) from bioprospecting work on the solid-state grown of *F. velutipes* supplied by the Institute of Microbiology, Chinese Academy of Sciences, Beijing, China.[44] The three fused cyclopentane ring cucumane backbone of flammulinol A was proposed from the ring rearrangement of protoilludane forming ceratopicane prior to cucumane backbone formation.[38]

8.5.1.3 Nitrogen-containing Compounds

An exhaustive chromatography protocol on an ethyl acetate crude extract of the rice-media-grown Chinese strain *F. velutipes* (at the Institute of Microbiology, Chinese Academy of Sciences, Beijing, China) successfully led to the purification and isolation of the first norsesquiterpene alkaloid **54**

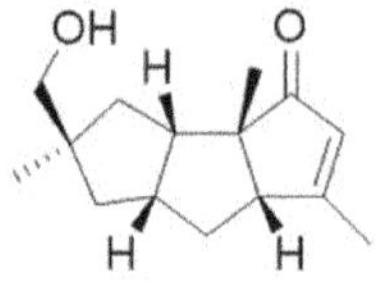

46

Figure 8.8 Molecular structures of cucumane sesquiterpenoids from *F. velutipes.*

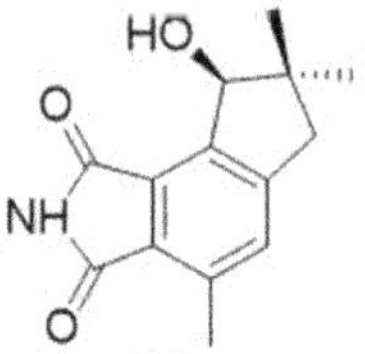

54

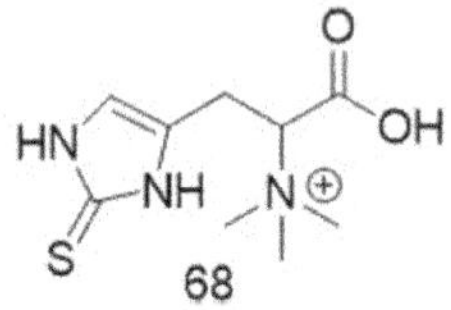

55

Figure 8.9 Molecular structures of alkaloids from *F. velutipes.*

(Figure 8.9). The molecular structure of the isolated white solid compound is derived from an isolactarane sesquiterpene, in which the presence of a cyclopentane ring was confirmed against flammulinolide G spectral data.[45] The naturally-occurring histidine derivative L-ergothioneine **55** (Figure 8.9) was isolated from different coloured *F. velutipes* species grown in China (snowy white, off-white, yellow, and snuff coloured *F. velutipes* grown at the farm of Zhejiang Academy of Agricultural Science) and determined based on liquid chromatography–mass spectrometry data analysis, with quantitative measurements in the range of 1.98–3.57 mg per g of dry mushroom.[46]

8.5.1.4 Monoterpenes

Monoterpenes were successfully isolated from the acetone extract of the fruiting body of *F. velutipes* grown on rice bran–sawdust media for almost 3 weeks. The terpenes were obtained as colourless oils, with their molecular structures unequivocally elucidated as (1*R*,2*R*,4*R*,8*S*)-(−)-*p*-menthane-2,8,9-triol **56** and its 8-epimer **57**, in yields of 14.2 and 5.0 mg, respectively, from an 8 kg fresh sample (Figure 8.10).[47]

8.5.1.5 Sterols

Ergosterol is a steroid component present in fungal cell membranes and is translated into fungal biomass; this can be used to estimate total biomass, although the content is ineffective for fungal taxa differentiation.[48] Through the saponification stage, a study of *F. velutipes* sample from Zhengdong Ecological Agriculture Development Center (Jiangsu, China) revealed ergosterols components that were characterized based on gas chromatography–

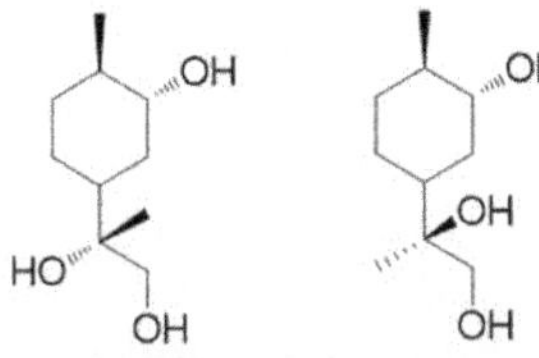

Figure 8.10 Molecular structures of monoterpenes from *F. velutipes*.

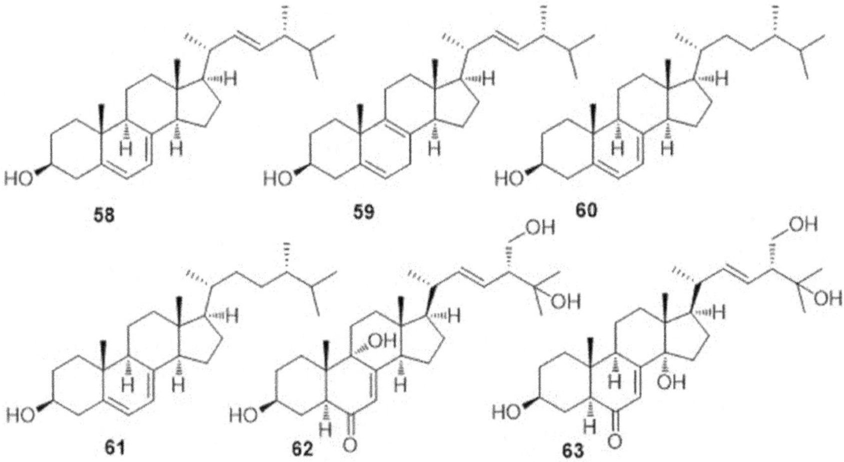

Figure 8.11 Molecular structures of sterols from *F. velutipes*.

mass spectrometric data analysis as ergosterol and its derivatives, namely ergosta-5,7,22-triene-3-ol (ergosterol) **58**, ergosta-5,8,22-triene-3-ol **59**, ergosta-5,7-diene-3-ol (22,23-dihydroergosterol) **60**, and ergosta-8(14)-ene-3-ol **61**. These ergosterols were found to constitute 54.78%, 11.27%, 27.94%, and 5.24% of total fungal steroids, respectively, with the two major components being successfully isolated through preparative HLPC.[49,50] Other ergosterol derivatives, namely 3,9,25-trihydroxy-24-hydroxymethylergosta-7,22-dien-6-one **62** and 3,14,25-trihydroxy-24-hydroxymethylergosta-7,22-dien-6-one **63** have been isolated from the fungal strain *F. velutipes* from fruiting bodies collected in the Yunnan–Guizhou plateau (Figure 8.11).[38]

8.5.2 Nutritional and Dietary Composition

Mushrooms are macrofungi species; spore-bearing fruiting bodies that can grow either above the soil (epigeous) or below the soil (hypogeous).[51] The species are rich in starches and proteins; however, they have a low-fat composition that is commonly composed of unsaturated fatty acids. The consumption of 100 g of mushroom provides around 22 calories.

In addition to their macronutrient value, mushrooms also contain important micronutrients, such as vitamin B-type molecules (*e.g.*, pantothenic acid, riboflavin, and niacin). Thus, mushrooms have been acknowledged as a vital nutrient source for promoting health and treating illnesses.[52,53] Since ancient times, mushrooms have been used as food and nutraceuticals by many Asian countries, including Japan, China, India, and Taiwan. The use of mushrooms has been well documented in a number of traditional medical practices, such as Japanese traditional medicine (Kampo), Indian traditional medicine (Ayurveda), traditional Chinese medicine (TCM), Tibetan traditional medicine (Amchi), and Korean medicine.[54] Historically, the Romans recognized mushrooms as the "Food of the Gods". In Greek civilization, mushrooms were believed to be a valuable food that could provide strength for warriors.[55]

Today, there is a high demand for edible mushrooms on the global market for daily consumption, accounting for around 34 billion USD production in 2013,[56] which increased to 35 billion USD in 2015, and reached 60 billion USD in 2021.[54] In recent years, the global consumption of cultivated mushrooms was estimated at 4.7 kg per capita per year.[57] Due to their healing properties, mushrooms are also regarded as the "elixir of life" in Chinese culture.[55] In general, mushrooms display all three functionalities of food: nutrition, taste, and physiological properties,[58] and are therefore considered to be both edible and medicinal species.[53] In addition to nutritional value, edible mushrooms have been acknowledged for their texture and aroma range (fruity, seafood, nutty, and curry).[59] The consumption of mushrooms is considered safe when no significant harmful effects are detected; however, several edible mushroom species have been re-classified as toxic.[60]

Flammulina velutipes, also known in different countries as enokitake, winter mushroom, golden needle mushroom, and velvet shank, holds high nutritional and medicinal value, as well as having a delicious taste. Several reports have shown that the dietary composition of *F. velutipes* includes carbohydrates, proteins, minerals, crude fiber, and vitamins. It also contains several bioactive secondary compounds, including terpenoids (*e.g.*, sesquiterpenoid enokipodins A–D), phenolic acids, and flavonoids (*e.g.*, kaempferol, apigenin, and epicatechin).[61] *F. velutipes* is one of the most commonly cultivated mushrooms and its production is predicted to increase in the near future, along with *Auricularia auriculajudae*, *Lentinula edodes*, *Agaricus bisporus*, and *Pleurotus ostreatus*.[57] The cultivated species of *F. velutipes* has a pure white bean sprout appearance with a velvety stem and a small snowy-white cap, while the wild species is usually orange to brown in appearance with a larger, shiny cap[62] (Figure 8.1).

8.5.2.1 Carbohydrates

Recent studies have suggested that carbohydrates generally constitute about one-half (estimated at 58% of mushroom dry matter) of the total weight of *F. velutipes*. However, the cultivated species has a slightly higher amount of carbohydrate than the wild type. *F. velutipes* is native to Asia, with some major producers, including China, Japan, and South Korea.[51] The carbohydrates in

F. velutipes can be broadly divided into monosaccharides (*i.e.*, glucose, fucose, ribose, mannose, galactose, and fructose), disaccharides (*i.e.*, sucrose and trehalose), and polysaccharides (*i.e.*, starch, chitin, and β-glucan).[9] Among the bioactive contents of mushrooms, mushroom polysaccharides are one of the promising nutritive substances that possess excellent nutritional and therapeutic properties. To date, a variety of polysaccharides have been isolated and characterized from various mushroom strains, thanks to the advanced development of agro-food technologies (advanced production and extraction techniques, and biomolecular tools (*e.g.*, mass spectrometry, chromatographic methods, nuclear magnetic resonance (NMR) spectroscopy, and DNA sequencing and fingerprinting).[54] In recent years, mushroom polysaccharides have attracted a lot of attention from both immunobiologists and chemists, and have become a research hotspot due to their fantastic prospects in the food and pharmaceutical industries.[63,64] Some mushroom polysaccharides have been subjected to clinical trials and are used in clinical settings to treat numerous ailments in Asia, including cancers.[65]

Accumulating evidence shows that *F. velutipes* polysaccharides extracted from fruiting bodies and mycelium exhibit some pharmacological properties and health-promoting effects, such as antimicrobial, antioxidant, antitumor, anti-inflammatory, immunomodulatory, and intestinal microbial regulation properties.[63,66] Nevertheless, since different extraction and purification methods are used to isolate the polysaccharide content from *F. velutipes*, the bioactivities exhibit certain differences.[64] Additionally, the beneficial values of compounds from *F. velutipes* can be greatly influenced by the growing environment, substrate type, maturity of the species at the harvesting stage, and post-harvest conditions, such as processing and storage methods. Thus, these variables have been linked to the variability in the bioactive composition data published despite the same *F. velupites* mushroom having been investigated.[9]

Wu *et al.* reported the polysaccharide components in *F. velutipes* using high-performance liquid chromatography (HPLC) and NMR spectroscopy. The authors found that the monosaccharide contents in *F. velutipes* included glucose, mannose, and xylose in a molar ratio of 3.5 : 0.8 : 1.4. Moreover, it was demonstrated that the polysaccharide components significantly reduced the CD4+, CD8+, ICAM-1, and MPO in the serum and colon of normal and burned male Wistar rats.[67] Xin *et al.* studied the effect of *F. velutipes* polysaccharides on an animal model of constipation. The results showed that treatment of *F. velutipes* polysaccharides increased intestinal peristalsis and contraction of the intestinal smooth muscle.[68]

8.5.2.2 Proteins and Amino Acids

Generally, mushrooms are rich in protein (19–35%) and contain all nine essential amino acids. Owing to the presence of essential amino acids, edible mushrooms have been proposed as a potential source of protein to combat protein deficiency and malnutrition.[58] The more abundant amino acids found in mushrooms are valine, glutamic acid, arginine, glutamine, and aspartic acid, while methionine is the minor constituent. The protein content in mushrooms is approximately four-fold higher than that of tomatoes and carrots, six-fold higher than that of oranges, and 12-fold higher than that of apples.[53,58] Mushrooms can produce different types of proteins and peptides, including fungal immunomodulatory proteins (FIP), lectins, ribonucleases, antimicrobial proteins, and laccases.[65] Some proteins in mushrooms, in general, possess immunomodulatory properties, either from proteins working alone or jointly with β-glucans.[69] Specifically, *F. velutipes* contains approximately 178.9–279.5 g per kg dry weight of protein. The protein composition of *F. velutipes* is similar to that of many green leafy vegetables, in the range of 200–300 g per kg dry weight, suggesting that *F. velutipes* may be valuable as an alternative protein source.[9] Furthermore, *F. velutipes* contains fungal immunomodulatory proteins that have been proposed against immunoprophylaxis in allergic diseases.[69]

Several studies have demonstrated that the most common amino acids found in *F. velutipes* are L-glutamic acid, L-alanine, glycine, and L-lysine, contributing to 2.6–3.0% of its dry weight. Moreover, other amino acids have been found in this fungus, including valine, leucine, isoleucine, methionine, threonine, and phenylalanine.[9] These amino acids, especially aspartic and glutamic acid, along with nucleotide compounds (5′-xanthosine monophosphate, 5′-guanosine monophosphate (5′-GMP), and 5′-inosine monophosphate (5′-IMP)), play a substantial role in influencing the pleasant taste of *F. velutipes*, which responsible for its umami or monosodium glutamate-like taste. Umami, also known as a "delicious savory taste" improves satiety and taste, and is recognized as the fifth basic taste after salty, bitter, sour, and sweet. Therefore, edible mushrooms have been used as an ingredient and additive replacement in a variety of food products, such as meat, flour, fat, and salt. Recently, several studies revealed the potential use of edible mushrooms as a meat replacement. For example, *F. velutipes* has been used as a meat substitute in goat meat nuggets. In this example, it was suggested that replacing 2–6% of goat meat with *F. velutipes* increased the water holding capacity, leading to a softer nugget texture.[70]

8.5.2.3 Dietary Fiber

Dietary fiber is part of the edible portion of analogous carbohydrate of *F. velutipes*. This component cannot be decomposed by digestive and/or enzymatic lysis in the human gastrointestinal tract. Dietary fiber generally possesses beneficial physiological properties, such as blood glucose and blood cholesterol attenuation.[51,71] Yeh and colleagues demonstrated that the total fiber content, the insoluble content, and the water-soluble content in *F. velutipes* were 29.34 mg per 100 g, 15.28 mg per 100 g, and 14.09 mg per 100 g, respectively. The authors also found that the fiber content extracted from *F. velutipes* accelerated the decomposition of cholesterol in hamsters, and therefore reduced triacylglycerol and low-density lipoprotein cholesterol levels.[51] Mushroom cell walls, in particular, non-digestible carbohydrates, are considered as a source of dietary fiber. It contains fibrillar and matrix component complex, which include chitin (a straight-chain (1→4)-β-linked polymer of *N*-acetyl-glucosamine) and polysaccharides, such as (1→3)-β-D-glucans and mannans, respectively. Studies have found that mushroom fiber has a broad range of health benefits, including lipid and blood glucose attenuation, immunomodulatory effects, and antitumour properties.[71]

8.5.2.4 Lipid and Fatty Acids

The importance and health benefits of unsaturated fatty acids for humans have been demonstrated in many studies. Fatty acids, especially the long-chain poly-unsaturated form, have been beneficial in reducing blood cholestrol and regulating cell physiology.[59] In general, lipid content in mushrooms is quite low, 1.75–15.5% per 100 g dry weight, and therefore mushrooms have been regarded as a healthy-low fat food.[72] Several studies have revealed the fatty acid composition of edible mushrooms and investigated their nutritional value in the human diet. Generally, mushrooms contain higher levels of polyunsaturated fatty acids than vegetables and foods of animal origin. Linoleic, linolenic, and oleic acids are the most important essential fatty acids found in mushrooms. These fatty acids have been reported from *F. velutipes* species across different countries, including Turkey, Israel, Portugal, and South Korea (Table 8.1).[59]

Ergönül *et al.* investigated the fatty acid compositions of six wild edible mushroom species (Basidiomycetes): *F. velutipes*, *Boletus reticulatus*, *Lactarius salmonicolor*, *Polyporus squamosus*, *Pleurotus ostreatus*, and *Russula anthracina* from different areas of Turkey. They found that the levels of saturated, monosaturated, and polyunsaturated in *F. velutipes* were 20.67, 18.58, and 60.65% of the total fatty acids (dry basis), respectively. In this study, 37 fatty acids were detected in *F. velutipes*, with the major constituents including palmitic acid (14.56%), *cis*-oleic acid (16.37%), and *cis*-linolenic acid (40.87%), as determined by gas-liquid chromatography equipped with flame ionization detection (GLC-FID).[72]

8.5.2.5 Micronutrients

Elements such as potassium, phosphorous, magnesium, calcium, sodium, and zinc are among the crucial ions to human bodily fluid and tissues, regulating numerous enzyme functions.[73] Several major elements found in edible mushrooms include potassium, phosphorus, chlorine, sulfur, magnesium, calcium, and sodium. Potassium is the most predominant element in mushrooms, and its accumulation increases in the order of: spore, spore-forming part, stipe, and cap. In addition, other elements have been reported from several edible mushrooms, including aluminum, cadmium, cesium, copper, chromium, iodine, iron, and zinc.[74] Indeed, the amounts of several trace elements (*e.g.*, potassium and phosphorous) are typically higher than in

Table 8.1 Percentages (in 100 g of total fatty acids) in *F. velutipes* collected from different countries.

Percentages (in 100 g of total fatty acids)			
Linoleic acid (%)	Linolenic acid (%)	Oleic acid (%)	Country
40.9	0.0	16.4	Turkey
51.2	13.0	10.7	Israel
56.3	1.0	15.1	Portugal
63.7	—	5.6	South Korea

most vegetables. It has also been reported that mushrooms are rich in vitamins B1, B2, and B12, C, D, E, niacin, and folate. Despite being grown in areas with low exposure to sunlight, mushrooms contain a high level of vitamin D, and when exposed to more sunlight or ultraviolet (UV)-B light, their vitamin D content, particularly D2, increases significantly.[58] It is noteworthy that wild-type mushrooms generally constitute a higher level of trace elements than cultivated species.[75] Smiderle *et al.* determined the mineral element content of cultivated *F. velutipes* obtained at the Public Municipal Market of Curitiba, the State of Paraná (PR), Brazil. The authors demonstrated that potassium, phosphorous, magnesium, and calcium were the major constituents found in *F. velutipes*, while other elements, such as sodium, manganese, iron, zinc, selenium, and lithium, were trace elements.[15]

8.6 Pharmacological Potential

F. velutipes has been reported to provide health benefits for humans as a source of antitumor, anticancer, anti-hypercholesterolemic, immunomodulatory, antioxidant, antimicrobial, anti-inflammatory, anti-neurological disorder, protective, anti-melanosis, and anti-obesity properties. Although it has numerous positive effects, there are also some adverse effects to consider.

8.6.1 Antitumor and Anticancer Properties

Edible and medicinal mushroom fruit bodies are well known as medicaments used to complement chemotherapy and radiotherapy by attenuating the numbers of side effects experienced by the patient, including nausea, bone marrow suppression, anaemia, and lowered resistance.[76] In fact, numerous studies have revealed that the diverse chemical components of mushrooms are cytotoxic to arrays of malignant cell lines.

Aqueous extracts of *F. velutipes* have indicated significant activities against a series of human breast cancer cell lines, including ER+, MCF-7, ER−, BT-20, and MDA-MB-231. According to the Annexin V-FITC assay, *F. velutipes* extract triggered fast apoptosis in MCF-7 and MDA-231 cells within 2 h.[77] The conditioned medium of human peripheral blood mononuclear cells treated with *F. velutipes* cold-water extracts significantly suppressed the proliferation of human leukemic U937 cells.[78] In addition, the acetone extract was more cytotoxic against MCF-7 (IC_{50} ranged from 17.7 to 38.36 μg mL^{-1}) and MDA-MB-231 (IC_{50} ranged from 114.5 to 184.2 μg mL^{-1}) than it was against MCF-10a (IC_{50} > 250 μg mL^{-1}).[79]

A polysaccharide component of the mushroom stipe was shown to significantly enhanced macrophage proliferation and phagocytic activity and inhibited the growth of cancer cells.[80] The polysaccharide showed significant antitumor effect against sarcoma SC-180 in a mouse model, with only 54.7–58.1% tumor cell proliferation at 24 mg per kg dose of treatment.[81,82] The polysaccharide extract might suppress cell division and reduce cell proliferation based on mitosis index, mitochondria activity, and cell proliferation study using a human hepatoma SMMC-7721 cell model.[82] Polysaccharides EA5, EA500, and EA501 isolated from the mushroom were able to inhibit solid ascites sarcoma 180 cells implanted in ICR mice by 84%, 84%, and 85%, respectively.[83] Administration of a hot water extract containing polysaccharide (EA6) (50 mg kg^{-1}) was able to extend the lives of mice intraperitoneally injected with ascites sarcoma 180.[84] Polysaccharides FVP-1 and FVP-2, isolated from *F. velutipes*, strongly inhibited the proliferation of BGC-823 cells in a concentration-dependent manner, with values of 78% and 95%, respectively, at a concentration of 200 g mL^{-1}.[85] The water-soluble polysaccharide of *F. velutipes* polysaccharide I-A can stimulate RAW264.7 macrophages to generate cytokines (interleukin-1, interleukin-6, and tumor necrosis factor-α) *in vitro*.[86]

Several antitumor proteins were isolated from an *F. velutipes* mycelium culture. Proflamin, a non-toxic mildly acidic glycoprotein antitumor agent, was discovered in mushroom mycelia and has a molecular weight of around 13 kDa and contains 90% protein and <10% carbohydrate. The protein is an effective antitumor agent against syngeneic malignancies, B-16 melanoma (B-16), and adenocarcinoma 755 (Ca-755), in which an oral dose of 10 mg kg^{-1} increased the median survival times of mice by 86% and 84%, respectively.[87] A hemagglutinin, with a molecular mass of 12 kDa, elicited a weak mitogenic response and inhibited proliferation of L1210 cells with an IC_{50} of 13 μM.[88] Protein FIP-fve is a fungal immunomodulatory protein, the oral treatment which (10 mg kg^{-1}) significantly increased the survival of BNL 1MEA.7R.1 (BNL) hepatoma-bearing mice as it triggered the innate and adaptive immune systems to improve the tumoricidal capability of macrophages and splenocytes from the mouse peritoneum.[89] FIP-fve inhibited the proliferation of A549 lung cancer cells in the MTT assay after 48 h of treatment. On flowcytometry, the FIP-fve treatment similarly induced cell cycle arrest but not apoptosis. FIP-fve suppressed A549 cell migration in a wound-healing assay and reduced the formation of filopodia fibers by reducing RACGAP1 expression and suppressing proliferation by activating the p53 pathway.[90]

Fungal sterol (FVS), ergosterol **58**, 22,23-dihydroergosterol **60**, and ergost-8(14)-ene-3-ol **61** are active against human glioma (U251) and human cervical carcinoma (HeLa) cell lines. FVS showed a significant inhibitory effect on U251 cells, with an IC_{50} value of 23.42 $\mu g\,mL^{-1}$ and was ineffective against the HeLa cells.[49]

Numerous novel bioactive compounds with anticancer properties have been discovered from *F. velutipes* cultures (Table 8.2). Wang *et al.*[44] successfully isolated flammulinolides A–G **47–53** and tested them against HeLa (human cervical adenocarcinoma), HepG2 (human hepatocellular liver carcinoma), and KB (human nasopharyngeal carcinoma) cell lines. The KB cell line was sensitive to flammulinolides A **47**, B **48**, and F **52**, indicated by IC_{50} values of 3.9, 3.6, and 4.7 μM, respectively. With an IC_{50} of 3.0 mM, flammulinolide C exhibited potent cytotoxicity against the HeLa cell line. Compounds flammulinolide A **47**, flammulinolide B **48**, and flammulinolide F **52** exhibited moderate cytotoxicity against the HepG2 cell line. Flammulinolide E **51** and flammulinolide G **53** showed moderate cytotoxicity against the HeLa and KB cell lines as well as flammulinolide C **49** against the KB cell line. Further work on solid culture yielded enokipodins E–J **16–21**, sterpurols A **43** and B **44**, 2,5-cuparadiene-1,4-dione **22**, enokipodin B **13**, enokipodin D **15**, and sterpuric acid, among which only enokipodin B **13**, D **15**, J **21**, and 2,5-cuparadiene-1,4-dione **22** showed cytotoxicity with moderate activity against HepG2 (liver cancer), MCF-7 (breast cancer), SGC7901 (stomach cancer), and A549 (lung cancer) cell lines.[44] The inhibitory activities of enokipodins A **12**, B **13**, C **14** and D **15**, produced by *F. velutipes*, against several cancer cell lines (HeLa, HL60, tsFT210, and tsNRK) has also been investigated. Enokipodins B **13** and D **15** had a higher antiproliferative effect than Enokipodins A **12** and C **13** against a HeLa, HL60, tsFT210, and tsNRK cell lines, with IC_{50} values of 4.9–18.7 μM.[41] The alkaloid sesquiterpene (*R*)-8-hydroxy-4,7,7-trimethyl-7,8-dihydrocyclopenta[*e*]isoindole-1,3(2*H*,6*H*)-dione **54** was isolated from solid culture and showed moderate cytotoxicity against KB cells, with an IC_{50} value of 16.6 μM.[45]

8.6.2 Antihypercholesterolemic and Antihyperglycemic Properties

Mushroom fiber of *F. velutipes* is able to increase hepatic LDL receptor mRNA concentrations and fecal cholesterol excretion while decreasing HDL cholesterol levels in a rat model. This impact may be related to decreasing ileal concentrations of apo A-I and apo A-IV mRNA or its action in increasing HDL receptor activity in adrenals/gonads.[91] Another *in vivo* study revealed that fiber promotes the breakdown of cholesterol and reduces the levels of TG (triacylglycerol), TC (total cholesterol), and LDL (low-density lipoprotein) in the blood and TG, TC, and phospholipids in the liver.[92]

In vitro studies were conducted on the hypoglycemic and hypolipidemic effects of *F. velutipes* compounds based on their inhibitory effect against HMG-CoA reductase, dipeptidyl peptidase-4 (DPP-4), and aldose reductase enzymes (Table 8.3), with the discovery that flammuspirone A **31**, flammuspirone C **33**, 7,13,14-trihydroxy-4-cadinen-15-oic acid methyl ester **42**,

Table 8.2 Anticancer activities of chemical constituents of *Flammulina velurtipes* against several cancer cell lines.[41,42,44,45,93]

	Cell lines (IC_{50}, μM)								
Compounds	HeLa	HepG2	KB	MCF-7	A549	SGC7901	HL60	tsFT210	tsNRK
Flammulinolide A **47**	>200	34.7	3.6						
Flammulinolide B **48**	>200	79.0	3.6						
Flammulinolide C **49**	3.0	>200	12.4						
Flammulinolide D **50**	>200	>200	>200						
Flammulinolide E **51**	59.5	>200	103.9						
Flammulinolide F **52**	>200	123	4.7						
Flammulinolide G **53**	25.8	>200	74.9						
Enokipodin A **12**	302.2						12.9	359.5	>403.0
Enokipodin B **13**	289.6	22.9		59.2	32.1	40.6	22.3	>378.6	>378.6
Enokipodin C **14**	4.9						1.6	8.9	8.9
Enokipodin D **15**	11.8	34.6		34.9	24.8	64.5	3.2	18.7	18.7
Enokipodin J **21**		23.4		33.0	36.8	47.5			
2,5-Cuparadiene-1,4-dione **22**		34.5		38.4	47.8	90.9			
(*R*)-8-Hydroxy-4,7,7-trimethyl-7,8-dihydrocyclopenta[*e*] isoindole-1,3 (2*H*,6*H*)-dione **54**			16.6						

Table 8.3 Antihypercholesterolemic and hyperglycemic secondary metabolites of *F. velutipes*.[93]

Compounds	Inhibition (IC_{50}, μM) HMG-CoA reductase	DPP-4	Aldose reductase
Flammufuranone A **29**	>150	>100	>150
Flammufuranone B **30**	>150	>100	>150
Flammuspirone A **31**	114.7	>100	>150
Flammuspirone B **32**	>150	>100	>150
Flammuspirone C **33**	77.6	75.9	>150
Flammuspirone D **34**	>150	83.7	>150
Flammuspirone E **35**	>150	83.7	>150
Flammuspirone F **36**	>150	>100	>150
Flammuspirone G **37**	>150	>100	>150
Flammuspirone H **38**	>150	79.7	>150
Flammuspirone I **39**	>150	>100	>150
Flammuspirone J **40**	>150	>100	>150
7,13,14-Trihydroxy-4-cadinen-15-oic acid methyl ester **42**	>55.5	80.5	>150
1,2,6,10-Tetrahydroxy-3,9-epoxy-14-nor-4(14)-eudesmane **41**	87.1	74.8	>150
Flamvelutpenoid E **27**	>150	>100	>150
Flamvelutpenoid F **28**	>150	>100	>150

and 1,2,6,10-tetrahydroxy-3,9-epoxy-14-nor-4(14)-eudesmane **41** inhibited HMG-CoA reductase, with IC_{50} values of 114.7, 77.6, 55.5, and 87.1 μM, respectively. The compounds flammuspirone C **33**, flammuspirone D **34**, flammuspirone E **35**, flammuspirone H **38**, 7,13,14-trihydroxy-4-cadinen-15-oic acid methyl ester **42**, and 1,2,6,10-tetrahydroxy-3,9-epoxy-14-nor-4(14)-eudesmane **41** demonstrated DPP-4 inhibiting action, with corresponding IC_{50} values of 75.9, 83.7, 70.9, 79.7, and 80.5 μM, but none of them showed any inhibitory activity against aldose reductase.[93]

8.6.3 Immunomodulatory Properties

Flammulina velutipes could be a valuable immunomodulatory pharmaceutical or nutraceutical. Carbohydrates obtained from *F. velutipes* biomass had a stimulating effect on macrophage-mature cells of the mononuclear phagocyte system that participate in the realization of inherent and acquired immunity mechanisms, and thus could be beneficial in enhancing the organism's immunity under various pathological conditions.[94] Cold water extract (CWE) from *F. velutipes* was revealed to be a potent inducer of monocyte-associated antigens, promoting TNF-α and IL-1β production in human blood peripheral mononuclear cells, suggesting that the mushrooms boosted cell-mediated immunity by activating T-lymphocytes, B-lymphocytes, and macrophages, and may act as an immunomodulator.[78] Another animal model study revealed the purified FvPB2 polysaccharide enhanced dose-dependent proliferation of mouse spleen cells and increased quantities of IgM and IgG produced by B cells.[95]

Proteins from *F. velutipes* were able to reduce IL-4 significantly and IFN- after stimulation with PMA/Ca-I.[96] The potential use of *F. velutipes* immunomodulatory protein (FIP-fve) as an adjuvant for tumor immunotherapy has been investigated, in which the immunoprophylactic effects of FIP-fve in Th2-mediated allergy anaphylaxis are connected with the activation of IFN-γ-releasing Th1 cells promoted by FIP-potential fve.[97] FIP-fve promotes fast elevation in calcium levels, as Ca^{2+} signaling is one of the FIP-fve-stimulated IFN-γ-events.[98] In an *in vivo* study, mice co-immunized with HPV-16 E7 and FIP-fve exhibited increased expression of HPV-16 E7-specific antibodies and increased HPV-16 E7-specific interferon (IFN)-c-producing CD4+ and CD8+ T cells. Assay of tumor protection showed that 60% of mice co-immunized with HPV-16 E7 and FIP-fve remained tumor-free after tumor challenge, in which FIP-fve may have contributed to the enhanced HPV-16 E7-specific T-cell immunity and higher IFN-c production of these T cells.[99]

8.6.4 Antioxidants

A methanol extract of *F. velutipes* has been shown to exhibit relatively low antioxidant activity.[100] Moreover, an investigation on a water extract at a concentration of 1 mg mL^{-1} showed potent antioxidant effect, with a 99.7% DPPH scavenging rate, 46.4% chelating ability of iron, and 68.1% ferric reducing antioxidant power.[92] Polysaccharide components showed exceptional antioxidant activity to eliminate hydroxyl, DPPH, and ABTS radicals, exhibited high reducing power and reduced malondialdehyde (MDA) formation in healthy mice liver homogenate caused by auto-oxidation and the Fe^{2+}–H_2O_2 system.[101] Extracts from waste culture medium of *F. velutipes* have been shown to exhibit strong actions against lipid oxidation, oxyMb antioxidation, and phenolic substances, primarily due to the antioxidative properties of the extracts.[102]

Antioxidant molecules isolated and characterised from *F. velutipes* are ergothioneine (ESH) and phenolic compounds.[103,104] ESH is highly deposited in the tip of the stipe and the bottom (base) of *F. velutipes*, with a concentration of around 60 mg per kg dry weight. The individual compounds of enokipodins B **13**, D **15**, J **21**, and 2,5-cuparadiene-1,4-dione **22** were evaluated against DPPH radicals and showed moderate antioxidant activity, with concomitant IC_{50} values of 78.6, 80.7, 154.2, and 116.5 μM, respectively.[42]

8.6.5 Antimicrobial and Antiparasitic Properties

Many studies on the antimicrobial effect of *F. velutipes* have shown moderate or weak activity, even no activity. An antibacterial study on a crude aqueous extract of *F. velutipes* showed a narrow antibacterial spectrum against Gram-negative bacteria, but it was extremely effective at preventing the growth of

Gram-positive bacteria.[100,105] An antimicrobial examination of FvP, FvP2, FvP-Fe, and FvP2-Fe polysaccharides against *S. aureus*, *E. coli*, and *B. subtilis* resulted in strong inhibition by FvP-Fe and FvP2-Fe, with no activity against Rhizopus and Aspergillus fungi.[101]

Ferreira *et al.*[106] initially proved that the edible fungus *F. velutipes* exhibits nematophagous activity. *F. velutipes* demonstrated a substantial nematicidal effect on *Panagrellus* sp. larvae, as determined by evaluating the nematicidal activity of an extract on the model nematode *Panagrellus* sp.

F. velutipes immunomodulatory protein (FIP-fve), a naturally occurring substance derived from *F. velutipes*, has been indicated to be a potential agent for viral prophylaxis therapy. A study indicated that FIP-fve inhibited RSV replication and provided substantial evidence that FIP-fve is an effective RSV infection suppressor.[107] Water-soluble polysaccharide (FVP1) extract demonstrated strong anti-HBV action by inhibiting the expression of hepatitis B surface antigen (HBsAg), hepatitis B e antigen (HBeAg), and hepatitis B virus (HBV) DNA replication.[108]

Antimicrobial evaluation of secondary metabolites indicated enokipodins A–D to exhibit antibacterial activity against Gram-positive bacteria (*Bacillus subtilis* and *Staphylococcus aureus*), but not against Gram-negative pathogens (*Escherichia coli* and *Pseudomona fluorescens*).[40] The minimum inhibitory doses (MIDs) of enokipodins A and C for *B. subtilis* were comparable to that of antibiotic penicillin G.[109] Enokipodins A, B, and C were also evaluated for anti-phytopathogenic potency against *Pyriculari oryzae* and *Candida albicans*, in which enokipodin B displayed the most significant antifungal activity. In addition, enokipodin B was also able to prevent spore germination. These enokipodins were also tested for antiparasitic activity against the *Plasmodium falciparum* strain 3D7, in which enokipodins A **12** and C **14** showed relatively selective antimalarial activity, with IC_{50} values of 9.7 and 4.2 μM, respectively.[41] Another study showed enokipodins B **13**, D **15**, I **20**, J **21**, and 2,5-cuparadiene-1,4-dione **22** to exhibit mild antibacterial activity against *B. subtilis* ($IC_{50} = 140.5–167.6$ μM), whereas enokipodins F **17**, G **18**, and I **20** showed weak antifungal activity against *Aspergillus fumigatus* ($IC_{50} = 229.1–235.1$ μM).[42]

The antibacterial activity of flamvelutpenoids A–D **23–26** was investigated against *Escherichia coli*, *Bacillus subtilis*, and *methicillin-resistant Staphylococcus aureus* (MRSA) with the compounds exhibiting weak antibacterial activity.[110] Flammulinolides A–G **47–53** were tested against *B. subtilis* and MRSA, with all of the compounds exhibiting only a moderate level of activity, with IC_{50} values of >250 μM.[44]

8.6.6 Anti-inflammatory Properties

FIP-fve could be an effective therapeutic agent for allergic asthma. In a mouse asthma model, oral doses of FIP-fve suppressed allergen ovalbumin (OVA)-induced chronic airway inflammation. FIP-fve decreased airway hyperresponsiveness and dramatically reduced the number of infiltrating inflammatory cells and Th2 cytokines in bronchoalveolar lavage fluid (BALF) and serum. FIP-fve also lowered serum concentrations of OVA-specific IgE and suppressed inflammatory cell infiltration in lung histopathology studies using hematoxylin and Liu's staining.[111] A further study revealed that oral FIP-fve exhibited anti-inflammatory activity on OVA-induced airway inflammation and an inhibiting effect on Th17 cells, decreasing airway remodelling and collagen expression.[112]

In Balb/c mice, FIP-fve demonstrated a reduction in Der m-induced airway AHR, airway inflammation, cell infiltration, and cytokine expression in bronchoalveolar lavage fluid. IL-1, IL-4, IL-5, IL-6, IL-7, IL-10, IL-13, IL-12 p70, IL-23, IL-27, the chemokine (C-X-C motif) ligand 1 (CXCL1), macrophage inflammatory protein-1 alpha (MIP-1), and TNF- were up-regulated by Der m and decreased by FIP-fve treatment. FIP-fve treatment was able to control asthma, inflammation, and the pathophysiology of the respiratory system as triggered by Der m.[113] A similar study revealed FIP-fve to significantly decrease HDM-specific immunoglobulin E in serum and decrease production of Th2 cytokines in bronchoalveolar lavage fluid and serum.[114] The inflammatory chemicals in *F. velutipes* are sensitive to heat, as a fresh sample showed considerably greater anti-inflammatory activity than processed fungus.[115]

8.6.7 Anti-neurological Disorder Properties

Flammulina velutipes polysaccharide (FvP) exhibited potential ability to improve learning, memory impairment, and cognitive function. FvP showed a moderate AchE inhibition effect, with an 18.51% inhibitory rate. These results suggested that FVP may have a potential application in enhancing cognitive impairment and memory (Yang *et al.*, 2011). Morris water maze and biochemical analyses of rat brains were utilized to evaluate the effects of FvP on scopolamine-induced impairments in learning and memory, in which FvP decreased the escape latency and total swimming distance of rats. Biochemical assay suggested that FVP significantly increased the activity of SOD and GSH-Px and neurotransmitter content. In addition, learning and memory-related signaling pathways were activated by FVP through an increase in connexin 36 and p-CaMK II expression. These results demonstrated that FVP is an effective agent against cognitive impairment progression.[116] Based on a rat model experiment, the effectiveness of FvP was significantly increased against Alzheimer's disease in combination with ginsenosides.[117] These findings provide scientific evidence to support the use of FVP as a safe and effective treatment for the prevention and treatment of Alzheimer's disease.

The flavonoids from *Flammulina velutipes* (FvF) identified contained arbutin, epicatechin, phillyrin, apigenin, kaempferol, and formononetin. FVF protects PC12 cells from neurotoxicity caused by H_2O_2, indicating that a neuroprotection pathway may be possible. FvF pretreated PC12 cells showed increases in cell survival, glutathione level, and activity of superoxide dismutase activity, while decreasing the release of lactate dehydrogenase and intracellular reactive oxygen species and malondialdehyde accumulation.[118]

Formononetin decreased levels of reactive oxygen species and malondialdehyde equivalent, whereas phillyrin increased glutathione and superoxide dismutase activity. All substances, except for apigenin, reduced the expression of BCL2-associated X protein and cytochrome-*C* and increased the expression of B-cell lymphoma-2 protein.[119]

8.6.8 Anti-melanosis Properties

An investigation of extract from *F. velutipes* against tyrosinase activity indicated that ethyl acetate and *n*-butanol extracts demonstrated potent tyrosinase-inhibiting activity.[120] A further study on the ethyl acetate extract revealed 1′,3′-dilinolenoyl-2′-linoleoylglycerol (LnLLn) to be the component with the highest tyrosinase inhibitory activity, with an IC_{50} value of 16.1 $\mu g\, mL^{-1}$.[121] Examination of melanosis-inhibiting effects of a crude water-soluble extract (ergothioneine (ESH)-rich extract) derived from the base and fruiting body waste on live crab, mature black tiger shrimp, and Pacific white shrimp indicated reduction in postharvest melanosis by decreasing phenoloxidase (PO) and prophenoloxidase (proPO) gene expression in hemocytes. Copper-chelating activity investigations have suggested that ESH may interact directly with Cu^{2+} at the putative binding sites of PPO and proPO, hence preventing shrimp melanosis.[122,123] Overall, ESH-rich or ethyl acetate extracts can be developed as an active ingredient for commercial whitening agents.

8.6.9 Anti-obesity Properties

The effect of the edible fungus *F. velutipes* as an anti-obesity agent has been explored based on its inhibitory effect on intracellular lipid accumulation. An acetone fraction of *F. velutipes* extracts indicated an inhibitory effect on adipose differentiation and intracellular lipid accumulation. Seven compounds, as methyl linoleate, methyl γ-linolenate, ergosterol **58**, ergosta-5,7-dien-3β-ol **60**, ergost-7-en-3β-ol, ergosterol peroxide, and 5,6-epoxy-24(*R*)-methylcholesta-7,22-dien-3β-ol were successfully isolated and characterized Ergosterol peroxide had the most potent inhibitory effect on intracellular lipid accumulation.[124] A mouse model study showed the anti-obesity activity of *Velutipes* polysaccharide (FvP) was related to its capability in increasing diversity and a change in fecal microbiota composition.[125]

8.7 Cultivation and Postharvest

8.7.1 Cultivation

Mushroom cultivation is an important economic activity. Wild and cultivated mushrooms have long been consumed due to their medicinal value and delicious taste.[126,127] It is estimated that around 3000 species are primary edible mushrooms within the kingdom of fungi, and 700 species are widely known as healthy therapeutic mushrooms.[128] However, only around

35 known species have been commercially produced.[129] The worldwide production of edible mushrooms for human consumption began in 1978. Since then, the production capacity has increased by around 35-fold (from approximately 1 billion kg in 1978 to 34 billion kg in 2013). In the same period, the world's population increased by only around 1.7-fold (from approximately 4.2 to 7.1 billion). Hence, the per capita consumption of edible mushrooms has increased significantly, exceeding 4.7 kg annually.[130] As a result, the economic value of world mushroom production in 2013 reached $63 billion, with cultivated mushrooms contributing around $34 billion.[131]

Today, mushrooms are cultivated by employing diverse technologies, ranging from simple to advanced techniques. It is reported that >100 countries cultivate commercial mushrooms on different production scales. The production of edible mushrooms increases every year, with around a 7% growth rate worldwide. Currently, the global consumption of mushrooms is approximately 12.74 million tons, and it is projected to increase to 20.84 million tons by 2026.[132] China is the leading mushroom producer, followed by Japan and the United States of America.[133] To date, the most produced edible mushrooms include *Flammulina velutipes* (Enokitake), *Pleurotus ostreatus* (Hiratake), *Lentinula edodes* (Shiitake), *Agaricus bisporus* (Tsukuritake), and *Auricularia auriculajudae* (Kikurage), and their production is expected to continuously increase in the future.[57] Due to its medicinal and nutritional value, the annual production of *F. velutipes* increased from about 143 000 metric tons in 1990 to 285 000 tons in 1997.[134] In 2013, the yearly production of *F. velutipes* was recorded at approximately 2.7 billion kg.[41] Several east Asian countries have been great suppliers of commercial mushrooms, with China being one of the main producers of *F. velutipes*, with an annual production exceeding 2 million tons, suggesting the high popularity of *F. velutipes* in the market.[135,136] In Japan, the annual production of *F. velutipes* is approximately 130 000 tons per annum.[137]

In general, the cultivation of edible mushrooms involves seven primary stages, as outlined in Figure 8.12.[132] Several factors, such as carbon : nitrogen ratio, pH, light, temperature, oxygen, moisture content, macronutrients (*e.g.*, magnesium, potassium, phosphorus, nitrogen, and carbon), and micronutrients (*e.g.*, iron, selenium, zinc, manganese, copper, and molybdenum) are crucial for efficient production of mushrooms.[133] The choice of raw materials for the production of the compost/substrate is the first important step during mushroom cultivation. Different substrate compositions can influence the chemical composition and nutritional value of the mushroom.[129] Edible mushrooms can be broadly classified as primary and secondary decomposers. Primary decomposers demand materials with lignin content and a high carbon : nitrogen ratio, but a low nitrogen content (*e.g.*, sawdust, sugarcane bagasse and straw). Secondary decomposers, however, demand materials with a low carbon:nitrogen ratio and high levels of cellulose, hemicellulose, and nitrogen (*e.g.*, manure and mixtures of numerous agricultural residues).[138] Alternatively, other agricultural materials can be used as a primary substrate for mushroom cultivation, including

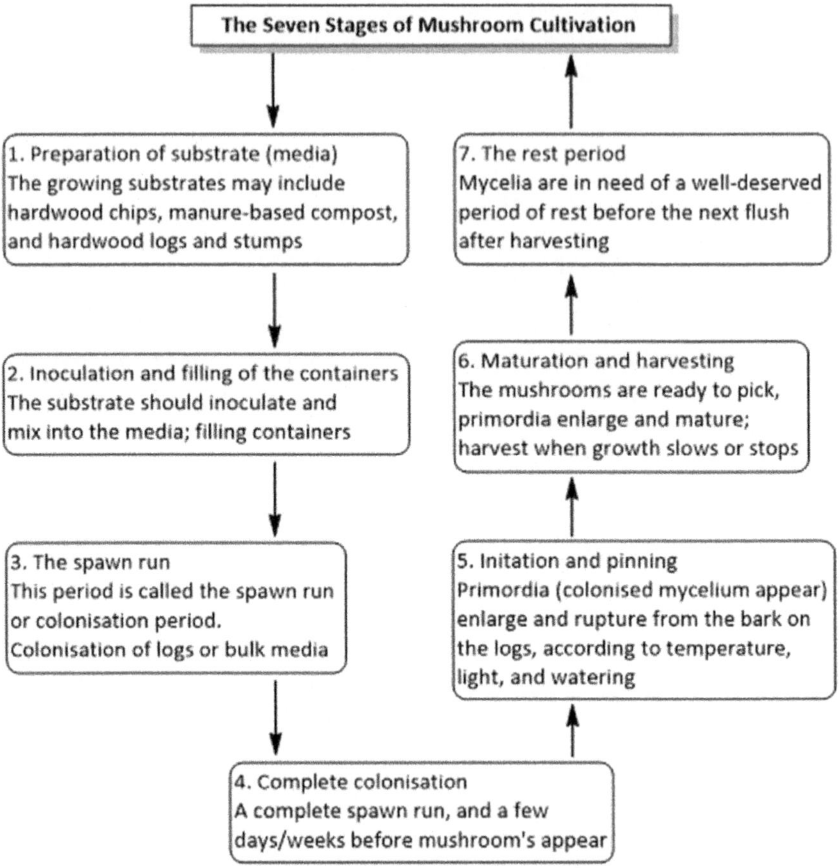

Figure 8.12 Cultivation stages of edible mushrooms. Reproduced from ref. 132, https://doi.org/10.3390/su14094941, under the terms of CC BY 4.0 License, https://creativecommons.org/licenses/by/4.0/.

soya, sunflower, kenaf, bulrush, and ramie stalks, banana leaves, palm fronds, and cotton seed hulls.[133]

The majority of cultivated mushrooms are decomposers or saprophytic fungi, which can produce some lignocellulosic enzymes, allowing them to grow on lignocellulosic substrates.[133] Historically, the first cultivation of *F. velutipes* was reported in China during the 8th century.[134] To date, several methods for *F. velutipes* cultivation on different substrates have been reported. *F. velutipes* is regarded as a broadly adapted species for diverse artificial cultivation media. Hiramori *et al.* cultivated *F. velutipes* on different compositions of fermented apple pomace (FAP) media: 0% (control group), 4.8%, 9.6%, and 14.4%. The results showed that the fruit body yield was positively correlated with increasing concentration of the FAP.[139] Harith *et al.* investigated the use of agro residues as an alternative carbon source in the fruiting substrate of *F. velutipes*. The authors demonstrated that paddy straw and palm empty fruit bunches in a ratio of 1 : 3 showed potential as a fruiting substrate, as indicated by mycelial growth and basidiocarp yield. However, the combination of paddy straw and palm-pressed fiber in a ratio of 1 : 1 was the best formulation as the fruiting substrate for *F. velutipes*.[140] Furthermore, it was reported that agro-industrial residues, such as coffee husk and spent coffee grounds, can potentially be used as a substrate for *F. velutipes* cultivation without any nutrient supplementation.[141]

More recently, spent mushroom substrate (SMS) has also been examined for its potential as a mushroom growing substrate. SMS is defined as leftover biomass (by-products) from mushroom industry activities, generated after the harvesting of mushrooms.[142] Some by-products, including degraded substrate and other agro-bio residues, such as chemical residues (potassium salts, superphosphate, and ammonium nitrate), composed of straw and gypsum, have been categorised as SMS.[80,143] SMS is considered to be of concern, as it is associated with a diverse range of environmental issues, such as groundwater contamination, and therefore hinders the future expansion of the mushroom industry.[144] It is estimated that 5 kg of SMS is generated for each kg of fresh mushroom produced,[145] and 10–50 million metric tons of SMS are expected to be produced annually on a global scale.[142] Ikeda and colleagues investigated the potential use of SMS-anaerobic digestion residue (ADR) as a medium base material for *F. velutipes* cultivation. This study revealed that the use of more than 50% SMS-ADR yielded 230–250 g fresh matter per bottle, which was comparable to the yield of the standard medium (containing 50% corn cob, 46% rice bran, and 4% fossil shells). Moreover, the levels of inorganics and amino acids in the fruiting bodies of *F. velutipes* grown in the SMS-ADR were the same as those harvested from the standard medium.[57]

8.7.2 Postharvest

It is worth noting that several post-cultivation factors, such as storage conditions and postharvest treatments could influence the nutritional value and chemical composition of edible mushrooms. Mushrooms continue to breathe and perform metabolic processes after being harvested. Thus, several chemical contents will gradually change during their storage.[146] Furthermore, due to their high water content, high respiratory rate, and lack of a cuticular structure, the quality of edible mushrooms quickly degrades over time. As a consequence, mushrooms are easily exposed to microbial decomposition, mechanical damage, weight loss, and enzymatic browning, and are therefore considered to have a short shelf life postharvest, which prevents their commercialisation in a fresh condition (1–3 days atroom temperature and 5–7 days if stored in a refrigerator).[147] Mushroom quality degradation is associated with several factors, including moisture, nutrient, and flavour losses; texture changes; discolouration; stipe elongation; and cup opening.[53]

F. velutipes is highly perishable due to the lack of a cuticle structure on the surface of itscap or stem, and its high moisture content.[148] The shelf life of *F. velutipes* is significantly affected by storage temperature. For example, a storage temperature of 1 °C generates a shelf life of 14–20 days, but when it is stored at 6 °C and 20 °C, the shelf life of *F. velutipes* decreases to 10 days and 2–3 days, respectively.[134]

Recently, the shelf life of various agricultural products has been predicted using mathematical equations. Niu *et al.* constructed a mathematical model to predict the shelf life of *F. velutipes* during its postharvest storage at different temperatures (4, 15, and 25 °C). The authors demonstrated that some variables, including decay index, whiteness value, weight loss rate, soluble protein content, and the *Pseudomonas* spp., can be used to construct a prediction model. From these results, the whiteness value was found to be the most accurate factor in predicting the shelf life of *F. velutipes*.[149]

Several innovations have been applied to extend the shelf life of *F. velutipes* after harvesting. Generally, the preservation techniques of edible mushrooms rely on packaging methods and storage temperature. These techniques can include dehydration, the use of chemical preservative agents and bioactive coatings, and modification of packaging atmospheres.[148] In general, the preservation methods can be classified into three categories: thermal (*i.e.*, drying, freezing), chemical (*i.e.*, washing, edible coating and film, ozone), and physical (*i.e.*, efficient packaging, irradiation, pulsed electric field, and ultrasound), which can be employed to halt and slow chemical deterioration as well as microbial growth. Among these methods, freezing is one of the best strategies by which to preserve the mushrooms; however, a long storage may results in a loss of amino acids and vitamins.[53,150]

Packaging of mushrooms using oriented polypropylene and polyethylene+polypropylene under vacuum could extend the shelf life of *F. velutipes* from 7 to 14 days.[151] Other methods, such as preservation using gamma, electron, and UV irradiation, have been found to delay microbial growth and, therefore, extend the shelf life of mushrooms.[152] Moreover, the pre-treatment of mushrooms using a non-thermal atmospheric pressure plasma jet and plasma-treated water significantly reduced the colony-forming unit number of *Escherichia coli* on the surfaces of mushrooms.[153] A study also showed that a combination of electrolysed water at different concentrations (5, 25, 50, and 100 $mg\,L^{-1}$) with passive atmosphere packaging was positively correlated with an extension in mushroom quality.[154] Donglu *et al.* investigated the effect of packaging material containing polyethylene combined with nano-Ag, nano-TiO_2, and nano-SiO_2, and attapulgite on the storage stability of *F. velutipes*. This study demonstrated that nanocomposite-based packaging material controlled oxygen and carbon dioxide levels, removed ethylene, and inhibited microorganism growth, leading to more effective preservation of mushrooms than regular polyethylene material.[155]

8.8 Safety, Toxicity, and Regulation

Compared to medicinal mushrooms, edible mushrooms are considered to be safer for consumption in large portions as they have nutritional benefits. *F. velutipes* was reported to increase creatine kinase (CK) plasma levels at

6–9 g per kg per day, which could be an indicator of cardiotoxicity.[60] This corresponds to the presence of cardiotoxin protein, known as flammutoxin, in the fresh fruiting body in which flammutoxin can be entirely eliminated through cooking.[156–158] A toxicity study also revealed a chance of liver damage caused by CK plasma elevation along with a bilirubin plasma increase at 3–6 g per kg per day, although transaminase activity was never reported.[60] These modes of toxicity manifested upon individual sensitivity to a daily consumption of 190–280 g of fresh *F. velutipes* over five days.[60] An attempt to identify enokitake allergens using serum from human patients showed that three enokitake protein bands (18 kDa, 39 kDa, and 50 kDa) interacted specifically with the patient's IgE.[159] A special allergic case of a 10 year-old girl who developed hypersensitivity reactions after consuming enokitake revealed that the patient's IgE reacted specifically with the band (75 kDa) of the enokitake protein.[160]

Wild mushroom foraging has a long history in some countries, such as Asian countries, Amazonian countries, Slavic countries, and Italy, although this cultural practice is uncommon in the United Kingdom.[16] In order to manage intoxication cases, 24 countries in Europe collectively released 268 different mushroom species consumed around Europe, although the guidelines and trade legislation vary among countries in terms of the mushroom species allowed for commerce.[161] Enokitake (*F. velutipes*) is commonly commercialized as a cultivated fungi in which the maximum residue level has become a basic requirement to state on food products.[162] Nevertheless, globalization has created another level of challenge to attenuate global outbreak related misconduct or contamination, such as the listeria cases in the USA reported in 2020.[163]

8.9 Conclusions and Future Perspectives

Due to its economic growth around the globe, wild *Flammulina velutipes* gathering has become a recreational activity. The fruiting body texture and taste of cultivated *F. velutipes* has surpassed those of its wild counterpart, with the cultivated product being in high demand. This makes enokitake one of the most important food commodities in commercialized cultivated mushrooms, worth billions of dollars annually. Historically, mushrooms have been consumed for their nutritional value; in modern society the mushroom is reported to have medicinal potential. Around 76 secondary metabolites have been characterized from the fruiting body of *Flammulina velutipes*, with several having notable pharmacological activity. Further research is necessary to ensure the nutritional medicinal value of the mushroom. In addition, HACCP system practice is necessary to warrant the value as well as safety of cultivated mushroom products as the global market for this mushroom is increasing every year.

References

1. The PLOS ONE Editors, *PLoS One*, 2018, **13**, e0199660.
2. B. Lowy, *Mycologia*, 1971, **63**, 983–993.

3. M. Cheek, E. Nic Lughadha, P. Kirk, H. Lindon, J. Carretero, B. Looney, B. Douglas, D. Haelewaters, E. Gaya, T. Llewellyn, A. M. Ainsworth, Y. Gafforov, K. Hyde, P. Crous, M. Hughes, B. E. Walker, R. Campostrini Forzza, K. M. Wong and T. Niskanen, *Plants, People, Planet*, 2020, **2**, 371–388.
4. H. Thatoi, S. K. Singdevsachan and J. K. Patra, in *Therapeutic, Probiotic, and Unconventional Foods*, ed. A. M. Grumezescu and A. M. Holban, Academic Press, 2018, pp. 79–99.
5. D. L. Barney, *Growing Mushrooms Commercially: Risks and Opportunities*, University of Idaho, College of Agriculture, Cooperative Extension System, Agricultural Experiment Station, 2000.
6. A. Rashid, A. H. Wani, M. Yaqub and S. Parveen, *Asian J. Pharm. Clin. Res.*, 2017, **10**, 399–405.
7. J. Živković, M. Ivanov, D. Stojković and J. Glamočlija, *J. Fungi*, 2021, **7**, 1–21.
8. Z. Xing and X. Zhao, in *Global Safety of Fresh Produce*, ed. J. Hoorfar, Woodhead Publishing, 2014, pp. 401–408.
9. C. Tang, P. C. Hoo, L. T. Tan, P. Pusparajah, T. M. Khan, L. H. Lee, B. H. Goh and K. G. Chan, *Front. Pharmacol.*, 2016, **7**, 474.
10. C. L. Schoch, S. Ciufo, M. Domrachev, C. L Hotton, S. Kannan, R. Khovanskaya, D. Leipe, R. Mcveigh, K. O'Neill, B. Robbertse, S. Sharma, V. Soussov, J. P. Sullivan, L. Sun, S. Turner, I. Karsch-Mizrachi, *Database (Oxford)*, 2020.
11. V. Sharma, S. Kumar and R. Tewari, *Flammulina velutipes, the culinary medicinal winter mushroom*, Directorate of Mushroom Research, Indian Council of Agricultural Research, 2009.
12. P. M. Wang, X. B. Liu, Y. C. Dai, E. Horak, K. Steffen and Z. L. Yang, *Mycol. Prog.*, 2018, **17**, 1013–1030.
13. V. M. Tereshina and A. S. Memorskaia, *Microbiology*, 2005, **74**, 279–283.
14. K. B. McKnight, J. R. Rohrer, K. M. K. Ward and K. H. McKnight, *Peterson Field Guide to Mushrooms of North America*, Houghton Mifflin Harcourt Publishing Company, 2nd edn, 2021.
15. F. Smiderle, E. Carbonero, G. Sassaki, P. Gorin and M. Iacomini, *Food Chem.*, 2008, **108**, 329–333.
16. U. Peintner, S. Schwarz, A. Mešić, P.-A. Moreau, G. Moreno and P. Saviuc, *PLoS One*, 2013, **8**, e63926.
17. S. Parnmen, N. Nooron, S. Leudang, S. Sikaphan, D. Polputpisatkul, O. Pringsulaka, S. Binchai and A. Rangsiruji, *Food Control*, 2021, **128**, 108182.
18. C. J. E. Schulp, W. Thuiller and P. H. Verburg, *Ecol. Econ.*, 2014, **105**, 292–305.
19. S. Egli, M. Peter, C. Buser, W. Stahel and F. Ayer, *Biol. Conserv.*, 2006, **129**, 271–276.
20. A. Butler, E. Ångman, Å. Ode Sang, I. Sarlöv-Herlin, A. Åkerskog and I. Knez, *J. Outdoor Recreation Tourism*, 2021, **33**, 100358.
21. S. Rezaeian and H. R. Pourianfar, *Waste Biomass Valorization*, 2017, **8**, 2631–2642.

22. K. Mori, in *Developments in Crop Science*, ed. P. J. Wuest, D. J. Royse and R. B. Beelman, Elsevier, 1987, vol. 10, pp. 455–459.
23. K. Nakamura, *Mushroom cultivation in Japan*, Asaki Publication House, Japan, 1981.
24. Statistics of Japan, https://www.e-stat.go.jp/dbview?sid=0003446942 (accessed June 2022).
25. F. Leifa, A. Pandey and C. Soccol, *Braz. Arch. Biol. Technol.*, 2001, **44**, 205–212.
26. C. Li and S. Xu, *Appl. Microbiol. Biotechnol.*, 2022, **106**, 3949–3955.
27. P. E. Mortimer, E. Boa, K. D. Hyde and H. Li, in *Encyclopedia of Mycology*, ed. Ó. Zaragoza and A. Casadevall, Elsevier, Oxford, 2021, pp. 405–417.
28. S. A. Pala, A. H. Wani and M. Y. Bhat, *Int. J. Med. Mushrooms*, 2013, **15**, 211–220.
29. A. Das, R. Banerjee, Y. Gadekar and S. Biswas, *Food*, 2020, **9**, 432.
30. A. M. Abdelshafy, T. Belwal, Z. Liang, L. Wang, D. Li, Z. Luo and L. Li, *Crit. Rev. Food Sci. Nutr.*, 2021, 1–21.
31. F. Ververidis, E. Trantas, C. Douglas, G. Vollmer, G. Kretzschmar and N. Panopoulos, *Biotechnol. J.*, 2007, **2**, 1214–1234.
32. M. B. Bahadori, C. Sarikurkcu, O. U. Yalcin, M. Cengiz and H. Gungor, *Microchem. J.*, 2019, **150**, 104172.
33. X. Lu, M. A. Brennan, J. Narciso, W. Guan, J. Zhang, L. Yuan, L. Serventi and C. S. Brennan, *LWT*, 2020, **118**, 108730.
34. M.-Y. Kim, P. Seguin, J.-K. Ahn, J.-J. Kim, S.-C. Chun, E.-H. Kim, S.-H. Seo, E.-Y. Kang, S.-L. Kim, Y.-J. Park, H.-M. Ro and I.-M. Chung, *J. Agric. Food Chem.*, 2008, **56**, 7265–7270.
35. A. K. Das, M. N. Islam, M. O. Faruk, M. Ashaduzzaman and R. Dungani, *S. Afr. J. Bot.*, 2020, **135**, 58–70.
36. T. Okuda and H. Ito, *Molecules*, 2011, **16**, 2191–2217.
37. M. A. Rahman, N. Abdullah and N. Aminudin, *Oxid. Med. Cell. Longevity*, 2015, **2015**, 403023.
38. Q. Tao, K. Ma, Y. Yang, K. Wang, B. Chen, Y. Huang, J. Han, L. Bao, X. B. Liu, Z. Yang, W. B. Yin and H. Liu, *J. Org. Chem.*, 2016, **81**, 9867–9877.
39. N. K. Ishikawa, K. Yamaji, S. Tahara, Y. Fukushi and K. Takahashi, *Phytochemistry*, 2000, **54**, 777–782.
40. N. K. Ishikawa, Y. Fukushi, K. Yamaji, S. Tahara and K. Takahashi, *J. Nat. Prod.*, 2001, **64**, 932–934.
41. A. Tabuchi, E. Fukushima-Sakuno, K. Osaki-Oka, Y. Futamura, T. Motoyama, H. Osada, N. K. Ishikawa, E. Nagasawa and K. Tokimoto, *Biosci., Biotechnol., Biochem.*, 2020, **84**, 876–886.
42. Y. Wang, L. Bao, X. Yang, L. Li, S. Li, H. Gao, X. S. Yao, H. Wen and H. W. Liu, *Food Chem.*, 2012, **132**, 1346–1353.
43. Y.-Q. Wang, L. Bao, X.-L. Yang, H.-Q. Dai, H. Guo, X.-S. Yao, L.-X. Zhang and H.-W. Liu, *Helv. Chim. Acta*, 2012, **95**, 261–267.
44. Y. Wang, L. Bao, D. Liu, X. Yang, S. Li, H. Gao, X. Yao, H. Wen and H. Liu, *Tetrahedron*, 2012, **68**, 3012–3018.

45. Z. Y. Xu, Z. A. Wu and K. S. Bi, *Chin. Chem. Lett.*, 2013, **24**, 57–58.
46. Z. Zhang, Q. Jin, G. Lv, L. Fan, H. Pan and L. Fan, *Int. J. Food Sci. Technol.*, 2013, **48**, 1057–1064.
47. Y. Hirai, M. Ikeda, T. Murayama and T. Ohata, *Biosci., Biotechnol., Biochem.*, 1998, **62**, 1364–1368.
48. A. Frainer and B. G. McKie, in *Advances in Ecological Research*, ed. S. Pawar, G. Woodward and A. I. Dell, Academic Press, 2015, vol. 52, pp. 169–200.
49. C. Yi, C. Sun, S. Tong, X. Cao, Y. Feng, C. K. Firempong, X. Jiang, X. Xu and J. Yu, *Int. J. Pharm.*, 2013, **448**, 44–50.
50. C. Yi, H. Zhong, S. Tong, X. Cao, C. K. Firempong, H. Liu, M. Fu, Y. Yang, Y. Feng, H. Zhang, X. Xu and J. Yu, *Int. J. Nanomed.*, 2012, 7, 5067–5078.
51. M. Y. Yeh, W. C. Ko and L. Y. Lin, *BioMed Res. Int.*, 2014, **2014**, 352385.
52. F. Mustafa, H. Chopra, A. A. Baig, S. K. Avula, S. Kumari, T. K. Mohanta, M. Saravanan, A. K. Mishra, N. Sharma and Y. K. Mohanta, *J. Fungi*, 2022, **8**, 1–21.
53. S. Marçal, A. S. Sousa, O. Taofiq, F. Antunes, A. M. M. B. Morais, A. C. Freitas, L. Barros, I. C. F. R. Ferreira and M. Pintado, *Trends Food Sci. Technol.*, 2021, **110**, 418–431.
54. Y. Mingyi, T. Belwal, H. P. Devkota, L. Li and Z. Luo, *Trends Food Sci. Technol.*, 2019, **92**, 94–110.
55. Y. K. Leong, F. C. Yang and J. S. Chang, *Carbohydr. Polym.*, 2021, **251**, 117006.
56. A. Das, C. M. Chen, S. C. Mu, S. H. Yang, Y. M. Ju and S. C. Li, *Pharmaceutics*, 2022, **14**, 1–17.
57. S. Ikeda, M. Yamauchi, T. Watari, M. Hatamoto, M. Yamada, S. Maki, H. Hara and T. Yamaguchi, *Environ. Technol. Innovation*, 2021, **24**, 102046.
58. A. K. Das, P. K. Nanda, P. Dandapat, S. Bandyopadhyay, P. Gullon, G. K. Sivaraman, D. J. McClements, B. Gullon and J. M. Lorenzo, *Molecules*, 2021, **26**, 1–25.
59. D. Sande, G. P. Oliveira, M. Moura, B. A. Martins, M. Lima and J. A. Takahashi, *Food Res. Int.*, 2019, **125**, 108524.
60. A. M. Mustonen, M. Määttänen, V. Kärjä, K. Puukka, J. Aho, S. Saarela and P. Nieminen, *Exp. Biol. Med.*, 2018, **243**, 639–644.
61. Y. Ding, S. Chen, H. Wang, S. Li, C. Ma, J. Wang, L. Cui and X.-Z. Tang, *J. Food Qual.*, 2021, **2021**, 1–8.
62. D. K. Banerjee, A. K. Das, R. Banerjee, M. Pateiro, P. K. Nanda, Y. P. Gadekar, S. Biswas, D. J. McClements and J. M. Lorenzo, *Foods*, 2020, **9**, 1–15.
63. V. S. P. Oetama, P. Hennersdorf, H. P. Saluz, M. A. Abdul-Aziz, G. Mrotzek and H. Haryanti, *Mar. Pollut. Bull.*, 2016, **110**, 718–725.
64. Y. Wang and H. Zhang, *Int. J. Biol. Macromol.*, 2021, **167**, 528–538.
65. X. Xu, H. Yan, J. Chen and X. Zhang, *Biotechnol. Adv.*, 2011, **29**, 667–674.

66. Y. Hao, X. Liao, X. Wang, S. Lao and W. Liao, *Int. J. Biol. Macromol.*, 2021, **185**, 582–591.
67. D. M. Wu, W. Q. Duan, Y. Liu and Y. Cen, *Int. J. Biol. Macromol.*, 2010, **46**, 100–103.
68. X. Xin, K. Zheng, Y. Niu, M. Song and W. Kang, *Open Chem.*, 2018, **16**, 155–162.
69. M. C. Garcia, P. Puchalska, C. Esteve and M. L. Marina, *Talanta*, 2013, **106**, 328–349.
70. E. Rangel-Vargas, J. A. Rodriguez, R. Dominguez, J. M. Lorenzo, M. E. Sosa, S. C. Andres, M. Rosmini, J. A. Perez-Alvarez, A. Teixeira and E. M. Santos, *Foods*, 2021, **10**, 1–19.
71. P. C. K. Cheung, *Food Sci. Hum. Wellness*, 2013, **2**, 162–166.
72. P. Gunc Ergonul, I. Akata, F. Kalyoncu and B. Ergonul, *Sci. World J.*, 2013, **2013**, 163964.
73. Y. Srikandace and D. S. G. Andayani, *Adv. Environ. Biol.*, 2015, **9**, 143–149.
74. P. Kalac, *J. Sci. Food Agric.*, 2013, **93**, 209–218.
75. F. A. Ayaz, H. Torun, A. Colak, E. Sesli, M. Millson and R. H. Glew, *Food Nutr. Sci.*, 2011, **02**, 53–59.
76. S. Patel and A. Goyal, *3 Biotech*, 2012, **2**, 1–15.
77. Y. H. Gu and J. Leonard, *Oncol. Rep.*, 2006, **15**, 417–423.
78. H. T. Ou, C. J. Shieh, J. Y. J. Chen and H. M. Chang, *J. Agric. Food Chem.*, 2005, **53**, 300–305.
79. C. I. Ukaegbu, S. R. Shah, A. H. Hazrulrizawati and O. R. Alara, *Beni-Suef Univ J. Basic Appl. Sci.*, 2018, **7**, 675–682.
80. G. T. Chen, Y. X. Fu, W. J. Yang, Q. H. Hu and L. Y. Zhao, *Int. J. Biol. Macromol.*, 2018, **107**, 2150–2156.
81. M. Y. K. Leung, K. P. Fung and Y. M. Choy, *Immunopharmacology*, 1997, **35**, 255–263.
82. S. M. Jiang, Z. M. Xiao and Z. H. Xu, *World J. Gastroenterol.*, 1999, **5**, 404.
83. T. Ikekawa, Y. Ikeda, Y. Yoshioka, K. Nakanishi, E. Yokoyama and E. Yamazaki, *J. Pharmacobio-Dyn.*, 1982, **5**, 576–581.
84. T. Ohkuma, K. Otagiri, T. Ikekawa and S. Tanaka, *J. Pharmacobio-Dyn.*, 1982, **5**, 439–444.
85. W. Yang, F. Pei, Y. Shi, L. Zhao, Y. Fang and Q. Hu, *Carbohydr. Polym.*, 2012, **88**, 474–480.
86. M. Wu, X. Luo, X. Xu, W. Wei, M. Yu, N. Jiang, L. Ye, Z. Yang and X. Fei, *J. Tradit. Chin. Med.*, 2014, **34**, 733–740.
87. T. Ikekawa, T. Miyano, A. Okura, H. Maruyama, K. Kawamura, Y. Sawasaki, K. Shiratori and K. Naito, *Jpn. J. Cancer Res.*, 1985, **76**, 142–148.
88. T. B. Ng, P. H. K. Ngai and L. Xia, *Mycologia*, 2006, **98**, 167–171.
89. H. H. Chang, K. Y. Hsieh, C. H. Yeh, Y. P. Tu and F. Sheu, *Int. Immunopharmacol.*, 2010, **10**, 239–246.
90. Y. C. Chang, Y. M. Hsiao, M. F. Wu, C. C. Ou, Y. W. Lin, K. H. Lue and J. L. Ko, *J. Agric. Food Chem.*, 2013, **61**, 12044–12052.
91. M. Fukushima, T. Ohashi, Y. Fujiwara, K. Sonoyama and M. Nakano, *Exp. Biol. Med.*, 2001, **226**, 758–765.

92. M. Y. Yeh, W. C. Ko and L. Y. Lin, *BioMed Res. Int.*, 2014, **2014**, 1–6.
93. Q. Tao, K. Ma, Y. Yang, K. Wang, B. Chen, Y. Huang, J. Han, L. Bao, X. B. Liu, Z. Yang, W. B. Yin and H. Liu, *J. Org. Chem.*, 2016, **81**, 9867–9877.
94. N. V. Kozhemyakina, E. P. Ananyeva, S. V. Gurina and V. A. Galynkin, *Appl. Biochem. Microbiol.*, 2010, **46**, 536–539.
95. W. H. Wang, J. S. Zhang, T. Feng, J. Deng, C. C. Lin, H. Fan, W. J. Yu, H. Y. Bao and W. Jia, *Sci. Rep.*, 2018, **8**, 1–13.
96. P. V. Jeurink, C. L. Noguera, H. F. J. Savelkoul and H. J. Wichers, *Int. Immunopharmacol.*, 2008, **8**, 1124–1133.
97. P. H. Wang, C. I. Hsu, S. C. Tang, Y. L. Huang, J. Y. Lin and J. L. Ko, *J. Agric. Food Chem.*, 2004, **52**, 2721–2725.
98. C. C. Ou, Y. M. Hsiao, W. J. Wu, G. J. Tasy, J. L. Ko and M. Y. Lin, *J. Agric. Food Chem.*, 2009, **57**, 11008–11013.
99. Y. Ding, S. V. Seow, C. H. Huang, L. M. Liew, Y. C. Lim, I. C. Kuo and K. Y. Chua, *Immunology*, 2009, **128**, e881–e894.
100. M. Karaman, E. Jovin, R. Malbaša, M. Matavuly and M. Popović, *Phytother. Res.*, 2010, **24**, 1473–1481.
101. Y. R. Dong, S. J. Cheng, G. H. Qi, Z. P. Yang, S. Y. Yin and G. T. Chen, *Carbohydr. Polym.*, 2017, **161**, 26–32.
102. H. N. D. Bao, Y. Ochiai and T. Ohshima, *Bioresour. Technol.*, 2010, **101**, 6248–6255.
103. M. Asatiani, V. Elisashvili, S. P. Wasser, A. Z. Reznick and E. Nevo, *Int. J. Med. Mushrooms*, 2007, **9**, 151–158.
104. H. N. D. Bao, H. Ushio and T. Ohshima, *J. Food Sci.*, 2009, **74**, C162–C169.
105. F. Bach, A. A. F. Zielinski, C. V. Helm, G. M. Maciel, A. C. Pedro, A. P. Stafussa, S. Ávila and C. W. I. Haminiuk, *LWT*, 2019, **107**, 214–220.
106. J. M. Ferreira, D. N. Carreira, F. R. Braga and F. E. de Freitas Soares, *3 Biotech*, 2019, **9**, 410.
107. Y. C. Chang, Y. H. Chow, H. L. Sun, Y. F. Liu, Y. T. Lee, K. H. Lue and J. L. Ko, *Antiviral Res.*, 2014, **110**, 124–131.
108. T. Zhang, J. Ye, C. Xue, Y. Wang, W. Liao, L. Mao, M. Yuan and S. Lian, *Carbohydr. Polym.*, 2018, **197**, 147–156.
109. N. K. Ishikawa, K. Miura, S. Tahara, K. Takahashi, K. Yamaji, H. Ishimoto, Y. Fukushi and S. Tahara, *Mycoscience*, 2005, **46**, 39–45.
110. Y. Q. Wang, L. Bao, X. L. Yang, H. Q. Dai, H. Guo, X. S. Yao, L. X. Zhang and H. W. Liu, *Helv. Chim. Acta*, 2012, **95**, 261–267.
111. Y. T. Lee, S. S. Lee, H. L. Sun, K. H. Lu, M. S. Ku, J. N. Sheu, J. L. Ko and K. H. Lue, *Cytokine*, 2013, **61**, 237–244.
112. Y. T. Lee, C. T. Wu, H. L. Sun, J. L. Ko and K. H. Lue, *J. Microbiol., Immunol. Infect.*, 2018, **51**, 598–607.
113. Y. C. Chang, Y. M. Hsiao, S. C. Hung, Y. W. Chen, C. C. Ou, W. T. Chang, K. H. Lue and J. L. Ko, *Biosci. Biotechnol. Biochem.*, 2015, **79**, 88–96.

114. P. Y. Chu, H. L. Sun, J. L. Ko, M. S. Ku, L. J. Lin, Y. T. Lee, P. F. Liao, H. H. Pan, H. L. Lu and K. H. Lue, *J. Microbiol. Immunol. Infect.*, 2017, **50**, 297–306.
115. D. Gunawardena, L. Bennett, K. Shanmugam, K. King, R. Williams, D. Zabaras, R. Head, L. Ooi, E. Gyengesi and G. Münch, *Food Chem.*, 2014, **148**, 92–96.
116. W. Yang, J. Yu, L. Zhao, N. Ma, Y. Fang, F. Pei, A. M. Mariga and Q. Hu, *J. Funct. Foods*, 2015, **18**, 411–422.
117. Y. Zhang, H. Li, X. Yang, G. Jin and Y. Zhang, *Int. J. Biol. Macromol.*, 2018, **112**, 788–795.
118. Q. Hu, J. Yu, W. Yang, B. M. Kimatu, Y. Fang, N. Ma and F. Pei, *Food Chem.*, 2016, **204**, 274–282.
119. Q. Hu, D. Wang, J. Yu, G. Ma, F. Pei and W. Yang, *J. Funct. Foods*, 2017, **37**, 586–593.
120. S. G. Jang, K. S. Jeon, E. H. Lee, W. S. Kong and J. Y. Cho, *J. Microbiol. Biotechnol.*, 2009, **19**, 681–684.
121. S. Y. Kim, W. S. Kong and J. Y. Cho, *Curr. Microbiol.*, 2011, **62**, 452–457.
122. A. B. Encarnacion, F. Fagutao, K. I. Shozen, I. Hirono and T. Ohshima, *Food Chem.*, 2011, **127**, 1594–1599.
123. A. B. Encarnacion, F. Fagutao, O. Jintasataporn, W. Worawattanamateekul, I. Hirono and T. Ohshima, *Food Res. Int.*, 2012, **45**, 232–237.
124. Y. Nobushi, Y. Hamada and K. Yasukawa, *J. Pharm. Nutr. Sci.*, 2013, **3**, 222–227.
125. R. Zhao, Q. Hu, G. Ma, A. Su, M. Xie, X. Li, G. Chen and L. Zhao, *J. Funct. Foods*, 2019, **56**, 255–264.
126. M. E. Valverde, T. Hernandez-Perez and O. Paredes-Lopez, *Int. J. Microbiol.*, 2015, **2015**, 376387.
127. G. M. Luna, L. Bongiorni, C. Gili, F. Biavasco and R. Danovaro, *Environ. Microbiol. Rep.*, 2010, **2**, 120–127.
128. A. R. Niazi and A. Ghafoor, *All Life*, 2021, **14**, 450–460.
129. S. M. S. Ab Rhaman, L. Naher and S. Siddiquee, *J. Fungi*, 2021, **8**, 1–17.
130. J. Mallerman, L. Papinutti and L. Levin, *J. Microbiol. Biotechnol.*, 2015, **25**, 57–65.
131. E. Fukushima-Sakuno, *J. Antibiot.*, 2020, **73**, 687–696.
132. H. El-Ramady, N. Abdalla, K. Badgar, X. Llanaj, G. Törős, P. Hajdú, Y. Eid and J. Prokisch, *Sustainability*, 2022, **14**, 4941.
133. N. Suwannarach, J. Kumla, Y. Zhao and P. Kakumyan, *Biology*, 2022, **11**, 1–27.
134. D. Handayani, R. A. Edrada, P. Proksch, V. Wray, L. Witte, R. W. M. Van Soest, A. Kunzmann and S. Soedarsono, *J. Nat. Prod.*, 1997, **60**, 1313–1316.
135. Q. Wang, M. Guo, R. Xu, J. Zhang, Y. Bian and Y. Xiao, *Front. Microbiol.*, 2019, **10**, 2845.
136. X. B. Liu, J. Li and Z. L. Yang, *Hereditas*, 2018, **155**, 3.
137. E. Manini, G. M. Luna, C. Corinaldesi, D. Zeppilli, G. Bortoluzzi, G. Caramanna, F. Raffa and R. Danovaro, *Microb. Ecol.*, 2008, **55**, 626–639.

138. D. Cunha Zied, J. E. Sánchez, R. Noble and A. Pardo-Giménez, *Agronomy*, 2020, **10**, 1239.
139. C. Hiramori, K. Koh, S. Kurata, Y. Ueno, S. Gamage, P. Huang and S. Ohga, *Adv. Microbiol.*, 2017, **07**, 719–728.
140. N. Harith, N. Abdullah and V. Sabaratnam, *Pesqui. Agropecu. Bras.*, 2014, **49**, 181–188.
141. Anonymous, *Bull. Pan Am. Health Organ.*, 1991, **25**, 267–273.
142. F. B. Muchena, C. Pisa, M. Mutetwa, C. Govera, W. Ngezimana and M. Serrano, *Int. J. Agron.*, 2021, **2021**, 1–9.
143. Z. Lou, Y. Sun, S. Bian, S. Ali Baig, B. Hu and X. Xu, *Chemosphere*, 2017, **169**, 23–31.
144. K. L. Chang, X. M. Chen, J. Sun, J. Y. Liu, S. Y. Sun, Z. Y. Yang and Y. Wang, *Environ. Technol.*, 2017, **38**, 1765–1769.
145. F. H. Mohd Hanafi, S. Rezania, S. Mat Taib, M. F. Md Din, M. Yamauchi, M. Sakamoto, H. Hara, J. Park and S. S. Ebrahimi, *J. Mater. Cycles Waste Manage.*, 2018, **20**, 1383–1396.
146. S.-Y. Tsai and Z.-Y. Chen, *ETP Int. J. Food Eng.*, 2019, 159–163.
147. K. Castellanos-Reyes, R. Villalobos-Carvajal and T. Beldarrain-Iznaga, *Foods*, 2021, **10**, 1–25.
148. D. Fang, W. Yang, Z. Deng, X. An, L. Zhao and Q. Hu, *J. Agric. Food Chem.*, 2017, **65**, 10368–10381.
149. Y. Niu, J. Yun, Y. Bi, T. Wang, Y. Zhang, H. Liu and F. Zhao, *Postharvest Biol. Technol.*, 2020, **167**, 111235.
150. K. Zhang, Y.-Y. Pu and D.-W. Sun, *Trends Food Sci. Technol.*, 2018, **78**, 72–82.
151. A. Blazejak and A. Schippers, *FEMS Microbiol. Ecol.*, 2010, **72**, 198–207.
152. A. Fernandes, A. L. Antonio, M. B. Oliveira, A. Martins and I. C. Ferreira, *Food Chem.*, 2012, **135**, 641–650.
153. S. Mitra, M. Veerana, E. H. Choi and G. Park, *Foods*, 2021, **10**, 1–16.
154. M. S. Aday, *LWT–Food Sci. Technol.*, 2016, **68**, 44–51.
155. F. Donglu, Y. Wenjian, B. M. Kimatu, A. M. Mariga, Z. Liyan, A. Xinxin and H. Qiuhui, *Innovative Food Sci. Emerging Technol.*, 2016, **33**, 489–497.
156. A. Harvey, *Handbook of Toxinology*, ed. W. T. Shier and D. Mebs, Marcel Dekker, New York, 1990.
157. J.-Y. Lin, Y.-J. Lin, C.-C. Chen, H.-L. Wu, G.-Y. Shi and T.-W. Jeng, *Nature*, 1974, **252**, 235–237.
158. J. Y. Lin, H. L. Wu and G. Y. Shi, *Toxicon*, 1975, **13**, 323–331.
159. K. Otsuji, K. Ohara, M. Nakamura, R. Amazumi, C. Higa, K. Kakazu and Y. Kondo, *Arerugi*, 2015, **64**, 63–67.
160. T. Shingaki, Y. Hiraguchi, M. Gen, S. Yoshmo, Y. Kumagai, Y. Ebishima, K. Owa, K. Otuji, Y. Kondou and Y. Suehiro, *Jpn. J. Allergol.*, 2017, **66**, 1240–1243.
161. M. Kotowski, *Slovenský národopis/Slovak Ethnology*, 2016, **64**, 169–178.
162. A. European Food Safety, *EFSA J.*, 2019, **17**, e05707.
163. J.-H. Yoon, D.-Y. Jeong, S.-B. Lee and S.-R. Kim, *Food Control*, 2021, **129**, 1–8.

CHAPTER 9

Fomes fomentarius *(L.) Fr.*

ARUN KUMAR GUPTA,*[a,e] MUZAMIL A. RATHER,[b] SHUVAM BHUYAN,[b] MUKESH S. SIKARWAR,[c] VIJAY KUMAR,[d] BINDU NAIK[e] AND POONAM MISHRA[a]

[a] Department of Food Engineering and Technology, Tezpur University, Assam-784028, India; [b] Department of Molecular Biology and Biotechnology, Tezpur University, Assam-784028, India; [c] Unit of Pharmaceutical Chemistry, Faculty of Pharmacy, AIMST University, Semeling, Kedah Darul Aman, Malaysia; [d] Department of Biosciences, Himalayan Institute of Medical Sciences, Swami Rama Himalayan University, Jolly Grant, Dehradun, Uttarakhand-248016, India; [e] Department of Food Science & Technology, Graphic Era Deemed to be University, Bell Road, Clement Town, Dehradun, Uttarakhand-248002, India

*Email: guptaarunkumar714@gmail.com

9.1 Introduction

Polypore fungus have long been utilized in folk medicine worldwide. The fruit body of *Fomes fomentarius* (L.) Fr. (*F. fomentarius*) has long been used in a variety of cultures. Tinder Fungus, sometimes known as Iceman Fungus, is a perennial necrotrophic parasitic fungus that can be found in the northern hemisphere in China, Japan, Europe, and North America. This fungus grows on a variety of host trees in Europe, including beech, birch, oak, linden, poplar, willow, and maple, except for rare alder and hornbeam species.[1,2] The fruit's body, which comes in a various sizes, resembles a horse's hoof. Depending on its source location, it might be anything from silvery-white and grey to nearly black. Low altitudes and latitudes experience mild weather.[3] A pioneering species is the tinder fungus.[4] In addition to the hyphae, it enters through the bark of trees. The white-rot host plant is produced by weak or healthy tree trunks. The fungus

Phytochemistry and Nutritional Composition of Significant Wild Medicinal and Edible Mushrooms: Traditional Uses and Pharmacology
Edited by Ajay Sharma, Garima Bhardwaj and Gulzar Ahmad Nayik

Published by the Royal Society of Chemistry, www.rsc.org

acts as a decomposer after the host tree dies.[3] Otzi, a 5000 year old Neolithic man, most likely employed this inedible species as tinder for both materials and spiritual purposes.[5] *F. fomentarius* (L.) Fr is well known to exhibit ethnographic hemostatic action. Additionally, recent pharmacological research has also shown that it can be used to prevent tumors and leukemia. Numerous secondary metabolites, including the linoleate of the tinder fungal sterol, protocatealdehyde, sitosterol, ergostanol, betulin, and cerevisterol, have been discovered in this fungus. These compounds have antioxidant, antithrombin, cytotoxic, anticancer, and anti-leukemia properties.[6–8] Because of their sturdy, lightweight, and resistant fruit bodies, they are intriguing candidates for bio-inspired and biodegradable applications. Scientific research on the nutritional, phytochemical, and pharmacological qualities of mushrooms has proliferated in recent years, leading to them flooding the market. This chapter covers the botanical classification, cultivation, sustainability, phytochemical and pharmacological properties, and biological activities of *F. fomentarius*.

9.2 Taxonomy

F. fomentarius is a basidiomycete white-rot fungus from the family Polyporaceae.

9.2.1 Taxonomic Tree

Domain: Eukaryota
Kingdom: Fungi
Phylum: Basidiomycota
Subphylum: Agaricomycotina
Class: Agaricomycetes
Subclass: Agaricomycetidae
Order: Polyporales
Family: Polyporaceae
Genus: Fomes
Species: *Fomes fomentarius*

9.3 Occurrence on Wood Substrate

Saprobic and parasitic, *F. fomentarius* is found throughout the year on the living or dead trunks, stumps, and logs of deciduous trees, solitary or in groups (Figure 9.1).

9.3.1 Dimensions

Hoof-shaped caps, up to 20 cm wide.

9.3.2 Upper Surface

The color ranges from mild dark grey to dark brown, and the furrows are concentrically furrowed, firm, and woody.

Figure 9.1 (a) *Fomes fomentarius* grown on beech; (b) upper and lower surface of mushroom; (c) section of mushroom showing bars (1 to 9); (d) bars cut into blocks (A to F). Reproduced from ref. 9, https://doi.org/10.1007/s00339-020-04270-2, under the terms of CC BY 4.0 license, http://creativecommons.org/licenses/by/4.0/.

9.3.3 Pore Surface

Pores 3–4 per mm; first cream-colored, then brownish.

9.3.4 Edibility

Inedible.

9.4 Taxonomic History

In 1753, Carl Linnaeus named *Boletus fomentarius* 'Hoof Fungus'. In 1821, Elias Magnus Fries renamed this fungus *Polyporus fomentarius.* Fomes was assigned as a genus by Fries in 1849, and the Flemish mycologist Jean Jacques Kickx moved Hoof Fungus to that genus in 1867; hence, its scientific name was established as *Fomes fomentarius*.

Synonyms: *Boletus fomentarius* L., *Polyporus fomentarius* (L.) Fr.

Other common names: Horse's Hoof Fungus (EN), tinder fungus, tinder polypore, ice man fungus.

9.5 Description

The appearance of *F. fomentarius* is considerably distinct from that of most other mushrooms. It has no stem or stipe, but it does have a flat bottom surface covered with pores, which are the openings of densely packed, spore-producing tubes. Also, the fruiting body is perennial, and its flesh is tough and woody.

9.6 Characteristics

The shape of the fruiting body of *F. fomentarius* is like the hoof of a horse, the reason for it being referred to as hoof fungus. As mentioned above, this fungus has no stem and gills; however, its flat lower surface is covered with pores. Moreover, the fungus has a perennial fruiting body with flesh that is tough and woody. Table 9.1 summarizes the characteristic features of *F. fomentarius*.

9.7 Resource Availability, Sustainability, and Distribution

- Geographical distribution: northern North America, Eurasia, Africa, Vihorlat Mountains (East Slovakia), China, and Japan. It may be found growing wild in parts of northern North America, Eurasia, and Africa.
- Polyporales hosts: various broad-leaved trees, including beech (*Fagus sylvatica*), birch (*Betula* spp.), cherry (*Prunus* spp.), poplar (*Populus* spp.), and oak (*Quercus* spp).[10]
- Beech, birch, oak, poplar, and maple are hosts for *F. fomentarius* throughout continental Europe; it is only found in birch in northern Scotland.[10]
- Horse chestnut and crack willow have also been identified as hosts.[11]

9.8 Medicinal Properties

Humans have used mushrooms as medicine for thousands of years. Hippocrates, a Greek physician who lived around 450 BCE, described the Amadou mushroom (*Phoms fomentarius*) as a powerful anti-inflammatory and wound healer. To help them live in the Alps of northern Italy about 5300 years ago, Otzi, the Ice Man, carried Amadou and a birch polypore in a pouch. Even though mushrooms have long been used by many civilizations,

Table 9.1 Characteristics of *F. fomentarius*.

S. no.	Parts	Characteristics
1.	Basidiocarp base	Soft white mottled mycelial core
2.	Basidiocarps	5–45 cm in diameter, 2–25 cm in height
3.	Shaped	Like horse hoof
4.	Outer part	Hard and woody
5.	Trama	Hard, cinnamon brown
6.	Odor	Like fruit while fresh
7.	Smell	Pleasant, mushroom-like, bitter taste
8.	Fruit bodies	Perennial, sessile, pileate, dimidiate to ungulate brackets, broadly attached to the substrate
9.	Hymenium	Porous, round, 2–3 cm in diameter, tubes 2–5 (8) mm long and gray-brown in color.
10.	Spores	$5.5–6 \times 18.5–19$ μm, elliptic-cylindrical, flat, hyaline

contemporary science has recently discovered that ancient humans always knew that mushrooms were a deep reservoir of potent healing.[12]

F. fomentarius has been handled with extreme care since the time of Hippocrates in the 5th century BC. The Okanagan–Colville indigenous people of British Columbia use this fungus to treat arthritis. After crushing and softening the fungus, a portion is applied to the skin over the diseased area and burned.[13] Surgeons, barbers, and dentists commonly employed it as a styptic, earning it the nickname "surgeons' agaric". In addition, in European, West Siberian, and Indian traditional medicine, an absorbent bandage containing tinder and iodine is used directly on wounds and burns. *F. fomentarius* is known as "Wundschwamm" or "Chirurgenschwamm" in German-speaking Alpine areas and is marketed in pharmacies as styptic bandages. Up to the 19th century, Austrian farmers utilized this absorbent wound tissue. Likewise, in Germany and Austria, *F. fomentarius* fruit bodies were utilized for traditional smoking activities around Easter.[14] Similar practices were also used by the Khanty people of West Siberia, who burned fruit bodies to generate smoke immediately after a person's death to prevent the deceased from having any effect on the living.[15]

F. fomentarius contains the active ingredient "fomitin", which is used to treat dysmenorrhea, hemorrhoids, and bladder issues.[16] It has also been reported that *F. fomentarius* is being used for pain relief in the management of cancers of the esophagus, stomach, and uterus.[17,18] Moreover, it is used in the management of glycosuria, alopecia, rheumatic piles, bladder disorders, colds, high blood pressure, eye diseases, dysmenorrhea, analgesics, abdominal pain, constipation, and cirrhosis.[12,19–22]

The exopolysaccharides (EPS) produced from the fungus have been reported to show a hypoglycemic effect in diabetic rats induced by streptozotocin (STZ), indicating that it could be used as an effective candidate for diabetes mellitus therapy.[22] Moreover, pro-inflammatory and immune-modulatory mediator activity, as well as anticancer activity against gastric cancer cells without toxicity to normal cells, has been reported.[23,24] EPS have been reported to increase humoral response and macrophage phagocytosis.[25] Similarly, intracellular polysaccharides (IPS) have also been demonstrated to have an anti-proliferative effect on human gastric cancer cell lines.[26] Tuberoside, a glycosidic triterpenoid from this fungus, has been reported to exhibit moderate apoptosis of lung, breast, and gastric cancer cells.[27]

Traditionally, *F. fomentarius* has been widely used in the field of medicine due to its perceived health benefits. It has been used to cauterize wounds since at least the fifth century; it was known as the "surgeons' agaric".[12] This fungus was traditionally used for therapeutic purposes in Northern Finland, Nepal, Europe, Western Siberia, Transylvania, Hungary, China, and India to treat rheumatism, aches, paralysis, hemorrhoids, bladder, and gastroenteric disorders, cough, hepatocirrhosis, oral ulcers, inflammation, constipation, cancer, and other ailments.[28,29]

The macrofungi host contains a myriad of bioactive compounds. Some of these bioactive compounds include phellandrene, hexadecenoic acid, octadecenoic acid, undecane, cymene, benzene dicarboxylic acid ester, β-glucan, daphnetin, and betulin. These compounds present in *F. fomentarius* render it with fungistatic, anti-nociceptive, antimicrobial, antioxidant, anti-inflammatory, and anti-cancer properties.[22,30–32] Indeed, it shows differential anti-microbial properties against *Pseudomonas aeruginosa, Serratia marcescens, Staphylococcus aureus, Bacillus subtilis, Absidia orchidis*, and *Aspergillus fumigatus*.[33,34] In contrast to attenuating the growth of bacterial strains, the antiviral activity of mycelia of *F. fomentarius* against influenza virus type A and herpes simplex virus type 2 has also been reported.[35] Further, culture filtrates of *F. fomentarius* have been found to be effective against the mechanical transmission of Tobacco Mosaic Virus (TMV) in tobacco, bell pepper, and tomato plants.[36] A comparative study on two fungal biopolymer complexes obtained from *F. fomentarius, viz.* a water-soluble melanin–glucan complex and an insoluble chitin–glucan–melanin complex, revealed the anti-infective properties of both complexes against *Candida albicans, Helicobacter pylori*, and HIV-1.[37]

Ethanolic extracts of *F. fomentarius* contain anti-diabetic fomentariol.[38] Aqueous extracts have also been reported to not only correct diabetes mellitus in streptozotocin (STZ)-induced diabetic rats but also contribute to alleviating the complications that arise along with the disease, such as ameliorating impaired diabetic kidney function.[39] *F. fomentarius* extract was shown to reduce blood glucose, enhance glutathione peroxidase activity, and dramatically reduce superoxide dismutase and catalase activities.

Nuclear factor-κB (NF-κB) regulates inflammation, immunoregulation, and cancer. The NF-κB complex is maintained in an inactive form in the cytoplasm by the inhibitory protein-κB (IκB). Concerning inflammatory diseases, it has been reported that ethyl acetate extracts of *F. fomentarius* biomass inhibit the degradation of IκB and moderately affect their phosphorylation.[40] Also, methanolic extracts of *F. fomentarius* have been shown to downregulate NF-κB in a dose-dependent manner, transactivating the expression of inducible nitric oxide synthase (iNOS), cyclooxygenase-2 (COX-2), and the cytokine tumor necrosis factor-α (TNF-α) in lipopolysaccharide (LPS) stimulated murine macrophages responsible for inflammatory and nociceptive responses.[41] However, a study later conducted identified the triterpene methyl 9-oxo-(10*E*,12*E*)-octadecadienoate purified from *F. fomentarius* to attenuate LPS-induced inflammatory responses in murine macrophages by inhibiting the signal transducer and activator of transcription activation, which led to the suppression of iNOS, COX-2, TNF-α and interleukin-6 (IL-6); albeit not affecting NF-κB, p38 and several kinases.[42]

In studies conducted on human breast cancer cells, an ethanol extract of *F. fomentarius* was reported to induce apoptosis, attenuate cell migration and expression of metalloproteinase-9, as well as the PI3K/AKT pathway, and upregulate E-cadherin expression.[31,43] Against murine sarcoma S180 cells, petroleum ether fractions of *F. fomentarius* showed cytotoxicity associated with apoptosis in addition to lowering immune organ toxicity.[44] Chloroform fractions of this fungus have also been shown to inhibit the proliferation of human leukemic cells *in vitro*.[45] The ergostane-type sterol 3β-*O*-glucopyranosyl-5α,6β-dihydroxy-ergosta-7,22*E*-diene, isolated from the ethanolic extract of the dried fruiting bodies of the polypore, have been shown to exhibit moderate *in vitro* antitumor activity against various carcinoma cells.[46] The water-extractable polysaccharide isolated from this fungus have been reported to induce cellular apoptosis by degrading DNA into multiple internucleosomal fragments on human lung carcinoma cells.[47] The EPS from mushrooms have been shown to exhibit several medicinal properties. For example, EPS from *F. fomentarius* has an antiproliferative effect on human gastric cancer cells in a dose and time-dependent manner. Besides EPS, ethanol extracts of mycelial biomass and intracellular polysaccharides also have a direct antiproliferative effect on human gastric cancer cell lines.[48] At non-cytotoxic concentrations, human gastric cancer cells may be more sensitive to doxorubicin-induced growth suppression if they are exposed to EPS.[49] A selenium (Se)-enriched culture medium of *F. fomentarius* incorporated Se into its EPS structure, which further improved its antibacterial activity against *S. aureus*, as well as its antioxidant activity and cytotoxic properties against cancer cells.[50]

9.9 Phytochemistry

Fomentarols A–D and sterols have been isolated from ethanolic extracts of *F. fomentarius* and moderate cytotoxic effects against human cancer cell lines were established[8,51] (Figure 9.2). Six compounds, paulownin, protocatechualdehyde, 4-(3,4-dihydroxyphenyl)-but-3-en-2-one,5α,8α-epidioxyergosta-6,22-dien-3β-ol, ergosta-7,22-dien-3β-ol, and β-sitosterol were identified from *F. fomentarius*.[6,52] The antidiabetic potential of *F. fomentarius* can be attributed to fomentariol. High-purity fomentariol has been isolated in ethanolic extract and standardized by high-resolution mass spectrometry, HPLC, and quantitative NMR spectroscopy.[38]

F. fomentarius emits a complex blend of compounds, with >10 sesquiterpenes. It releases octan-3-one, *cis*-furanoid linalool oxide, β-phellandrene, β-myrcene, and several sesquiterpene hydrocarbons.[53,54] Nine compounds have been identified by means of GC–MS in a petroleum ether extract of the mushroom. Compounds such as ergosta-7,22-dien-3-palmitate, stearic acid, palmitic acid, ergosta-7,22-dien-3-one, ergosta-7,22-dien-3-one, dimethyl acetal,[7,55] and sterols have also been isolated. The primary components of the essential oil of the mushroom, n-hexadecanoic acid and octadecanoic acid, demonstrate cytotoxicity against various types of cancer cells.[44] See Figure 9.3.

A novel type of alkyl succinic acid was isolated and identified as fomentaric acid from a petroleum ether extract of *F. fomentarius*. This compound was found to be a long chain aliphatic acid.[56,57] In a basic aqueous extract of *F. fomentarius*, phenolics, carbohydrates, minerals, and crude fat were found, along with a low quantity of proteins and chitin. Beta-glucans accounted for one-third of the total carbohydrates, with alpha-glucans accounting for the remainder. The extract had a high concentration of acid-resistant ultraviolet (UV)-absorbing phenolic components, such melanins, lignins, and humic acids. The strong antioxidant activity of the extract tested

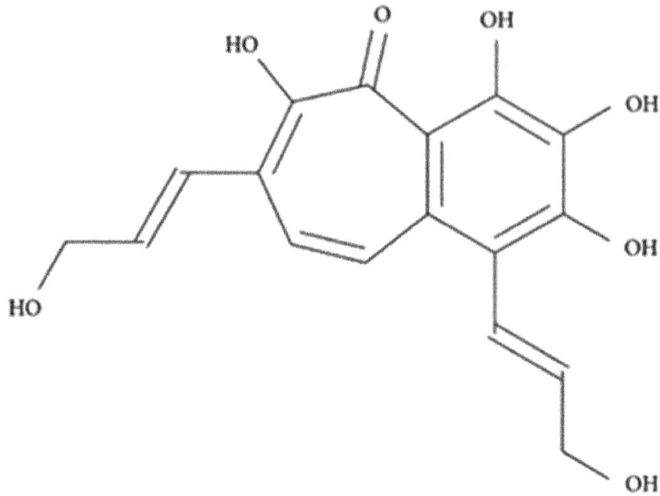

Figure 9.2 Fomentariol. Reproduced from ref. 38, https://doi.org/10.1155/2018/2434691, under the terms of the CC BY 4.0 license, https://creativecommons.org/licenses/by/4.0/.

in vitro might be attributed to these components. Moderate sun-protective capability was observed against UV-B rays, while a moderate amount of protection against UV-A rays was also predicted.[58–62]

Eleven compounds were identified as 3,4-dihydroxybenzoic acid ethyl ester, 3,4-dihydroxybenzaldehyde, 4-(3,4-dihydroxyphenyl)-3*E*-buten-2-one, 2,4-dihydroxy-3,5-dimethylacetophenone, *p*-hydroxybenzaldehyde, 4-hydroxyacetophenone, ergosta-7,22-diene-3β-ol, ergosta-7,22-diene-3-one, 4-(4-hydroxyphenyl)but-3-en-2-one (9), 1,5-bis-(3,4-dihydroxyphenyl)-1,4-pentadien-3-one, and 3,4-dihydroxyacetophenone. These above-mentioned compounds showed α-glucosidase inhibition, acetylcholinesterase inhibition, anti-oxidative, and antimicrobial activity.[59]

Using a combination of heating to reflux, thin layer chromatography, and silica gel column chromatography, two chemicals were isolated from petroleum ether and chloroform extracts of *F. fomentarius*. MS, ^{1}H-NMR, ^{13}C-NMR, and DEPT were used to identify the compounds [4,6,8(14),22(23)-tetraen-3-one-ergostane] and [7(8) 22(23)-diene-3-one-ergostane]. [3,3-Dimethyl-(8-hydroxy-3-octenyl)cyclohexanol] was identified as one of the compounds extracted from the methanol extract using HPLC. The findings revealed that the supernatant and precipitates of the petroleum ether extract, ergosterol and ergosterol peroxide, and the methanol extract of the *F. fomentarius* body may all prevent tumour formation.[60]

Three novel phenyl-ethanediols, (1*R*)-(3-ethenylphenyl)-1,2-ethanediol, (1*R*)-(3-formylphenyl)-1,2-ethanediol, and (1*R*)-(3-acetophenyl)-1,2-ethanediol, were identified in the extracts of the fungi. Spectroscopic technologies, such as comprehensive 2D NMR studies, were used to reveal their structures.[60,61]

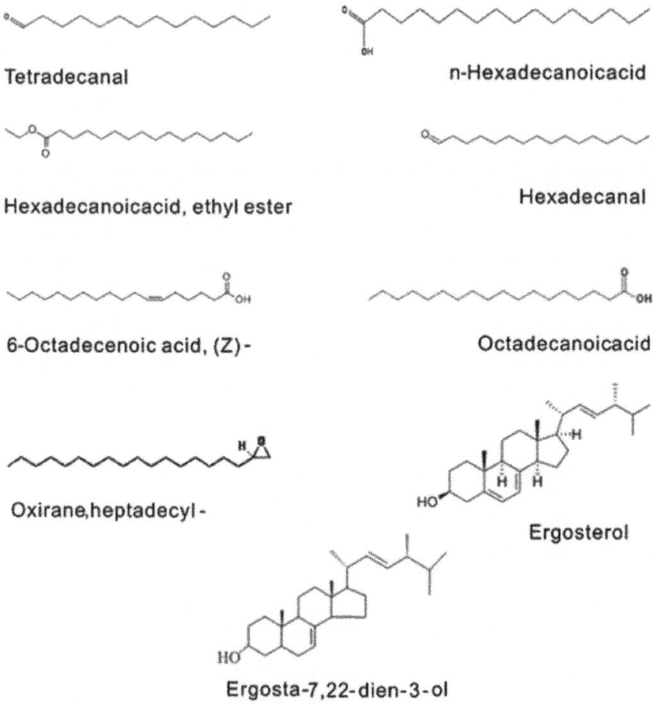

Figure 9.3 Compounds in *Fomes fomentarius* petroleum ether fraction. Reproduced from ref. 44, https://doi.org/10.1371/journal.pone.0109599, under the terms of the CC BY 4.0 license, https://creativecommons.org/licenses/by/4.0/.

Researchers have developed a novel chemical test to alter the composition of milled *F. fomentarius* fruiting bodies. Following repeated acidic–alkali treatment, the amount of chitin increased 13-fold while the amount of glucan decreased 14-fold. The polymer containing equivalent numbers of *N*-acetyl-D-glucosamine and D-glucosamine units after the first and second alkali treatments was identified as chitosan.[58,62]

Fomentariol was isolated from *F. fomentarius* and tested in murine macrophages activated with lipopolysaccharides to determine its anti-inflammatory properties. The findings implied that fomentariol, one of the bioactive chemicals that mediates the physiological benefits of *F. fomentarius*, differently modulates the inflammatory responses caused by lipopolysaccharides in macrophages.[63]

By employing DEAE-sepharose FF and Concanavalin A-sepharose 4B, five anti-complementary polysaccharides (MFKF-NP, MFKF-AP1, and MFKF-AP2) were isolated from hot-water extracts of *F. fomentarius*. The results showed that MFKF-AP1 appears to be a complement activator in both the conventional and alternative pathways.[62–64]

F. fomentarius is a fungus that lives in soil. The fruiting body of the fungus is made up of microscopic, long fibrous cells that are held together by sturdy and elastic cell walls. The primary components of the cell walls are glucan, chitin, and melanin-like compounds. Hemicellulose clusters of mannan, galactan, xylan, arabinan, and rhamnan, as well as uronic and glucuronic acids, have been discovered to be minor components of *F. fomentarius*.[62,65] The cell walls of *F. fomentarius* might be potentiated by being more bioavailable and accessible for certain receptors, immune cells, intestinal microbiota, and other dietary supplements, and *F. fomentarius* could be an excellent choice to replace various important fungi.[62,66] Corn gluten meal (CGM) nutritional supplements contain chitin, beta 1.3/1.6 D glucan, and melanin complex, all of which are derived from this fungus, and can be combined with other dietary supplements or dietary fiber.[67]

9.10 Pharmacological Potential

Figure 9.4 and Table 9.2 present some of the important pharmacological properties of *F. fomentarius.*

9.11 Cultivation and Harvesting

For *in vitro* mycelial cultures of *F. fomentarius*, Chen *et al.*[49] optimized a submerged culture medium at 6.0 pH that was inoculated with *F. fomentarius* inoculum and incubated it for 7 days. Here, *F. fomentarius* was initially inoculated on a PDA medium and incubated for 7–8 days at 25 °C. 10 mm^2 of punched-out grown colonies were transferred to a seed culture medium which was maintained at the same temperature for 4 days and later reinoculated in the optimal culture medium. Henning *et al.*[76] cultivated *F. fomentarius in vitro* on a liquid medium by inoculating the fungal complete medium (CM) with shredded colonies of 10 days old *F. fomentarius* from malt extract agar followed by incubation at 25 °C in the dark for 18–20 days. The matured colonies were harvested and dried at 50 °C for 2 days. *F. fomentarius* may also be cultivated on a solid lignocellulose medium. To inoculate sterile millet grains enriched with calcium sulphate dihydrate, the developed fungal mycelium was utilized as spawn to inoculate hemp shives hydrated with deionized water in a sterile polypropylene culture bag after being incubated for 14 days in the dark. Kneading was used to combine overgrown millet spawn with damp hemp shives. The bags were heat-sealed and incubated in the dark for 7 days. The bags were then mixed, and incubated for another 7 days. The overgrown culture was then crushed using a sterile shredder and transferred into a box which was maintained at 25 °C for another 19 days. The grown culture was then dried at 50 °C for 3 days and harvested by carefully stripping off the pure surface mycelium.

Traditionally, *F. fomentarius* fruit bodies are harvested for tinder, textiles, and medicine. These macro-fungi can yield up to three well-developed fruiting bodies in a year. After the fruiting bodies have grown to a fairly

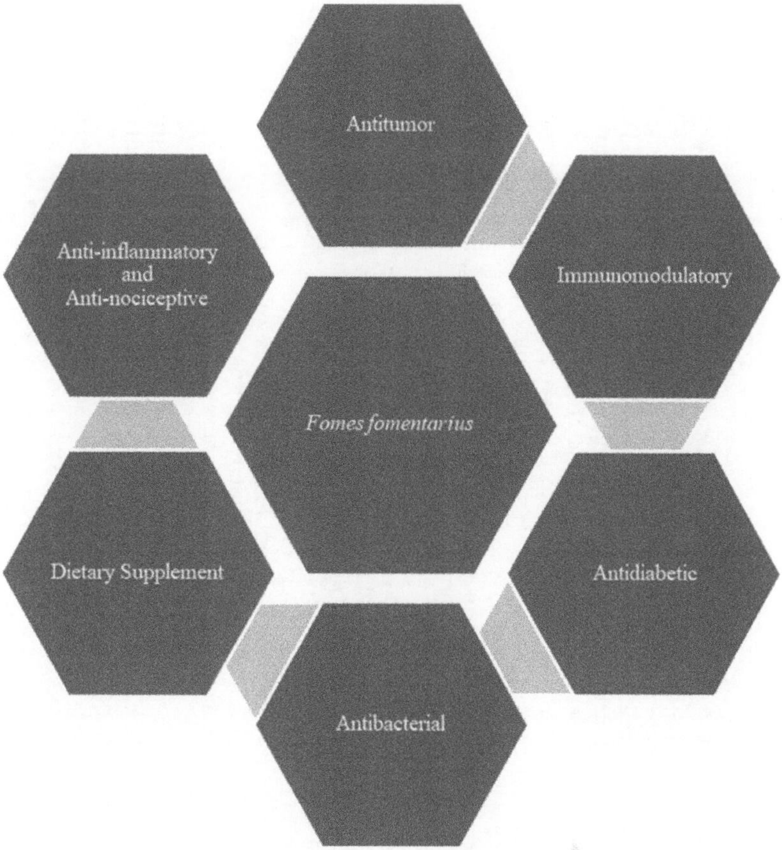

Figure 9.4 Biological activities of *F. fomentarius* (L.) Fr.

large size, the fungus is pried off the trunks of hardwood trees with the help of a sharp knife, axe or a stake. Under warm and moist conditions, new fruit bodies regrow on the same tree over a period of 2–3 weeks. The exterior of *F. fomentarius*, which is hard and woody, is trimmed off and discarded while the inner softer part is retained for use.[77,78]

9.12 Safety, Toxicity, and Regulation

Albeit inedible, there have been no reports of toxicity of this fungus and therefore there are no apparent underlying concerns of safety.

9.13 Trade, Storage and Transport

The use of *F. fomentarius* as tinder dates back to the Neolithic era.[79] Other products reported to be made from it include hats, bags, belts, cleaning

Table 9.2 The biological activities related to the medicinal uses of *Fomes fomentarius* (L.) Fr.

S. no.	Biological activities related to medicinal uses	Observations	Ref.
1.	Antibacterial and cytotoxic activities	*Staphylococcus aureus, Staphylococcus epidermidis, Micrococcus luteus, Bacillus subtilis, Enterococcus faecalis, Escherichia coli, Klebsiella pneumoniae, Pseudomonas aeruginosa, Salmonella abony*, and 10 strains of *Helicobacter pylori* were all potent against budding fungal extracts in aqueous extracts. Methanol and aqueous extracts were likewise active against a gastric cancer cell line, and all four extracts were active against a tested HeLa cell line with substantial activity. Healthy MRC-5 cells were unaffected by the extracts.	68
2.	Biological function in gastric cancer intervention	An ethanol extract of mycelia and intracellular polysaccharides had a direct antiproliferative effect on human gastric cancer cell lines in a dose-dependent manner. However, a human normal gastric cell line was less susceptible to it.	26
3.	Apoptotic properties of medicinal mushrooms in a human lung carcinoma cell line	Water-extractable polysaccharide induces cell death in lung carcinoma cells in a dose-dependent manner. It also induces apoptosis by increasing LDH release, disruption of cellular morphology, and induction of nuclear damage.	47
4.	Exopolysaccharide and its antitumor effects	Ethanol extracts of mycelial biomass and IPS also have a direct antiproliferative effect on human gastric cancer cell lines and at non-cytotoxic concentrations, human gastric cancer cells may be more sensitive to doxorubicin-induced growth suppression if they are exposed to EPS.	49
5.	Effects on streptozotocin-induced diabetic rats	The administration of an aqueous extract to STZ-induced diabetic rats reduced blood glucose levels by reducing the rise in serum insulin levels, as well as the rise in total cholesterol and triglyceride levels in the diabetic rats' serum and liver.	22
6.	Biological properties of exopolysaccharide by bioincorporating selenium	In all cell lines, exopolysaccharides showed a much stronger inhibitory impact at a concentration of 200 g mL^{-1}. The percentages of Se-treated 5637, A549, and KYSE 30 cells were 10%, 15%, and 15%, respectively. Because of their antioxidant properties, Se-polysaccharides can be utilized as a supplement to suppress cancer cell growth and promote immune system performance.	50
7.	Compositions and anti-tumor activity	GC–MS analysis of the fraction revealed nine primary chemical components. *In vitro* cytotoxic efficacy against murine sarcoma was dosage and duration dependent for the petroleum ether fraction. Apoptosis was shown to be linked to cytotoxic effects. The extract inhibited tumour development caused by S180 cells *in vivo* and decreased immune organ damage, indicating that it possesses anti-tumor efficacy.	44

8.	Anti-inflammatory and anti-nociceptive effects	In LPS-stimulated RAW 264.7 macrophages, methanol extract substantially reduced NO, prostaglandin E2, and tumour necrosis factor-production. A methanol extract lowered the levels of inducible NO synthase and cyclooxygenase-2 in a dose-dependent fashion.	69
9.	Antimicrobial effect of various formulations against hospital isolates	Against Candida species, the extracts were more effective than antibiotics. Because of the protecting properties of cetyl alcohol and sodium lauryl sulfate, it was more effective. Due to the stabilizing effects of the polymers, oil, and alcohols included in these preparations, emulsion, paste, and cream were more successful than the ethanol extract, aqueous extract, and gel formulation.	70
10.	Bioactive polysaccharides and investigation of antibacterial and antitumor activities	The Taguchi technique was used to optimize the biomass and polysaccharide production of an Iranian pharmaceutical. The growth of *S. aureus* and *E. coli* was inhibited by this polysaccharide by 50% and 25%, respectively, according to biological tests. Its antioxidant activity also doubled, to 16.11%. A549 and 5637 cancer cells showed a cell vitality of 45%, whereas MKN-45 and AGS cancer cells exhibited cell viabilities of 66% and 62%, respectively.	71
11.	Antiproliferation activities of a chloroform fraction in K562 cells	In K562 cells, a chloroform fraction caused cell membrane damage, influenced intracellular [Ca^{2+}], promoted ROS production, and triggered cell death.	72
12.	Tinder fungus as a dietary supplement	Glucan, chitin, and melanin-like compounds are found in the cell walls of fungi. Cell walls have the potential to become crucial in fungi employed as dietary supplements since they are bio-available and accessible to certain receptors, immunological cells, gut bacteria, and so on.	62
13.	Biological potential *via* dry and wet milling	In DPPH, FRAP, and ABTS experiments, the suspensions show good antioxidant activity. For dry and wet milled samples, IC_{50} values of 0.02–0.04% w:v and 0.04% w:v were reported in the DPPH test, respectively. In the FRAP test, dry and wet milling samples exhibited IC_{50} values of 0.02% w:v and 0.04% w:v, respectively, whereas the ABTS assay led to IC_{50} values of 0.04% w:v and 0.005% w:v, respectively.	73
14.	Cytotoxic activity	According to the findings, the ethanol extract showed the maximum efficacy in BSLA, while the ethyl acetate fraction exhibited the highest cell cytotoxicity. Palmitic acid, in combination with other chemicals identified in *F. fomentarius*, was discovered to have cytotoxic effect against colon cancer cells. This trait might offer a biological foundation for future clinical testing.	74
15.	Immunomodulatory effects of polysaccharides	Fungus polysaccharides increase the release of TNF-alpha, IFN-gamma, and IL-2 by mice immunocytes, as well as humoral immune response and macrophage phagocytosis.	75

cloths, tablecloths, *etc.*, which have been sold on the international market and at various world exhibitions since the late 1800s. These naturally occurring macro-fungi have not yet been exploited to their full potential. *F. fomentarius* fruit bodies could offer us a new direction to strive for an alternative to using animal hides in textiles and furniture. Moreover, it also has to its name a myriad of uses in pharmacology and ethnomedicine.[77,78]

The fruit bodies are traditionally kept smooth and moist in a nylon bag or a wooden box for 1–2 weeks after harvesting. The soft inside sections of the fruit bodies are sliced into shapes, which are then stretched and smoothed out by hand using tiny circular motions before drying. These fungus are dried and kept in a hemp sac for lengthy periods of time. Before use, these stored fungi are soaked in hot water and then trimmed for 2 days.[77] Dry milling of *F. fomentarius* into a powder also enables its longer storage and transport.[73]

9.14 Conclusions and Future Perspectives

Mushrooms play a crucial role in a variety of ecosystems; however, they are frequently collected from artificial cultures to avoid natural gathering. The biological potential of wild mushrooms might lead to the development of new medications. The continual emergence of drug-resistant bacteria, viruses, and cancer cells, as well as adverse effects posed by currently used medicines, are the most pressing challenges in medicine. The identification and study of naturally-sourced chemicals, as well as their mechanisms of action, are both urgently needed. The majority of the current research on bioactive chemicals found in wild mushrooms has been conducted on animal models or *in vitro*. The majority of these substances are used as adjuvant treatments and are for symptom alleviation rather than being a cure. This chapter, highlights the potential of *F. fomentarius,* a rich source of bioactive compounds with appreciable biological properties. As a result, *F. fomentarius* must be regarded not only as a delicacy but also as a functional food, and, in certain circumstances, a medicinal agent. It has been proven from the scientific literature that they are edible. As a result, edible mushrooms that contain bioactive chemicals have a great deal of promise as a therapeutic treatment. Clinical studies and analysis of the efficacy of these compounds are required, as are better clinical research procedures. More study into the discovery of new compounds and their bioactive potential is needed to address the rising demand for treatments and medications. Many rare and endemic wild mushroom species have limited habitats and may contain chemicals that hold medicinal promise.

References

1. L. Ryvarden and R. L. Gilbertson, European Polypores, *Synopsis Fungorum 6*, Fungiflora, Oslo, Norway, 1993, vol. 1–2, p. 745.
2. S. Yamashita, T. Hattori and H. Abe, Host preference and species richness of wood-inhabiting aphyllophoraceous fungi in a cool temperate area of Japan, *Mycologia*, 2010, **102**(1), 11–19.

3. F. W. Schwarze, J. Engels and C. Mattheck, *Fungal strategies of wood decay in trees*, Springer Science & Business Media, 2000, p. 95.
4. J. Heilmann-Clausen and L. Boddy, Inhibition and stimulation effects in communities of wood decay fungi: exudates from colonized wood influence growth by other species, *Microb. Ecol.*, 2005, **49**(3), 399–406.
5. R. Pöder and U. Peintner, Laxatives and the Ice Man, *Lancet*, 1999, **353**(9156), 926.
6. W. Feng and J. S. Yang, Studies on chemical constituents of *Fomes fomentarius* (L. Ex. Fr.), *Chin. Pharm. J.*, 2010, **45**(20), 1528–1530.
7. T. Z. Huang, D. Y. Du, Y. Q. Chen, B. Yuan, X. Y. Ju, Y. J. Feng, L. Wang and J. H. Jiang, Chemical constituents and antitumor activity of fruiting body of *Fomes fomentarius*, *Mycosystema*, 2012, **5**, 775–783.
8. Y. Zang, J. Xiong, W. Z. Zhai, L. Cao, S. P. Zhang, Y. Tang, J. Wang, J. J. Su, G. X. Yang, Y. Zhao and H. Fan, Fomentarols A–D, sterols from the polypore macrofungus *Fomes fomentarius*, *Phytochemistry*, 2013, **92**, 137–145.
9. C. Müller, S. Klemm and C. Fleck, Bracket fungi, natural lightweight construction materials: hierarchical microstructure and compressive behavior of *Fomes fomentarius* fruit bodies, *Appl. Phys. A*, 2021, **127**(3), 1.
10. CABI; EPPO, Distribution Maps of Plant Diseases, (Edition 1), CAB International Wallingford UK, 2002, DOI: 10.1079/DMPD/20066500867.
11. J. Pitt, Notes on the spread of *Fomes fomentarius* in West Kent, *Field Mycol.*, 2019, **20**(2), 58–59.
12. U. Grienke, M. Zöll, U. Peintner and J. M. Rollinger, European medicinal polypores–A modern view on traditional uses, *J. Ethnopharmacol.*, 2014, **154**(3), 564–583.
13. C. Hobbs, *Medicinal Mushrooms: An Exploration of Tradition, Healing and Culture*, Botanica Press, Santa Cruz, 1995.
14. R. Rutalek, Ethnomykologie–Eine Übersicht, *Österr Z Pilzkd.*, 2002, **11**, 79–94.
15. M. Saar, Fungi in Khanty folk medicine, *J. Ethnopharmacol.*, 1991, **31**(2), 175–179.
16. S. Killermann, Ehemaliger apothekerpilz, *Z. Pilzkd.*, 1938, **22**, 11–13.
17. R. G. Wasson, *Soma: divine mushroom of immortality*, Harcourt Brace and World, New York, 1969.
18. J. Ying, X. Mao, Q. Ma, Y. Zong and H. Wen, *Leones of Medicinal Fungi from China*, Science Press, Beijing, 1987, p. 389.
19. M. Kuo, *Fomes fomentarius*, Retrieved from the MushroomExpert.Com website, October 2, 2019.
20. H. Kim and M. J. Song, Analysis of traditional knowledge for wild edible mushrooms consumed by residents living in Jirisan National Park (Korea), *J. Ethnopharmacol.*, 2014, **153**(1), 90–97.
21. X. Kang, A. Kirui, A. Muszyński, M. C. Widanage, A. Chen, P. Azadi, P. Wang, F. Mentink-Vigier and T. Wang, Molecular architecture of fungal cell walls revealed by solid-state NMR, *Nat. Commun.*, 2018, **9**(1), 1–2.

22. J. S. Lee, Effects of *Fomes fomentarius* supplementation on antioxidant enzyme activities, blood glucose, and lipid profile in streptozotocin-induced diabetic rats, *Nutr. Res.*, 2005, **25**(2), 187–195.
23. R. G. Jayasooriya, C. H. Kang, M. J. Seo, Y. H. Choi, Y. K. Jeong and G. Y. Kim, Exopolysaccharide of Laetiporus sulphureus var. miniatus downregulates LPS-induced production of NO, PGE2, and TNF-α in BV2 microglia cells via suppression of the NF-κB pathway, *Food Chem. Toxicol.*, 2011, **49**(11), 2758–2764.
24. M. J. Seo, B. W. Kang, J. U. Park, M. J. Kim, H. H. Lee, Y. H. Choi and Y. K. Jeong, Biochemical characterization of the exopolysaccharide purified from Laetiporus sulphureus mycelia, *J. Microbiol. Biotechnol.*, 2011, **21**(12), 1287–1293.
25. H. L. Gao, L. S. Lei, C. L. Yu, Z. G. Zhu, N. N. Chen and S. G. Wu, Immunomodulatory effects of *Fomes fomentarius* polysaccharides: an experimental study in mice, *Nan fang yi ke da xue xue bao = J. South. Med. Univ.*, 2009, **29**(3), 458–461.
26. W. Chen, Z. Zhao and Y. Li, Simultaneous increase of mycelial biomass and intracellular polysaccharide from *Fomes fomentarius* and its biological function of gastric cancer intervention, *Carbohydr. Polym.*, 2011, **85**(2), 369–375.
27. H. Zhang, B. Feng, K. Liu and S. Guo, Isolation and purification of two triterpenoids from the Chinese medicinal plant Fomes officinalis Ames, *Asian J. Chem.*, 2013, **25**(11), 6130–6132.
28. J. Gáper, S. Gáperová, P. Pristas and K. Naplavova, Medicinal value and taxonomy of the tinder polypore, *Fomes fomentarius* (Agaricomycetes): a review, *Int. J. Med. Mushrooms*, 2016, **18**(10), 851–859.
29. P. Hassanzadeh, M. Kharaziha, M. Nikkhah, S. R. Shin, J. Jin, S. He, W. Sun, C. Zhong, M. R. Dokmeci, A. Khademhosseini and M. Rolandi, Chitin nanofiber micropatterned flexible substrates for tissue engineering, *J. Mater. Chem. B.*, 2013, **1**(34), 4217–4224.
30. L. Prasad, A. Kundu and D. Bahukhandi, Comparative analysis of volatile fractions of *Fomes fomentarius* and F. rhabarbarinus, *Indian Phytopathol.*, 2018, **71**(1), 25–31.
31. S. O. Lee, M. H. Lee, K. R. Lee, E. O. Lee and H. J. Lee, *Fomes fomentarius* ethanol extract exerts inhibition of cell growth and motility induction of apoptosis via targeting AKT in human breast cancer MDA-MB-231 cells, *Int. J. Mol. Sci.*, 2019, **20**(5), 1147.
32. L. A. Bojin, A. F. Serb, M. C. Pascariu, A. Moaca, R. Kostici, V. L. Purcărea, M. Penescu, M. V. Ivan, M. Georgescu and E. Sisu, Assessment of antioxidant properties of different *fomes fomentarius* extracts, *Farmacia*, 2020, **68**, 322–328.
33. I. Suay, F. Arenal, F. J. Asensio, A. Basilio, M. Angeles Cabello, M. Teresa Díez, J. B. García, A. González del Val, J. Gorrochategui, P. Hernández and F. Peláez, Screening of basidiomycetes for antimicrobial activities, *Antonie van Leeuwenhoek*, 2000, **78**(2), 129–140.
34. P. Dresch, K. Rosam, U. Grienke, J. M. Rollinger and U. Peintner, Fungal strain matters: colony growth and bioactivity of the European medicinal

polypores *Fomes fomentarius*, Fomitopsis pinicola and Piptoporus betulinus, *AMB Express*, 2015, **5**(1), 1–4.
35. T. Krupodorova, S. Rybalko and V. Barshteyn, Antiviral activity of Basidiomycete mycelia against influenza type A (serotype H1N1) and herpes simplex virus type 2 in cell culture, *Virol. Sin.*, 2014, **29**(5), 284–290.
36. M. Aoki, M. Tan, A. Fukushima, T. Hieda, S. Kubo, M. Takabayashi, K. Ono and Y. Mikami, Antiviral substances with systemic effects produced by Basidiomycetes such as *Fomes fomentarius*, *Biosci., Biotechnol., Biochem.*, 1993, **57**(2), 278–282.
37. O. F. Seniuk, L. F. Gorovoj, G. V. Beketova, H. O. Savichuk and P. G. Rytik, Kucherov II, Prilutskay AB, Prilutsky AI. Anti-infective properties of the melanin–glucan complex obtained from medicinal tinder bracket mushroom, *Fomes fomentarius* (L.: Fr.) Fr. (Aphyllophoromycetideae), *Int. J. Med. Mushrooms*, 2011, **13**(1), 7–18.
38. N. Maljurić, J. Golubović, M. Ravnikar, D. Žigon, B. Štrukelj and B. Otašević, Isolation and determination of fomentariol: Novel potential antidiabetic drug from fungal material, *J. Anal. Methods Chem.*, 2018, 1–19.
39. V. Meyer, E. Y. Basenko, J. P. Benz, G. H. Braus, M. X. Caddick, M. Csukai, R. P. De Vries, D. Endy, J. C. Frisvad, N. Gunde-Cimerman and T. Haarmann, Growing a circular economy with fungal biotechnology: a white paper, *Fungal Biol. Biotechnol.*, 2020, **7**(1), 1–23.
40. R. D. Petrova, J. Mahajna, A. Z. Reznick, S. P. Wasser, C. M. Denchev and E. Nevo, Fungal substances as modulators of NF-κB activation pathway, *Mol. Biol. Rep.*, 2007, **34**(3), 145–154.
41. J. K. Park, K. W. Park, K. S. Shin, C. M. Lee, S. J. Seok, J. B. Kim, B. S. Koo, B. S. Han and S. H. Yoon, Isolation and chemical analysis of potent anti-complementary polysaccharides from fruiting bodies of the *Fomes fomentarius*, *Microbiol. Biotechnol. Lett.*, 2013, **41**(2), 198–206.
42. J. H. Choe, Y. J. Yi, M. S. Lee, D. W. Seo, B. S. Yun and S. M. Lee, Methyl 9-Oxo-(10E,12E)-octadecadienoate Isolated from *Fomes fomentarius* Attenuates Lipopolysaccharide-Induced Inflammatory Response by Blocking Phosphorylation of STAT3 in Murine Macrophages, *Mycobiology*, 2015, **43**(3), 319–326.
43. K. Kulasinski, R. Guyer, D. Derome and J. Carmeliet, Water adsorption in wood microfibril-hemicellulose system: Role of the crystalline–amorphous interface, *Biomacromolecules*, 2015, **16**(9), 2972–2978.
44. Y. Zhang, Y. Xiao, P. Wang and Q. Liu, Compositions and anti-tumor activity of pyropolyporus fomentarius petroleum ether fraction *In Vitro* and *In Vivo*, *PLoS One*, 2014, **9**(10), e109599.
45. Y. Zhang, P. Wang, Y. Xiao, X. Wang, S. Yang and Q. Liu, Chemical compositions and antiproliferation activities of the chloroform fraction from Pyropolyporus fomentarius in K562 cells, *Hum. Exp. Toxicol.*, 2015, **34**(7), 732–743.
46. R. R. Mvondo, P. Meukam, J. Jeong, D. D. Meneses and E. G. Nkeng, Influence of water content on the mechanical and chemical properties of tropical wood species, *Results Phys.*, 2017, **7**, 2096–2103.

47. S. H. Kim, R. Jakhar and S. C. Kang, Apoptotic properties of polysaccharide isolated from fruiting bodies of medicinal mushroom *Fomes fomentarius* in human lung carcinoma cell line, *Saudi J. Biol. Sci.*, 2015, **22**(4), 484–490.
48. J. Ruiz-Herrera, *Fungal cell wall: structure, synthesis, and assembly*, CRC Press, 1991.
49. W. Chen, Z. Zhao, S. F. Chen and Y. Q. Li, Optimization for the production of exopolysaccharide from *Fomes fomentarius* in submerged culture and its antitumor effect in vitro, *Bioresour. Technol.*, 2008, **99**(8), 3187–3194.
50. H. Alvandi, A. Hatamian-Zarmi, B. E. Hosseinzadeh, Z. B. Mokhtari-Hosseini, E. Langer and H. Aghajani, Improving the biological properties of *Fomes fomentarius* MG835861 exopolysaccharide by bioincorporating selenium into its structure, *Carbohydr. Polym. Technol. Appl.*, 2021, **2**, 100159.
51. J. Breitenbach and F. Kränzlin, in *Fungi of Switzerland, Vol. 2. Non gilled fungi-Heterobasidiomycetes, Aphyllophorales, Gasteromycetes*, 1986.
52. I. Akata, B. Ergönül and F. Kalyoncu, Chemical compositions and antioxidant activities of 16 wild edible mushroom species grown in Anatolia, *Int. J. Pharmacol.*, 2012, **8**(2), 134–138.
53. L. Barros, B. A. Venturini, P. Baptista, L. M. Estevinho and I. C. Ferreira, Chemical composition and biological properties of Portuguese wild mushrooms: a comprehensive study, *J. Agric. Food Chem.*, 2008, **56**(10), 3856–3862.
54. J. Fäldt, M. Jonsell, G. Nordlander and A. K. Borg-Karlson, Volatiles of bracket fungi Fomitopsis pinicola and *Fomes fomentarius* and their functions as insect attractants, *J. Chem. Ecol.*, 1999, **25**(3), 567–590.
55. M. Gucia, G. Jarzyńska, E. Rafał, M. Roszak, A. K. Kojta, I. Osiej and J. Falandysz, Multivariate analysis of mineral constituents of edible Parasol Mushroom (Macrolepiota procera) and soils beneath fruiting bodies collected from Northern Poland, *Environ. Sci. Pollut. Res.*, 2012, **19**(2), 416–431.
56. J. Falandysz, T. Kunito, R. Kubota, M. Gucia, A. Mazur, J. J. Falandysz and S. Tanabe, Some mineral constituents of Parasol Mushroom (Macrolepiota procera), *J. Environ. Sci. Health, Part B*, 2008, **43**(2), 187–192.
57. P. Singh and S. Bangaswami, Fomentaric acid, a novel type of alkyl succinic acid from *Fomes Fomentarius* (L.) Fr, *Tetrahedron Lett.*, 1967, **8**(2), 149–152.
58. R. Wan and M. Eghbalian, Multiscale model for damage-fluid flow in fractured porous media, *Int. J. Multiscale Comput. Eng.*, 2016, **14**(4), 367–387.
59. L. Ran, M. Qingyun, K. Fandong, X. Qingyi, D. Qiong and Z. Youxing, Analysis of Chemical Constituents from the Fruiting Bodies of *Fomes fomentarius*, *J. Trop. Biol.*, 2018, **04**, 445–451.
60. J. Li, Study on Chemical Constituent and Pharmacological Activity of Fomes Fomentarius, Jilin Agricultural University, 2011, p. 68. Available at: https://oversea.cnki.net/KCMS/detail/detail.aspx?dbcode=CMFD&dbname=CMFD2011&filename=1011150555.nh&uniplatform=OVERSEAS_EN&v=FzlaQIIXhivgccoXVpGsbh1ckw7mCby2LKsbi9GqDfgpYtkea4EFxhdbKleVUic7.

61. J. Y. Zhao, J. H. Ding, Z. H. Li, Z. J. Dong, T. Feng, H. B. Zhang and J. K. Liu, Three new phenyl-ethanediols from the fruiting bodies of the mushroom *Fomes fomentarius*, *J. Asian Nat. Prod. Res.*, 2013, **15**(3), 310–314.
62. L. Kalitukha and M. Sari, Fascinating vital mushrooms. Tinder fungus (*Fomes fomentarius* (L.) Fr.) as a dietary supplement, *Int. J. Res. Stud. Sci. Eng. Tech.*, 2019, **6**, 1–9.
63. D. W. Seo, Y. J. Yi, M. S. Lee, B. S. Yun and S. M. Lee, Differential modulation of lipopolysaccharide-induced inflammatory cytokine production by and antioxidant activity of fomentariol in RAW264. 7 cells, *Mycobiology*, 2015, **43**(4), 450–457.
64. L. Ma, H. Chen, W. Zhu and Z. Wang, Effect of different drying methods on physicochemical properties and antioxidant activities of polysaccharides extracted from mushroom Inonotus obliquus, *Food Res. Int.*, 2013, **50**(2), 633–640.
65. S. Zheng, C. Li, T. B. Ng and H. X. Wang, A lectin with mitogenic activity from the edible wild mushroom Boletus edulis, *Process Biochem.*, 2007, **42**(12), 1620–1624.
66. J. Z. Yin and L. X. Zhou, Analysis of nutritional components of 4 kinds of wild edible fungi in Yunnan, *Food Res. Dev.*, 2008, **29**, 133–136.
67. R. Oliver and R. W. D.-I. Corn, *Gluten meal dietary supplements comprises chitin, beta D glucan and melanin complex, which are produced from* fomes fomentarius *mushroom product, where the dietary supplements can be used with other dietary supplements or dietary fiber*, DE202006016129U1, 2006, https://patents.google.com/patent/DE202006016129U1/en.
68. M. Kolundžić, NĐ Grozdanić, M. Dodevska, M. Milenković, F. Sisto, A. Miani, G. Farronato and T. Kundaković, Antibacterial and cytotoxic activities of wild mushroom *Fomes fomentarius* (L.) Fr., Polyporaceae, *Ind. Crops Prod.*, 2016, **79**, 110–115.
69. Y. M. Park, I. T. Kim, H. J. Park, J. W. Choi, K. Y. Park, J. D. Lee, B. H. Nam, D. G. Kim, J. Y. Lee and K. T. Lee, Anti-inflammatory and anti-nociceptive effects of the methanol extract of *Fomes fomentarius*, *Biol. Pharm. Bull.*, 2004, **27**(10), 1588–1593.
70. G. Gedik, G. Dülger, A. S. Hülya, A. Özyurt and A. S. Ahmet, The antimicrobial effect of various formulations obtained from *Fomes fomentarius* against hospital isolates, *Mantar Dergisi*, 2019, **10**(2), 103–109.
71. H. Alvandi, A. Hatamian-Zarmi, B. Ebrahimi Hosseinzadeh and Z. B. Mokhtari-Hosseini, Optimization of production conditions for bioactive polysaccharides from *fomes fomentarius* and investigation of antibacterial and antitumor activities, *Iran. J. Med. Microbiol.*, 2020, **14**(6), 596–611.
72. Y. Zhang, P. Wang, Y. Xiao, X. Wang, S. Yang and Q. Liu, Chemical compositions and antiproliferation activities of the chloroform fraction from Pyropolyporus fomentarius in K562 cells, *Hum. Exp. Toxicol.*, 2015, **34**(7), 732–743.

73. A. K. Darkal, M. M. Zuraik, Y. Ney, M. J. Nasim and C. Jacob, Unleashing the biological potential of *Fomes fomentarius* via dry and wet milling, *Antioxidants.*, 2021, **10**(2), 303.
74. A. Shahvaroughi Farahani, M. Kurepaz-mahmoodabadi and A. Manayi, Evaluation of cytotoxic activity of *Fomes fomentarius* extracts against brine shrimp larvae and colon adenocarcinoma cell line (SW742), *Adv. Pharmacol. Ther. J.*, 2021, **1**(2), 77–85.
75. H. L. Gao, L. S. Lei, C. L. Yu, Z. G. Zhu, N. N. Chen and S. G. Wu, Immunomodulatory effects of *Fomes fomentarius* polysaccharides: an experimental study in mice, *Nan fang yi ke da xue xue bao = J. South. Med. Univ.*, 2009, **29**(3), 458–461.
76. L. M. Henning, U. Simon, A. Abdullayev, B. Schmidt, C. Pohl, T. Nunez Guitar, C. Vakifahmetoglu, V. Meyer, M. F. Bekheet and A. Gurlo, Effect of *fomes fomentarius* cultivation conditions on its adsorption performance for anionic and cationic dyes, *ACS Omega*, 2022, **7**(5), 4158–4169.
77. N. Papp, K. Rudolf, T. Bencsik and D. Czégényi, Ethnomycological use of *Fomes fomentarius* (L.) Fr. and Piptoporus betulinus (Bull.) P. Karst. in Transylvania, Romania, *Genet. Resour. Crop Evol.*, 2017, **64**(1), 101–111.
78. Happy. *Fomes Fomentarius*, or Hoof Fungus – Harvesting, Uses and Benefits. Retrieved from the HubPages. Website: https://discover.hubpages.com/health/Fomes-fomentarius-or-Hoof-Fungus-Harvesting-Uses-and-Benefits, (2020, November).
79. U. Peintner, R. Pöder and T. Pümpel, The iceman's fungi, *Mycol. Res.*, 1998, **102**(10), 1153–1162.

CHAPTER 10

Ganoderma lucidum *(Lingzhi Mushroom): Its Medicinal Uses, Biomolecules and Therapeutic Applications*

RINCHEN YANGZOM[a] AND PHURPA WANGCHUK*[b]

[a] National Biodiversity Centre, Serbithang, Ministry of Agriculture and Forests, Thimphu, Bhutan; [b] Centre for Molecular Therapeutics, Australian Institute of Tropical Health and Medicine, James Cook University, Building E4, McGregor Rd, Smithfield, Cairns, QLD 4878, Australia
*Email: phurpa.wangchuk@jcu.edu.au

10.1 Introduction

Ganoderma is the largest genus, with over 300 species belonging to the family Ganodermataceae;[1,2] it is probably the most morphologically complex genus of polypores.[3,4] *Ganoderma lucidum* is one of these species, which was initially described as *Boletus lucidus* Curtis.[5–8] It is a macrofungus and used in many indigenous medicines of China, Japan, and other Asian countries.[9] Recently, this medicinal mushroom, has gained popularity due to its chemical constituents and medicinal properties for treating cancer, as well as promoting health and longevity. It is popularly known as the "mushroom of immortality", "king of herbs", "healing fungus", "herb of longevity", "herb of spiritual potency", "elixir of life", "wonder mushroom", "magic mushroom", "the symbol of traditional Chinese medicine", "supernatural mushroom", "ten-thousand-year mushroom", and "celestial herb".

Phytochemistry and Nutritional Composition of Significant Wild Medicinal and Edible Mushrooms: Traditional Uses and Pharmacology
Edited by Ajay Sharma, Garima Bhardwaj and Gulzar Ahmad Nayik

Published by the Royal Society of Chemistry, www.rsc.org

Morphologically, it has been described as a perennial, large-shelf mushroom or bracket fungus, with a woody texture and a shiny surface (Figure 10.1). Its fruiting body (pileus) is shiny, kidney or fan shaped, elongated or irregular with a thick margin that becomes brown as it matures. Its stipe (stem) is a long, corky, shiny, smooth, chocolate brown with a varnished crust, found on the trunks of both living and dead trees. It has basidiospores, which are double-walled spores with an ellipsoid shape and truncated apex and yellow to brown inner layers and obovate pores with a rugose surface that are cream become ochreous.[1,10–13] Owing to several factors, such as geographical locations, climatic conditions, inter-hybridization or the natural genetic development of individual species,[14,15] there is enormous variation in its morphological characteristics. For this reason, it is recommended to use combined methods, involving molecular, chemotaxonomic and macro–micro-morphological studies to develop a reliable genus speciation and also differentiate amongst species.[16] Secondary metabolite profiles, such as the content of polysaccharides and triterpenoids, have proven to be useful tools in differentiating the species of Ganoderma growing in various regions of the worlds. For example, less polysaccharide (1.1 to 5.8%) and triterpenic acid content was reported in European *G. lucidum* compared to the East Asian species.[7,15]

10.2 Ecological Distribution

The *Ganoderma* species are ecologically significant wood-degrading fungi that live in hardwood and coniferous forests with a global distribution in terrestrial ecosystems ranging from tropical to temperate geographical regions. The genus of white-rot fungi with a cosmopolitan distribution causes

Figure 10.1 Huge *Ganoderma lucidum* growing on the roots of a tree in a warm broadleaved forest in Bhutan.

basal stem rot, as well as root and butt rots through decomposition of lignin and cellulose in a variety of tree species.[17–19] The majority of the populations of *Ganoderma* species are found in tropical and subtropical regions of Asia, Africa, Europe, and America, while small populations are also found in temperate zones.[11,20] While *G. lucidum* is known to be distributed throughout the world, it has not reported in several countries (Figure 10.2).

10.3 Ethnomycological Uses

Ethnomycology refers to studying human interaction, selection and historical uses of mushrooms found in the environment.[22] Known as Lingzhi in China and Reishi in Japan, *Ganoderma lucidum* has been used since ancient times in China, Japan, and Korea as bush food/medicine. The ancient Taoists believed this fungus to be the "elixir of eternal youth". The emperors of the early dynasties of China and Japan revered the mushroom as an important herbal medicine and ate the mushrooms and drank them in special teas to attain greater vitality and longevity.[23] This is portrayed in paintings, drawings, carvings, carpet design, jewelry, perfumes bottles, and other creative artworks in China, as well as in Chinese traditional furniture.[24–26] During the Ming Dynasty in China in the 16th century, the mushroom was described in the "Compendium of Materia Medica", and later in the Pharmacopoeia of the People's Republic of China and American Herbal Pharmacopoeia.[27,28] These *materia medica* and pharmacopoeia describe the therapeutic properties of *G. lucidum* as a tonic, relaxant, expectorant, revitalizer or energizer, with immunomodulatory, hepatoprotective, cardioprotective, bacteriostatic, memory enhancing, anti-asthmatic, anti-insomnia, anti-dizziness, and anti-aging properties. However, these health benefits are largely based on traditional uses and cultural mores, and have been rarely substantiated by scientific investigation. Nevertheless,

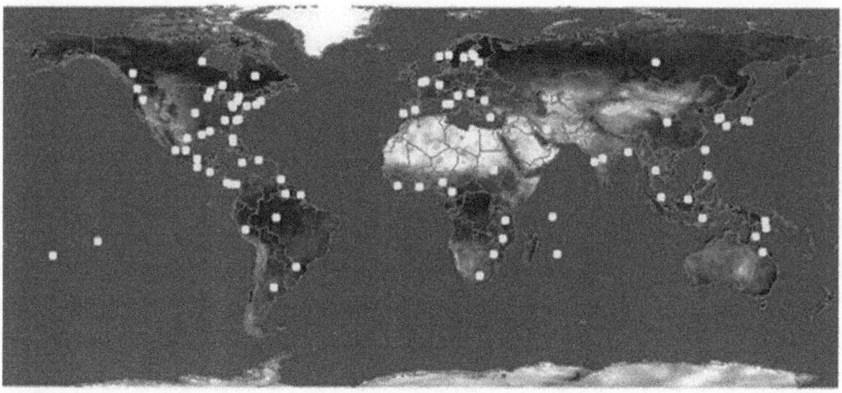

Figure 10.2 Worldwide distributions of *Ganoderma lucidum*. Reproduced with permission from ref. 21.

G. lucidum is the most prescribed traditional medicine in China,[27] and it is gaining popularity in other Asian countries.

Globally, *G. lucidum* is also highly regarded in many folk medicines and is used as a dietary supplements. It is known to be prized for medicinal rather than for its nutritional value. There are many expensive commercial products in the forms of dietary supplements, tea, coffee and powder, which have become popular due to the claimed benefits of anti-cancer properties. In northern Namibia, it is used for veterinary purposes.[29] In Bhutan, many wild *Ganoderma* spp. grow in the forest and are claimed to be recorded as used in Tibetan and Bhutanese Sowa Rigpa medicines, even though they are not currently used.

Ganopoly products are applied to enhance wound healing, heal carcinomas, treat postburn infections, and to prevent injuries in skin flap ischemia reperfusion.[30] In China, Korea, the USA, and European countries, *G. lucidum* is mixed with other ingredients to produce cosmetics, including skin lightening (whitening) products, and is used in antiaging *Ganoderma* extract-based formulations.[31] Ganoderma extracts have been used to remove toxins and help repair skin damage caused by excessive exposure to UV radiation and free radicals.[32,33] Also, such products are made for hair growth in men and to provide hair with sun protection and prevent color fading.[26,34] Lately, studies have demonstrated the use of *G. lucidum* extracts to enrich wines by influencing their sensory profiles and chemical compositions or for promoting the health properties of beer and brandy.[35,36]

10.4 Cultivation

The cultivation of *G. lucidum* is undertaken globally to meet its growing demand on the international market. Lingzhi was popular amongst very wealthy people in the ancient times and was collected from the wild from its natural habitat, and only recently has its cultivation increased due to its worldwide consumption as a herbal medicine or dietary supplement.[26,37] A spore separation method developed by the Institute of Microbiology, under the Chinese Academy of Sciences, Beijing, in the late 1960s was a breakthrough in the cultivation of this fungus, which made the cultivation of Lingzhi popular in China, Japan, and Korea.[25] The growth of the fungus on solid substrate/state, fixed liquid medium or through submerged methods has been reported in the literature.[23] The common substrates used in its commercial production are grains, cork residues, olive oil press cakes, sawdust, wood logs, sunflower seed hull, tea waste, cotton seed husk, residues of farm crops, corn cobs, wheat straw while wood log, short basswood segment, tree stumps, sawdust bags, and bottles. Compared to commercial substrate-based cultivation, natural log cultivation is known to yield superior quality mushrooms and the products attract the best rates on the southeast Asian markets.[26] For successful mushroom cultivation with intact genotypic, phenotypic, and physiological characteristics, it is essential to observe proper seed culture under viable storage conditions.[38]

10.5 Production and Marketing

The annual production of Lingzhi products was over 110 000 tons by the largest manufacturer and exporter in China, with a total estimated annual market price of US$2.5 billion.[23,25] The USA is reported to be the biggest market for Ganoderma and its products.[39] The dietary supplement industry, with global sales of $109 billion, gained traction, and was expected to double by 2020. The Lingzhi industry represents 2% of global dietary supplement sales.[18]

Commercial products, including herbal medicines and dietary supplements or nutraceuticals, are produced from the fruitbodies, mycelia or spores of the mushroom.[9,18,40] These products are made into different dosage forms, such as capsules, pills, powder, decoctions, tinctures, and tea. More than 100 Lingzhi-based products are currently marketed worldwide (Table 10.1).[41,42] *G. lucidum* is also considered to be an economically cost effective and eco-friendly myco-remediation product for removing toxic metals from contaminated soil and polluted water.[12,43,44]

10.6 Nutritional Value

In recent times, *G. lucidum* has been recognized for its exceptional source of nutrition and medicinal properties as a complete healthy food package.[31]

Table 10.1 Products containing *G. lucidum* that are available on the market.

Products/applications	Bioactivity	Ref.
Cosmetics formulation	Boost collagen, improve elasticity, provide hydration, anti-inflammatory	31, 32 and 34
Cosmeceutical skincare product	Anti-tyrosinase activity (anti-wrinkle and hyperpigmentation agent), anti-oxidant activity, anti-inflammatory	32 and 45
Yogurt and fermented milks	Improve nutritional value	35 and 46
Beer production	Improve sensory properties (taste, body, liveliness)	47
Shiraz wine product	Influenced sensory profiles and chemical compositions	35
Korean rice wine (Yakju)	Angio tensin I-converting enzyme, ACE inhibitory activity	35, 36 and 48
Brandy	Improve sensory characteristics and health properties of spirit drinks	48
Hair growth products for men and other hair cosmetics	Provide sun protection, prevent color fading	26 and 34
Improve skin vigor and related problems	Anti-tyrosinase activity, skin anti-aging, anti-oxidant activity, anti-inflammatory properties	30 and 45
Tea production	Enrich with bioactive compounds and improve flavor	49
Myco-remediation/bioremediation product	Removal of heavy metals from abandoned battery slag (strong affinity for toxic metals)	12, 43 and 44

Most mushrooms contain about 90% water by weight and the remaining 10% comprises proteins, fats, carbohydrates, fiber, vitamins, and minerals, while *G. lucidum* mostly comprises fiber (58%) followed by carbohydrates (26–28%). *G. lucidum* is one such mushroom known for promoting health, vigor, and longevity, and is used to produce functional food and beverages. For example, yogurt enriched with β-glucans from *G. lucidum* showed an increase in 17% carbohydrate, 23% protein, and a 17-fold surge in fat content.[46] Moreover, the application of *G. lucidum* extracts in Korean rice wine (Yakju) has been shown to enhance its functional properties, with its ACE (Angio tensin I-converting enzyme) inhibitory activity increased by 10% compared to wine without *G. lucidum* extract, imparting the wine with anti-hypertensive properties.[36] In addition, health-promoting properties have been introduced into commercial Pilsner beer[47] and brandy using *G. lucidum*.

10.7 Toxicity and Health Hazards

Some studies suggest caution when using *G. lucidum* or its extracts due to its toxicity.[52] However, several studies have confirmed it to be safe for consumption. For example, Wachtel-Galor and team[53] have shown that ingestion of this woody fungus has no toxic effects on the liver, kidneys or DNA. Similarly, Smina *et al.*[54] indicated that the toxicity of the total triterpenes is not significant. Dewi *et al.*[55] indicated that oral administration of the polysaccharide peptides of *G. lucidum* up to a dose of 1200 mg per kg body weight per day is nontoxic to the liver, which was also supported by another study[56] that used three different doses (300, 600 and 1200 mg per kg body weight) of mushroom extracts and found them to be safe to the renal organs and their functioning. Further, through laser-Induced breakdown spectroscopic (LIBS) analysis, Sharif and team[57] reported that *G. lucidum* is safe for consumption and a recent study demonstrated low hepatic and renal toxicity in a *G. lucidum* extract.[58] However, a case of *G. lucidum* inducing acute pancreatitis in an elderly lady was reported,[59] which warrants deeper understanding about the relationship between acute pancreatitis and *G. lucidum*.

10.8 Phytochemical Composition and Pharmacological Activity

It was reported that the tumour growth inhibition effects of ethanol extracts of *G. lucidum* are stronger that those of aqueous extracts.[65] Studies have determined the anti-obesity and microbiota-modulating effects of aqueous extracts of *G. lucidum*,[66] with significant radio-protective activity observed in normal cells,[51] as well as neurodegenerative effects and gamma ray irradiation enhancement in the recovery of cellular and immunocompetence.[67] Moreover, ethanolic extracts have shown anti-inflammatory effects,[68] anti-renal cancer,[67] and antimicrobial activities,[69] while methanolic extracts were found to increase cellular autophagy and the formation of autophagosomes in gastric cancer cell lines.[67] Sleep-prolonging effects and enhanced sleep quality upon the administration of *G. lucidum* extract was demonstrated in rats.[70] Studies have shown that the primary active component extracted from *G. lucidum* using organic solvent is triterpene, while with water extraction the main components are polysaccharides.[51]

G. lucidum extracts contain > 400 bioactive compounds,[7,15,18,33,34,60–62] and of these, >240 secondary metabolites have been isolated. The pharmacologically active constituents isolated from *G. lucidum* belong to the chemical classes of polysaccharides, steroids, proteins, sterols, alkaloids, phenols, nucleotides, fatty acids, vitamins, minerals, lectins, lactones, adenosine, germanium, glycopeptides and peptidoglycans.[45,53,61,63,64] However, the content of each of these components varies depending upon the strains used, the growing location, harvesting time, post-harvest processing and storage conditions, and the production methods used to produce the fungus.[33,64]

10.8.1 Triterpenes

Terpenes are the largest group of secondary metabolites and the most diverse set of natural derivatives of the branched C5 carbon skeleton of isoprene; they are widely distributed throughout the plant kingdom. Triterpenoids are a subclass of terpenes, with a basic C30 hydrocarbon skeleton derived from squalene. Triterpenoids have complex and highly oxidized chemical structures with molecular weights in the range of 400–600 kDa.[28] 240 triterpenes have been isolated from the fruiting bodies, mycelia, spores, and culture media of *G. lucidum*, of which 112 are C30 lanostanes (ganoderic acids), 55 are C30 lanostanes (aldehydes, alcohols, esters, glycosides, lactones, and ketones), 27 are C27 lanostanes (lucidenic acids), 18 are C27 lanostanes (alcohols, lactones, and esters), 3 are C24 and C25 lanostanes (lucidone A, B, and C), 1 is a meroterpenoid (Lingzhiol), 23 are steroids and 1 is a benzofuran (ganofuran B) and the strong bitterness of *G. lucidum* is derived from these triterpenoid compounds.[27,61,64] The major triterpenes isolated are ganoderic and lucidenic acids, with other important triterpenes including ganoderals, ganodermic acids, and ganoderiols.[63,64]

G. lucidum is valued for its antihypertensive, antihistaminic, anticancer, anti-oxidative, anti-HIV, anti-inflammatory, anti-angiogenic, hypocholesterolemic, hepatoprotective and hypolipidemic activity due to its triterpenoid content.[45,62,71] Triterpenoids are becoming increasingly recognized in cancer treatments because of their various biological actions, including cell cycle arrest reducing the rate of proliferation, cytotoxicity, regulation of cancer metastasis, suppressing inflammation, inhibition of invasion, and mediating oxidative stress in cancers (Table 10.2).[51,72] A study has shown the highest oxidative effects of terpene-rich fractions of *G. lucidum* containing ganoderic acids A, B, C and D, lucidenic acid B and ganodermanontriol in their maximum concentrations as compared to polysaccharide rich fractions.[62] It has been demonstrated that triterpenoids, namely lucidumol B, ganoderic acid beta, ganolucidic acid A, ganodermanontriol, and ganodermanondiol, exhibit important anti-HIV properties, with ganodermanontriol being reported to also be effective against Epstein Barr, influenza, and Dengue viral infections.[73] Triterpenoids, with their antidiabetic properties, are known for their inhibition by aldose reductase.[74]

10.8.2 *G. lucidum* Polysaccharides, GLPS (Ganopolys)

Ganopoly are macromolecules joined by glycosidic bonds with a molecular mass of >500 kDa. The major bioactive ganopolys isolated are (1→3), (1→6)-α/β-glucans, α-D-glucans, α-D-mannans, and polysaccharide–protein complexes. Glucose, galactose, xylose, arabinose, mannose, fucose, and rhamnose constitute the polysaccharide molecules and various bioactive peptidoglycans, namely ganoderans A, B, and C, have also been extracted from *G. lucidum.*[64,75,76] To date, >200 ganopolys have been extracted from the fruiting bodies, mycelia, and liquid cultures of *G. lucidum*,[11,62,77] and the quantity of water-insoluble polysaccharides was found to be higher than that of water-soluble polysaccharides.[62] GLPS is widely used in tumour therapies for its immunomodulating activity, inhibiting tumour growth through induction of apoptosis and anti-angiogenic effects (Table 10.2).[67,72,78] It was reported that anticancer efficacies improved when GLPS is used with other anticancer agents.[51] in particular, β-D-glucans are known to have greater antitumor activity and good absorption capacity compared to other ganopolys.[45,62,63,71] However, the size of molecules, the degree of branching, number of branches, side chain lengths, and water solubility effect the biological activities of glucans.[11,74] β-Glucan has been revealed to be a potent antioxidant against atherosclerosis.[79] Moreover, ganderans A, B and C, glycoproteins, and heteropolysaccharides with a hydrophilic nature and high molecular weights exhibit significant antitumor effects.[62] GLPS with a high molecular weight can be used as a prebiotic agent to prevent metabolic disorders related to gut dysbiosis and obesity in obese individuals.[80] A proteoglycan, mainly composed of D-glucose, D-mannose, and D-galactose, extracted from fruiting bodies of *G. lucidum* induces antihyperlipidemic, antidiabetic, and antioxidant activities, in addition to hepatoprotective effects.[9,63,80]

10.8.3 Proteins and Peptides

In 1989, the first fungal immunomodulatory protein was isolated from the mycelial extracts of *G. lucidum*. This was Lingzhi-8 (LZ-8), a polypeptide that consists of 110 amino acids with an acetylated amine ending and a 12 kDa molecular mass. LZ-8 inhibits tumour growth, controls immune responses through inducing cell and cytokine production in rodent macrophages and T-cells, and also exhibits important therapeutic effects on gastric and lung cancers.[61,63,64,67,81] A recent study reported that a recombinant immunomodulatory protein (rLZ-8) improved chemotherapy-induced neutropenia – a condition that is related to abnormally low levels of neutrophils in the blood, which is the most common complication of myelosuppressive anti-cancer therapy.[82]

Table 10.2 Common therapeutic activities of *G. lucidum* – mainly studies conducted over the past decade (2010–2021).

Therapeutic actions/effects – cancer treatments	Bioactive compounds	Ref.
Suppress the development of colorectal adenomas	GL extract (GLE)	86
Enhance the sensitivity of epithelial ovarian cancer cells to cisplatin	GL extract (GLE)	96
Inhibit tumour growth and improve anti-tumour immunostimulatory activity	GL extract (GLE)	58
Anti-inflammatory breast cancer	GL extract (GLE)	97
Anti-proliferative activity and repress cell migration and invasiveness	GL extract (GLE)	98
Anti-cancer and anti-inflammatory properties	GL extract (GLE)	99
Regulate the Rac/lamellipodin pathway in breast cancer metastasis	GL extract (GLE)	100
Anticancer activity	GL extract (GLE)	101
Activate immunity and antitumor activity on breast cancer cells	GLPS (*G. lucidum* polysaccharide)	102
Enhance antitumor effects	GLPS (*G. lucidum* polysaccharide)	76
Proliferation and migration of oral squamous cell carcinoma	GLPS (*G. lucidum* polysaccharide)	103
Anti-tumour effects in brain tumour cells	Water extract	104
Inhibit Wnt/β-catenin signaling to suppress osteosarcoma cell growth	GL	105
Suppress tumour growth by inducing apoptosis	*G. lucidum* spore oil (GLSO)	106
Anticancer adjuvant in breast cancer	Extract from sporoderm-breaking spores	107
Induce activation and maturation of human dendritic cells, inhibit proliferation, promote apoptosis of lung cancer cells, increase WBCs in the treatment of chemotherapy-induced neutropenia	Lingzhi-8 (LZ-8)	108
Stimulate dendritic cells and act as an adjuvant in a cancer DNA-based vaccine	Lingzhi-8 (LZ-8)	109
Antiproliferative action	β-1,3-Glucan, mycelial extracts	110
Induces autophagy of a gastric adenocarcinoma cell line – human gastric tumour cells	Methanolic extract	111
Anti-proliferation effects (induces DNA destruction, apoptosis in breast cancer cells, and G1 cell cycle arrest)	Ethanol-soluble and acidic component (ESAC)	112
Anti-cancer properties	Ergosterol peroxide	113
Apoptosis induction	Ganoderic acids (GAs)	114

Table 10.2 (*Continued*)

Therapeutic actions/effects – cancer treatments	Bioactive compounds	Ref.
Other therapeutic properties		
Reduce the negative effects of COVID-19	GL	91
Hypnotic effects (promote sleep in rats)	GL extract	70
Anti-inflammatory efficacy against chronic inflammatory diseases such as atherosclerosis	GL	115
Anti-obesity and microbiota-modulating effects	Aqueous extract of *Ganoderma lucidum* mycelium (WEGL), ganopoly	80
Anti-inflammatory activities	Lingzhi-8 (LZ-8)	81
Antioxidative activity (induce oxidative stress)	Total triterpenes	116
Improves aerobic endurance, lower body flexibility and velocity (improves the physical fitness of women with fibromyalgia)	GL	88
Improves the function of macular pre-ganglionic elements	Macuprev® (containing 600 mg of GL with other ingredients)	93
Anti-ulcer properties through its free radical scavenging activity	Ethanolic extract	117
Anti-aging effects	Ganodermasides A and B (ergosterol derivatives)	118
Protective effect against liver injury (hepatoprotective effects)	Proteoglycans	9
Anti-diabetes mellitus type 2, inhibition of the cholesterol synthesis pathway, alpha-glucosidase inhibitor	Ganoderol B	119
Anti-oxidant capacity	GLPS (*G. lucidum* polysaccharide)	120
Anti-diabetic effect	GLPS (*G. lucidum* polysaccharide)	121
Antioxidant and hepatoprotective effects	GLPS (*G. lucidum* polysaccharide)	89
Inhibition of cervical cancer cell lines and cytotoxicity	GLTs (GL triterpenoids)	122
Anti-aging and hepatoprotective effects	PSP (polysaccharide peptide)	89

Dyslipidemia effect (antihyperlipidemic activity)	GL	87
Antidiabetic effects and improves dyslipidemia	GL capsules	87
Decreases atherosclerosis (cardiovascular effect)	GL capsules	90
Antioxidant activity	PSP (polysaccharide peptide)	90
Antidiabetic effect	GLE	99
Antimicrobial activity (antibacterial/ antifungal)	Ethanol extracts	69
Antibacterial activity	GL extracts	123
Ameliorative effects against	GL extracts	124
Epstein Barr and influenza viruses, a novel drug against dengue viral infection	Ganodermanontriol	73
Ameliorates respiratory tract infection symptomatology	GL + elderberry	92
Progresses cognitive damage, inhibits apoptosis in hippocampus tissues and cells in Alzheimer's disease and improves neuronal damage	GLTs (GL triterpenoids)	125
Prevents neutropenia in tumour chemotherapy	Recombinant LINGZHI-8 (rLZ-8)	108
Immunoregulatory fungal protein	LINGZHI-8 (LZ-8)	82
Induces endoplasmic reticulum stress-mediated autophagic apoptosis	Recombinant LINGZHI-8 (rLZ-8)	126
Promotes epidermal growth to induce apoptosis of lung cancer cells and prevent growth	Recombinant LINGZHI-8 (rLZ-8)	127
Improves the therapeutic activities of a DNA vaccine against MBT-2 tumour in mice.	Recombinant LINGZHI-8 (rLZ-8)	128
Anticancer potential	Sporoderm-breaking spores extract	107
Inhibit colorectal cancer	Ethanol extracts of sporoderm-broken spores	129

Ganodermin is another protein that has been extracted from the fruiting bodies of *G. lucidum*, which has a molecular mass of 15 kDa and exhibits antifungal activity.[64] Some other bioactive proteins extracted from the fruiting bodies of *G. lucidum* are LZP-1, LZP-2, and LZP-3.[63] The polysaccharide–protein complex known as GLPP consists of fructose, rhamnose, galactose, xylose, and glucose, connected *via* b-glycosidic bonds, and exhibits antioxidant properties.[77]

10.8.4 Enzymes, Nucleosides/Tides, and Amino Acids

Amylase, carboxyl proteinase, cellulases, lignin-modifying enzymes, α-galactosidase, isozymes, and laccase are some enzymes that have been isolated from the Lingzhi mushroom.[63] More than 18 amino acids have been reported to be present in *G. lucidum* extracts, of which the most abundant are glutamic and aspartic acids, glycine, and leucine. Leucine has been shown to exhibit hypoglycemic and antioxidant effects.[23,76]

Cytidine, adenosine, inosine, hypoxanthine, guanosine, adenine, thymidine, guanine, thymine, uridine and uracil have also been extracted from the mycelia and fruiting bodies of *G. lucidum*. Adenosine and its derivatives have been shown to inhibit platelet aggregation.[11,63,76]

10.8.5 Vitamins and Essential Minerals

Vitamins B1, B2, B6, β-carotene, C, D, E, Niacin (nicotinic acid – vitamin B3), and pro-vitamin D2 have been found in *G. lucidum*.[28,62,63,76]

Potassium, sulfur, calcium, phosphorus, magnesium are the major mineral components of *G. lucidum*, while iron, zinc, copper, sodium, manganese, and strontium are found in small amounts, with silicon, carbon, chromium, lead, cobalt, aluminum, arsenic, nickel, and molybdenum among the other minerals identified in this fungus.[62,76,83] A selenium content of ≤72 μg Se per g dry weight of the fungus and[84] and a germanium content of 489 μg per g dry weight of the fungus have been reported as being present in the fruiting bodies of wild *G. lucidum*, which possess immunopotentiating, antitumor, antioxidant, and antimutagenic effects,[83] with others including heavy metal detoxification and free-radical scavenging.[63]

10.8.6 Alkaloids, Fiber, and Steroids

Choline and betaine are alkaloids that have been isolated from the spores of *G. lucidum*.[85] The crude fiber content in *G. lucidum* is 59%, with β-glucans (1→3) mainly forming the indigestible part of the plant food.

G. lucidum has been reported to contain 20 types of sterols, with their skeletons divided into cholesterols and ergosterols.[76] Ergosteryl esters and ergosterol (pro vitamin D2) from the fruiting bodies of *G. lucidum* exhibit anti-cancer, anti-inflammatory, and anti-angiogenesis effects.[11,81]

10.9 Clinical Trials and Therapeutic Applications

The numbers of preclinical (*in vivo*, *in vitro* or *ex vivo*) studies have significantly increased over the past four decades, with the primary objectives being: (i) to discover new pharmacologically bioactive compounds, (ii) elucidate the underlying mechanisms of the biological activities/effects, (iii) identify any potential side effects, (iv) characterize biological compounds, and (v) study their functions and therapeutic applications. Only a limited number of preclinical studies have translated into clinical trials.[15] A study of 123 patients with colorectal adenomas treated with *G. lucidum* extract of 1.5 g per day for a year suggested the suppression of the development of colorectal adenomas, which are precancerous lesions of the large bowel.[86] In a controlled cross-over trial, 26 patients suffering from dyslipidemia and/or hypertension treated with 1.44 g of *G. lucidum* daily for 12 weeks showed improvements in the dyslipidemia of diabetes.[87] Upon its administration at a dose of 6 g per day for six weeks, *G. lucidum* was shown to promote a significant improvement in velocity, lower body flexibility, and aerobic endurance in 64 women suffering from fibromyalgia.[88] A study established the anti-aging, hepatoprotective, antioxidative, and activities of *G. lucidum* by suppressing oxidative stress effectively in 42 healthy subjects who took the fungus in capsule form (225 mg) for six successive months.[89] In a clinical trial, the polysaccharide peptide (PsP) of *G. lucidum* was shown to exhibit antioxidant propertied against atherosclerosis in patients suffering from angina in stable and serious conditions. 37 serious and 34 stable patients were treated with 750 mg of PsP per day for 90 days.[90]

G. lucidum and its potential in the treatment of COVID-19 infections was determined in a clinical study of 150 COVID-19 patients treated with a dose of 300 mg per kg of body weight daily for 10 days. The results showed that *G. lucidum* plays a significant role in reducing the effects of COVID-19.[91] A study demonstrated a natural way to curb the severity and length of respiratory tract infections in a group of 60 women and men aged over 65, a particularly vulnerable population, through treatment with a combination of 1.5 g of elderberry and 0.5 g of *G. lucidum* extracts or a placebo for 14 weeks.[92] The effectiveness of Macuprev® containing 600 mg of *G. lucidum* plus other ingredients was tested in 30 patients suffering from transitional age-related macular degeneration for six months. An improvement in the role of the macular pre-ganglionic elements was observed, without related choroidal and retinal ultra-structural changes upon Macuprev® supplementation.[93]

A single-case study showed promising results in a patient (50 years of age) who self-medicated with *G. lucidum* for three months to combat the symptoms of Parkinson's disease, with the study recommending future clinical research on a larger sample size for validation of the results.[94] Moreover, a study that investigated *G. lucidum* efficacy in 60 patients with gynecological cancers showed stabilization of the disease, although it was not possible to ensure *G. lucidum* use in treating gynecological cancers as the study was incomplete. In an another study, an attempt to explore the efficacy and safety of *G. lucidum* spore powder in treating Alzheimer's disease resulted in no promising efficacy after a 6 week treatment in 42 patients; a larger sample size and longer treatment duration may be the key to the success of such clinical trials in the future.[95] The small amount of clinical research data may be due to a small sample size, lack of data for replication, short duration of treatment, inappropriate research approaches or irrelevant statistical models, which hinder the use of *G. lucidum* as a potential conventional drug.[15]

10.10 Biodiscovery Potential and Conservation Aspects

G. lucidum has become popular worldwide as a significant natural source of pharmaceuticals, nutraceuticals, and cosmeceuticals with immense biodiscovery potential. *G. lucidum* products are the most sought-after natural remedies in the treatment of diseases, including numerous cancers. Dietary supplements containing *G. lucidum* are widely used as therapeutic agents and there has been a boom in the use of *G. lucidum* in the cosmetics industry. Further investigations could lead to the discovery of new therapeutic metabolites, nutraceuticals, and pharmacological formulations.

Patents have been published that provide additional valued information and highlights applications for developing Lingzhi-based industries.[131] According to Google patents,[132] from over 4000 patented products or procedures associated with *G. lucidum*, 32 products associated with the therapeutic properties/effects of *G. lucidum* were published in the period 2021–2022 (a duration of a little over two years). This indicates that the number of patents on Lingzhi is rapidly increasing. In a patent taken from Google patents,[132] a liquid-state fermentation broth of *G. lucidum* was prepared for the amelioration, prevention or treatment of Alzheimer's disease or other relevant symptoms induced by Alzheimer's disease. In another patent,[132] a *G. lucidum* mycelium antihypertensive peptide and its preparation method were described, combining the technical fields of polypeptide biotechnology and antihypertensive drugs.

Another patent[132] details a *G. lucidum* spore powder that exhibits an immunization effect and its preparation process, which involves the use of spore powder and tussah pupa (including various pupae) as well as a number of food raw materials for promoting the growth of healthy cells, removing free radicals, and improving the growth of cells of organism immunity, improving the organism immunity function, dealing with the invasion of foreign microorganisms (viruses), and increasing the variety of special diet foods.

Another patent[132] relates to a simple and effective experimental method of synthesizing an anti-tumor *G. lucidum* spore oil lycopene compound preparation, showing the experimental research steps of the antiproliferative action of LZFQ on four types of tumor cells, the influence of LZFQ on the tumour quality of A549 tumor-bearing mice, and the influence of LZFQ with different concentrations on the apoptosis of A549 cells.

An invention was detailed in a patent related to the technical fields of polypeptide biotechnologies and antihypertensive drugs and a *Ganoderma lucidum* mycelium antihypertensive peptide and a preparation method thereof were described;[132] the antihypertensive peptide in the patent has the characteristics of being low in molecular weight and easily absorbed by the human body, and could be used as a food additive for healthcare or as an antihypertensive drug.

A patent[132] describes the preparation of food and cosmetic raw materials, and discloses an antioxidant *Ganoderma lucidum* extract and its preparation.

Moreover, a compound *Ganoderma lucidum* spore oil extract for preventing cerebral apoplexy and a preparation method and application thereof was invented.[132] A lucid *Ganoderma lucidum* fruiting body extract for preventing cerebral apoplexy and preparation method and application thereof were patented to overcome the limitations of the efficacy of *Ganoderma lucidum* extract in the market.[132]

A *Ganoderma lucidum* spore oil vitamin E soft capsule was also patented,[132] which was prepared from 50–55 parts by mass of capsule core material liquid and 45–50 parts by mass of capsule shell glue liquid.

The preservation of the wild natural strain of *G. lucidum* is considered to be important for the mass cultivation of this fungus for commercial purposes, which is possible, unlike for a few other high-priced fungi, such as the caterpillar fungus (*Ophiocordyceps sinensis*), matsutake (*Tricholoma matsutake*) or King mushroom (*Boletus edulis*). As consumers still prefer the wild natural strain of *G. lucidum*, proper harvesting, conservation strategies, and plans are key to sustaining wild populations and to protect mushroom diversity for local uses. Human encroachment into habitats and deforestation has led to habitat loss, resulting in a loss of mushroom diversity. To conserve biodiversity is to maintain species richness and preserve reservoirs of valuable genes for direct benefits and sustainability; hence, mushroom diversity in the wild is vital to contributing to genetic resources for creating diverse varieties in artificial breeding programs as per consumer demand.[130]

10.11 Conclusions and Future Prospects

Recent research has reported that >400 phytochemicals are found in *G. lucidum* and some 240 secondary metabolites/triterpenes and around 200 primary bioactive compounds have been extracted from its mycelia, spores, fruiting bodies, and culture media. The vast chemical and structural diversity of *G. lucidum* makes it an important medicine and food supplement, and more new products are increasingly receiving attention globally. Triterpenoids and polysaccharides are the two major constituents of *G. lucidum*, which exhibit pharmacological and therapeutic effects. Triterpenoids show antihypertensive, antihistaminic, anticancer, anti-oxidative, anti-HIV, anti-inflammatory, anti-angiogenic, hypocholesterolemic, hepatoprotective and hypolipidemic activity, while polysaccharides have antitumor properties through anti-immunomodulatory and anti-angiogenesis activities. However, clinal, *in vitro* and *in vivo* research needs to be enhanced to recognize any probable side effects, to elucidate the underlying molecular mechanisms of the pharmacological activities, and to implement appropriate research approaches and statistical models. Further, a scientific evaluation system is required for quality control, which is essential to ensure the identity, value, and safety of *G. lucidum* products or drugs available on the market. As natural products are preferred by the consumer for a healthier life, it is imperative to establish more scientific evidence to understand mushrooms better. With the demand for natural products and drugs for improving health and vigor, bioprospecting potential should be explored.

Acknowledgements

We appreciate our families for always providing us with their full support.

References

1. S. Bhosle, K. Ranadive, G. Bapat, S. Garad, G. Deshpande and J. Vaidya, *Mycosphere*, 2010, **1**(3), 249–262.
2. S. Baby, A. J. Johnson and B. Govindan, *Phytochemistry*, 2015, **114**, 66–101.
3. L. Ryvarden, *Mycologia*, 2000, **92**, 180–191.
4. S. Jargalmaa, J. A. Eimes, M.-S. Park, J. Y. Park, S.-Y. Oh and Y. W. Lim, *PeerJ*, 2017, **5**, e3596.
5. J.-M. Moncalvo, H.-H. Wang and R. S. Hseu, *Mycologia*, 1995, **87**, 223–238.
6. A. M. Gottlieb and J. E. Wright, *Fungal Biol.*, 1999, **103**, 661–673.
7. F. Hennicke, Z. Cheikh-Ali, T. Liebisch, J. G. Maciá-Vicente, H. B. Bode and M. Piepenbring, *Phytochemistry*, 2016, **127**, 29–37.
8. T. R. Kinge, A. M. Mih, M. D. S. Neves, D. R. T. D. Sousa, M. D. P. S. B. C. Ferreira, M. Z. M. Frota, J. V. B. Souza and J. L. L. Lozano, *J. Yeast Fungal Res.*, 2015, **6**(2), 17–24.
9. S. Wachtel-Galor, J. Yuen, J. A. Buswell and I. F. F. Benzie, in *Herbal Medicine: Biomolecular and Clinical Aspects*, ed. I. F. F. Benzie and S. Wachtel-Galor, Boca Raton (FL), 2011.
10. M. P. Vishwakarma, R. P. Bhatt and S. Gairola, *Int. J. Med. Aromat. Plants*, 2011, **1**, 33–40.
11. M. Siwulski, K. Sobieralski, I. Golak-Siwulska, T. SokóB and A. T. Kara, *Herba Pol.*, 2015, **61**, 105–118.
12. M. I. Mafia, F. M. Aminuzzaman, M. S. M. Chowdhury and J. F. Tanni, *J. Agric. Nat. Resour.*, 2020, **3**(2), 160–174.
13. A. Bijalwan, M. Dobriyal, T. Thakur, M. Thakur and K. Bahuguna, *Biodiversity Int. J.*, 2020, **4**, 202–209.
14. Y.-J. Park, O.-C. Kwon, E.-S. Son, D.-E. Yoon, W. Han, Y. B. Yoo and C.-S. Lee, *Mycobiology*, 2012, **40**, 71–75.
15. R. Ahmad, M. Riaz, A. Khan, A. Aljamea, M. Algheryafi, D. Sewaket and A. Alqathama, *Phytother. Res.*, 2021, **35**, 6030–6062.
16. S. G. Hong and H. S. Jung, *Mycologia*, 2004, **96**, 742–755.

17. M. P. A. Coetzee, S. Marincowitz, V. G. Muthelo and M. J. Wingfield, *IMA Fungus*, 2015, **6**, 249–256.
18. A. L. Loyd, B. S. Richter, M. A. Jusino, C. Truong, M. E. Smith, R. A. Blanchette and J. A. Smith, *Front. Microbiol.*, 2018, **9**, 1557.
19. V. Fryssouli, G. I. Zervakis, E. Polemis and M. A. Typas, *MycoKeys*, 2020, **75**, 71–143.
20. S. Zhao, Q. Gao, C. Rong, S. Wang, Z. Zhao, Y. Liu and J. Xu, *J. Fungi*, 2020, **6**, 269.
21. https://www.discoverlife.org/, Global Biodiversity Information Facility, Systematic Mycology and Microbiology Laboratory (15th March 2022).
22. R. Sitotaw, E. Lulekal and D. Abate, *J. Ethnobiol. Ethnomed.*, 2020, **16**, 1–14.
23. B. S. Sanodiya, G. S. Thakur, R. K. Baghel, G. B. Prasad and P. S. Bisen, *Curr. Pharm. Biotechnol.*, 2009, **10**, 717–742.
24. M. L. Roberts, Doctoral dissertation, Swinburne University of Technology, 2004.
25. S. Li, C. Dong, H. Wen and X. Liu, *Mycology*, 2016, **7**, 74–80.
26. K. K. Hapuarachchi, W. A. Elkhateeb, S. C. Karunarathna, C. R. Cheng, A. R. Bandara, P. Kakumyan, K. D. Hyde, G. Daba and T.-C. Wen, *Mycosphere*, 2018, **9**, 1025–1052.
27. X. Zhou, J. Lin, Y. Yin, J. Zhao, X.-F. Sun and K. Tang, *Am. J. Chin. Med.*, 2007, **35**(4), 559–574.
28. C. Sharma, N. Bhardwaj, A. Sharma, H. S. Tuli, P. Batra, V. Beniwal, G. K. Gupta and A. Sharma, *J. Herb. Med.*, 2019, 100268.
29. N. P. Kadhila-Muandingi, Doctoral dissertation, The University of Namibia, 2010.
30. Z. Yin, B. Yang and H. Ren, *Adv. Exp. Med. Biol.*, 2019, **1182**, 311–321.
31. O. Taofiq, S. A. Heleno, R. C. Calhelha, M. J. Alves, L. Barros, A. M. González-Paramás, M. F. Barreiro and I. C. F. R. Ferreira, *Food Chem. Toxicol.*, 2017, **108**(Pt A), 139–147.
32. L. Jiang, *Global J. Res. Anal.*, 2016, **4**, 124–125.
33. O. Taofiq, A. M. González-Paramás, A. Martins, M. F. Barreiro and I. C. F. R. Ferreira, *Ind. Crops Prod.*, 2016, **90**, 38–48.
34. Y. Wu, M.-H. Choi, J. Li, H. Yang and H.-J. Shin, *Cosmetics*, 2016, **3**, 22.
35. A. N. H. Nguyen, D. L. Capone, T. E. Johnson, D. W. Jeffery, L. Danner and S. E. P. Bastian, *Foods*, 2019, **8**, 538.
36. J. H. Kim, D. H. Lee, S. H. Lee, S. Y. Choi and J. S. Lee, *J. Biosci. Bioeng.*, 2004, **97**, 24–28.
37. K. S. Bishop, C. H. Kao, Y. Xu, M. P. Glucina, R. R. M. Paterson and L. R. Ferguson, *Phytochemistry*, 2015, **114**, 56–65.
38. S. S. Veena and M. Pandey, *Mycosphere*, 2010, **1**, 191–194.
39. K. Perumal, Shri AMM Murugappa Chettiar Research Centre, 2009.
40. S. Yeung, Q. Chen, Y. Yu, B. Zhou, W. Wu, X. Li, Y. Huang and Z. Wang, *Acta Chromatogr.*, 2021, 100–113.
41. G. Venturella, V. Ferraro, F. Cirlincione and M. L. Gargano, *Int. J. Mol. Sci.*, 2021, **22**, 634.

42. J. Frljak, A. H. Mulabećirović, S. Isakovi, E. Karahmet and A. Toroman, *Open J. Prev. Med.*, 2021, **11**, 9–22.
43. C. Gupta, R. M. Balakrishnan, U. Priyanka and A. Pugazhendhi, *Biocatal. Agric. Biotechnol.*, 2019, 101239.
44. A. R. Ipeaiyeda, C. O. Adenipekun and O. Oluwole, *Cogent Environ. Sci.*, 2020, **6**, 1847400.
45. M. Abate, G. Pepe, R. Randino, S. Pisanti, M. G. Basilicata, V. Covelli, M. Bifulco, W. Cabri, A. M. D'Ursi, P. Campiglia and M. Rodriquez, *Pharmaceuticals*, 2020, **13**, 224.
46. A. M. Vanegas-Azuero and L. F. Gutierrez, *J. Dairy Sci.*, 2018, **101**, 1020–1033.
47. I. Leskosek-Cukalovic, S. Despotovic, N. S. Lakj, M. Nilai, V. Nedovj and V. V. Tfaevj, *Food Res. Int.*, 2010, **43**, 2262–2269.
48. S. Pecić, M. Veljović, S. Despotovic, I. Leskošek-Čukalović, V. Tešević, N. Nikićević and M. Nikšić, *Food Technologists, Biotechnologists and Nutritionists*, 2011, 89.
49. W.-F. Bai, X.-Y. Guo, L.-Q. Ma, L. Guo and J. Lin, *Indian J. Microbiol.*, 2012, **53**, 70–76.
50. H. Yang and L. Zhang, *Food Chem.*, 2009, **112**, 1–5.
51. C. Kao, A. C. Jesuthasan, K. S. Bishop, M. P. Glucina and L. R. Ferguson, *Funct. Foods Health Dis.*, 2013, **3**, 48–65.
52. S. K. Gill and M. J. Rieder, *Can. J. Clin. Pharmacol.*, 2008, **15**(2), e275–e285.
53. S. Wachtel-Galor, B. Tomlinson and I. F. Benzie, *Br. J. Nutr.*, 2004, **91**, 263–269.
54. T. P. Smina, J. Mathew, K. K. Janardhanan and T. P. Devasagayam, *Environ. Toxicol. Pharmacol.*, 2011, **32**, 438–446.
55. S. C. Dewi, D. Sargowo, M. A. Widodo, T. A. Wihastuti, T. Heriansyah, M. T. Hartanto, I. D. A. Pambayun, S. Bakhri, G. D. Wicaksono S, N. Q. A'ini, D. P. Sahudi, S. H. Hantoko, J. Indra Jaya, K. Octariny and A. I. Ardhi, *J. Hypertens.*, 2015, **33**, e30.
56. T. A. Wihastuti, D. Sargowo, M. A. Widodo, T. Heriansyah, S. Soeharto, K. W. Anita and N. Q. A'ini, *Natl. J. Physiol., Pharm. Pharmacol.*, 2015, **5**, 296–302.
57. S. Sharif, G. Mustafa, H. Munir, C. M. Weaver, Y. Jamil and M. Shahid, *J. Food Nutr. Res.*, 2016, **4**, 703–708.
58. R. Zhao, Q. Chen and Y. M. He, *Sci. Rep.*, 2018, **8**, 12680.
59. A. Yüksel, E. Gezer, Z. Z. Gümüş and Ö. Tuncer, *Turk. J. Fam. Med. Primary Care*, 2017, **11**, 288–291.
60. M. F. Ahmad, *Biomed. Pharmacother.*, 2018, **107**, 507–519.
61. S. Chen, J. Xu, C. Liu, Y. Zhu, D. R. Nelson, S. Zhou, C. Li, L. Wang, X. Guo, Y. Sun, H. Luo, Y. Li, J. Song, B. Henrissat, A. Levasseur, J. Qian, J. Li, X. Luo, L. Shi, L. He, L. Xiang, X. Xu, Y. Niu, Q. Li, M. V. Han, H. Yan, J. Zhang, H. Chen, A. Lv, Z. Wang, M. Liu, D. C. Schwartz and C. Sun, *Nat. Commun.*, 2012, **3**, 913.
62. D. Cor, Z. Knez and M. Knez Hrncic, *Molecules*, 2018, **23**, 649.
63. M. F. Ahmad, *Plant and Human Health, Volume 2*, 2019.

64. S. W. Chan, B. Tomlinson, P. Chan and C. W. K. Lam, *Pharm. Biol.*, 2021, **59**, 1161–1171.
65. V. K. Chaturvedi, S. Agarwal, K. K. Gupta, P. W. Ramteke and M. P. Singh, *3 Biotech*, 2018, **8**, 1–20.
66. C.-J. Chang, C.-S. Lin, C.-C. Lu, J. Martel, Y.-F. Ko, D. M. Ojcius, S.-F. Tseng, T.-R. Wu, Y.-Y. M. Chen, J. D. Young and H.-C. Lai, *Nat. Commun.*, 2015, **6**, 7489.
67. Y. Cao, X.-W. Xu, S. Liu, L. Huang and J. Gu, *Front. Pharmacol.*, 2018, **9**, 1217.
68. S. Cheng and A. D. Sliva, *Integr. Cancer Ther.*, 2015, **14**(3), 249–257.
69. S. G. Jonathan and F. E. Awotona, *Afr. J. Biomed. Res.*, 2010, **13**, 131–139.
70. X. Y. Cui, S. Y. Cui, J. Zhang, Z. J. Wang, B. Yu, Z. F. Sheng, X. Q. Zhang and Y. H. Zhang, *J. Ethnopharmacol.*, 2012, **139**, 796–800.
71. M. F. Ahmad, *Biomed. Pharmacother.*, 2018, **107**, 507–519.
72. Z. Xu, X. Chen, Z. Zhong, L. Chen and Y. Wang, *Am. J. Chin. Med.*, 2011, **39**, 15–27.
73. S. Bharadwaj, K. E. Lee, V. D. Dwivedi, U. Yadava, A. Panwar, S. J. Lucas, A. Pandey and S. G. Kang, *Sci. Rep.*, 2019, **9**, 19059.
74. K. WjDska, W. M. Czka, K. Gabryelska and M. Grabarczyk, *Molecules*, 2019, **24**, 4075.
75. I. C. F. R. Ferreira, S. A. Heleno, F. S. Reis, D. S. Stojkovi, M. Queiroz, M. H. Vasconcelos and M. D. Sokovj, *Phytochemistry*, 2015, **114**, 38–55.
76. Y. Yang, H. Zhang, J. Zuo, X. Gong, F. Yi, W.-C. Zhu and L. Li, *Biomed. Dermatol.*, 2019, **3**, 1–17.
77. Á. Trigos and J. S. Medellín, *Revista mexicana de micología*, 2011, **34**, 63–83
78. S. Cheng and D. Sliva, *Integr. Cancer Ther.*, 2015, **14**, 249–257.
79. K. K. Oh, M. Adnan and D. H. Cho, *J. Food Biochem.*, 2021, **45**(9), e13906.
80. C. J. Chang, C. S. Lin, C. C. Lu, J. Martel, Y. F. Ko, D. M. Ojcius, S. F. Tseng, T. R. Wu, Y. Y. Chen, J. D. Young and H. C. Lai, *Nat. Commun.*, 2015, **6**, 7489.
81. S.-J. Chen, H. H. Lin, W.-C. Huang, P.-J. Tsai, W. Chen, D.-C. Chen and L.-T. Chuang, *Food Agric. Immunol.*, 2017, **28**, 1393–1407.
82. X. Lei, C. Zhi, W. Huang, X. Sun, W. Gao, X.-L. Yin, X. Zhang, C. Liang, H. Zhang and F. Sun, *Front. Pharmacol.*, 2020, **11**, 956.
83. S. Elmansy, *Microb. Biosyst.*, 2019, 45–57.
84. J. Falandysz, *J. Environ. Sci. Health, Part C: Environ. Carcinog. Ecotoxicol. Rev.*, 2008, **26**, 256–299.
85. M. F. Ahmad, *J. Ethnopharmacol.*, 2020, **260**, 113047.
86. S. Oka, S. Tanaka, S. Yoshida, T. Hiyama, Y. Ueno, M. Ito, Y. Kitadai, M. Yoshihara and K. Chayama, *Hiroshima J. Med. Sci.*, 2010, **59**(1), 1–6.
87. S. Chen, J. Xu, C. Liu, Y. Zhu, D. R. Nelson, S. Zhou, C. Li, L. Wang, X. Guo and Y. Sun, *Nat. Commun.*, 2012, **3**, 1–9.
88. D. Collado Mateo, F. Pazzi, F. J. Dominguez Munoz, J. P. Martin Martinez, P. R. Olivares, N. Gusi and J. C. Adsuar, *Nutr. Hosp.*, 2015, **32**, 2126–2135.

89. H. F. Chiu, H. Y. Fu, Y. Y. Lu, Y. C. Han, Y. C. Shen, K. Venkatakrishnan, O. Golovinskaia and C. K. Wang, *Pharm. Biol.*, 2017, **55**, 1041–1046.
90. D. Sargowo, N. Ovianti, E. Susilowati, N. Ubaidillah, A. W. Nugraha, K. S. Proboretno, M. Failasufi, F. Ramadhan, H. Wulandari, Y. Waranugraha and D. H. Putri, *Indian Heart J.*, 2018, **70**, 608–614.
91. M. M. O. Al-Jumaili, F. Al-Dulaimi and M. A. Ajeel, *Syst. Rev. Pharm.*, 2020, **11**, 537–541.
92. C. Gracián-Alcaide, J. A. Maldonado-Lobón, E. Ortiz-Tikkakoski, A. Gómez-Vílchez, J. Fonollá, J. L. López-Larramendi, M. Olivares and R. Blanco-Rojo, *Appl. Sci.*, 2020, **10**, 8259.
93. M. Parravano, M. Tedeschi, D. Manca, E. Costanzo, A. Di Renzo, P. Giorno, L. Barbano, L. Ziccardi, M. Varano and V. Parisi, *Adv. Ther.*, 2019, **36**, 2493–2505.
94. K. P. C. Kuypers, *J. Med. Food*, 2021, **24**, 766–773.
95. G. H. Wang, L. H. Wang, C. Wang and L. H. Qin, *Medicine*, 2018, **97**, e0636.
96. S. Zhao, G. Ye, G. Fu, J. X. Cheng, B. B. Yang and C. Peng, *Int. J. Oncol.*, 2011, **38**, 1319–1327.
97. I. J. Suarez-Arroyo, R. Rosario-Acevedo, A. Aguilar-Perez, P. L. Clemente, L. A. Cubano, J. Serrano, R. J. Schneider and M. M. Martínez-Montemayor, *PLoS One*, 2013, **8**, e57431.
98. I. J. Suarez-Arroyo, T. J. Rios-Fuller, Y. R. Feliz-Mosquea, M. Lacourt-Ventura, D. J. Leal-Alviarez, G. Maldonado-Martinez, L. A. Cubano and M. M. Martinez-Montemayor, *J. Cancer*, 2016, **7**, 500–511.
99. A. Barbieri, V. Quagliariello, V. Del Vecchio, M. Falco, A. Luciano, N. J. Amruthraj, G. Nasti, A. Ottaiano, M. Berretta, R. V. Iaffaioli and C. Arra, *Nutrients*, 2017, **9**, 210.
100. A. Acevedo-Diaz, G. Ortiz-Soto, I. J. Suarez-Arroyo, A. Zayas-Santiago and M. M. M. Montemayor, *Nutrients*, 2019, **11**, 1116.
101. M. M. Martínez-Montemayor, T. Ling, I. J. Suárez-Arroyo, G. Ortiz-Soto, C. L. Santiago-Negrón, M. Y. Lacourt-Ventura, A. Valentín-Acevedo, W. H. Lang and F. Rivas, *Front. Pharmacol.*, 2019, **10**, 115.
102. L. Zhao, Y. Dong, G. Chen and Q. Hu, *Carbohydr. Polym.*, 2010, **80**, 783–789.
103. Z.-W. Zeng and K. Xiao, *Adv. Polym. Technol.*, 2020, **2020**, 1–6.
104. A. Y. Cheng, Y. C. Chien, H. C. Lee, Y. H. Hsieh and Y. L. Yu, *Molecules*, 2020, **25**, 3585.
105. Q.-H. Zhang, Q.-X. Hu, D. Xie, B. Chang, H.-G. Miao, Y.-G. Wang, D.-Z. Liu and X.-D. Li, *Integr. Cancer Ther.*, 2019, **18**, 1534735419890917.
106. C. Jiao, W. Chen, X. Tan, H. Liang, J. Li, H. Yun, C. He, J. Chen, X. Ma, Y. Xie and B. B. Yang, *J. Ethnopharmacol.*, 2020, **247**, 112256.
107. J. Su, L. Su, D. Li, O. Shuai, Y. Zhang, H. Liang, C. Jiao, Z. Xu, Y. Lai and Y. Xie, *Front. Immunol.*, 2018, **9**, 1765.
108. X. Lei, C. Zhi, W. Huang, X. Sun, W. Gao, X. Yin, X. Zhang, C. Liang, H. Zhang and F. Sun, *Front. Pharmacol.*, 2020, **11**, 956.
109. T. T. Chu, I. F. Benzie, C. W. Lam, B. S. Fok, K. K. Lee and B. Tomlinson, *Br. J. Nutr.*, 2012, **107**, 1017–1027.

110. J. Jiang and D. Sliva, *Int. J. Oncol.*, 2010, **37**(6), 1529–1536.
111. F. S. Reis, R. T. Lima, P. Morales, I. C. Ferreira and M. H. Vasconcelos, *Molecules*, 2015, **20**, 17872–17882.
112. G. Wu, Z. Qian, J. Guo, D. Hu, J. Bao, J. Xie, W. Xu, J. Lu, X. Chen and Y. Wang, *Am. J. Chin. Med.*, 2012, **40**, 631–642.
113. Q. P. Wu, Y. Z. Xie, Z. Deng, X. M. Li, W. Yang, C. W. Jiao, L. Fang, S. Z. Li, H. H. Pan, A. J. Yee, D. Y. Lee, C. Li, Z. Zhang, J. Guo and B. B. Yang, *PLoS One*, 2012, **7**, e44579.
114. B.-J. You, M.-H. Lee, N. Tien, M.-S. Lee, H.-C. Hsieh, L.-H. Tseng, Y.-L. Chung and H.-Z. Lee, *PLoS One*, 2013, **8**, e53616.
115. K. K. Oh, M. Adnan and D. H. Cho, *J. Food Biochem.*, 2021, **45**, e13906.
116. T. P. Smina, J. Joseph and K. K. Janardhanan, *Redox Rep.*, 2016, **21**, 254–261.
117. K. A. Rony, J. Mathew, P. P. Neenu and K. K. Janardhanan, *Indian J. Nat. Prod. Resour.*, 2011, 2(1), 19–27.
118. Y. Weng, L. Xiang, A. Matsuura, Y. Zhang, Q. Huang and J. Qi, *Bioorg. Med. Chem.*, 2010, **18**, 999–1002.
119. S. Fatmawati, K. Shimizu and R. Kondo, *Phytomedicine*, 2011, **18**, 1053–1055.
120. W. Zhao, X. Jiang, W. Deng, Y. Lai, M. Wu and Z. Zhang, *Food Chem. Toxicol.*, 2012, **50**, 303–309.
121. R. Pan, J. Lou and L. Wei, *Exp. Ther. Med.*, 2021, **22**, 820.
122. C.-R. Cheng, Q.-X. Yue, Z.-Y. Wu, X.-Y. Song, S.-J. Tao, X.-H. Wu, P.-P. Xu, X. Liu, S.-H. Guan and D.-A. Guo, *Phytochemistry*, 2010, **71**, 1579–1585.
123. S. Quereshi, A. Pandey and S. Sandhu, *J. Sci. Res.*, 2010, **3**, 9–13.
124. M. S. Hossen, M. M. Billah Prince, E. M. Tanvir, M. A. Z. Chowdhury, M. A. Rahman, F. Alam, S. Paul, M. Saha, M. Y. Ali, N. C. Bhoumik, N. Karim, S. H. Gan and M. I. Khalil, *Evidence-Based Complementary Altern. Med.*, 2018, **2018**, 6254929.
125. N. Yu, Y. Huang, Y. Jiang, L. Zou, X. Liu, S. Liu, F. Chen, J. Luo and Y. Zhu, *Oxid. Med. Cell. Longevity*, 2020, **2020**, 9894037.
126. C. Liang, H. Li, H. Zhou, S. Zhang, Z. Liu, Q. Zhou and F. Sun, *Oncol. Rep.*, 2012, **27**, 1079–1089.
127. T.-Y. Lin, H.-Y. Hsu, W. Sun, T.-H. Wu and S.-M. Tsao, *Int. J. Cancer*, 2017, **140**, 2596–2607.
128. C.-C. Lin, Y. L. Yu, C.-C. Shih, K.-J. Liu, K. L. Ou, L.-Z. Hong, J. D. C. Chen and C.-L. Chu, *Cancer Immunol. Immunother.*, 2011, **60**, 1019–1027.
129. K. Li, K. Na, T. Sang, K. Wu, Y. Wang and X. Wang, *Oncol. Rep.*, 2017, **38**, 2803–2813.
130. Y. L. Chien, C. T. Ho, B. H. Chiang and L. S. Hwang, *Food Chem.*, 2011, **126**, 1586–1592.
131. L. Wang, J. Li, J. Zhang, Z.-M. Li, H. Liu and Y.-Z. Wang, *RSC Adv.*, 2020, **10**, 42084–42097.
132. G. patents, Google patents, https://patents.google.com/.

CHAPTER 11

Lentinula edodes *(Berk.) Pegler*

KASHIF AMEER,*[a] MARUF AHMED,[b] MIAN ANJUM MURTAZA[a] AND SHAHIDA ANUSHA SIDDIQUI[c,d]

[a] Institute of Food Science and Nutrition, University of Sargodha, Sargodha, 40100, Pakistan; [b] Department of Food Processing and Preservation, Hajee Mohammad Danesh Science & Technology University, Dinajpur, Bangladesh; [c] Technical University of Munich Campus Straubing for Biotechnology and Sustainability, Essigberg 3, 94315 Straubing, Germany; [d] German Institute of Food Technologies (DIL e.V.), Prof.-von-Klitzing Str. 7, 49610 D-Quakenbrück, Germany
*Emails: kashifameer89@gmail.com, kashif.ameer@uos.edu.pk

11.1 Introduction

Mushrooms have been reported specifically in traditional and alternative medicinal modalities, such as traditional Chinese medicine (TCM) for thousands of years. Furthermore, dried extracts obtained from mushroom fruiting bodies have a significant share of the herbal medicine market of western countries.[1,2] The marketing and advertisement of such medicinal products in Europe and North America is regulated in a poor manner, which allows nutraceutical companies to make false and fictitious statements pertaining to the medicinal significance of mushrooms.[3] Therefore, several issues must be taken into account objectively for assessment of the medicinal benefits and traditional uses of mushrooms in alternative and traditional medicines.[4] The medicinal use of shiitake mushrooms (*Lentinula edodes*) increased during the Ming Dynasty period (1368–1644) after a method was developed for cultivating mushrooms on logs.[4] A tonic based on shiitake mushrooms has been reported in being helpful to relieve pain, aches, and fatigue attributed to the aging process.[5]

Phytochemistry and Nutritional Composition of Significant Wild Medicinal and Edible Mushrooms: Traditional Uses and Pharmacology
Edited by Ajay Sharma, Garima Bhardwaj and Gulzar Ahmad Nayik

Published by the Royal Society of Chemistry, www.rsc.org

The scientific name for the shiitake mushroom is *Lentinula edodes* (Berk.) Pegler, which is one of the most widely cultivated mushrooms globally.[6] This mushroom exhibits medicinal significance and has been regarded as a healthy food for centuries in East Asia, especially in Japan, Korea, and China.[7] Shiitake mushrooms have even gained popularity as medicinal ingredients in nutritional and medicinal products in North America and Europe.[8] In the published literature, shiitake mushrooms have been reported to exhibit several health benefits owing to their chemical constituents and bioactive compounds.[9] The specific health benefits rendered by the shiitake mushroom include its beneficial effect on cardiovascular health and protective effects against intestinal worms and lung diseases.[10] The major bioactive constituents found in Shiitake mushrooms include the polysaccharide lentinan [β-(1→3)-D-glucan]. Moreover, the mycelium and culture extracts of shiitake mushrooms, such as LEM, LAP, and KS-2, have been reported to demonstrate strong anti-tumor activity when administrated orally or by injection to both humans and animals.[11–13] Lentinan has also shown potent efficacy against various parasitic, viral, and bacterial infections.[14] LEM is also found to effect in treating AIDS; moreover, HIV infection inhibition in cultured human T cells and AZT effect enhancement against *in vitro* viral replication have been reported.[15] Water-soluble lignins extracted from the EP3 and EPS4 extracts of the shiitake mushroom have also been shown to exhibit immunomodulatory and antiviral effects.[16] Mushroom-derived polyacetylene compounds, including centinamycins A and B, exhibit potent antibacterial effects.[17] Another compound, eritadenine exhibits a cholesterol-lowering effect and antiviral properties.[18]

Traditional Chinese medicine advocates the symptomatic treatment of cardiovascular diseases in individuals suffering from arteriosclerosis.[19] The treatment modalities, including herbal medicine and acupuncture, are designed for restoration and balance of energy flow in the human body.[20] Western medicine addresses heart disease in a different manner, using prescription drugs to reduce blood pressure and serum cholesterol as well as making exercise and dietary recommendations.[21] The exploitation of combination therapies in traditional Chinese medicine is usually adjusted for individual patients, making the measurement of the particular efficacy of shiitake mushrooms an arduous task in clinical trials.[22] As China has embraced modern medicinal philosophy, the global demand for herbal products has reached annual sales of $50 billion in terms of nutritional supplements (Figure 11.1).[23]

11.1.1 Taxonomic Classification

In 1877, the scientific taxonomy of the shiitake mushroom (*L. edodes*) was first described as *Agaricus edodes* by the English cryptogamist and plant pathologist Miles Joseph Berkeley.[24] Moreover in 1976, it was classified in the genus Lentinula by the British mycologist David Pegler. In the division *Basidiomycota*, the shiitake mushroom belongs to the class *Agaricomycetes*, in the order *Agaricales* in the *Omphalotaceae* family and is ranked under the genus *Lentinula*. This mushroom is also known by the other common names

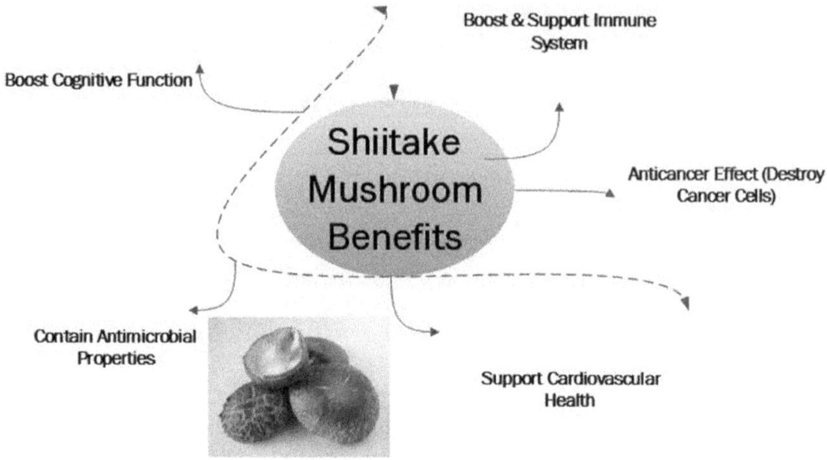

Figure 11.1 Health benefits of the shiitake mushroom.

of black mushroom, oakwood mushroom, black forest mushroom, golden oak mushroom, or sawtooth oak mushroom.[24,25]

11.1.2 Distribution

Shiitake mushrooms have been reported to grow in deciduous trees and their decaying wood parts. In particular, maple, beech, chestnut, mulberry, oak, sweet gum, and poplar are the main tree plants that serve as of the growth habitat of the shiitake mushroom.[26] Natural distribution regions of shiitake mushroom include the warm and moist climate of Southeast Asia. The shiitake mushroom represents 25% of the total global mushroom production. An overview of the global shiitake mushroom market indicates that China, Italy, Poland, Netherland, and Romania are the leading producers of shiitake mushrooms, with market shares of 28.09%, 9.46%, 6.84%, 6.27% and 5.36%, respectively. In India, the leading region for mushroom production is Uttar Pradesh, followed by Kerala and Tripura.[27]

11.2 Traditional Uses

Shiitake mushrooms are traditionally used in fresh and dried forms for various purposes in East Asian countries. In Japan, miso soup is prepared using shiitake mushrooms and is employed as a fundamental cuisine for a variety of vegetarian dashi. Shiitake mushrooms are utilized as a chief ingredient in simmered and steamed dishes. In China, shiitake mushrooms are utilized in a method of cooking known as sautéing.[27]

11.3 Nutritional and Chemical Composition of the Shiitake Mushroom

Raw shiitake mushrooms are a good source of energy. A 100 g serving of shiitake mushroom, provides 34 kcal, carbohydrates (7%), protein (2%), moisture (90%), sugar (2.4 g), dietary fiber (2.5 g), calcium (2 mg), magnesium (20 mg), phosphorus (112 mg), potassium (304 mg), sodium (9 mg), zinc (1.0 mg), and fat (<1%). The detailed nutritional composition of the shiitake mushroom is presented in Table 11.1. Moreover, raw shiitake mushroom has been reported to provide 20% or more of vitamins B on basis of daily value. Similarly, the shiitake mushroom also provides vitamin D2 when its ergosterol is exposed to the ultraviolet B rays of the sun (Table 11.1).[27,28]

Several experiments have been conducted in recent years to elucidate the bioactivity of the chemical compounds of the shiitake mushroom and the Memorial Sloan Kettering Cancer Center (MSKCC) has reported promising results.[29] In the cell walls of accumulated hyphae, the β-(1→3)-D-glucan lentinan is found as the polysaccharide and forms the chemical composition of shiitake mushroom fruiting bodies.[30,31] Research has shown that lentinan acts as an adjuvant and plays a significant contributory role in the stimulation of antibody production in mice injected with the hepatitis B vaccine.[32] Other *in vivo* experiments have demonstrated that lentinan exhibits immune-stimulating potential at the cellular level to attack cancer cells as well as virus-infected cells.[33] In another study, the findings of clinical trials conducted by Japanese researchers suggested that advanced stomach cancer patients who were treated after exposure to chemotherapy survived longer when lentinan was added to their treatment.[34,35] These results endorsed the anticancer effects of shiitake mushrooms and encouraged further study. Shiitake mushrooms are ranked as the second most popular mushroom on the global market on the basis of consumer demand.[36] The nutritional

Table 11.1 Nutritional composition of the Shiitake mushroom (per 100 g dry weight basis).

Nutritional composition		
1	Fat	0.5 g
2	Carbohydrate	6.8 g
3	Fiber	2.5 g
4	Protein	2.2 g
5	Calories	34 kcal
6	Niacin	4 mg
7	Pantothenic acid	1.5 mg
8	Vitamin B6	0.2 mg
9	Riboflavin	0.2 mg
10	Manganese	0.2 mg
11	Phosphorus	112 mg
12	Selenium	5.7 μg
13	Copper	0.1 mg
14	Zinc	1 mg
15	Potassium	304 mg
16	Magnesium	20 mg
17	Vitamin D	18 IU
18	Iron	0.4 mg

components of the shiitake mushroom are bioactive polysaccharides, including β-D-glucan, xylomannan, heteroglycan, lentinan, and eritadenine; free sugars, such as glycerol, fibrehalose, mannitol, mannose, arabitol, arabinose, and trehalose; vitamins, such as B1, B12, and D2; and dietary fiber (Figure 11.2a and b).[37–39] Numerous bio-components of the shiitake mushroom exert their pharmacological activity against cardiovascular disease, hyperlipidemia, hypertension, depressed immunity, cancer and hepatic disorders.[40,41]

11.4 Cultivated Shiitake (*Lentinula edodes*) Yield

Cultivated shiitake mushrooms are regarded as one of the most commonly utilized edible mushrooms and represent an important forestry product in Japan.[42] In 2018, in Japan, the total recorded production of shiitake mushrooms was 88 832 tons, accounting for around 19% of total mushroom production.[43,44] Cultivated shiitake mushrooms are classified as a white-rod fungus belonging to Basidiomycetes, which grow naturally in callus and the wood logs of broad-leaved trees.[45] Wood logs on the forest floor and sawdust substrates are usually the common cultivation places of shiitake mushroom.[46] Although indoors sawdust is the most common medium for the cultivation of shiitake mushrooms, the traditionally employed log-cultivated method is the most significant method overall for shiitake mushroom cultivation.[47] In Japan, oak logs are mainly employed for shiitake mushroom cultivation. Generally, these logs are usually 1–15 cm long and are cut to pieces of almost 1 m in length in winter and autumn, while inoculation of shiitake mushroom spawn is carried out during early spring.[48,49] After completion of inoculation, logs are stacked in a horizontal manner to facilitate the growth of mycelia over several months (known as spawn run).[50] Wild fungi also colonize the logs and affects shiitake mushroom production. To date, several fungal species (almost 50) have been identified as shiitake mushroom pests during log cultivation.[51]

11.5 *In vitro* Anti-atherosclerotic Activity

Food bio-components, which exhibit anti-hypercholesterolemic and antioxidant activities, have been advocated as a natural remedy for pathophysiological complications arising due to elevated oxidative stress.[37,52] Atherosclerosis is classified as a cardiovascular disease which is associated with high consumption of saturated fats and low-density lipoproteins in oxidized form (ox-LDL) being excessively produced.[53,54] Management of hyperlipidemia is recommended by the International Atherosclerotic Society, with an emphasis on controlling LDL levels to reduce the incidence rate of atherosclerosis.[55] It has been reported that LDL oxidation in terms of

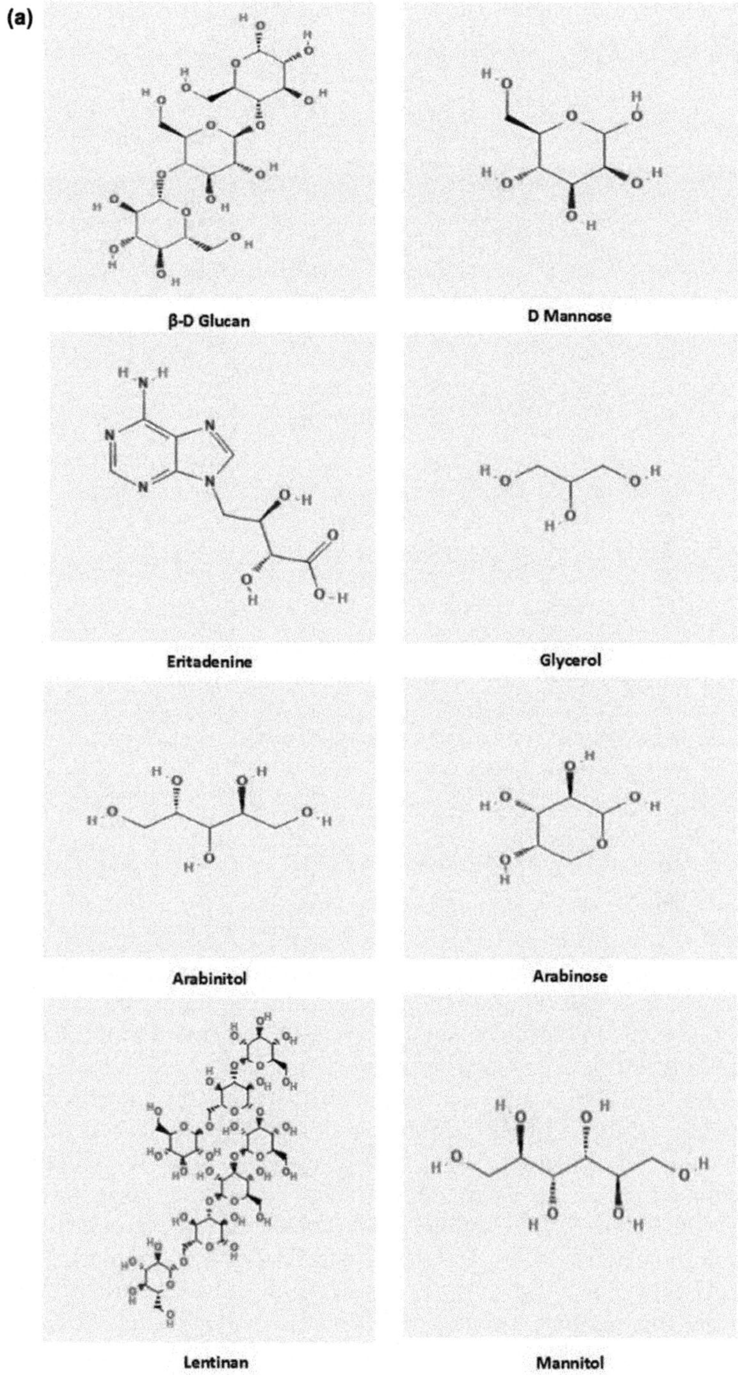

Figure 11.2 Phytochemical medicinal significance of the shiitake mushroom.

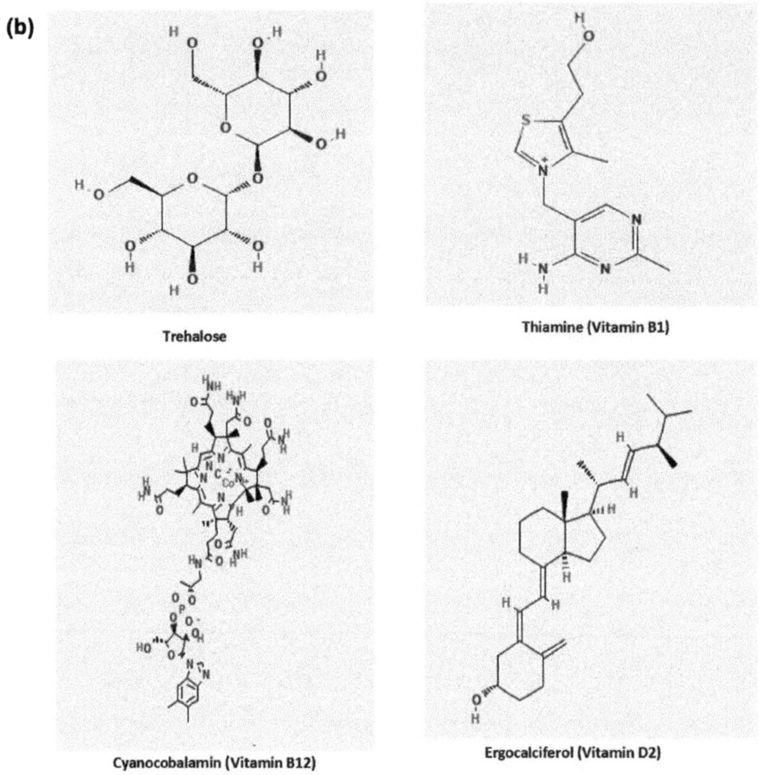

Figure 11.2 (Continued)

reaction kinetics shows a lag period of slow oxidation followed by a corresponding rise in an enhanced oxidation period.[56,57] Conjugated dienes are formed during the lag period of slow oxidation. Malondialdehyde (MDA) production has been reported to take place during the lag period.[58] The biocomponents of the shiitake mushroom exhibit a mediating effect, with a corresponding rise in the lag time of oxidation leading to decreased levels of MDA. Moreover, the hexane fraction of shiitake mushroom extract has been reported to exhibit excellent anti-hypercholesterolemic activity owing to the production of conjugated dienes.[59,60] In the case of LDL oxidation, conjugated diene formation has been reported to occur in conjunction with MDA production through the breakage of double bonds present in lipid hydroperoxide molecules.[61] The MDA levels exhibit positive correlation to that of LDL oxidation. The hexane solvent fraction of the shiitake mushroom extract contained the smallest amount of MDA (21.57 μM per mg LDL), which is evident of the fact that the shiitake mushroom extract has the highest biocomponent activity to withstand LDL oxidation.[62] The elevated level of total cholesterol is described in published literature as one of the risk factors. The hexane fraction of shiitake mushroom extract inhibited HMG Co-A reductase activity (44.26%) in comparison with that of a synthetic statin named pravastatin (87.64%).[63] The shiitake mushroom is regarded as being among the top-ranked edible and medicinal mushrooms with respect to inhibitory effect on HMG-Co-A reductase activity.[64] The hexane fraction of shiitake mushroom extract contains bio-functional components that have anti-atherosclerotic effects, with the main identified compounds being butyric acid, oleic acid, linoleic acid, ergosterol, and α-tocopherol (vitamin E). The scavenging of free radicals has been reported by vitamin E (α-tocopherol) molecules through the mechanism of hydrogen atom transfer (HAT).[65] α-Tocopherol reacts with free radicals, including lipid peroxyl molecules, and interrupts further peroxidation hydroperoxide production as well as α-tocopherol radicalized formation. Both these intermediary products exhibit harm to the human body due to them being unreactive.[66] These α-tocopherol radicals may either react with peroxyl molecules to change to functionally unreactive intermediate products or may revert back to their original vitamin E form.

As mentioned above, the hexane fraction of shiitake mushroom extract is also known for the presence of ergosterol in its composition and its role as a membrane antioxidant after ingestion has already been elucidated in the published literature.[67] The inhibitory effect rendered by ergosterol peroxide from the hexane fraction of shiitake mushroom extract was reported to be higher than the inhibitory activities of thiourea and α-tocopherol. The supplementation of ergosterol has also been carried out in one research study with oleic acid in *Saccharomyces cerevisiae*, with this combination of ergosterol with oleic acid leading to reduced ROS production, lower oxidative stress, and also less resultant damage at the cellular and subcellular levels.[68] Ergosterol as well as ergosterol peroxide combination has been reported to cause suppression of LPS-induced inflammatory responses through inhibition of the transcriptional activities of C/EBPb and NF-κB as well as MAPK phosphorylation. The antioxidant effects of the shiitake mushroom have also been measured by means of DPPH radical scavenging activity and reducing capacity assay (Folin–Ciocalteu testing), with these antioxidant effects being ascribed to the presence of various bioactive compounds.[54] However, the antioxidant activity of the mushroom extract is dependent on the species type and type of solvent employed for extraction. In the case of the polar solvent based bio-separation of mushroom bioactive components, phenolic acids and polyphenolic compounds have been shown to exhibit radical scavenging activities, whereas triterpenoids have been shown to exhibit metal chelating activities. In the case of non-polar solvent, such as hexane extraction, ergosterol has been identified as one of the principal antioxidant compounds.[69] Other known antioxidant compounds include 5,6-dihydroxy ergosterol and structurally similar components, such as 7,22-ergostadienol and ergosta-7,22-dien-3-ol, in the hexane fraction have been linked to playing a contributory role in chelating transition metal ions such as Fe^{2+}, thus exhibiting oxidative stress and LDL lowering effects.[67]

11.6 Lipid Peroxidation Inhibitory Effect of the Shiitake Mushroom

Fatty acids are regarded as one of the most important bio-functional components of edible and medicinal mushrooms, including the shiitake mushroom. The hexane fraction of the shiitake mushroom has been reported to contain linoleic, oleic, and butyric acids. Oleic acid is a monounsaturated fatty acid.[70] These fatty acids have been reported to exhibit anti-atherosclerotic effects after incorporation in various diets, especially the Mediterranean diet. The shiitake mushroom, in terms of its medicinal effects, exhibits anti-atherosclerotic effects, the mechanism of which involves the reduction in the expression of vascular cell adhesion molecule 1 (VCAM-1) as well as NF-κb activity and dampening in endothelial activity.[71] Resultantly, the monocytes adhere to the endothelial walls and this adherence is minimized in conjunction with a slowing down of the associated pro-atherogenic and pro-inflammatory processes. Butyric acid is a short chain fatty acid found in shiitake mushroom extract. Bioactive component rich shiitake mushroom extract has been show to confer anti-atherosclerotic effects through modulation of NF-κb and VCAM-1 activities.[72]

Linoleic acid exists chemically in the form of 9,12-octadecadienoic acid and it is classified as one of the most abundant fatty acids reported in medicinal and edible mushrooms, including the shiitake mushroom.[73] Linoleic acid along with other constitutional and stereoisomers are collectivity termed as "conjugated linoleic acid (CLA)" and have become an active topic of research with respect to anti-atherosclerotic effects. It has been shown by previous researchers on anti-atherosclerotic effects of CLA in an animal model that arteriosclerosis progression decreased slowly due to inhibitory effects.[74] Dietary supplementation of CLA extracted from the shiitake mushroom may lower serum LDL levels, triglycerides, and the progression of arteriosclerosis progression. The proposed CLA mechanism involves enhanced expression of nuclear transcription factors in vascular smooth muscle cells, such as peroxisome proliferator activated receptor γ (PPAR-γ).[75] Some *in vitro* studies have reported the pro-atherogenic and pro-oxidative role of linoleic acid. However, *in vivo* evidence has yet to be established by conducting further research. In humans, consuming a linoleic acid rich diet which has been linked to a reduced incidence of risk factors that contribute to coronary heart disease. The hexane fraction of the shiitake mushroom comprises ergosterols that exhibit an inhibitory effect towards HMG Co-A reductase.[76] This has also been endorsed by several *in vivo*, *in vitro*, and *ex vivo* studies published in the literature. Ergosterol from shiitake and *Sarcodon aspratus* has been reported to cause downregulation of the expression of LDL receptor (LDLR) and genes pertaining to HMG-CoA reductase in RAW 264.7 cells.[77]

11.7 Immunomodulatory Properties of the Shiitake Mushroom

A wide range of medicinal mushrooms exhibiting immunomodulatory effects have been reported in traditional Chinese medicine and Indian medical systems for more than 2000 years. Moreover, there is an increasing tendency among health conscious consumers regarding the medicinal use of mushrooms as a nutraceutical. The shiitake mushroom is a rich source of β-glucans and other bioactive components.[78] The β-glucans from the shiitake mushroom act as immunomodulators and activate both the adaptive and innate immune systems. These β-glucans aid in the proliferation and activation of innate immune system components such as natural killer (NK) cells, macrophages, and neutrophils and also play a role in developing and releasing cytokines.[79] These cytokines subsequently help to promote B cell antibody production and boost the differentiation of T cells into T helper (Th1) and Th2 cells owing to mediation of humoral and cell immunity.[80] Due to their large molecular weight, mushroom polysaccharides, including β-glucan, do no exhibit the capacity to infiltrate immune cells and directly activate them. Polysaccharides stimulate the immune system *via* various cell receptors, such as complement receptor-3 (CR3), dectin-1, lactosylceramide (LacCer), and toll-like receptor (TLR)2.[81] Mushroom polysaccharide effectiveness is usually determined by evaluating their affinity for immune cell receptors. In such cases, the suitability of the polysaccharides is usually influenced by their low affinity for invulnerable cell receptors. These low affinity polysaccharides exhibit immunomodulatory effects. Furthermore, these polysaccharides have been reported to exhibit an immune system activation mechanism.[82] Conversely, low molecular weight polysaccharides enter the immune cells and lead to cellular activation owing to their simple structures and shape, and hence modulate the immunity. In the case of immunologically active polysaccharides, the degree of branching number (DB) is in the range of 20–40%. Even with highly-researched lentinan, a DB level of 32% exhibited maximal anticancer and immunomodulatory activities, whereas it was found that DB levels from 32 to 40% exhibited a negative relationship with biological activities.[83] These immunomodulatory components of edible mushrooms, including the shiitake mushroom, promote B cell growth to produce antibodies, leading to T cell dissociation to partner T cells, which help to combat humoral and cell invulnerability.[84] Lectins have also been classified as immunomodulatory compounds, along with terpenoids, terpenes, immunomodulatory polysaccharides, and immunomodulatory proteins (FIPs). All of these compounds have been reported to reduce cytokine production. In addition to straight chain and branched polysaccharides, polysaccharide confirmation may lead to a change in bioactivity owing to changes in structural stability.[85] Mushroom polysaccharides exhibiting a random coil shape have been reported to exhibit immunomodulatory and anticancer effects. The most common strategy used to enhance the biological activity of polysaccharides is *via* their chemical modification. This technique is commonly utilized to manufacture immunomodulators. Modification strategies include hydroxylation, formyl-methylation, sulfation, amino-ethylation, and carboxymethylation.[86] This inclusion of chemical groups in polysaccharide structures may lead to enhancement in the probable interactions between immune cell receptors and modified polysaccharides through hydrostatic or hydrogen linkages, hence leading to boosting of immunological response. *Lentinus edodes* exhibits sulfated cell wall glucans, which impart it with strong anticancer and immunomodulatory properties compared to native polysaccharides. The improved properties due to sulfation could be related to greater solubility, as was reported in the case of hyper-branched glucan TMb3.[42,84,87] Resultantly, the chemical composition, branching, molecular weight, and chemical modification of mushroom-derived polysaccharides may have a significant influence on immunomodulatory activity. Recently, a novel class of protein compounds that exhibit immunomodulatory effects have been reported in the literature, known as fungal immunomodulatory proteins (FIPs). These FIPs in conjunction with toll-like receptors stimulate antigen-presenting cells and lead to the generation of IL-12 and NO-like cytokines.[88] By stimulating p38/MAPK phosphorylation, this may cause stimulated differentiation and proliferation of T helper cells (Th0) to Th1 and Th2 cells, hence activating B cells and macrophages. These compounds exhibit immunomodulatory activity through p38/MAPK phosphorylation, regulating the expression of the intercellular adhesion of molecules on the surface of T cells.[89]

There are several published reports available which report on the effectiveness of the immune boosting properties of edible mushrooms as a possible remedy to treat SARS-CoV-2 (COVID-19) patients. The β-glucans extracted from the edible shiitake mushroom have been found to exhibit potential to reduce key cytokines in the case of cytokine storm being detected in patients suffering from COVID-19.[90] The β-glucans, as well as carbohydrates, present in the cell walls of the shiitake mushroom have been utilized as a nutraceutical to treat heart disease and excessive cholesterol. Furthermore, among medicinal mushrooms, the shiitake mushroom because of its immunomodulatory properties may help to relieve asthma and lung infections by strengthening the immune system tolerance to respiratory diseases.[38] Edible mushroom consumption may offer physiologically plausible immunomodulatory activity against SARS-CoV-2. Interaction with the immune cells on receptors is one of the modulatory ways through which mushroom interact, similar to that of the gut microbiome.[91] For instance, T-lymphocytes exhibit receptors that bind to mushroom polysaccharides. This is one of the pathways through which mushrooms may alter the response of immune cells to combat SARS-CoV-2. Extracts derived from the shiitake mushroom have been demonstrated to exhibit an inhibitory effect in the case of influenza *in vivo*.[92]

It has been suggested in the literature that mushroom consumption is linked with an improvement in immunity; however, to date, very few studies have been carried out in this area. In one such clinical study, whole mushrooms were fed to humans as food.[91] It has also been reported in the literature that whole mushrooms were also consumed by humans as a source of food.

11.8 Antiviral Effects of the Shiitake Mushroom

Antiviral properties of the shiitake mushroom have been reported by several published studies. For instance, in the case of infectious hematopoietic necrosis virus (IHNV), the antiviral activity of lentinan was assessed. Lentinan (LNT-1) is prepared from the shiitake mushroom by obtaining powder from the mycelia of the *L. edodes* mushroom.[92] The analysis showed that LNT-1 exhibits a molecular weight of 3.79×10^5 Da with a β-(1–3)-glucan backbone and β-(1–6)-glucosyl side-branching units. LNT-1 has been shown to exhibit antiviral efficacy against IHNV whereas the antiviral mechanism was because of immediate inactivation as well as viral replication limitation.[43] Alongside *L. edodes*, *Coriolus versicolor*, *Agaricus bisporus*, *Pleurotus abalonus*, and *Pleurotus citrinopileatus* contain anti-HIV compounds in their compositions. Polysaccharides, lentin, lectin, polysaccharopeptides, peptides, ubiquitin-like protein, and laccase have been reported to primarily target HIV reverse transcriptase in conjunction with reduction in viral entrance, viral enzymes, replication, viral protein production, cellular proteins, and improve immunity against HSV-1, HSV-2, influenza A virus, HCV, HIV, FCV, and EV71.[93] Polysaccharides, lentin, lectin, and laccase from *L. edodes* have been found to exhibit IC_{50} values in the range of 0.1–2.2 M against HIV-1 compared to those of anti-HCV drugs, such as AZT, dideoxycytidine, and dideoxyinosine, which exhibit IC_{50} values in the range of 0.03–0.5 M, 0.5–1.5 M, and 2–10 M against HIV-1, respectively. The antiviral mechanism of mushroom compounds has been well established in previous research against enveloped viruses; however, there is still a need to investigate the antiviral mechanisms of non-enveloped viruses, such as enteroviruses and NoV.[94] It could be inferred that mushroom-derived bioactive metabolites could be utilized as antiviral options against RNA and DNA viruses and the SARS nCoV that causes COVID-19. The cap and stem parts of *L. edodes* contain total glucan contents of 20.54 and 26.54 g per 100 g dry weight in their chemical compositions. The total glucan is further classified into α-glucan and β-glucan.[95] The α-glucan contents in the cap and stem of *L. edodes* (shiitake) are reported to be 0.76 and 1.45 g per 100 g dry weight, respectively, whereas the β-glucan contents of the cap and stem parts are 19.78 and 25.31 g per 100 g dry weight, respectively.[94]

11.9 Shiitake Mushroom Use in Synthesizing Vegetable-based Edible Films

An aqueous extraction of the shiitake mushroom allows an aqueous extract full of ingredients to be employed for the development of edible films.

In such an instance, salt (1.5%), corn starch (3%) and the sodium salt of carboxy methyl cellulose (CMC-Na, 0.5%), as well as gelatin (0.5%), were added together to prepare a vegetable-based paper derived from the shiitake mushroom. As the shiitake mushroom exhibits unique flavor characteristics, the researchers added salt in an appropriate amount. Salt addition not only helped to mitigate the taste issues associated with the originally added materials, but also improved the flavor profile of the shiitake mushroom as well as the taste of the final product.[96] As raw materials, cabbage and jujube were utilized for preparing vegetable-based paper, and other secondary ingredients, such as agar, CMC, compound thickener, and sodium alginate were mixed together in a ratio of 5 : 5 : 6. The other components such as starch, isolated soy protein, and sodium alginate were mixed together, in the case of jujube, in a mixing ratio of 5 : 4 : 4. The researchers reported unique physical properties of the components used as raw materials. The addition of various additives in a specific ratio as thickener replacer was carried out in order to retain the taste of the jujube paper.[97]

11.10 Toxicity/Side Effects of the Shiitake Mushroom

The shiitake mushroom is classified as the second most popular edible mushroom and, as aforementioned, it is a rich source of a chemical compound known as lentinan. When the shiitake mushroom is prepared or consumed in a regular serving quantity, it is mostly harmless and do not cause any kind of toxic and side effects.[28,98] However, when the shiitake mushroom is taken for medicinal purposes and utilized in excessive amounts or in an uncooked form, then it may cause possible side effects, such as stomach pain, sunlight sensitivity, skin edema or may lead to breathing issues and skin reactions. In the case of skin reactions, rarely, the shiitake mushroom in slightly cooked or raw forms may lead to allergenicity and this allergic reaction is known as shitake dermatitis, characterized by streaky pruriginous, micro-papular, and erythematous rashes occurring all over the human body, including on the scalp and face, which usually appear 24 h after consumption of the fungus; further exposure to the ultraviolet rays in sunlight may worsen the skin issues.[99] This skin allergy is thought to be probably caused by lentinan; however, proper cooking of the shiitake mushroom may result in the elimination of this allergenicity.[100–110]

11.11 Conclusions

A wide range of mushrooms exhibit numerous bioactivities and health benefits. Shiitake is an edible mushroom that originates from East Asia, the cultivation and consumption of which is recorded throughout the world. Medicinal and edible mushrooms have been used for centuries to treat ailments in traditional and alternative medicines and oriental therapies.

Contemporary research findings endorse the health benefits of edible mushrooms. Modern clinical practices in Korea, Japan, and China rely on mushroom-derived extracts and supplements. Mushrooms have been researched for their phytochemistry, nutritional, and medicinal properties. For medicinal effects, mushrooms have been consumed by health-conscious consumers worldwide for cancer prevention, to combat cardiovascular diseases, for blood circulation improvement, and for their immunomodulatory and cholesterol-lowering properties. Some medicinal mushrooms such as the shiitake mushroom have been utilized for centuries to treat emotional and physical stress, chronic hepatitis, and gastric ulcers. Shiitake mushrooms exhibit antioxidant, anti-inflammatory, antibacterial, antifungal, anticarcinogenic, and immunomodulatory activities. Important bio-components in the shiitake mushroom include vitamin E, butyric acid, ergosterol, and oleic and linoleic acids, which endow the shiitake mushroom with the properties that make it an important medicinal mushroom to act as an anti-atherosclerotic agent. These therapeutic benefits of the shiitake mushroom support that it should be used to develop functional foods and nutraceuticals.

References

1. A. J. Kora, *Bull. Natl. Res. Cent.*, 2020, **44**, 1-9.
2. K. I. Alzand, M. S. M. Bofaris and A. Ugis, *World J. Pharm. Res.*, 2019, **8**, 31–46.
3. R. K. Saini, A. Rauf, A. A. Khalil, E. Y. Ko, Y. S. Keum, S. Anwar, A. Alamri and K. R. Rengasamy, *S. Afr. J. Bot.*, 2021, **141**, 344–356.
4. D. Yadav and P. S. Negi, *Food Res. Int.*, 2021, **148**, 110599.
5. R. M. Chugh, P. Mittal, N. Mp, T. Arora, T. Bhattacharya, H. Chopra, S. Cavalu and R. K. Gautam, *Front. Pharmacol.*, 2022, **13**, 925387.
6. S. Li, A. Wang, L. Liu, G. Tian, S. Wei and F. Xu, *J. Food Meas Charact.*, 2018, **12**, 2012–2019.
7. S. Chen, C. Liu, X. Huang, L. Hu, Y. Huang, H. Chen, Q. Fang, N. Dong, M. Li, W. Tang and S. Nie, *J. Funct. Foods*, 2020, **66**, 103791.
8. L. Xu, X. Fang, W. Wu, H. Chen, H. Mu and H. Gao, *Food Chem.*, 2019, **285**, 406–413.
9. I. F. Olawuyi and W. Y. Lee, *Int. J. Food Sci. Technol.*, 2019, **54**, 2321–2328.
10. S. L. Ooi, S. C. Pak, P. S. Micalos, E. Schupfer, C. Lockley, M. H. Park and S. J. Hwang, *Molecules*, 2021, **26**, 2539.
11. K. Lee, H. Y. Lee, Y. Choi, H. S. Kim and J. Jeong, *Food Sci. Technol. Res.*, 2019, **25**, 115–122.
12. R. Salwan, S. Katoch and V. Sharma, in *Fungi in Sustainable Food Production*, ed. X. Dai, M. Sharma and J. Chen, Springer, Cham, 2021, ch. 2, pp. 165–180.
13. Z. Chen, X. Fang, W. Wu, H. Chen, Y. Han, H. Yang and H. Gao, *Food Chem.*, 2021, **364**, 130398.

14. R. Gaitán-Hernández, D. López-Peña, M. Esqueda and A. Gutiérrez, *Int. J. Med. Mushroom*, 2019, **21**, 841–850.
15. S. Sharma and S. Prakash, *J. Indian Soc. Periodontol.*, 2019, **23**, 216–219.
16. H. Zhu, M. Hu, D. Wang, G. Xu, X. Yin, X. Liu, M. Ding and L. Han, *Biomed. Pharmacother.*, 2020, **126**, 110049.
17. X. Wang, X. He, C. Sun, M. Peng, Q. Zhang, C. S. Brennan, W. Guan, F. Wang and N. Zhang, *Int. J. Food Sci. Technol.*, 2022, **57**, 6613–6620.
18. D. R. Byron, R. M. S. José, R. R. Felipe, M. V. J. Kelly and C. Dagoberto, *Afr. J. Food Sci.*, 2018, **12**, 374–382.
19. B. Duran-Rivera, F. Rojas-Rodas, W. Silva-Lopez, C. Gomez-Suarez and D. Castro-Restrepo, *Bionatura*, 2020, **5**, 1183–1188.
20. Z. S. Amin, S. Abbas, A. Munir, M. M. A. Qadri, M. Raza, M. Danish, S. Khan, B. Hafey, M. K. Hanif and H. Kashif, *J. Pharm. Res. Int.*, 2022, **34**, 18–35.
21. S. Im, Y. G. Chun, S. Liang and J. A. Han, *Korean J. Food Sci. Technol.*, 2018, **50**, 8–13.
22. E. A. Erdoğan Eliuz, *Int. J. Environ. Health Res.*, 2021, **32**, 1828–1841.
23. H. Kaur, L. M. Nyochembeng, P. Banerjee, E. Cebert and S. R. Mentreddy, *J. Agric. Sci.*, 2019, **11**, 1–19.
24. D. Pegler, *Kavaka*, 1975, **3**, 11–20.
25. D. S. Hibbett and M. J. Donoghue, *Conserv. Biol.*, 1996, **10**, 1321–1327.
26. D. J. Royse, L. C. Schisler and D. A. Diehle, *Interdiscip. Sci. Rev.*, 1985, **10**, 329–335.
27. Data Bridge Market Research. Market–Global Industry Trends and Forecast to 2029|Data Bridge Market Research. Global Shiitake Mushroom Cultivation Market–Industry Trends and Forecast to 2029, 2022. Available from: https://www.databridgemarketresearch.com/reports/global-shiitake-mushroom-cultivation-market.
28. P. G. Miles and S. T. Chang, Lentinula — A Mushrooming Mushroom, in *Mushrooms: Cultivation, Nutritional Value, Medicinal Effect, and Environmental Impact*, CRC Press, Boca Raton, 2004, pp. 120–160.
29. M. Antonelli, D. Donelli and F. Firenzuoli, Lentinan for Integrative Cancer Treatment: An Umbrella Review, in: Proceedings, 2021, vol. 79, pp. 1–8.
30. D. Lemanne and V. Maizes, *J. Altern. Complementary Med.*, 2018, **24**, 902–909.
31. W. Xiaokang, N. P. Brunton, J. G. Lyng, S. M. Harrison, S. T. Carpes and K. Papoutsis, *Food Biosci.*, 2020, **36**, 100619.
32. X. Yang, Y. Zhang, Y. Kong, J. Zhao, Y. Sun and M. Huang, *Int. J. Food Prop.*, 2019, **22**, 1100–1111.
33. U. Pandya, U. Dhuldhaj and N. S. Sahay, *Nat. Prod. Res.*, 2019, **33**, 2668–2680.
34. K. M. S. U. Imam, Y. Xie, Y. Liu, F. Wang and F. Xin, *ACS Food Sci. Technol.*, 2021, **1**, 551–558.
35. P. A. Ayeka, *J. Evidence-Based Complementary Altern. Med.*, 2018, **2018**, 1–11.

36. T. Kobayashi, M. Oguro, M. Akiba, H. Taki, H. Kitajima and H. Ishihara, *J. For. Res.*, 2020, **25**, 269–275.
37. M. A. Rahman, N. Abdullah and N. Aminudin, *Saudi J. Biol. Sci.*, 2018, **25**, 1515–1523.
38. K. Arunachalam, S. P. Sasidharan and X. Yang, *Food Chem. Adv.*, 2022, **1**, 100023.
39. H. Zhao, L. Wang, M. Brennan and C. Brennan, *J. Futur. Foods*, 2022, **2**, 18–24.
40. G. Venturella, V. Ferraro, F. Cirlincione and M. L. Gargano, *Int. J. Mol. Sci.*, 2021, **2**, 1–31.
41. M. Zeb and C. H. Lee, *Molecules*, 2021, **26**, 251.
42. S. Zhao, Q. Gao, C. Rong, S. Wang, Z. Zhao, Y. Liu and J. Xu, *J. Fungi*, 2020, **6**, 1–37.
43. F. Shahzad, D. Anderson and M. Najafzadeh, *Nutrients*, 2020, **12**, 1–13.
44. G. Baktemur, E. Kara, M. Yarar, N. Yilmaz, E. Ağçam and A. Akyildiz, *Not. Bot. Horti Agrobot.*, 2022, **50**, 12553.
45. F. Atila, *Sci. Hortic.*, 2019, **9**, 263–268.
46. S. Gao, Z. Huang, X. Feng, Y. Bian, W. Huang and Y. Liu, *Sci. Rep.*, 2020, **10**, 1–8.
47. N. F. Azman, M. Yamauchi, M. Yamada, S. Ikeda, T. Yamaguchi, F. Yagi and H. Hara, *J. Mater. Cycles Waste Manag.*, 2019, **21**, 336–344.
48. T. Bito, F. Teng, N. Ohishi, S. Takenaka, E. Miyamoto, E. Sakuno, K. Terashima, Y. Yabuta and F. Watanabe, *Mycoscience*, 2014, **55**, 462–468.
49. K. Ogawa and T. Yashima, *Cellulose*, 2020, **27**, 9605–9621.
50. S. G. Shin, H. S. Lee and S. Lee, *J. Asia Pac. Entomol.*, 2012, **15**, 174–181.
51. T. Ohnuki, Y. Aiba, F. Sakamoto, N. Kozai, T. Niizato and Y. Sasaki, *Sci. Rep.*, 2016, **6**, 1–6.
52. J. Cui and Y. Chisti, *Biotechnol. Adv.*, 2003, **21**, 109–122.
53. C. Ponnusamy, V. V. S. Uddandrao, S. P. Pudhupalayam, S. Singaravel, T. Periyasamy, P. Ponnusamy, P. Prabhu, V. Sasikumar and S. Ganapathy, *Biosci. Biotechnol. Res. Asia*, 2022, **19**, 1–11.
54. N. A. Amirullah, N. Zainal Abidin and N. Abdullah, *Food Res. Int.*, 2018, **105**, 517–536.
55. C. Tang, P. C. X. Hoo, L. T. H. Tan, P. Pusparajah, T. M. Khan, L. H. Lee, B. H. Goh and K. G. Chan, *Front. Pharmacol.*, 2016, **7**, 474.
56. B. Muszyńska, P. Pazdur, J. Lazur and K. Sułkowska-Ziaja, *Med. Int. Rev.*, 2017, **28**, 189–195.
57. S. Kumar, *Pharma. Innov. J.*, 2022, **11**, 1–11.
58. I. Golak-Siwulska, A. Kałuzewicz, S. Wdowienko, L. Dawidowicz and K. Sobieralski, *Herba. Polonica. Sciendo*, 2018, **64**, 71–81.
59. T. Yamada, T. Oinuma, M. Niihashi, M. Mitsumata, T. Fujioka, K. Hasegawa, H. Nagaoka and H. Itakura, *J. Atheroscler. Thromb.*, 2002, **9**, 149–156.

60. A. Pop, A. Păucean, S. M. Man, M. S. Chis, M. Mihai and S. Muste, in *Wild Mushrooms*, ed. S. B. Dhull, A. Bains, P. Chawla and P. K. Sadh, CRC Press, 2022, pp. 259–315.
61. S. Ramesh, M. Majrashi, M. Almaghrabi, M. Govindarajulu, E. Fahoury, M. Fadan, M. Buabeid, J. Deruiter, R. Clark, V. Mulabagal and D. C. Agrawal, in *Overview of Therapeutic Efficacy of Mushrooms*, ed. D. C. Agrawal and M. Dhanasekaran, Medicinal Mushrooms Springer, Singapore, 2019, pp. 103–141.
62. R. Z. Pasko, T. G. Timm, G. G. de Lima, C. V. Helm, E. A. de Lima, G. S. Henriques and L. B. B. Tavares, *Int. J. Med. Mushrooms*, 2022, **24**, 53–66.
63. I. Diallo, F. Boudard, S. Morel, M. Vitou, C. Guzman, N. Saint, A. Michel, S. Rapior, L. Traoré, P. Poucheret and F. Fons, *Int. J. Med. Mushrooms*, 2020, **22**, 535–546.
64. A. Jedinak, S. Dudhgaonkar, Q. L. Wu, J. Simon and D. Sliva, *Nutr. J.*, 2011, **10**, 1–10.
65. M. P. Singh, S. N. Rai, S. K. Dubey, A. T. Pandey, N. Tabassum, V. K. Chaturvedi and N. B. Singh, *Crit. Rev. Biotechnol.*, 2022, **42**, 913–930.
66. M. H. Z. Abidin, N. Abdullah and N. Z. Abidin, *Int. J. Food Prop.*, 2017, **20**, 1251–1261.
67. S. Paisansak, P. Sangtanoo, P. Srimongkol, T. Saisavoey, O. Reamtong, K. Choowongkomon and A. Karnchanatat, *J. Food Sci. Technol.*, 2021, **58**, 85–97.
68. X. Lu, M. A. Brennan, W. Guan, J. Zhang, L. Yuan and C. S. Brennan, *Foods*, 2021, **10**, 731.
69. S. M. Badalyan, A. Barkhudaryan and S. Rapior, *Int. J. Med. Mushrooms*, 2021, **23**, 21–31.
70. L. M. Cheung and P. C. K. Cheung, *Food Chem.*, 2005, **89**, 403–409.
71. N. Zhang, H. Chen, Y. Zhang, L. Ma and X. Xu, *J. Sci. Food Agric.*, 2013, **93**, 3107–3313.
72. H. Y. Fu, D. E. Shieh and C. T. Ho, *J. Food Lipids*, 2002, **9**, 35–43.
73. T. Rahman and M. Choudhury, *Bang. J. Med. Biochem.*, 2013, **5**, 24–32.
74. H. Maeda, T. Katsuki, T. Akaike and R. Yasutake, *Jap. J. Cancer Res.*, 1992, **83**, 923–928.
75. S. Pil-Nam, K. M. Park, G. H. Kang, S. H. Cho, B. Y. Park and H. Van-Ba, *LWT*, 2015, **62**, 62–68.
76. M. Zhu, P. Nie, Y. Liang and B. Wang, *Carbohydr. Polym.*, 2013, **95**, 644–648.
77. H. Van Ba, H. W. Seo, S. H. Cho, Y. S. Kim, J. H. Kim, J. S. Ham, B. Y. Park and S. P. Nam, *Food Control*, 2016, **70**, 201–209.
78. B. Mallard, D. N. Leach, H. Wohlmuth and J. Tiralongo, *PLoS One*, 2019, **14**, e0224740.
79. L. Zhou, Q. Zhang, Y. Zhang, J. Liu and Y. Cao, *Int. Immunopharmacol.*, 2009, **9**, 455–462.
80. X. Li, H. Zhang and H. Xu, *Int. J. Biol. Macromol.*, 2009, **45**, 377–380.

81. J. Nisar, I. Mustafa, H. Anwar, M. U. Sohail, G. Hussain, M. I. Ullah, M. N. Faisal, S. A. Bukhari and A. Basit, *Int. J. Med. Mushrooms*, 2017, **19**, 981–990.
82. X. Xu, H. Yan and X. Zhang, *J. Agric. Food Chem.*, 2012, **60**, 11560–11566.
83. G. Accardi, D. Majo, A. Roszczyk, J. Turło, Z. R. Zago and B. Kaleta, *Int. J. Mol. Sci.*, 2022, **23**, 8980.
84. K. Kumar, R. Mehra, R. P. F. Guiné, M. J. Lima, N. Kumar and R. Kaushik, *Foods*, 2021, **10**, 2996.
85. C. Dong, T. Yang and T. Lian, *Int. J. Med. Mushrooms*, 2014, **16**, 485–495.
86. F. Zhu, B. Du, Z. Bian and B. Xu, *J. Food Compos. Anal.*, 2015, **18**, 165–173.
87. S. Rathee, D. Rathee, D. Rathee, V. Kumar and P. Rathee, *Rev. Bras. Farmacogn.*, 2012, **22**, 459–474.
88. H. A. El Enshasy and R. Hatti-Kaul, *Trend Biotechnol.*, 2013, **31**, 668–677.
89. V. Rezvani, H. R. Pourianfar, S. Mohammadnejad, A. Madjid Ansari and L. Farahmand, *Appl. Microbiol. Biotechnol.*, 2020, **104**, 6855–6871.
90. M. A. Rahman, M. S. Rahman, N. M. B. Bashir, R. Mia, A. Hossain, S. K. Saha, A. J. Kakon and N. C. Sarker, *Int. J. Med. Mushrooms*, 2021, **23**, 1–11.
91. K. K. Hapuarachchi and T. C. Wen, Therapeutic Potential of Medicinal Mushrooms: Insights into Its Use Against Covid-19, in *Biology, Cultivation and Applications of Mushrooms*, ed. A. Arya and K. Rusevska, Springer, Singapore, 2022, pp. 27–46.
92. N. Suwannarach, J. Kumla, K. Sujarit, T. Pattananandecha, C. Saenjum and S. Lumyong, *Molecules*, 2020, **25**, 1800.
93. T. V. Teplyakova and T. A. Kosogova, *Int. J. Med. Mushrooms*, 2016, **18**, 375–386.
94. T. Teplyakova and T. Kosogova, *J. Pharm. Pharmacol.*, 2015, **3**, 357–371.
95. M. Takehara, K. Kuida and K. Mori, *Arch. Virol.*, 1979, **59**, 269–274.
96. H. Kumar, K. Bhardwaj, R. Sharma, E. Nepovimova, N. Cruz-Martins, D. S. Dhanjal, R. Singh, C. Chopra, R. Verma, K. A. Abd-Elsalam and A. Tapwal, *J. Fungi*, 2021, **7**, 427.
97. E. Tavassoli-Kafrani, H. Shekarchizadeh and M. Masoudpour-Behabadi, *Carbohyd. Polym.*, 2016, **137**, 360–374.
98. Healthline. Why Shiitake Mushrooms Are Good For You, [cited 2022 Aug 20]. Available from: https://www.healthline.com/nutrition/shiitake-mushrooms#side-effects.
99. D. Boels, A. Landreau, C. Bruneau, R. Garnier, C. Pulce, M. Labadie, L. De Haro and P. Harry, *Clin. Toxicol.*, 2014, **52**, 625–628.
100. M. Hérault, J. Waton, A. C. Bursztejn, J. L. Schmutz and A. Barbaud, *Ann. Dermatol. Venereol.*, 2010, **137**, 290–293.
101. J. H. Maeng, H. M. Shahbaz, K. Ameer, Y. Jo and J. H. Kwon, *J. Food Process Eng.*, 2017, **40**, e12421.

102. C. C. Zhao, K. Ameer and J. B. Eun, *LWT*, 2021, **143**, 111149.
103. H. G. Lee, Y. Jo, K. Ameer and J. H. Kwon, *Food Sci. Biotechnol.*, 2018, **27**, 1607–1617.
104. K. Ameer, S. W. Bae, Y. Jo, H. G. Lee, A. Ameer and J. H. Kwon, *Food Chem.*, 2017, **229**, 198–207.
105. Z. Wu, K. Ameer and G. Jiang, *J. Food Sci. Technol.*, 2021, **58**, 62–73.
106. G. Jiang, K. Ameer, H. Kim, E. J. Lee, K. Ramachandraiah and G. P. Hong, *Foods*, 2020, **9**, 1227.
107. G. Jiang, K. Ameer and J. B. Eun, *J. Food Sci. Technol.*, 2020, **57**, 4551–4561.
108. K. Ameer, H. M. Shahbaz and J. H. Kwon, *Comp. Rev. Food Sci. F. Saf.*, 2017, **16**, 295–315.
109. K. Ramachandraiah, K. Ameer and G. Jiang, *Sci. Total Environ.*, 2022, **844**, 157234.
110. Z. Wu, K. Ameer, C. Hu, A. Bao, R. Wang, W. Tang, N. Chaudhary and G. Jiang, *Food Sci. Tech.*, 2021, **42**, 1–13.

CHAPTER 12

Leucopaxillus giganteus (Sowerby) Singer (Giant Leucopax)

FARAK ALI,[a,b] SHAHNAZ ALOM,[a,b] ARPITA PAUL,[b] NILAYAN GUHA[b] AND Md. KAMARUZ ZAMAN*[b]

[a] Girijananda Chowdhury Institute of Pharmaceutical Sciences-Tezpur, Dekargaon, Sonitpur-784501, Assam, India; [b] Department of Pharmaceutical Sciences, Faculty of Science and Engineering, Dibrugarh University, Dibrugarh-786004, Assam, India
*Email: kzaman71@dibru.ac.in

12.1 Introduction

Mushrooms are not only considered to be part of a balanced diet but are also a key ingredient in the development of drugs and dietary supplements.[1,2] Foods rich in nutrients, such as minerals, amino acids, vitamins, crude fiber, proteins, and lipids, are necessary for the body to function normally. Mushrooms are an important source of nutrition and can therefore be included in a balanced diet.[3] Along with dried mushrooms, the mycelia of mushrooms also possess high nutritional value and serve as a functional food for many tribes or communities. Owing to their excellent bioactive properties, people are inclined to consume mushrooms and their mycelia, and as a result their production has increased and driven the commercialization of local edible species. Simultaneously, the pharmaceutical industry has also become very focused on the *in vitro* production of mycelia. It has been reported that the growth rate of cancers can be successfully suppressed

Phytochemistry and Nutritional Composition of Significant Wild Medicinal and Edible Mushrooms: Traditional Uses and Pharmacology
Edited by Ajay Sharma, Garima Bhardwaj and Gulzar Ahmad Nayik

Published by the Royal Society of Chemistry, www.rsc.org

by consuming mushrooms regularly in the diet.[4] In Asia and its subcontinent, mushrooms have long been used as a traditional food and medicine.[5] Most edible mushrooms found in forests possess antioxidant activity; therefore, regular diet-based consumption of mushrooms can help in the management of various diseases/disorders caused by oxidative stress, such as cancer, diabetes, aging, *etc.* These mushrooms also exhibit immunomodulatory activity.[4,6–8]

L. giganteus (Sowerby) Singer (*L. giganteus*) is an edible mushroom belonging to the family Tricholomataceae, order *Agaricales*, and phyla Basidiomycete[7] that is native to Portugal. It is known as the giant leucopax or giant funnel because a fully matured fungus can have a cap diameter of 45 cm, which is much bigger than those of other species of mushrooms. Wild *L. giganteus* usually grows in hedgerows or damp wooded edges, grassy roadside verges, parklands, or permanent pastures. Currently, 80–85% of important bioactive compounds are obtained from the fruiting bodies of mushrooms that are harvested either from farms or from the wild. The cultivation of fruiting bodies of *L. giganteus* is not done due to it being a time-consuming and difficult process. As a result, around 15% of bioactive compounds are obtained from the mycelia of *L. giganteus*.[9]

12.2 Taxonomy

12.2.1 Taxonomic History

The taxonomic history of *Leucopaxillus giganteus* is presented in Figure 12.1.

12.2.2 Taxonomic Classification

Kingdom: Fungi
Phyla: Basidiomycota
Class: Agaricomycetes
Order: Agaricales
Family: Tricholomataceae
Genus: *Leucopaxillus*
Species: *Leucopaxillus giganteus*

12.2.3 Vernacular Names

The common name assigned is the giant leucopax or the giant funnel.

12.3 Distribution and Relationship

L. giganteus generally forms fairy rings in grasslands. The fruiting bodies of *L. giganteus* start to develop in summer and autumn. This fungus is a saprophyte, meaning that it derives its nutrients from dead and decomposed organic matter. This species of fungus is distributed throughout the

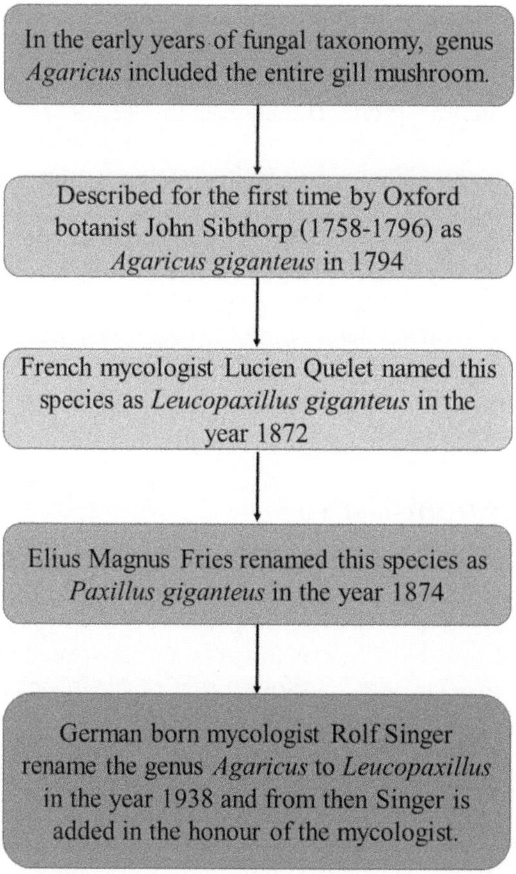

Figure 12.1 Representation of the taxonomic history of *Leucopaxillus giganteus.*

temperate zones of the northern hemisphere, mainly found in the grasslands of Braganca, northeast Portugal, as well as throughout northern Europe.[7] Moreover, it can also be found in the grass pastures of the black sea region of Turkey, the northwest region of Rajasthan, and the Binchuan area of Dali.[10,11] *L. giganteus* is also found in the temperate zone of the northern hemisphere in North America, Britain, and Europe. It can also be found in the Himalayan region of Asia. The highest numbers of this species can be found in European countries and North America, whereas in India, leucopax can be found in the Himalayan region.

12.4 Morphology Characteristics

The identification of mushrooms is a tedious job since most mushrooms often look similar. *L. giganteus* can be easily identified by its size and certain morphology characteristics, such as cap, stems, gills, spores, and some microscopic characteristics. The cap of *L. giganteus* is usually like the top part an umbrella, fleshy, and a fragile part of the mushroom. The color of the *L. giganteus* cap is ivory and it has a velvet, flat surface and margins that point downwards and eventually tend towards a funnel shape. The size of the cap varies from 15–30 cm. The stem is the slender yet quite tough part of the mushroom over which the cap resides. The dimensions of the stem vary from 4–6 cm in height and 2–3 cm in diameter and it has a creamy white buff color that eventually turns brown upon maturation. The stems of this fungus are devoid of a ring-like structure, and there is a bed of white colored mycelia present at its base.[12] The gills of the fungus are cream-colored, narrow, and closely packed at a young age, but darken during its maturation. Spores of 6–9 × 4–5 μm can be seen on the cap as well as on the stem, which are oblong, disc-shaped, translucent, and exhibit an even surface when visualized using a light microscope. The odor and taste are of *L. giganteus* are pleasant but not distinctive.[13]

12.5 Traditional Uses

Wild edible mushrooms have been an integral part of the diets of many ethnic groups since time immemorial due to their high nutritional value. Due to their high medicinal value and nutritive importance, wild mushrooms have also been greatly exploited by tribes or foragers. Besides medicinal mushrooms, other mushroom species are also found that result in poisoning when consumed.[14] Medicinal mushrooms possess health-promoting properties as they contain a very low lipid content, *trans*-isomers of unsaturated fatty acids, and different phytochemicals, such as sterols, chitosan, triterpenes, and phenolic compounds.[15] *L. giganteus* is one such important mushroom that has a high nutritional value and is used as a functional food. The mycelium and fruiting bodies of mushrooms are used as a dietary supplement for treating different ailments, such as obesity, asthma, and cardiovascular diseases, as well enhancing immunity and strengthening bones.

12.6 Cultivation

As the consumption of mushrooms is increasing daily, so is their cultivation. Harada *et al.* developed a method for cultivating *L. Giganteus* in the Sugi (*Cryptomeria japonica*) forests of Japan. Sugi forests are being continuously abandoned in Japan due to the low price of Sugi trees, so to compensate for this a method has been developed that involves the use of this abandoned land in the cultivation of giant mushrooms. As mushrooms require moist soil with decaying matter, abandoned forests are the best choice for cultivation. Various strains of *L. giganteus* were obtained from Myco Bank in Mie, Hyogo, and Fukuoka of Japan, and the mycelia of each strain were extracted on PYGA medium (4% w:v glucose, 0.3% w:v yeast extract, 0.3% w:v polypeptone, 0.1% w:v $MgSO_4{\cdot}7H_2O$, 0.1% w:v KH_2PO_4 and 1.5 w:v agar; pH 5.5) from the basidiospores of fruiting bodies. The strains were prepared and inoculated with media and buried in spring and autumn. Investigation of the inoculated land was conducted after six months and hyphae were observed only in the inoculated land. After two years, the fruiting bodies of *L. giganteus* were harvested and 3.8–13.4 kg were recovered per inoculated area. Every year, 173–1103 fruiting bodies are collected, each with an approximate weight of around 19 ± 8 g. To support the cultivation and distribution of nutraceutically important *L. giganteus,* its genetic sequence is being collected in Russia, France, Italy, China, and the United States and kept in the database of Gene Bank.[16]

12.7 Detailed Extraction of Major Phytochemicals

Clitocine and some polysaccharides have been isolated from the fruiting body of *L. giganteus*. The freshly collected *L. giganteus* mushrooms (2.7 kg) were extracted with 95% ethanol (6 L), filtered, and column chromatography was carried out over a silica gel column using a CH_2Cl_2–MeOH (10 : 0–0 : 10) eluent. Four fractions were obtained, of which the fourth fraction was further sub-fractioned using CH_2Cl_2–MeOH (1 : 9). The collected sub-fractions were purified using a reverse-phase C18 (ODS, 50 μM) column, and, finally, separation was carried out on an HPLC system using MeOH–H_2O (5 : 95) to yield the compound clitocine. The purity of the compound was found to be more >95%.[17]

Niu *et al.* (2021) extracted the dried and powdered fruiting body of *L. giganteus* using distilled water in a water bath (extraction temperature 80 °C, extraction time 4.5 h). Centrifugation was then carried out at 4000 rpm for 10 min to separate the supernatant. After that, ethanol was added in a four-fold volume to the supernatant, which was kept overnight at 4 °C to precipitate the polysaccharide. After this time, the obtained precipitate was centrifuged at 8000 rpm for 10 min and lyophilized to obtain the crude polysaccharide. The isolated polysaccharides were further characterized using FT-IR spectroscopy, SEM, TGA, Congo red measurements, and XRD.[18] Later in the same year, Niu *et al.* (2021) isolated a polysaccharide from a cold-water extract of *L. giganteus*. For this, the crushed and powdered fruiting bodies of the fungi (1 : 20 w : v) were soaked in water at 4 °C for 24 h. The resulting extract was centrifuged and frozen before being thawed again. Then, a three-fold volume of absolute ethanol was added, and the mixture was stored overnight. After this time, the liquid was further centrifuged at 8000 rpm and distilled water was used to precipitate the polysaccharide fraction. This fraction was then purified by column chromatography using a Sephadex G-200 column and lyophilized. Ion chromatography revealed that the isolated polysaccharides comprised fructose, galactose, glucose, xylose, mannose, and glucuronic acid. In addition, the presence of traces of uronic acid was confirmed using the *m*-hydroxy biphenyl method. Also, FT-IR and NMR spectroscopy results were similar to the results of ion chromatography. SEM images showed the presence of irregular small fragments with a rough and loosely arranged sheet-like stacked structure.[19]

12.8 Quantitative Data Representation of Phytochemicals

Quantitative analysis has shown that the moisture and energy contents of *L. giganteus* were 3.40 g per 100 g and 20.38 ± 1.71 kcal, respectively. In addition, the proximate chemical composition of this wild edible mushroom was – total fat (0.14 ± 0.00 g per 100 g), crude protein (2.87 ± 0.19 g per 100 g), ash (0.18 ± 0.01 g per 100 g), and carbohydrates (1.91 ± 0.24 g per 100 g). Moreover, the polyunsaturated fatty acid (PUFA) content (46.67 ± 0.47%) was higher than that of the monounsaturated fatty acid (MUFA) content (34.08 ± 0.46%). An appreciable amount of sugar (0.64 ± 0.00 g per 100 g) was also found to be present in the mushroom. Mannitol (0.14 ± 0.00 g per 100 g) and trehalose (0.50 ± 0.00 g per 100 g) were the main sugars detected.[20] The presence of amino acids was also studied and the percentages of essential and non-essential amino acids were found to be 37.9% and 62.1%, respectively. The main amino acids detected were aspartic acid (21.1 ± 0.31 μg mg^{-1}), threonine (13.2 ± 0.95 μg mg^{-1}), serine (12.9 ± 1.86 μg mg^{-1}), glutamic acid (37.6 ± 2.26 μg mg^{-1}), proline (8.92 ± 0.66 μg mg^{-1}), glycine (10.8 ± 0.87 μg mg^{-1}), alanine (16.1 ± 1.40 μg mg^{-1}), valine (10.4 ± 10.4 μg mg^{-1}), isoleucine (14.3 ± 0.80 μg mg^{-1}), leucine (17.2 ± 1.13 μg mg^{-1}), tyrosine (4.82 ± 0.58 μg mg^{-1}), phenylalanine (10.1 ± 0.70 μg mg^{-1}), histidine (9.38 ± 0.82 μg mg^{-1}), lysine (13.3 ± 1.14 μg mg^{-1}), tryptophan (3.58 ± 0.33 μg mg^{-1}), arginine (15.1 ± 1.42 μg mg^{-1}), cysteine (3.60 ± 0.00 μg mg^{-1}), and methionine (4.05 ± 0.00 μg mg^{-1}).[21]

Furthermore, organic acids were reported to be present in *L. giganteus*, such as oxalic, *cis*-aconitic, citric, and fumaric acids. Amongst them, citric acid was present in the highest amount (90.0–99.6%). Additionally, the presence of alkaloids was detected in *L. giganteus* in 2008 by Ribeiro *et al.* Further, a methanol extract of *L. giganteus* was reported to contain phenols (6.29 ± 0.20 mg g^{-1}), ascorbic acid (0.13 ± 0.0069 mg g^{-1}), β-carotene (1.88 ± 0.090 μg g^{-1}) and lycopene (0.69 ± 0.034 μg g^{-1}).[1] Similarly, another study reported phenols (15.28 mg g^{-1}), ascorbic acid (0.97 mg g^{-1}), flavonoids (2.23 mg g^{-1}), β-carotene (1.46 μg g^{-1}), and lycopene (34.92 μg g^{-1}) as the major bioactive constituents present in *L. Giganteus*.[22] On the contrary, in 2008 Ribeiro *et al.* failed to detect the presence of phenols in the mushroom.[23]

12.9 Phytochemistry

All the investigations conducted on *L. giganteus* have resulted in the isolation of several compounds, the structures have been verified and elucidated from their different extractions. A novel compound, (*R*)-4-ethoxy-2-hydroxy-4-oxobutanoic acid (**1**), and seven compounds (**2–7**) were isolated, the structures of which are shown in Figure 12.2.[24,25]

(*R*)-4-Ethoxy-2-hydroxy-4-oxobutanoic acid (**1**) was isolated in the form of a pale-yellow oil; molecular formula was confirmed to be $C_6H_{10}O_5$ using

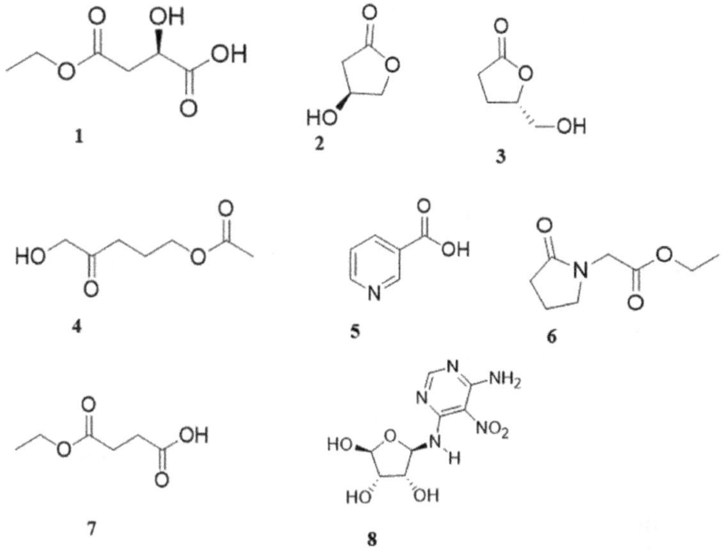

Figure 12.2 Structures of the compounds obtained from *Leucopaxillus giganteus*.

high-resolution electrospray ionization mass spectrometry, and its structure was revealed using DEPT, COSY, HMQC, and HMBC NMR spectra. One methyl, two methylene, one methine, and two tetra-substituted carbons were observed in the structure from the DEPT experiment. A COSY correlation (H-2/H-3) and HMBC correlations (H-2/C-1, C-3; H-3/C-2, C-4) determined the presence of a malic acid moiety. COSY (H-1′/H-2′) and HMBC (H-1′/C-4, 2′; H-2′/C-1′) correlations both indicated that the carboxylic acid at the C-4 position was esterified. By comparing the particular rotation of **1** with that of the synthetic compound and its enantiomer, the absolute configuration of **1** was determined. However, no particular rotation or CD data have been published in the literature; therefore, the absolute configuration of the isolated chemical has yet to be determined.[26–28] As a result of our findings, it was determined that compound (**1**) is a new chemical (Figure 12.2). Compound (**6**) was confirmed to be ethyl-2-(2-oxopyrrolidin-1-yl)acetate, which was separated from a natural source for the first time. Compound (**3**) was identified as (*S*)-5-(hydroxymethyl)-dihydrofuran-2(3*H*)-one.[29] Compound (*S*)-4-hydroxydihydrofuran-2(3*H*)-one (**2**) was isolated and found to be (*R*)-(+)-β-hydroxy-gamma-butyrolactone with the molecular formula $C_4H_6O_3$.[30] Compound (**4**) was isolated and identified as catathelasmol D with the molecular formula $C_7H_{12}O_4$.[31] Compound (**5**) was isolated and found to be nicotinic acid with the molecular formula $C_6H_5NO_2$.[32] Monoethyl succinate was found to be compound (**7**) with the molecular formula $C_6H_9ClO_4$.[24,33] 6-Amino-5-nitro-4-(ribofuranosylamino)pyrimidine, a natural compound commonly known as clitocine was isolated as compound (**8**), from *L. giganteus*.[34]

12.10 Pharmacological Activities

The presence of some important phytoconstituents, such as clitocine, in *L. giganteus*, makes this species pharmacologically important as it exhibits antioxidant, antimicrobial, and anticancer properties, amongst others.[19,35,36]

12.10.1 Antioxidant Activity

L. giganteus exhibits strong antioxidant activity[1] due to the high number of polyphenolic compounds present in its mycelium.[37] Free radical scavenging capacity is dependent on the carbon source used for the energy and growth of the mycelium. Barros *et al.* in their experiments showed that to increase the phenolic content and synthesis antioxidant activities, monosaccharides such as glucose and fructose were far better than disaccharides such as sucrose. They also reported that the aldohexose glucose showed the most prominent effect on increasing the antioxidant activity by increasing the phenolic content and decreasing the EC_{50} value of *L. giganteus*. The mycelium of *L. giganteus* showed DPPH (2,2-diphenyl-1-picrylhydrazyl) radical scavenging activity, inhibition of hemolysis, and reduced power activity by inhibiting free radical formation and protecting against damage caused by oxidative stress of the cell membrane of erythrocytes in the case of hemolysis.[35] Barros *et al.* investigated the antioxidant and hemolysis activity of *L. giganteus* using a linoleate-β carotene model and found that a methanolic extract of *L. giganteus* showed significant DPPH radical scavenging activity, reducing the power activity due to a high phenolic content. Although the extract showed hemolysis activity, it was lower than the antioxidant activity. Nitrogen sources can also increase the antioxidant activity of the mycelium of *L. giganteus*. Among all the known nitrogen sources, ammonium hydrogen phosphate was found to be a suitable source for increasing the phenol and flavonoid content during the growth stage of the mycelium of *L. giganteus,* which is responsible for its significant antioxidant properties.[20]

12.10.2 Antimicrobial Activity

Alongside its antioxidant activity, *L. giganteus* also exhibits strong antimicrobial activity.[38] Alves *et al.* determined the antimicrobial activity of an aqueous methanolic extract of *L. giganteus* using a microdilution method. From their experiments, they found that *L. giganteus* effectively prevents the growth of *Proteus mirabilis* and *Escherichia coli* and has the ability to kill *Streptococcus pyrogens*, *Streptococcus agalactiae*, and *Pasteurella multocida*.[39] The active compound clitocybin is extracted from *L. giganteus* at the commercial level for the production of clitocybin antibiotics, which are effective antimicrobial agents. From their experiments, Barros *et al.* proved that the presence of a nitrogen source during the growth of the mycelium of *L. giganteus* increases its phenolic and flavonoid content and also enhances its antioxidant and antimicrobial properties. They also reported that a methanolic extract of the mycelium of *L. giganteus* was found to inhibit gram-positive bacteria, such as *S. aureus*, *B. subtilis,* and *B. cereus*. Therefore, *L. giganteus* can be used as a promising antimicrobial agent against human respiratory tract infections and gastrointestinal infections caused by some bacterial agents.[1,39] *L. giganteus* can also inhibit the formation of bacterial biofilm that is mainly formed by multi-resistant bacteria. Alves *et al.* reported that *L. giganteus* inhibits the biofilm formation of *E. coli*, *Pr. Mirabilis*, *A. baumannii*, and *P. aeruginosa*. Among all these bacterial biofilms, *L. giganteus* showed the highest inhibition potential against *E. coli*.[40]

12.10.3 Antitumor Activity

L. giganteus exhibits prominent antitumor activity due to the presence of clitocine, an active phytoconstituent.[41] Clitocine exhibits anti-proliferative activity against human cervical cancer cell (HeLa) cells in a dose and time-dependent manner. Ren *et al.* reported that clitocine inhibits HeLa proliferation *via* the induction of apoptosis. The process of apoptosis was induced by various pathways, such as caspase-3, caspase-8, and caspase-9 activation, the reduction of Bcl-2 levels and the rise of Bax levels, the secretion of cytochrome *C* (Cyt *C*) from mitochondria to the cytosol, *etc.*[15,17] Multidrug resistance (MDR) is a primary drawback associated with combating cancer as it decreases the effectiveness of many anticancer drugs. Clitocine has the potential to inhibit MDR in both MES-SA/Dx5 and R-HepG2 cells by inhibiting the overexpression of P-glycoprotein. Clitocin inhibits doxorubicin-induced MDR by inhibiting the NF-κB p65 protein levels and suppressing P-glycoprotein activation. Consequently, it is an effective treatment for cancer chemotherapy.[42] Niu *et al.* reported that the polysaccharides present in *L. giganteus* have excellent organ-protective effects as well as antitumor effects and can also prevent H22 solid tumor cell proliferation in a dose-dependent manner. The polysaccharides of *L. giganteus* can induce apoptosis *via* the mitochondrial pathway and cause S-phase arrest of tumor cells.[19] The polysaccharide galactomannoglucan found in *L. giganteus* is mainly responsible for its antitumor and organ-protective effects.[43]

12.11 Extinction and Conservation

In recent years, the mushroom trade has boomed due to their scientific and socio-economic importance. However, a lack of proper conservation techniques and legislation to protect different mushroom species and their habitat have led to the extinction of certain species. According to 'The national red list of North Macedonia', *L. Giganteus* is categorized as an endangered species. Therefore, measures should be undertaken to protect the fungi population, such as creating public awareness regarding their ecology and diversity. Additionally, strict and efficient regulations of the conservation and trade control of mushrooms should be implemented.[44]

12.12 Conclusions and Future Perspectives

Mushrooms and the phytoconstituents present in them can serve as promising candidates for the development of pharmaceuticals to treat diseases with minimal side effects. Oxidative stress is a key player in the progression of chronic illnesses, such as diabetes, cancer, neurodegenerative diseases, amongst others. *L. giganteus* shows significant antioxidant activity; therefore, it can attenuate the damage caused by oxidative stress. This fungus mainly contains clitocybin and galactomannoglucan, which exhibit significant anticancer activity. In addition, *L. giganteus* extracts often exhibit antimicrobial activity against respiratory tract and gastrointestinal infections. However, most of these studies have been limited to *in vitro* models; therefore, in-depth studies at the pre-clinical and clinical levels are necessary to understand the underlying mechanism involved. Further, analysis of the polysaccharide composition of *L. giganteus* will give information regarding the dietary fiber content of the mushroom, which ultimately underpins its positive health benefits. Moreover, the identification and isolation of bioactive compounds present in *L. giganteus* will help in the further exploration of its therapeutic potential and encourage research in the advancement of efficient therapies for the management of ailments.

Acknowledgements

The authors would like to offer sincere thanks to Dr Kamaruz Zaman for constant guidance, motivation, drawing an outline, communication, *etc.* The authors wish to offer their heartfelt gratitude to both research institutions for providing the opportunity to carry out this work.

References

1. L. Barros, M. J. Ferreira, B. Queirós, I. C. F. R. Ferreira and P. Baptista, *Food Chem.*, 2007, **103**, 413–419.
2. L. M. Cheung and P. C. K. Cheung, *Food Chem.*, 2005, **89**, 403–409.
3. J. A. Buswell, Y. J. Cai, S. T. Chang, J. F. Peberdy, S. Y. Fu and H. S. Yu, *World J. Microbiol. Biotechnol.*, 1996, **12**, 537–542.
4. A. T. Borchers, C. L. Keen and M. E. Gershwin, *Exp. Biol. Med.*, 2004, **229**, 393–406.
5. R. Chang, *Nutr. Rev.*, 1996, **54**, S91.
6. L. M. Cheung and P. C. K. Cheung, *Food Chem.*, 2005, **89**, 403–409.
7. L. Barros, P. Baptista, L. M. Estevinho and I. C. F. R. Ferreira, *Food Chem.*, 2007, **105**, 179–186.
8. D. Agrahar-Murugkar and G. Subbulakshmi, *Food Chem.*, 2005, **89**, 599–603.
9. U. Lindequist, T. H. J. Niedermeyer and W.-D. Jülich, *Evidence-Based Complementary Altern. Med.*, 2005, **2**, 285–299.
10. F. A. Ayaz, H. Torun, A. Colak, E. Sesli, M. Millson and R. H. Glew, *Food Nutr. Sci.*, 2011, **02**, 53–59.

11. H. Liu, J. Zhang, T. Li, Y. Shi and Y. Wang, *Biol. Trace Elem. Res.*, 2012, **147**, 341.
12. Leucopaxillus giganteus, Giant Funnel, identification, [cited 2022 Mar 17]. Available from: https://www.first-nature.com/fungi/leucopaxillus-giganteus.php.
13. Giant funnel (Leucopaxillus giganteus) - JungleDragon, [cited 2022 Mar 17]. Available from: https://www.jungledragon.com/specie/10739/giant_funnel.html.
14. M. Thakur, *Biol. Macrofungi*, 2018, 379–394.
15. V. K. Chaturvedi, S. Agarwal, K. K. Gupta, P. W. Ramteke and M. P. Singh, *3 Biotech*, 2018, **8**, 334.
16. E. Harada, N. Tada, I. Kamei, T. Morizono and Y. Matsuda, *J. For. Res.*, 2021, **26**, 386–394.
17. G. Ren, Y. Zhao, L. Yang and C.-X. Fu, *Cancer Lett.*, 2008, **262**, 190–200.
18. L.-L. Niu, Y.-R. Wu, H.-P. Liu, Q. Wang, M.-Y. Li and Q. Jia, *J. Food Meas. Charact.*, 2021, **15**, 2842–2853.
19. L. Niu, Y. Wu, H. Liu, Q. Wang, M. Li and Q. Jia, *Chem. Biodiversity*, 2021, **18**, e2001010.
20. L. Barros, P. Baptista, D. M. Correia, S. Casal, B. Oliveira and I. C. F. R. Ferreira, *Food Chem.*, 2007, **105**, 140–145.
21. F. A. Ayaz, L. T. Chuang, H. Torun, A. Colak, E. Sesli and J. Presley, *et al.*, *Int. J. Food Sci. Nutr.*, 2011, **62**, 328–335.
22. H. T. Feleke and A. Doshi, *Indian J. Nat. Prod. Resour.*, 2017, **8**, 254–262.
23. B. Ribeiro, P. B. Andrade, P. Baptista, L. Barros, I. C. F. R. Ferreira and R. M. Seabra, *et al.*, *J. Agric. Food Chem.*, 2008, **56**, 4769–4774.
24. I. Y. Malya, J. Wu, E. Harada, M. Toda, C. N. D'Alessandro-Gabazza and T. Yasuma, *et al.*, *Biosci., Biotechnol., Biochem.*, 2020, **84**, 1332–1338.
25. R. Dai, M. Liu, W. N. Nik Nabil, Z. Xi and H. Xu, *Molecules*, 2021, **26**, 1113.
26. X. L. Yang, M. Y. Jiang, K. L. Hsieh and J. K. Liu, *Chin. J. Nat. Med.*, 2009, **7**, 119–122.
27. S. Yang, T. Shen, L. Zhao, C. Li, Y. Zhang and H. Lou, *et al.*, *Fitoterapia*, 2014, **93**, 168–174.
28. H.-P. Xiong, J.-L. Mi, J.-M. Le, Z.-J. Wu and W.-S. Chen, *Chem. Nat. Compd.*, 2017, **53**, 791.
29. M. S. Abdel-Kader, A. M. Al-Taweel and K. S. El-Deeb, *Nat. Prod. Sci.*, 2008, **14**, 56–61.
30. J. Wu, M. Tsujimori, H. Hirai and H. Kawagishi, *Biosci., Biotechnol., Biochem.*, 2011, **75**, 783.
31. L. Zhang, Y. Shen, H. J. Zhu, F. Wang, Y. Leng and J. K. Liu, *J. Antibiot.*, 2009, **62**, 239–242.
32. K. Arpha, C. Phosri, N. Suwannasai, W. Mongkolthanaruk and S. Sodngam, *J. Agric. Food Chem.*, 2012, **60**, 9834–9841.
33. N. Komoto, S. Ikegami and S. Tamura, *Agric. Biol. Chem.*, 1972, **36**, 2547–2553.

34. J. Sun, C. A. Yeung, N. N. Co, T. Y. Tsang, E. Yau and K. Luo, *et al.*, *PLoS One*, 2012, **7**, e40720.
35. L. Barros, I. C. F. R. Ferreira and P. Baptista, *Food Sci. Technol. Int.*, 2008, **14**, 47–55.
36. M. Alves, I. Ferreira, J. Dias, V. Teixeira, A. Martins and M. Pintado, *Planta Med.*, 2012, **78**, 1707–1718.
37. L. Barros, M. Dueñas, I. C. F. R. Ferreira, P. Baptista and C. Santos-Buelga, *Food Chem. Toxicol.*, 2009, **47**, 1076–1079.
38. L. Barros, B. A. Venturini, P. Baptista, L. M. Estevinho and I. C. F. R. Ferreira, *J. Agric. Food Chem.*, 2008, **56**, 3856–3862.
39. M. J. Alves, I. C. F. R. Ferreira, A. Martins and M. Pintado, *J. Appl. Microbiol.*, 2012, **113**, 466–475.
40. M. J. Alves, I. C. F. R. Ferreira, I. Lourenço, E. Costa, A. Martins and M. Pintado, *Pathogens*, 2014, **3**, 667–679.
41. H. Fortin, S. Tomasi, J. G. Delcros, J. Y. Bansard and J. Boustie, *ChemMedChem*, 2006, **1**, 189–196.
42. J. Sun, C. A. Yeung, N. N. Co, T. Y. Tsang, E. Yau and K. Luo, *et al.*, *PLoS One*, 2012, **7**, e40720.
43. S. Wasser, *Appl. Microbiol. Biotechnol.*, 2002, **60**, 258–274.
44. Giant funnel – National Red List, [cited 2022 Mar 28]. Available from: http://redlist.moepp.gov.mk/giant-funnel/.

CHAPTER 13

Macrolepiota procera

KAMALPREET KAUR,[a] GURPREET KAUR,*[b] AND PARDEEP KAUR SANDHU[b]

[a] Department of Chemistry, Mata Gujri College, Fatehgarh Sahib, Punjab, India; [b] Department of Zoology, Mata Gujri College, Fatehgarh Sahib, Punjab, India
*Email: gpreet.scorpio@gmail.com

13.1 Introduction

Mushrooms have been utilized by humans since the prehistoric era for food and pharmaceutical uses; around 1069 edible mushroom species have been documented. Edible mushroom farming and production are on the rise, especially in Europe, America, and Asia. However, wild mushrooms have been shown to possess more fiber and bioactive constituents than farmed mushrooms.[1,2] The nutritional value of high-grade mushrooms is nearly comparable to that of milk, which explains their high nutritional relevance. Mushrooms, both cultivated and wild, contain carbohydrates, protein, minerals, fats, fiber, and vitamins.[3] It has been shown that mushrooms contain a high level of bioactive compounds that can be used for the cure of a variety of ailments and to slow down the aging process.[4–6] Mushrooms are regarded as superfoods because they contain bioactive compounds that can be used in nutraceutical research for extending human life. Terpenoids, polyphenols, flavonoids, polysaccharides, and alkaloids are present in the fruiting bodies of many mushrooms, and they all have diverse pharmacological properties.[7,8] Most mushrooms exhibit biological attributes, such as anticancer, antiviral, antimicrobial, antibiotic, immunomodulatory, and blood lipid lowering properties.[9,10] Various research studies have demonstrated the ethnomedicinal usefulness of many edible mushrooms, and the United States

Phytochemistry and Nutritional Composition of Significant Wild Medicinal and Edible Mushrooms: Traditional Uses and Pharmacology
Edited by Ajay Sharma, Garima Bhardwaj and Gulzar Ahmad Nayik

Published by the Royal Society of Chemistry, www.rsc.org

National Cancer Institute has selected mushrooms as a reservoir of novel therapeutic agents for the cure of cancer.[3] Despite their culinary qualities, edible mushrooms gathered from contaminated soils may contain harmful metals, such as arsenic, mercury, and other contaminants.[11–13] This chapter aims to cover the phytochemistry, nutritional aspects, and pharmacological potential of *Macrolepiota procera* (*M. procera*).

M. procera, is a saprophytic basidiomycete fungus in the Agaricaceae family. During the autumn, this culinary mushroom species grows alone or in groups along meadows, roadsides, and woodland clearings. It has a velvety brown cap that is roughly 30 cm in diameter and an extremely fibrous stem that makes it inedible. *M. procera* contains mannitol, lepiotan, glucose, trehalose, glycerol, and almost 20 amino acids.[14] The fruiting bodies of this mushroom have been extensively utilized in the production of functional foods and medications. *M. procera* shows anticancer, antioxidant, antiproliferative, and antimicrobial properties. Many studies have reported that this mushroom has the potential to absorb lethal and trace metals, particularly in its edible portion.[15–17]

13.2 Taxonomy

M. procera (Scop.: Fr.) Singer is a basidiomycete fungus belonging to the order Agaricales, with a big, conspicuous fruiting body similar to a parasol (Figure 13.1). This saprophytic mushroom is commonly referred to as a shaggy parasol, field parasol, and parasol mushroom and it cultivates alone or in tiny, dispersed groups at the edges of woods or along forest borders.[18] *Macrolepiota* was divided into two categories by Singer:[19] *Macrolepiota* and *Macrosporae*. Bon[20] split the genus *Macrolepiota* into divisions according to the presence or lack of clamp connections. Three significant clades were discovered using molecular phylogenetic analysis: *Macrolepiota*, *Macrosporae*, and *Volvate*. The pileus of these fungi are ovoid, convex, and 10–21 cm in diameter. The lamellae are thin, white, free, compactly packed when young, whitish to creamish in color when grown, ranging in length from 2 to 3 cm. The stem is highly fibrous, subcylindrical, and acquires a maximum height prior to the formation of the cap, coated with brown squamules in asymmetrical bands. The annulus is pale and membranous on the upper side, brownish underneath and around 4 cm below the stem apex. Basidiocarps range in size from modest to giant. Basidiospores are thick-walled, whitish smooth, hyaline, elliptical to ovoid, dextrinoid, and congophilous. The pleurocystidia is missing, basidia thin-walled, clavate, four spored. The cheilocystidia are hyaline, thin-walled, clavate to subclavate, in bunches forming a sterile edge.[21,22]

13.3 Distribution

M. procera (Scop.) Singer is extensively present in temperate and sub-tropical zones, such as North America, Europe, Africa, and Asia.[23] This mushroom cultivates wildly in Cedrus and Pinus forests in various zones in Portugal

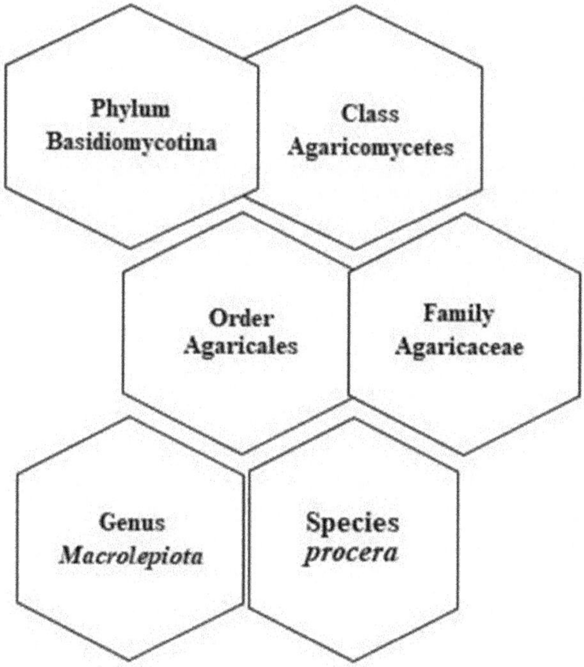

Figure 13.1 Taxonomy of *Macrolepiota procera.*

and Morocco.[24] *M. procera* is also found throughout the northwestern Himalayas, from Garhwal to Himachal Pradesh.[25] Kumar *et al.*[22] demonstrated that this parasol mushroom is found in the Nagaland state of India. *M. procera* is the only species that has been found in Pakistan. The parasol mushroom is saprotrophic and produce fruiting bodies from June to November in gardens, forests, and on meadows and pastures. It favors calcareous and sandy soils in warm climates.[26–28]

13.4 Traditional Uses

Mushrooms have been characterized as therapeutic foods for the treatment of hypercholesterolemia, hypertension, diabetes, and cancer, in addition to being utilized as nutraceuticals.[29] *M. procera* is used traditionally in many countries as food and medicine. This pileus of this mushroom is particularly valued by locals and is commonly known by various names, such as kandi koir, tatmour, and badi chattri. It is preferred due to the flavor and scent of freshly cooked mushrooms that have been baked and broiled.[30] In Europe, *Amanita phalloides* and *Chlorophyllum molybdites* are often confused with the parasol mushroom. *M. procera* is often grown in kitchen gardens due to its popularity and adaptability. This mushroom is very good at accumulating toxic metals and micronutrients in its cap and stipe.[31,32]

13.5 Phytochemistry of *M. procera*

Mushrooms are not only known for their characteristic flavor and taste but are also a rich source of diverse varieties of biomolecules.[33–36] Chen *et al.*[37] carried out an investigation on the chemical profile and biological activity of *M. procera* and reported 12 novel lanostane class of triterpenoidal compounds, including lepiotaprocerins A–L (Figure 13.2). The triterpenoid compounds were verified by the combined use of nuclear magnetic resonance (NMR) spectroscopies, high-resolution electrospray ionization mass spectroscopy (HRESIMS) analysis, and X-ray diffractometry (XRD) analysis. Interestingly, six lepiotaprocerins (A–F) were found to contain a rigid and infrequently found 1-ene-1,11-epoxy moiety, which is generally unknown for the lanostane class of triterpenoids.

Zurga *et al.*[38] reported a single domain protein called 'β-trefoil lectin' (Table 13.1), purified from *M. procera* by employing affinity chromatography. The molecular mass of the lectin isolated from *M. procera* was found to be 16 kDa and its isoelectric point (pI) was detected at a pH value of 4.8. Electrospray ionization mass spectrometry (ESI-MS) analysis indicated only one peak at *m*/*z* of 15 831 Da. The crystal structure revealed that the lectin isolated from *M. procera* is a ricin B like lectin and it contains three homologous domains of amino acid sequence, designated as α, β and γ-types. The α-site revealed strong affinity for carbohydrates, including lactose, galactose, and lacto-*N*-biose, whereas the second binding site, which was a putative γ-site located in a γ-repeat, had lower affinity for galactose. The third novel binding site present in the lectin protein was for hydrophobic moieties. The isolated lectin showed excellent nematocidal activity, demonstrating its significant role in defending mushrooms against parasitic attack.

Moreover, Nowak *et al.*[39] examined the presence of polysaccharides in 53 different varieties of wild mushrooms, including *M. procera*. Maceration followed by sonication and removal of protein by employing Savage reagent yielded 15.70% polysaccharides in *M. procera*.

Barros *et al.*[40] investigated the individual phenolic mixtures responsible for antioxidant and other bioactivities present in a methanolic extract of various dried Portuguese wild mushrooms, including *M. procera*, by high-performance liquid chromatography (HPLC) combined with a photo-diode array and mass spectrometry analysis. The total phenols in *M. procera* was found to be 3.17 $mg\,g^{-1}$ of extract.

A thorough investigation of the phytoconstituents present in *M. procera* was performed by Ciric *et al.*[41] They investigated the comparative nutritional composition of three Serbian *Macrolepiota* species, *viz. M. procera*, *M. rhacodes*, and *M. mastoidea*. All the mushrooms were observed to be a good source of carbohydrates, but the highest carbohydrate content (60.3 g per 100 g) was that of *M. procera*, followed by *M. mastoidea* (51.58 g per 100 g), and the lowest carbohydrate content was that of *M. rhacodes* (40.9 g per 100 g). Moreover, *M. procera* contained the lowest protein content of the three tested samples.

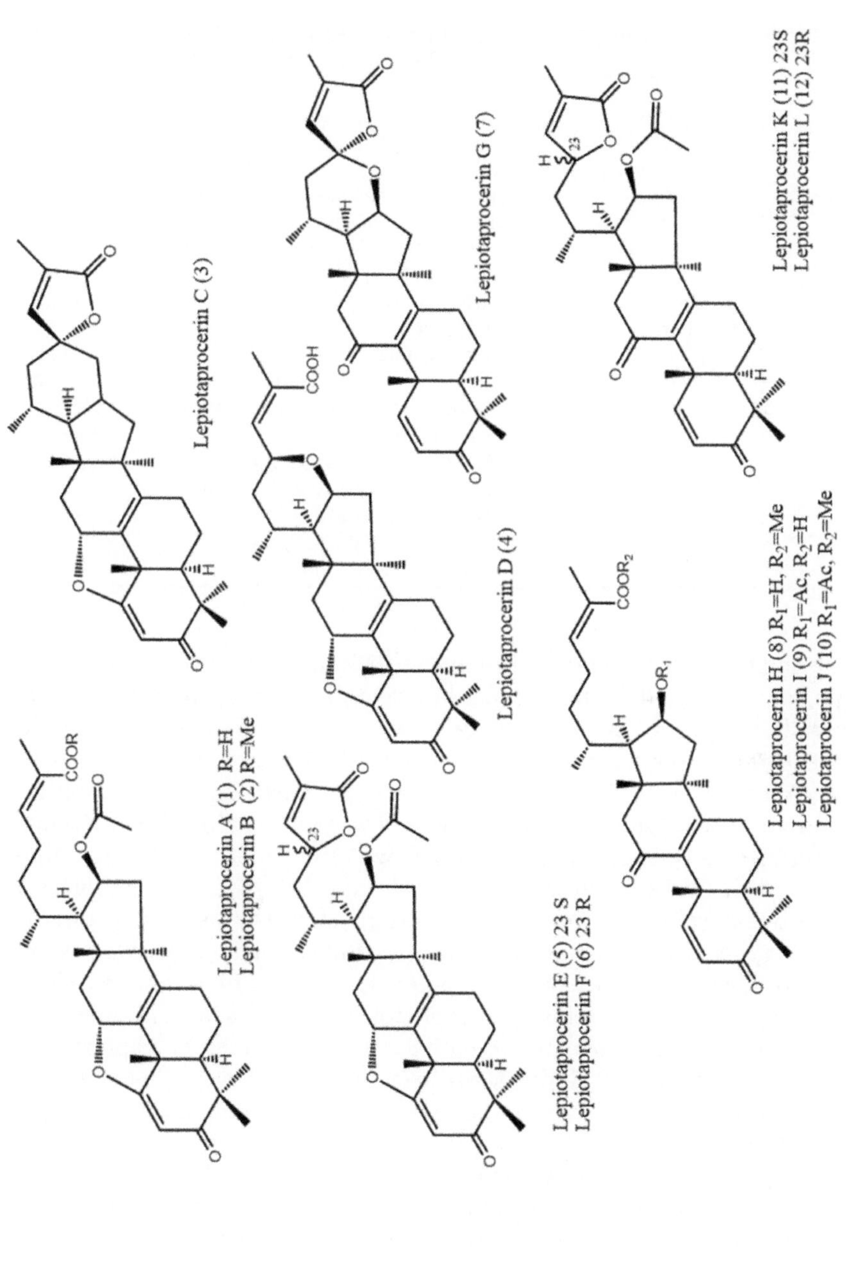

Figure 13.2 Lanostane triterpenoids of the *Macrolepiota procera* mushroom.

Table 13.1 Phytoconstituents present in *Macrolepiota procera* collected from different regions.

Macrolepiota procera	Country origin	Phytoconstituents	Ref.
	Poland	Protocatechuic acid and caffeic acid	43
	Portugal	Tocopherols, fructose, trehalose, mannitol and melezitose	40
		Cinnamic acid	
	Slovenia	β-Trefoil lectin protein	38
	Turkey province	Linoleic acid, oleic acid, palmitic acid	49
	Serbia	Trehalose, mannitol, linoleic acid, palmitic acid, citric acid, oleic acid, tocopherols, hydroxybenzoic acid, fumaric acid	41
	Poland	Lepiotaprocerins	37
	Turkey	β-Carotene and lycopene	42
		Fumaric acid	36
	Northern Morocco	Cinnamic acid, paraben, syringic acid, coumaric acid, glycerol, threonine and tryptophan	14
	Portugal	D-Galactose, threonine, tryptophan, protocatechuic acid, cinnamic acid, paraben, syringic acid	14
	Northern India	Phenolics, flavonoids, alkaloids, lycopene and β-carotene	21

Its ash and fat content were very low in comparison to earlier reports available in the literature. Trehalose and mannitol were found to be the major sugars present in the *Macrolepiota* mushrooms, with the highest content found in *M. procera*. Moreover, fructose and melezitose, which were found to be present in *M. procera* species cultivated in Portugal, were absent in the same species taken from Serbia. Oxalic acid and malic acid were present in higher amounts in *M. procera* in contrast to for the other two varieties of mushrooms. Additionally, citric acid was present only in *M. procera* (Figure 13.3). *M. procera* contained relatively higher concentrations of phenolic acids, such as coumaric and hydroxybenzoic acids, in contrast to *M. mastoidea* and *M. rhacodes*. Moreover, linoleic acid was present in considerable amounts in *M. procera*, along with palmitic and oleic acids. β-Tocopherol was present as the dominant component in *M. mastoidea* and *M. rhacodes* and α-tocopherol was present in relatively low amounts in all the *Macrolepiota* species. However, δ-tocopherol was present only in *M. procera*. In addition to this, Aytar *et al.* estimated the β-carotene and lycopene content of a methanolic extract of the parasol mushroom;[42] spectrophotometric analysis indicated the presence of β-carotene, lycopene, and total phenolic content at 0.091 $\mu g\,mL^{-1}$, 0.059 $\mu g\,mL^{-1}$ and 36.25 mg gallic acid equivalents (GAE) per gram of extract, respectively. Likewise, Babita *et al.*[21] explored nutraceutical components present in a methanolic extract of *M. procera* from northern regions in India. Phenolic

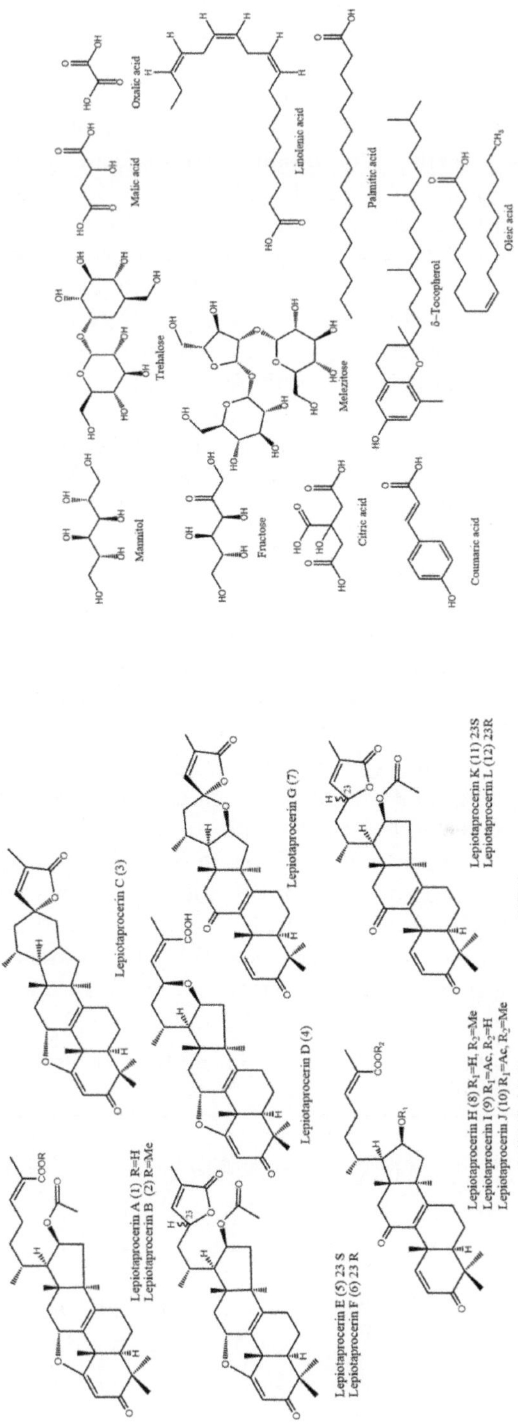

Figure 13.3 Major phytoconstituents in the *Macrolepiota procera* mushroom.

compounds were identified as the predominant component, followed by flavonoids. However, smaller amounts of alkaloids, lycopene, and β-Carotene were present.

M. procera extract was analyzed for the identification and quantification of phytoconstituents with the help of liquid chromatography and mass spectrometry methods.[43] The predominant phenolic acid present in *M. procera* was found to be protocatechuic acid (5.19 mg per kg dry weight). In addition to this, caffeic acid was detected as a trace amount. The total phenolic content of *M. procera* was found to be 8.21 mg of GAE per gram of dried mushroom. Likewise, Cayan *et al.*[36] employed the HPLC-diode array detection (DAD) method for the analysis of bio-active constituents in *M. procera*. *M. procera* was found to contain fumaric (1.97 $\mu g\,g^{-1}$), protocatechuic, caffeic, gallic, hydroxycinnamic, and cinnamic acids. Similarly, Taofiq *et al.*[44] reported a cinnamic acid content of 522 $\mu g\,g^{-1}$ in an ethanolic extract of *M. procera* from Portugal. The differences observed in the chemical profile of *M. procera* are attributable to growing techniques and conditions, extraction techniques, processing methods, conditions of storage, and exposure to UV radiation, pollution, and extreme weather conditions.[36,45,46]

Erbiai *et al.*[14] reported the comparative analysis of the chemical profile of *M. procera* obtained from northern Morocco (MR) and Portugal (PL). Moreover, it was determined that the chemical compositions and the amounts of active bio-components are also influenced by the geographical location and ecological environment of the mushrooms. The chemical composition of significant phyto-compounds was assessed with the help of liquid chromatography–mass spectrometry (LC–MS) and gas chromatography–mass spectrometry (GC–MS) analysis. The predominant components present in *M. procera* precured from MR and PL were detected to be protocatechuic acid, cinnamic acid, and paraben. Slightly higher concentrations of these chemical constituents were present in *M. procera* from Portugal. Additionally, the constituents present in smaller amounts in *M. procera* from MR and PL were detected to be gallic, syringic, ferulic, hydroxybenzoic, and vanillic acids (Table 13.2). A methanolic extract of *M. procera* mushroom revealed the presence of 51 and 35 biomolecules in

Table 13.2 Phenolic composition of *M. procera* from northern Morocco and Portugal.[14]

Phenolic compounds	*M. procera* (μg per g of dry weight) Portugal	*M. procera* (μg per g of dry weight) Northern Morocco
Ferulic acid	6.36	5.11
Cinnamic acid	90.60	81.93
Paraben	17.87	15.30
Gallic acid	16.41	11.58
p-Hydroxybenzoic acid	16.76	9.99
p-Coumaric acid	3.07	2.04
Syringic acid	0.40	0.44
Protocatechuic acid	125.5	93.52
Vanillic acid	8.42	8.61

M. procera (PL) and *M. procera* (MR) mushrooms, with sugars being the most common component, present in 53.71% and 59%, respectively. Glycerol (24.43%) was found to be the predominant sugar in the *M. procera* sample procured from the MR variety, whereas D-galactose was the predominant sugar present in *M. procera* from PL, at 38.43% and 24.43%, respectively. The amino acids threonine and tryptophan were found to be the key compounds present in the MR and PL varieties of *M. procera*, found at 1.26% and 2.82%, respectively. In other studies, glutamic acid was reported to be the chief amino acid present in the *M. procera* mushroom. Moreover, linolelaidic and succinic acids were found to be the major fatty and organic acids present in *M. procera* (MR). Urea was present in *M. procera* (MR) but it was absent in *M. procera* (PL). The major sugars of *M. procera* were observed to be mannose, mannitol, glucose, and trehalose, present in quantities of 11.09, 2.42, 10.78, and 0.11 g per 100 g, respectively.[47]

Fernandes *et al.*[48] explored the influence of processing techniques, *viz.* drying, freezing, and gamma irradiation on the biochemical profile and antioxidant potential of *M. procera*. The amounts of fatty acids, sugars, and tocopherols were found to be largely influenced by the nature of the preservation technique. The predominant tocopherols determined in *M. procera* were α, β, γ, and δ-tocopherols, present in 2.5–5.2, 26–77, 09–43, and 65–91 μg per 100 g dry weight of fresh, frozen, dried and irradiated *M. procera*. Fructose, trehalose, and mannitol were the major carbohydrates detected in fresh and processed *M. procera* mushrooms. Yilmaz *et al.*[49] employed the use of gas chromatography to quantify the compositions of various fatty acids in *M. procera*. The predominant fatty acid present in *M. procera* was found to be linoleic acid, *i.e.*, 18 : 2 present at 62.44%. *M. procera* contained more unsaturated fatty acids (81.95%) than saturated fatty acids (15.90%), and can therefore can be used as a low-fat food for people with high cholesterol. The ratio of unsaturated fatty acids to saturated fatty acids in *M. procera* was as high as 5 : 15, and the most dominant polyunsaturated fatty acid (PUFA) and linoleic acid were present at 62.44% in the *M. procera*. Palmitic and oleic acids were present at 10.95% and 17.40%, respectively. Even-carbon and single-carbon fatty acids were also detected to be present at 94.98% and 2.41%.

13.6 Proximate Composition of *M. procera*

Wild mushrooms are an important food due to their organoleptic, nutritional, and medicinal properties. The most important nutritional components present in *M. procera* include moisture, crude fat, ash, crude protein, fiber, and total carbohydrates.[50–52] The nutritional significance of mushrooms is assessed by their dry weight, which mostly lies in the range of 5–15%. Moreover, the moisture content of mushrooms directly influences their nutritional profile.[51,53] *M. procera* mushrooms contain high carbohydrate and protein content but low fat and ash content. The major components investigated in *M. procera* include total carbohydrates (66.78 g per 100 g), proteins (24.22 g per 100 g), ash (5.37 g per 100 g), and fat (2.23 g per 100 g). The protein and carbohydrate contents estimated by Beluhan *et al.*[47] were in agreement with the protein content of 23.9 g per 100 g and carbohydrate content of 68.4 g per 100 g as described previously by Ouzouni and Riganakos.[54] The total energy contributed by 100 g dry weight of *M. procera* was estimated to be 1629.51 kJ. In another study, Kumari *et al.*[21] analyzed the proximate composition of wild edible macrolepiotoid mushrooms belonging to three different species, *viz. Macrolepiota procera*, *M. rhacodes*, and *M. dolichaula* from north-west India on a dry weight basis. Nutritional studies afforded much lower values of the different components. In addition to carbohydrates (60.82%), protein (19.95%), moisture (8.5%), crude fat (3.4%), and ash (1.93%), 5.1% crude fibers were also reported by the same research group in dried *M. procera* samples. Proximate analysis also estimated that 353.7 kcal of energy was provided by 100 g of *M. procera*.

Mushrooms have a shelf life of 1–3 days and are easily perishable due to post-harvest modifications.[55,56] To increase the shelf life of mushrooms, processing methods such as drying and freezing are used. In terms of preservation, drying is the cheapest method, whereas freezing is the most efficient method. Freezing not only retains the nutritional value of mushrooms but also maintains their color, texture, taste, and fragrance. Due to the conversion of water into ice during freezing, the enzymatic and microbial activities of mushrooms are slowed significantly.[57] Cooking drastically changes the nutritional and chemical profiles of mushrooms. The effects of different processing techniques, such as drying, freezing, and cooking, on the proximate chemical composition, such as the moisture, ash, fat, protein, and carbohydrate content of *M. procera* was evaluated.[58] Cooking reduced the nutrient concentrations more effectively than freezing and drying. The moisture content in cooked *M. procera* (64.36 g per 100 g fresh weight) was much lower than that in dried and frozen mushrooms and no substantial differences were observed in the moisture content of dried and frozen mushrooms (90.01 and 89.85 g per 100 g fresh weight, respectively). Carbohydrates were found to be the major macronutrient present in *M. procera*. As per expectations, the carbohydrate content in cooked *M. procera* was very low (16.40 g per 100 g dry weight) in comparison to 80.38 and 79.28 g per 100 g dry weight in dried and frozen mushrooms, respectively. The ash content was considerably reduced on cooking (5.73 g per 100 g dry weight) in a cooked sample in contrast to 9.86 and 9.16 g per 100 g dry weight in dried and frozen *M. procera*, respectively. The protein, fat, and reducing sugar content showed a significant increase in cooked mushrooms. The protein content varied from 11.85 g per 100 g dry weight in cooked mushrooms to 7.62 and 9.36 g per 100 g dry weight in dried and frozen mushrooms, respectively. Cooking leads to the hydrolysis of polysaccharides (starch and chitin), which release sugars and increase the reducing sugar content. The fat content showed a remarkable increase in cooked mushrooms (57.43 g per 100 g dry weight) in comparison to dried mushrooms (1.45 g per 100 g dry weight). Moisture is lost while cooking, which changes the concentration of nutrients. The loss of nutrients while cooking is also due to (i) various chemical reactions, (ii) thermal degradation, (iii) increased solubility in cooking medium, and (iv) constituent interactions. The highest energy was provided by cooked *M. procera* (623.67 kcal per 100 g dry weight), while dried and frozen samples contributed much lower energy (365.01 and 379.87 kcal per 100 g dry weight, respectively). Moreover, the energy contribution was higher in frozen samples in comparison to those dried as freezing provides better stability of nutrients than the harsh conditions used in drying.[29,59]

Likewise, Fernandes *et al.*[48] investigated the effect of preservation treatments, including freezing, drying, and gamma irradiation, on the nutritional composition of *M. procera* in comparison to the nutritional parameters of fresh mushrooms. The major components present in *M. procera* were found to be carbohydrates, protein, fat, ash, and moisture, present in ranges of 60–70 g per 100 g dry weight, 19.0–28.6 g per 100 g dry weight, 2.2–3.6 g per 100 g dry weight, 7.6–10.3 g per 100 g dry weight, and 85.8–87.7 g per 100 g fresh weight, respectively, in fresh, frozen, dried, and gamma irradiated samples. Maximum protein content was observed in frozen *M. procera* as low temperature prohibited the degradation of protein. Additionally, gamma irradiation also hindered fat loss due to the prevention of lipid oxidation due to the availability of less oxygen.[60] Aydin *et al.*[61] also investigated the influence of various processing techniques, such as canning, drying, and freezing, on the compositional profile of *M. procera*. The moisture content of fresh and frozen mushroom samples was almost the same. The maximum moisture was noticed for canned mushrooms, whereas the lowest value was observed for dried mushrooms. The ash content was the highest in dried *M. procera* followed by fresh and frozen samples and the lowest value of 0.37 g was detected for the canned sample. The ash content in the dried sample was greater than that reported by Barros *et al.*[62] The fat and protein contents were the highest in dried samples and the lowest in canned *M. procera* samples. Likewise, the energy contributed by a dried sample was noted to be the highest (872.87 kJ per 100 g) and lowest energy (165.42 kJ per 100 g) was provided by canned samples.

13.7 Mineral Composition

M. procera is a rich reservoir of potassium (K), magnesium (Mg), and selenium (Se).[63–66] However, *M. procera* also contains toxic metals such as Cd and Pb.[67,68] It was observed that mushrooms found in contaminated soil accumulate more Pb compared to those in uncontaminated soil.[69] Gucia *et al.*[70] demonstrated the presence of elements such as K, Mg, P, Fe, Ca, Al, Fe, Na, Zn, Mn, and Cu in the cap and stem of *M. procera* (Figure 13.4) from different regions of northern Poland using various spectroscopic techniques. Moreover, significant amounts of Cd, Pb, and Hg were detected in the different parts of *M. procera* (Figures 13.5 and 13.6).

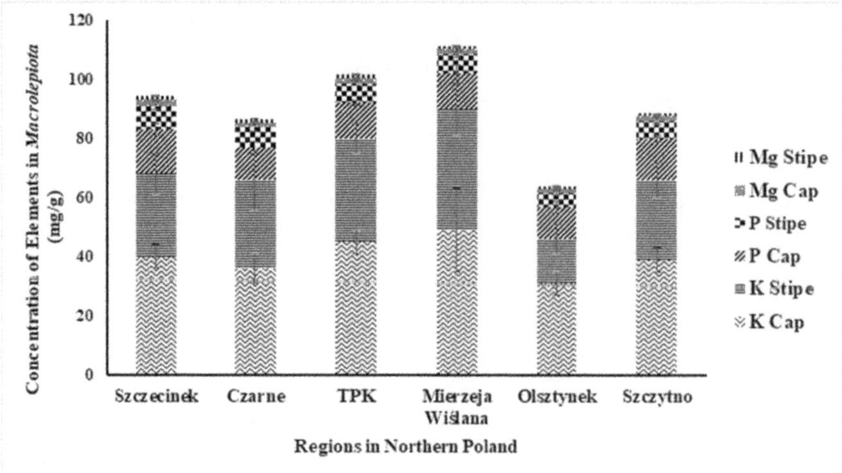

Figure 13.4 Concentration of essential elements in *Macrolepiota procera* in different regions of northern Poland (adapted from ref. 70).

In another investigation, Durdic *et al.*[71] demonstrated the amount of Pb to be in the ranges of 0.12–6.65 mg kg^{-1} and 0.036–2.48 mg kg^{-1} in the cap and stipe, respectively, of *M. procera* from Serbia. In addition to this, Janco *et al.*[72] described that higher amounts of the heavy metals As, Cd, and Hg were observed in the cap of *M. procera* from eight different regions of Slovakia (Figure 13.7).

In another study, the concentrations of Hg were found to be in the regions of 0.04–6.29 mg kg^{-1} and 0.03–2.83 mg kg^{-1} in the cap and stipe, respectively, of *M. procera* from 22 regions of Slovakia.[73] Moreover, Kosanic *et al.*[74] monitored the concentrations of Zn, Pb, Mn, Ni, Cu, Cd, Cr, Fe, and Co in *Macrolepiota* mushrooms collected from Serbia. All the metals were found to be within the tolerable daily intake, except for Cd. In another study, the quantities of Cu and Zn were detected to be predominant in *Macrolepiota* mushrooms found in the regions of southern Spain and northern Morocco. The toxic metals Cr, As, Cd, and Hg in the parasol mushroom were observed to be above the tolerable daily intake.[75,76] All of these findings indicate that *Macrolepiota* mushrooms accumulate enormous quantities of metals and should be limited in the diet to avoid risk to human health.

13.8 Pharmacological Activities

Various investigations have described that the parasol mushroom has numerous biological attributes, such as antioxidant, anticancer, antiparasitic, anti-inflammatory, and antimicrobial properties (Table 13.3). The parasol

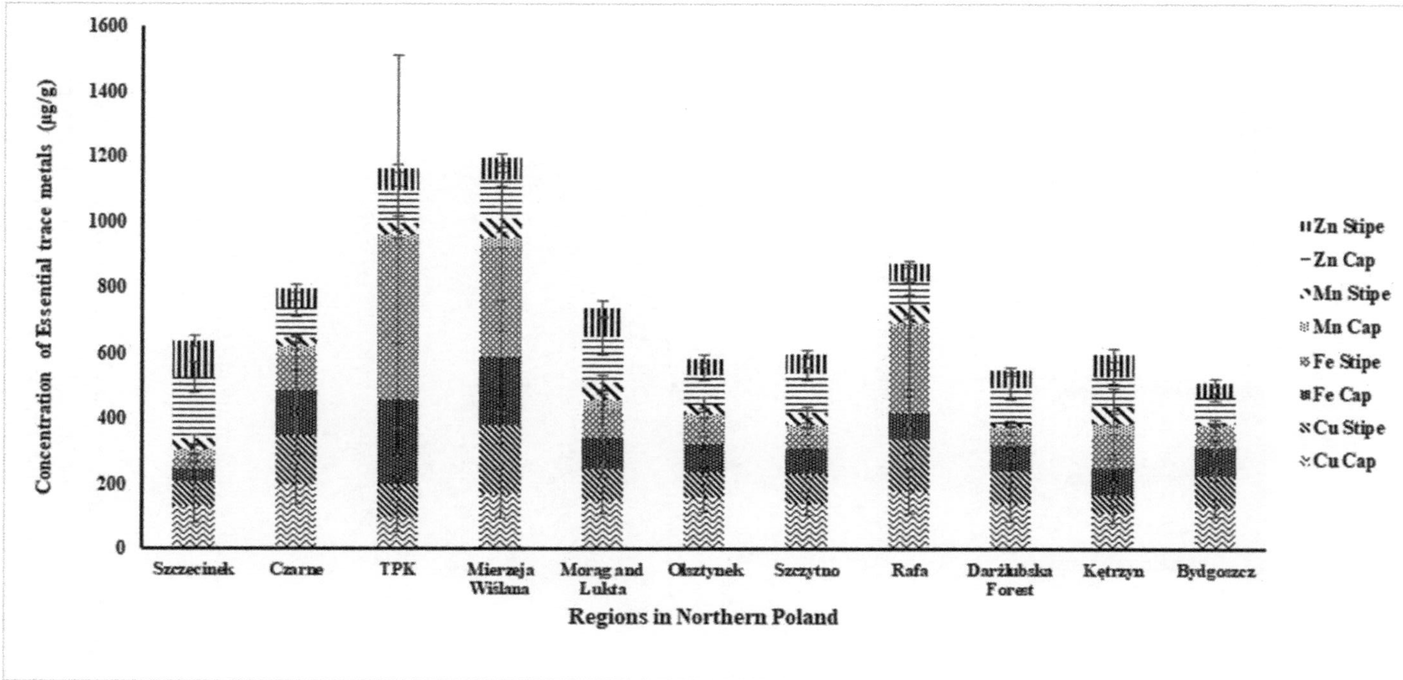

Figure 13.5 Concentration of the essential trace metals in *Macrolepiota procera* in different regions of northern Poland (adapted from ref. 70).

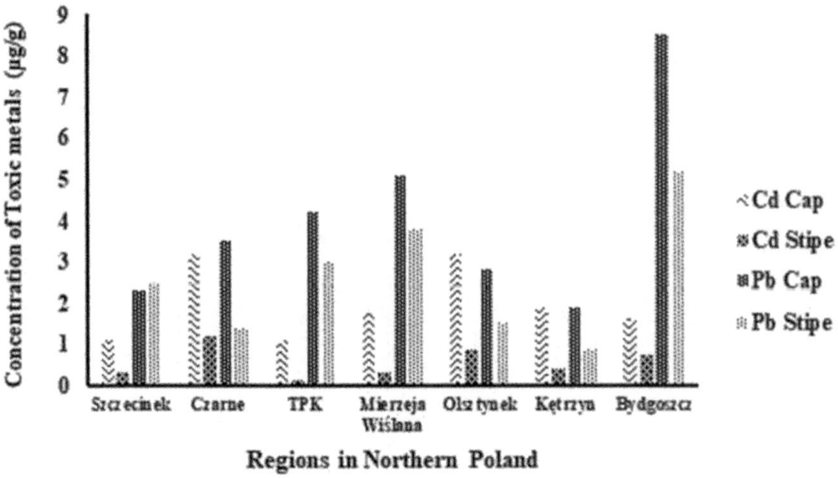

Figure 13.6 Concentration of cadmium and lead in *Macrolepiota procera* in different regions of northern Poland (adapted from ref. 70).

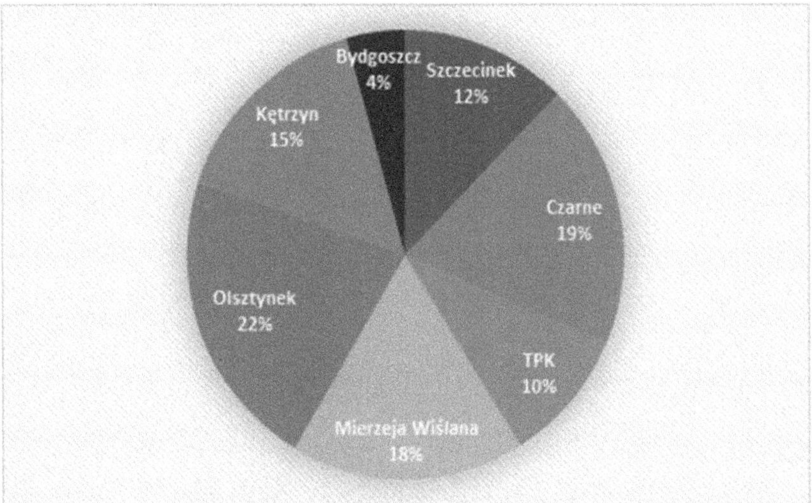

Figure 13.7 Concentration of mercury in *Macrolepiota procera* in different regions of northern Poland (adapted from ref. 70).

mushroom serves as a repository for a large number of diverse bioactive compounds, which can be used as substitutes for prevalent synthetic drugs. This section covers the major pharmacological potential of mushrooms, which helps in alleviating deadly infections.

Table 13.3 Pharmacological potential of the *M. procera* mushroom.

S. no	Biological activity	Results	Ref.
1	Anticancer	$IC_{50} = 29.39\ mg\ mL^{-1}$	74
	HeLa cell line	$25.55\ mg\ mL^{-1}$	79
	A549	$68.49\ mg\ mL^{-1}$	
	LS174	$6.18\ \mu g\ mL^{-1}$	77
	A549	$2\ mg\ mL^{-1}$	
	HepG2	>400	41
	MCF-7 (breast carcinoma)	>400	78
	Lung and cervical cancer cells	>400	
	MCF 7	$256\ \mu g\ mL^{-1}$	
	COLO-205	$349\ \mu g\ mL^{-1}$	
2	Antioxidant	92.72% at $200\ mg\ L^{-1}$	89
		$1.19\ mg\ mL^{-1}$ (MPE from northern Morocco)	14
		$1.31\ mg\ mL^{-1}$ (MPE from Portugal)	14
		$0.191\ mg\ mL^{-1}$	42
		$181.53\ mg\ g^{-1}$	33
		$3.7\ mg\ mL^{-1}$	41
		IC_{50} $311.40\ mg\ mL^{-1}$	74
		$752\ \mu g\ mL^{-1}$	78
3	Antimicrobial	99.99% at $500\ mg\ L^{-1}$	89
		$0.75\ mg\ mL^{-1}$ (*Staphylococcus aureus*)	41
		$2\ mg\ mL^{-1}$ (*Escherichia coli*)	
		0.35 (*Bacillus cereus*)	
		9 mm inhibition Zone (*E. faecalis, K. pneumoniae*)	42
4	Anti-inflammatory	Inhibition of NO production	44
5	Amylase inhibition	$0.17\ mmol\ g^{-1}$ extract	33
6	Glucosidase inhibition	$0.24\ mmol\ g^{-1}$ extract	33
7	Acetylcholinesterase activity	$0.83\ mg\ g^{-1}$ extract	33
	Butyrylcholinesterase activity	$0.86\ mg\ g^{-1}$ extract	
8	Apoptosis	28% apoptosis in cells	41
		↑Bax, caspase 9, PTEN, PUMA, p53	78
		Chromatin condensation and shrinkage of cells, membrane blebbing in the cells	
9	Antifungal	$1.50\ mg\ mL^{-1}$ (*Aspergillus niger*)	41
		$0.50\ mg\ mL^{-1}$ (*Trichoderma viride*)	
		$0.75\ mg\ mL^{-1}$ (*Penicillium funiculosum*)	
10	Immunomodulatory	↑NK cells and Th1 cytokines	82
11	Antitubercular activity	$50\ \mu g\ mL^{-1}$	37

13.8.1 Anticancer Activity

Mushrooms contain an endless supply of active constituents with positive health effects and no known negative adverse effects, and they have the potential to be key therapeutic agents in the fight against cancer.[77] *M. procera* extract at a concentration of 50 mg mL^{-1} serves as a potent glucose phosphate dehydrogenase inhibitor and shows significant effect against lung cancer cells, with an inhibitory concentration of 6.2 μg mL^{-1}. Molecular docking studies have shown that *M. procera* contains active constituents such as paraben, quercetin, and syringic acid, which exhibit efficient binding potential with glucose phosphate dehydrogenase. Early investigations also described the protective efficacy of *M. procera* extract against colon and lung cancer.[78,79] In another investigation, triterpenoids of *M. procera* extract demonstrated remarkable protective efficacy against mammalian cancer cell lines.[37] Moreover Ciric *et al.*[41] reported that *M. procera* extract selectively killed cancerous cells and did not reveal any toxic effects on hepatic primary cells, which shows that the extracts are safe to healthy cells. In another study, *M. procera* showed remarkable inhibition against lung cancer cells and inhibitory concentration was observed to be 2 mg mL^{-1} after 72 h. The levels of CDK6, Cyclin Dl, Bcl-2, Akt, and NOXA genes were found to be diminished and the levels of Caspases, Bax, PUMA, PTEN, p21, and p53 augmented in cells treated with extract in contrast to their expression in control cells. Treatment of *M. procera* extract led to 28% apoptosis in cells and decreased invasion in cancer cells.[77] Moreover, Han *et al.*[80] reported that *M. procera* extract showed more inhibitory activity against tumor metastasis in contrast to mice treated with a tumor vaccine alone. The tumor HK vaccine in combination with *M. procera* extract exhibited a strong Th1 response and caused the induction of cytotoxic T lymphocyte (CTL) response in murine splenocytes in comparison to the tumor vaccine alone. This study highlighted that *M. procera* extract could be a potent adjuvant of antitumor immune response.

13.8.2 Immunomodulatory Activity

Mushrooms have huge potential for the induction of humoral and cellular immune response. Various mushrooms assist in the rejuvenation of the immune system, which is suppressed by cancer treatments such as radiation and chemotherapy, and intensify antimutagenic and anticancer activity. Thus, mushrooms serve as potent agents for immunotherapy in the cure of cancer and other immunosuppressive diseases.[81] *M. procera* extract exhibits potential immunostimulatory benefits in terms of enhancing the Th1 type of cytokines such as IL-12 through the induction of CD3+ and CD4+ T lymphocytes. An aqueous extract of *M. procera* was shown to hamper tumor metastasis in colon cells compared to a tumor control group. Administration of *M. procera* extract enhanced the levels of IL-12 in peritoneal macrophages and resulted in an induction in tumoricidal activity. Treatment with *M. procera* extract markedly amplified the natural killer (NK) cell cytotoxicity to tumor cells. The *M. procera* extract in combination with cisplatin exhibited prolonging of the lifespan of tumor-bearing mice. *M. procera* extract stimulates the non-specific immune response and can be utilized as a potent agent in the treatment of cancer.[82]

13.8.3 Anti-inflammatory Activity

Anti-inflammatory medication is found to be effective against various fatal infections and it has been reported that the parasol mushroom assists in preventing early neoplastic development and malignant conversion. *M. procera* extract has been found to contain active anti-inflammatory lepiotaprocerin compounds. Treatment with lepiotaprocerins showed marked reduction in the levels of nitric oxide in murine macrophages in contrast to the positive control.[37] In another study, *M. procera* extract exhibited inhibitory action on the induction of nitric oxide in the macrophages. It has been observed that *M. procera* extract comprises potent compounds such as coumaric acid, *p*-hydroxybenzoic, and cinnamic acid, all of which are linked with the treatment of inflammatory illnesses.[41] Similarly Taofiq *et al.*[44] demonstrated the strong anti-inflammatory potential of *M. procera* extract due to its coumaric and cinnamic acid content.

13.8.4 Antioxidant Activity

In biological organisms, oxidative damage is caused by the generation of an excess of reactive oxygen species (ROS), which causes impairment in macromolecules. Such as DNA, RNA, lipids, and proteins, resulting in tissue damage.[83] The overproduction of ROS molecules has been detected in the etiology of chronic illnesses like diabetes, ageing, cardiovascular disease, cancer, and arthritis.[84,85] The antioxidant potential of mushroom extracts has also become a valuable foundation for the pharmaceutical and food sectors, replacing synthetic antioxidant ingredients and leading to the production of goods that contain natural products. Methanolic *M. procera* extract showed robust lipid peroxidation repression and DPPH radical scavenging action, as tested by TBARS and DPPH assays.[86] Similarly, Barros *et al.*[58] revealed the EC_{50} values of *M. mastoidea* and *M. procera* mushrooms toward DPPH scavenging activity to be 8.2 $mg\,mL^{-1}$ and 5.3 $mg\,mL^{-1}$, respectively. Similarly, Popescu *et al.*[87] also demonstrated that *M. procera* extract exhibited good antioxidant effects due to its significant polyphenol content. In another investigation, *M. procera* extract exhibited strong antioxidant activity and its IC_{50} value was observed to be 14.73 $mg\,mL^{-1}$.[33] Moreover, Kosanic *et al.*[74] displayed that *M. procera* extract unveiled antioxidant action, with an IC_{50} value of 0.311 $mg\,mL^{-1}$. Similarly, Fernandes *et al.*[88] described that different extraction methods for *M. procera* extract showed antioxidant activity in the range of 2.7–7.9 $mg\,mL^{-1}$. In another investigation, the IC_{50} of *M. procera* extract for DPPH free radical scavenging effects was found to be 0.191 $mg\,mL^{-1}$, which indicated the huge antioxidant potential of the mushroom.[42] Moreover, the maximum DPPH scavenging potential of *M. procera* silver nanoparticles (Mp-AgNPs) was observed to be 92.72%, at a concentration of 200 mg L^{-1}.[89]

13.8.5 Antimicrobial Activity

Methanolic extract of *M. procera* has been shown to exhibit huge protective efficacy against Gram positive and negative bacteria and yeasts.[90] Similarly, Kosanic *et al.*[79] reported that *M. procera* extract exhibited potent inhibition against several bacterial and fungal strains (Table 13.4). In another study, *M. procera* extract from Poland revealed moderate microbicidal activity against several microorganisms.[43] In another investigation, Dyakov *et al.*[91] showed that *M. procera* extract did not exhibit any effect against methicillin-resistant strain *Staphylococcus aureus* (MRSA). In contrast, Circu *et al.*[41] revealed the potent activity of *M. procera* extract against MRSA. Moreover, Cuvalova *et al.*[92] demonstrated that *M. procera* and *Laetiporus* extracts displayed antibiofilm activity against *S. aureus*. In another investigation, *M. procera* extract revealed protective efficacy against *K. pneumoniae* and *E. faecalis*, with a 9 mm inhibition zone diameter.[42] In another study, the protective efficacy of MPE was tested against chili anthracnose caused by *Colletotrichum capsica.* Treatment of MPE displayed the maximum colony inhibition and 80% decline in the spore production of *C. capsica* at 1000 ppm.[93] Moreover, the lanostane triterpenoids of *M. procera* extract showed protective potential against *Mycobacterium tuberculosis* at a concentration of 50 μg mL^{-1}.[37] Likewise, Gonca[89] demonstrated that Mp-AgNPs triggered single-strand breakage in the DNA of *E. coli*. The Mp-AgNPs presented 66% and 73% antibiofilm efficacy against *P. aeruginosa* and *S. aureus*, respectively. The Mp-AgNPs at a concentration of 500 mg L^{-1} displayed potent inhibitory activity against *E. coli* of 99.99%.

Table 13.4 Antimicrobial potential of the *M. procera* mushroom.[74]

Pathogenic organism	Minimum inhibitory concentration of *M. procera* (mg mL^{-1})
Bacillus cereus	5
Candida albicans	5
Proteus mirabilis	10
Bacillus subtilis	5
Alternaria alternata	10
Aspergillus niger	20
Bacillus subtilis	5
Penicillium expansum	20
Proteus mirabilis	10
Trichoderma viride	10
Fusarium oxysporum	5
Cladosporium	10
Bacillus cereus	5

Thus, Mp-AgNPs, in spite of their exorbitant price, are an ideal alternative antimicrobial agent due to their eco-friendliness and non-toxicity.

13.8.6 Other Therapeutic Benefits

The macrocypins present in *M. procera* extract block the endopeptidases of papain and cysteine cathepsins, which exhibit both exopeptidase and endopeptidase activity.[94] The MPE Lectin (MpL) has toxic effects on the worm *Caenorhabditis elegans*, showing that MpL plays a role in defending fruiting bodies against parasitic infections.[38]

13.9 Toxicological Risks

According to the World Health Organization criteria for hazardous metals, the permissible daily consumption amounts of cadmium (Cd) and lead (Pb) for an adult are 0.06 mg and 0.21 mg, respectively. The tentative daily consumption for the metals Cu, Fe, and Zn for an adult is in the range of 3–60 mg. It was observed that the concentration of Cd in *M. procera* extract was detected to be three times higher than permissible and it was found to be approximately 0.19 mg.[74] Similarly Sarikurkcu *et al.*[31] described that *M. procera* extract has the ability to accumulate extraordinary quantities of metals and its consumption should be avoided by humans due to potential health risks. However according to EU standards, the allowed quantity of lead in mushrooms is 0.3 $mg\,kg^{-1}$ of fresh mushrooms.[17]

According to dietary intake evaluation, the intake of *M. procera* once a week might be acceptable, but eating this fungus more than once a week could result in harmful metal concentrations that exceed the approved daily eating limits for humans. Various studies have demonstrated the amounts of As, Hg, Cd, and Pb in *M. procera* taken from provinces in Poland and Slovakia to be above the permissible daily dietary intake.[70,72] However *M. procera* is a strong reservoir of essential elements, such as like calcium, sodium, iron, potassium, magnesium, copper, and zinc. Several studies have recommended the sporadic or occasional eating of parasol mushroom to be safe.[17]

13.10 Need for Cultivation

Amongst wild mushrooms that are of culinary interest, *M. procera* is an interesting delicacy for human consumption and a more comprehensive knowledge about their nutritional and medicinal value has led to its increased consumption worldwide. The growth of this fungus in the wild is slow, restricted by area and season. It is a significant source of income as well as food for the people living in the areas where it grows. However, the collection of this fungus from the wild should not compromise its conservation. The successful cultivation of this valuable mushroom can not only help to preserve the species in the wild but also ensures maximum growth for food security.

The cultivation of *M. procera* can boost the economy, especially that of rural people by providing them with business opportunities.[95] Wild mushrooms are active accumulators of toxic elements from polluted environments and present a risk to the health of consumers. The geographical location and ecological conditions also affect the quality and quantity of important biopharmaceuticals obtained from mushrooms. Mushrooms grown under controlled conditions are safe for human consumption and can be improved for biopharmaceutical purposes. Moreover, cultivation ensures guaranteed unadulterated products of a particular species.[96]

13.10.1 Methods of Cultivation

M. procera can be cultivated on solid substrate or in submerged medium. In the former method, a natural or inert synthetic substrate is used for the conversion of plant materials, commonly agricultural residues, into fungal mycelium and fruiting bodies. However, this is time consuming and challenges are faced in the constant production of biomass with respect to yield and quality.[96] Peksen and Kiber[97] investigated the cultivation of *M. procera* from Turkey and found that wheat was the most suitable grain for the generation of spawn. Mycelial growth was observed in substrates prepared from agricultural residues used in different combinations, but fructification did not occur. In another investigation, Sharma *et al.*[98] commercially cultivated *M. procera* on compost prepared using a short composting method. The time between spawning and casing was 18–20 days when the compost bag was incubated at 27 °C. After application of the casing layer, the primordial mature to fruiting body was 16–30 cm in length and 30 g in weight. Moreover, the submerged culture of mushrooms has more industrial potential than solid-substrate culture as the former is rapid and leads to efficient growth of mycelium and the production of bioactive constituents.[99,100]

However, the optimization of the culture conditions in the submerged culture depends on the species being cultivated.[101]

13.10.2 Critical Conditions for the Cultivation of *M. procera*

pH, temperature, culture medium, source of carbon (C) and nitrogen (N), C : N ratio, availability of nutrients, and genetic potential are the critical factors in the cultivation of fungi. The optimization of these critical factors for the growth of mycelium and formation of fruiting bodies is important. There are limited studies on the optimization of the culture conditions of *M. procera*. Shim *et al.* [102] investigated the critical conditions for the artificial cultivation of *M. procera* and found the maximal mycelia growth to be achieved at pH 7 and 30 °C, with the most suitable carbon source being maltose and glycine as the best nitrogen source. However, the maximum density of the mycelium was reported using ammonium phosphate. The most suitable C : N ratio was 10 : 1 if the basal medium was supplemented with 1% glucose. Moreover, Peksen and Kiber[95] cultivated *M. procera* from Turkey and determined the optimum culture conditions to be pH 6.0 and a temperature of 20–25 °C. Growth was found to be significantly reduced at 15 °C and 30 °C due to decreased food absorption and denaturation of enzymes, respectively. The maximum mycelial growth was obtained at a combination of pH 6 and 20 °C and at pH 7.0 and 25 °C. Among the seven carbon sources used, along with base Sabouraud media, *i.e.*, Dextrose, maltose, lactose, mannitol, xylose, sucrose and glucose, the maximum diameter of the colony was observed in glucose. Yeast and peptones were shown to be the most beneficial nitrogen sources. However, dextrose was the most efficient carbon source in the submerged culture of *M. procera*, whereas peptones and malt extract were key nitrogen sources for the formation of maximal mycelia biomass.[101] Additionally, Erbiai *et al.*[14] successfully cultivated the mycelium of *M. procera* from Northern Morocco and Portugal on potato dextrose agar at 25 °C.

13.11 Preservation and Processing of *M. procera*

Mushrooms have a very short shelf life as they are rich in moisture, rapid enzymatic activity, and a high rate of respiration. In the absence of physical protection measures for reducing loss of water and the attack of microbes, they undergo many damaging post-harvest changes, such as browning, elongation of the stipe, opening of the cap, and loss of texture (Figure 13.8).

Thus, *M. procera* needs to be preserved to maintain its unique texture and flavor. The increase in its shelf life will also help to market the produce at distance, hence increasing its profitability. The common methods of preserving *M. procera* are drying, freezing, canning, and irradiation. However, as *M. procera* has great nutraceutical value, the preservation and processing technologies used should not chemically modify the harvested product. The goal should be to enhance the shelf life of this important fungus while keeping its nutritional and therapeutic worth as much as possible.[55] Kumar *et al.*[103] reported the formation of different value-added edible products, such as soup, pickles, and bakery products at the commercial level; however, the utilization of *M. procera* in such products is yet to be explored.

13.11.1 Preservation Technologies for *M. procera*

The reported preservation technologies for *M. procera* are drying, canning, freezing, and irradiation (Figure 13.9). The comparative assessment of different methods with respect to their effect on morphological, physicochemical, organoleptic, and medicinal properties are mentioned.

13.11.1.1 Drying

Drying is the most common and cheapest method of preserving fungi.[104,105] Moreover, it decreases the moisture content of mushrooms and minimizes loss by microbial degradation.[106] The dried products can be stored at a low temperature and in take up less space than fresh mushrooms. Moreover,

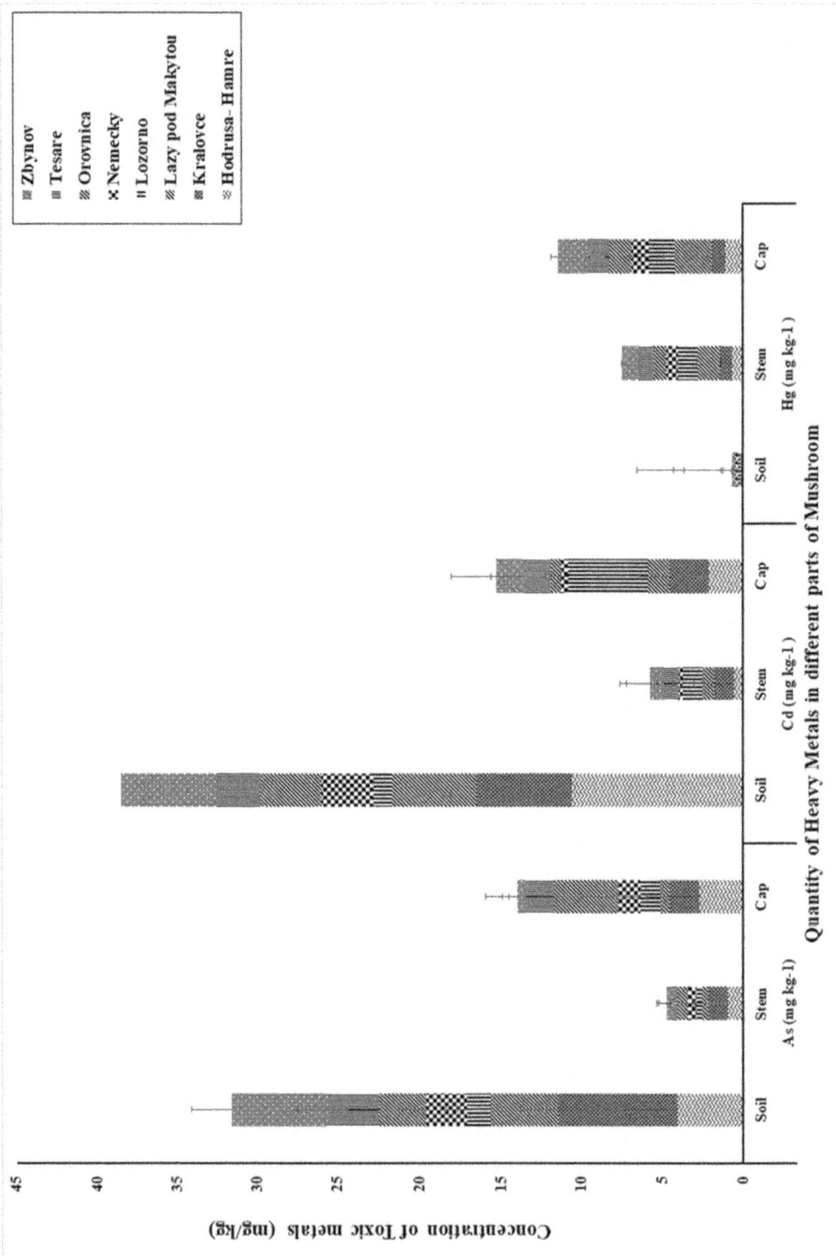

Figure 13.8 Concentration of toxic metals in *Macrolepiota procera* in different regions of Slovakia (adapted from ref. 72).

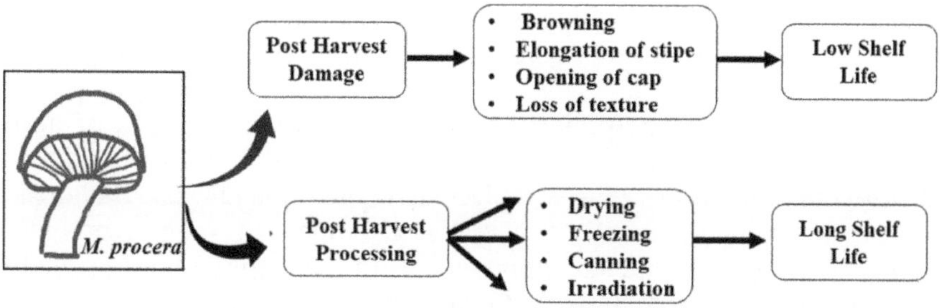

Figure 13.9 Post-harvest processing of *Macrolepiota procera.*

loss of water increases the nutritional value of mushrooms.[107] The heat used in the drying process increases the antioxidant properties of *M. procera* as its oxidative enzymes become inactivated.[48,58] However, drying changes the chemical properties of dried samples, as they contain more saturated fatty acids than fresh samples. *M. procera* dried at 30 °C has a high β-tocopherol content and mannitol as the main sugar, while fresh mushrooms have more δ-tocopherol and trehalose as the main sugar. A high temperature in drying also leads to a loss in organoleptic properties. This also increases the ability of microbes to survive for longer periods in dry foods.[108]

13.11.1.2 Freezing

Freezing is the most efficient method of preserving mushrooms as it retains not only the nutritional value but also the organoleptic properties of the fungi. The conversion of water to ice in the process of freezing reduces microbial activity.[109] Samples frozen at −20 °C have high δ-tocopherol and fatty acid content like that of fresh samples, and mannitol is the predominant sugar in dried mushrooms.

Freezing can increase the reducing power of *M. procera*, which leads to the increment in its antioxidant activity. Further freezing can cause the disruption of cell walls and the easy extraction of intracellular compounds.[48]

13.11.1.3 Canning

In the canning process, mushrooms are boiled in boiling water for 30 minutes and canned after draining while still hot. Aydin *et al.*[61] found that canning was the least effective method of preserving mushrooms in comparison to drying and freezing, as the mushrooms contained higher moisture and lower ash, crude protein, fat, and energy content. The pileus component of the mushrooms also darkened during storage.

13.11.1.4 Irradiation

Irradiation of food *via* the control of dose and irradiation time is the safest way of increasing its shelf life. In 1991, the ICGFI recommended the maximum dose for the irradiation of mushrooms to be 3 kGy. There are two methods of irradiating mushrooms: gamma and electron beam radiation.[48]

13.11.1.4.1 Gamma Irradiation. Gamma irradiation is the most appropriate processing method for the maintenance of the chemical profile of *M. procera*. Irradiation reduces molecular oxygen and the lipid autooxidation process, which prevents fat content loss. Moreover, higher contents of monounsaturated fatty acids, alpha and gamma tocopherol, and trehalose have also been observed in irradiated mushrooms. Irradiated samples have been shown to the greatest thiobarbituric acid reactive substance inhibition in a TBARS assay.[48] Alves *et al.*[110] investigated the effect of both gamma and electron beam irradiation on *M. procera* for its antimicrobial potential against clinical isolates of Gram-positive and Gram-negative bacteria, including resistant strains, and found that a dose of ≤2 kGY was appropriate for the preservation of the fungus without compromising its antimicrobial activity.

13.11.1.4.2 Electron Beam Irradiation. Drying induces many undesirable changes in mushrooms products that can be attenuated by electron beam irradiation. Irradiation can prevent the discoloration of mushrooms[103] and is an effective method by which to preserve their texture, taste, and aroma. Electron beam sources are used commonly for thin samples in low density.[55,111] Cardoso *et al.*[112] demonstrated that electron beam radiation improved the number of organic acids in mushrooms and also lessened the negative effects of drying. Samples of *M. procera* dried at 30 °C were electron beam irradiated at doses of 1.0 kGy and 6.0 kGy and were monitored after 0, 6, and 12 months of storage. The organoleptic qualities of the mushrooms were retained, demonstrating the efficacy of the electron beam irradiation, which may be used as a supplementary method to drying as it has decontamination capabilities.[113]

13.12 Conclusions and Future Perspectives

This chapter highlights the importance of *M. procera* as an edible and medicinal mushroom. A huge number of research studies have indicated the occurrence of pharmacologically active compounds in *M. procera. M. procera* is a rich source of various important compounds but it also contains hazardous substances. Therefore, the dietary intake of *M. procera* should be limited to one or two meals per week. The biological properties of this parasol-like mushroom may be of interest to a number of enterprises in the pharmaceutical and nutraceutical domains. In the future, studies are required to explore the value-added bioactive components of *M. procera* and its utilization as a functional food. Additionally, studies should be carried out on the cultivation, toxicity, and therapeutic usefulness of the parasol mushroom.

References

1. P. Kalač, *Food Chem.*, 2009, **113**, 9–16.
2. A. Srikram and S. Supapvanich, *Agric. Nat. Resour.*, 2016, **50**, 432–436.
3. S. M. Ayodele and J. A. Okhuoya, *S. Afr. J. Sci.*, 2009, **105**, 158–190.
4. M. Kozarski, A. Klaus, D. Jakovljevic, N. Todorovic, J. Vunduk, P. Petrović, M. Niksic, M. M. Vrvic and L. Van Griensven, *Molecules*, 2015, **20**, 19489–19525.
5. H. M. Hsieh and Y. M. Ju, *Appl. Microbiol. Biotechnol.*, 2018, **102**, 4987–4994.
6. W. A. Elkhateeb, G. M. Daba, P. W. Thomas and T. C. Wen, *Egypt. Pharm. J.*, 2019, **18**, 88–101.
7. M. Akyuz and S. Kirbag, *EurAsian J. BioSci.*, 2009, **3**, 58–63.
8. N. Nowacka, R. Nowak, M. Drozd, M. Olech, R. Los and A. Malm, *PLoS One*, 2015, **10**, 0140355.
9. P. Bobek and S. Galbavy, *Br. J. Biomed. Sci.*, 2001, **58**, 164.
10. Y. Sun, Y. Shao, Z. Zhang, L. Wang, A. M. Mariga, G. Pang, C. Geng, C. T. Ho, Q. Hu and L. Zhao, *J. Food Drug Anal.*, 2014, **22**, 463–467.
11. J. Falandysz, *Food Chem.*, 2017, **214**, 96–101.
12. M. H. Rashid, M. M. Rahman, R. Correll and R. Naidu, *Int. J. Environ. Res. Public Health*, 2018, **15**, 919.
13. I. Mirończuk-Chodakowska, K. Socha, M. E. Zujko, K. M. Terlikowska, M. H. Borawska and A. M. Witkowska, *Int. J. Environ. Res. Public Health*, 2019, **16**, 3614.
14. E. H. Erbiai, L. P. da Silva, R. Saidi, Z. Lamrani, J. C. Esteves da Silva and A. Maouni, *Biomolecules*, 2021, **11**, 575.
15. J. Alonso, M. A. García, M. Pérez-López and M. J. Melgar, *Arch. Environ. Contam. Toxicol.*, 2003, **44**, 0180–0188.
16. I. Širić, M. Humar, A. Kasap, I. Kos, B. Mioč and F. Pohleven, *Environ. Sci. Pollut. Res.*, 2016, **23**, 18239–18252.
17. A. K. Kojta, M. Gucia, G. Krasińska, M. Saba, I. C. Nnorom and J. Falandysz, *Pol. J. Environ. Stud.*, 2016, **25**(6), 2445–2460.
18. H. J. Cho, H. Lee, M. S. Park, C. Kim, K. Wisitrassameewong, A. Lupala, K. H. Park, M. J. Kim, J. J. Fong and Y. W. Lim, *Mycobiology*, 2019, **47**, 368–377.
19. R. Singer, *The Agaricales in Modern Taxonomy*, Sven Koeiltz Scientific Books, Koenigstein, Germany, 1986.
20. M. Bon, F. Medjebeur-Thrun, and A. Einhellinger, *Die Großpilzflora von Europa. 3. Lepiotaceae*, IHW-Verlag, 1996.
21. B. Kumari and N. S. Atri, *Int. J. Pharm. Pharm. Sci.*, 2014, **6**, 200–204.
22. R. Kumar, S. Pandey, R. R. Rishi and K. Giri, *Asian J. Agric.*, 2017, **1**, 6–8.
23. M. J. Melgar, J. Alonso and M. A. Garcia, *Food Chem. Toxicol.*, 2016, **88**, 13–20.
24. M. Haimed, A. Nmichi, A. O. Touhami and A. Douira,*J. Anim. Plant Sci.*, 2013, **18**, 2723–2749.

25. K. C. Semwal, S. L. Stephenson, V. K. Bhatt and R. P. Bhatt, *Mycosphere*, 2014, **5**(3), 440–461.
26. S. Ahmad, S. H. Iqbal, and A. N. Khalid, *Fungi of Pakistan*, Mycological Society of Pakistan, Uni. Punjab, Monograph, 1997.
27. E. C. Vellinga, *Aust. Syst. Bot.*, 2003, **16**(3), 361–370.
28. L. M. Rizal, K. D. Hyde, E. Chukeatirote and S. Chamyuang, *Mycosphere*, 2015, **6**, 414–420.
29. P. Manzi, A. Aguzzi and L. Pizzoferrato, *Food Chem.*, 2001, **73**, 321–325.
30. G. Steinhauser and V. Steinhauser, *J. Food Prot.*, 2016, **79**(11), 1995–1999.
31. C. Sarikurkcu, B. Tepe, M. S. Kocak and M. C. Uren, *Food Chem.*, 2015, **175**, 549–555.
32. M. Mędyk, M. Grembecka, J. Brzezicha-Cirocka and J. Falandysz, *J. Environ. Sci. Health*, 2017, **52**, 577–583.
33. I. Akata, G. Zengin, C. M. N. Picot and M. F. Mahomoodally, *S. Afr. J. Bot.*, 2019, **120**, 95–99.
34. S. Wei, *Int. J. Med. Mushrooms*, 2008, **10**(4), 315–324.
35. K. Sulkowska-Ziaja, B. Muszynska, P. Motyl, P. Pasko and H. Ekiert, *Int. J. Med. Mushrooms*, 2012, **14**(4), 385–393.
36. F. Çayan, G. Tel-Çayan, E. Deveci and M. E. Duru, *J. Food Process. Preserv.*, 2021, **45**(9), 15695.
37. H. P. Chen, Z. Z. Zhao, Z. H. Li, Y. Huang, S. B. Zhang, Y. Tang, J. N. Yao, L. Chen, M. Isaka, T. Feng and J. K. Liu, *J. Agric. Food Chem.*, 2018, **66**(12), 3146–3154.
38. S. Zurga, J. Pohleven, M. Renko, S. Bleuler-Martinez, P. Sosnowski, D. Turk, M. Künzler, J. Kos and J. Sabotič, *FEBS J.*, 2014, **281**(15), 3489–3506.
39. R. Nowak, N. Nowacka-Jechalke, M. Juda and A. Malm, *Eur. J. Nutr.*, 2018, **57**(4), 1511–1521.
40. L. Barros, M. Dueñas, I. C. Ferreira, P. Baptista and C. Santos-Buelga, *Food Chem. Toxicol.*, 2009, **47**(6), 1076–1079.
41. A. Ciric, I. Kruljević, D. Stojković, Â. Fernandes, L. Barros, R. C. Calhelha, I. C. Ferreira, M. Soković and J. Glamočlija, *Food Funct.*, 2019, **10**(12), 7678–7686.
42. E. C. Aytar, İ. Akata and A. Ç. I. K. Leyla, *Mantar Dergisi*, 2020, **11**(2), 121–128.
43. N. Nowacka, R. Nowak, M. Drozd, M. Olech, R. Los and A. Malm, *LWT-Food Sci. Technol.*, 2014, **59**(2), 689–694.
44. O. Taofiq, R. C. Calhelha, S. Heleno, L. Barros, A. Martins, C. Santos-Buelga, M. J. R. Queiroz and I. C. Ferreira, *Food Res. Int.*, 2015, **76**, 821–827.
45. M. Naczk and F. Shahidi, *J. Pharm. Biomed. Anal.*, 2006, **41**(5), 1523–1542.
46. C. Dinçer, İ. Tontul, İ. B. Çam, K. S. Özdemir, A. Topuz, H. Ş. Nadeem, S. T. AY and R. S. Göktürk, *Turk. J. Agric. For.*, 2013, **37**(5), 561–567.
47. S. Beluhan and A. Ranogajec, *Food Chem.*, 2011, **124**(3), 1076–1082.

48. Â. Fernandes, L. Barros, J. C. Barreira, A. L. Antonio, M. B. P. Oliveira, A. Martins and I. C. Ferreira, *LWT-Food Sci. Technol.*, 2013, **54**(2), 493–499.
49. N. Yilmaz, I. Türkekul, S. Bulut and F. Sahin, *Asian J. Chem.*, 2013, **25**(3), 1216.
50. L. Barros, P. Baptista, D. M. Correia, S. Casal, B. Oliveira and I. C. Ferreira, *Food Chem.*, 2007, **105**(1), 140–145.
51. P. Manzi, L. Gambelli, S. Marconi, V. Vivanti and L. Pizzoferrato, *Food Chem.*, 1999, **65**(4), 477–482.
52. Y. L. Lee, S. Y. Jian and J. L. Mau, *LWT-Food Sci. Technol.*, 2009, **42**(2), 594–598.
53. P. Mattila, P. Salo-Väänänen, K. Könkö, H. Aro and T. Jalava, *J. Agric. Food Chem.*, 2002, **50**(22), 6419–6422.
54. P. Ouzouni and K. Riganakos, *Acta Aliment.*, 2007, **36**(1), 99–110.
55. Â. Fernandes, A. L. Antonio, M. B. P. Oliveira, A. Martins and I. C. Ferreira, *Food Chem.*, 2012, **135**(2), 641–650.
56. L. Barros, R. C. Calhelha, J. A. Vaz, I. C. Ferreira, P. Baptista and L. M. Estevinho, *Eur. Food Res. Technol.*, 2007, **225**(2), 151–156.
57. W. Haiying, Z. Shaozhi and C. Guangming, *LWT-Food Sci. Technol.*, 2007, **40**(6), 1112–1116.
58. L. Barros, P. Baptista, D. M. Correia, J. Sá Morais and I. C. Ferreira, *J. Agric. Food Chem.*, 2007, **55**(12), 4781–4788.
59. P. Manzi, S. Marconi, A. Aguzzi and L. Pizzoferrato, *Food Chem.*, 2004, **84**(2), 201–206.
60. W. W. Nawar, *Food Rev. Int.*, 1986, **2**(1), 45–78.
61. E. Aydin, I. B. Gurbuz, H. Karahan and C. Basdar, *J. Food Process. Preserv.*, 2017, **41**(2), 12802.
62. L. Barros, M. J. Ferreira, B. Queiros, I. C. Ferreira and P. Baptista, *Food Chem.*, 2007, **103**(2), 413–419.
63. J. Falandysz and M. Gucia, *Environ. Geochem. Health*, 2008, **30**(2), 121–125.
64. M. Gucia, G. Jarzyńska, A. K. Kojta and J. Falandysz, *J. Environ. Sci. Health*, 2012, **47**(2), 81–88.
65. E. Kułdo, G. Jarzyńska, M. Gucia and J. Falandysz, *Chem. Pap.*, 2014, **68**(4), 484–492.
66. V. Stefanović, J. Trifković, S. Djurdjić, V. Vukojević, Ž. Tešić and J. Mutić, *Environ. Sci. Pollut. Res.*, 2016, **23**(21), 22084–22098.
67. P. Kalac and L. Svoboda, *Food Chem.*, 2000, **69**(3), 273–281.
68. M. T. Schlecht and I. Säumel, *Environ. Pollut.*, 2015, **204**, 298–305.
69. J. Falandysz and R. Treu, *J. Environ. Sci. Health B*, 2017, **52**(3), 147.
70. M. Gucia, G. Jarzyńska, E. Rafał, M. Roszak, A. K. Kojta, I. Osiej and J. Falandysz, *Environ. Sci. Pollut. Res.*, 2012, **19**(2), 416–431.
71. S. Durdic, V. Stanković, S. Ražić and J. Mutić, *Environ. Sci. Pollut. Res.*, 2021, **28**(42), 59003–59014.
72. I. Janco, M. Šnirc, M. Hauptvogl, H. Franková, N. Čeryová, J. Štefániková and J. Árvay, *J. Microbiol. Biotechnol. Food Sci.*, 2021, **11**(2), 4764.

73. I. Jančo, M. Šnirc, M. Hauptvogl, L. Demková, H. Franková, V. Kunca, T. Lošák and J. Árvay, *J. Fungi*, 2021, **7**(9), 772.
74. M. Kosanić, B. Ranković, A. Rančić and T. Stanojković, *J. Food Drug Anal.*, 2016, **24**(3), 477–484.
75. M. Mleczek, M. Siwulski, A. Budka, P. Mleczek, S. Budzyńska, M. Szostek, N. Kuczyńska-Kippen, P. Kalač, P. Niedzielski, M. Gąsecka and P. Goliński, *Chemosphere*, 2021, **263**, 128095.
76. M. Barea-Sepúlveda, E. Espada-Bellido, M. Ferreiro-González, H. Bouziane, J. G. López-Castillo, M. Palma and G. F. Barbero, *J. Food Compos. Anal.*, 2022, **108**, 104419.
77. M. Seçme, O. Kaygusuz, C. Eroglu, Y. Dodurga, O. F. Colak and P. Atmaca, *Int. J. Med. Mushrooms*, 2018, **20**(11), 1075–1086.
78. S. Arora, S. Goyal, J. Balani and S. Tandon, *Int. J. Med. Mushrooms*, 2013, **15**(3), 301–314.
79. R. Zara, A. Rasul, T. Sultana, F. Jabeen and Z. Selamoglu, *Saudi J. Biol. Sci.*, 2022, **29**(5), 3372–3379.
80. K. H. Han, D. H. Kim, K. Y. Song, K. H. Lee, T. B. Kang and T. J. Yoon, *Korean J. Pharmacogn.*, 2012, **43**(1), 32–38.
81. P. A. Ayeka, *Evidence-Based Complementary Altern. Med.*, 2018, **2018**, 7271509.
82. D. H. Kim, K. H. Han, K. Y. Song, K. H. Lee, S. Y. Jo, S. W. Lee and T. J. Yoon, *Korean J. Pharmacogn.*, 2010, **41**(2), 115–121.
83. J. A. Vaz, L. Barros, A. Martins, C. Santos-Buelga, M. H. Vasconcelos and I. C. Ferreira, *Food Chem.*, 2011, **126**(2), 610–616.
84. M. L. Circu and T. Y. Aw, *Free Radical Biol. Med.*, 2010, **48**(6), 749–762.
85. E. M. Jeong, M. Liu, M. Sturdy, G. Gao, S. T. Varghese, A. A. Sovari and S. C. Dudley Jr, *J. Mol. Cell. Cardiol.*, 2012, **52**(2), 454–463.
86. A. A. F. Zielinski, C. W. I. Haminiuk and T. Beta, *LWT-Food Sci. Technol.*, 2016, **65**, 897–907.
87. M. L. Popescu, T. Costea, I. Nencu, L. E. Duţu and C. E. Gird, *Power*, 2016, **5**, 27.
88. Â. Fernandes, J. Barreira, A. L. Antonio, M. B. P. P. Oliveira, A. Martins and I. C. Ferreira, *Food Bioprocess Technol.*, 2014, **7**(6), 1606–1617.
89. S. Gonca, *Mersin Üniversitesi Tıp Fakültesi Lokman Hekim Tıp Tarihi ve Folklorik Tıp Dergisi*, 2022, **12**(1), 198–208.
90. T. Ozen, C. Darcan, O. Aktop and I. Turkekul, *Comb. Chem. High Throughput Screening*, 2011, **14**(2), 72–84.
91. M. Y. Dyakov, O. V. Kamzolkina, O. V. Shtaer, N. A. Bis'ko, N. L. Poedinok, O. B. Mikhailova, O. V. Tikhonova, T. E. Tolstikhina, B. F. Vasil'eva and O. V. Efremenkova, *Microbiology*, 2011, **80**(2), 274–285.
92. A. Čuvalová, I. Strapáč, L. Handrová and V. Kmeť, *Ann. Univ. Paedagog. Crac. Stud. Math.*, 2018, 17–23.
93. P. Phadungpran, W. Pongnak and K. Soytong, *J. Agric. Technol.*, 2015, **11**(8), 2457–2461.

94. J. Sabotic, T. Popovič, V. Puizdar and J. Brzin, *FEBS J.*, 2009, **276**, 4334–4345.
95. A. Pekşen and B. Kibar, *Acta Sci. Pol. Hortorum Cultus*, 2020, **19**(1), 11–20.
96. M. L. Gargano, L. J. van Griensven, O. S. Isikhuemhen, U. Lindequist, G. Venturella, S. P. Wasser and G. I. Zervakis, *Plant Biosyst. Int. J. Dealing Aspects Plant Biol.*, 2017, **151**(3), 548–565.
97. A. Pekşen and B. Kibar, *Uluslararası Tarım ve Yaban Hayatı Bilimleri Dergisi*, 2017, **3**(2), 68–79.
98. V. P. Sharma, S. Kumar, and S. Sharma, *ICAR-Directorate of Mushroom Research*, Chambaghat, Solan, Himachal Pradesh, India, 2020.
99. B. C. Lee, J. T. Bae, H. B. Pyo, T. B. Choe, S. W. Kim, H. J. Hwang and J. W. Yun, *Enzyme Microb. Technol.*, 2004, **35**(5), 369–376.
100. Z. Zhang, W. Shen, D. Liu and J. Li, *Afr. J. Microbiol. Res.*, 2011, **5**(21), 3452–3461.
101. A. Pekşen and B. Kibar, *Anadolu J. Agric. Sci.*, 2016, **31**(1), 16–24.
102. S. M. Shim, Y. H. Oh, K. R. Lee, S. H. Kim, K. H. Im, J. W. Kim, U. Y. Lee, J. O. Shim, M. J. Shim, M. W. Lee and H. S. Ro, *Mycobiology*, 2005, **33**(1), 15–18.
103. A. Kumar, and A. Satpathy, in *Biology, Cultivation and Applications of Mushrooms*, Springer, Singapore, 2022, pp. 561–577.
104. S. G. Walde, V. Velu, T. Jyothirmayi and R. G. Math, *J. Food Eng.*, 2006, **74**(1), 108–115.
105. S. K. Giri and S. Prasad, *J. Food Eng.*, 2007, **78**(2), 512–521.
106. S. V. Jangam, C. L. Law and A. S. Mujumdar, *Drying of Foods, Vegetables and Fruits*, 2011, vol. 2, ISBN: 978-981-08-7985-3.
107. W. Cao, Y. Nishiyama and S. Koide, *Biosyst. Eng.*, 2003, **85**(3), 331–337.
108. M. Maskan, *J. Food Eng.*, 2000, **44**(2), 71–78.
109. G. Jaworska and E. Bernaś, *Food Chem.*, 2009, **113**(4), 936–943.
110. M. J. Alves, Â. Fernandes, J. C. Barreira, I. Lourenço, D. Fernandes, A. Moura, A. R. Ribeiro, J. Salgado, A. Antonio and I. C. Ferreira, *J. Appl. Microbiol.*, 2015, **118**(3), 592–598.
111. IAEA-International Atomic Energy Agency, Dosimetry for food irradiation. Technical Report Series 490, Vienna, 2002.
112. R. V. Cardoso, M. Carocho, Â. Fernandes, J. C. Barreira, S. Cabo Verde, P. M. Santos, A. L. Antonio, A. M. Gonzaléz-Paramás, L. Barros and I. C. Ferreira, *J. Food Sci.*, 2021, **86**(6), 2276–2287.
113. Â. Fernandes, J. C. Barreira, A. L. Antonio, M. B. P. Oliveira, A. Martins and I. C. Ferreira, *Food Chem.*, 2014, **149**, 91–98.

CHAPTER 14

Russula virescens *(Schaeff.) Fr.*/Russula adusta *Krombh.*/ Russula delica *Fr.*/Russula nigricans *(Bull.) Fr.*/Russula cyanoxantha

MANMEET KAUR (0000-0001-8120-4774),* SHIVANI SHARMA, JASPREET KAUR AND HARPREET SINGH SODHI

Department of Microbiology, Punjab Agricultural University, Ludhiana-141004, India
*Email: manmeet1-mb@pau.edu

14.1 Introduction

The Russulaceae family includes the Basidiomycota genus of Russula fungi. It is estimated that there are 750 different Russula species. Russula mushrooms have been found in 17 northeastern provinces of Thailand and some of them have been used as food, including *Russula (R.) virescens* (Khai-Dong in Thai); *R. helios* (ko-rang-som in Thai); *R. medullata* (Ko-Kab-Yang in Thai); *R. alboareolata* (Nam-Pang in Thai); and *R. monspeliensis* (Khai-Na-Lae-Paeng in Thai). The following Russula mushrooms have a documented history of use in traditional medicines to treat a variety of diseases, as shown in Table 14.1.[1]

Several species of the genus *Russula* Pers are included in the Compactae Fr section. The carpophores of members of this group are thick, with compact

Phytochemistry and Nutritional Composition of Significant Wild Medicinal and Edible Mushrooms: Traditional Uses and Pharmacology
Edited by Ajay Sharma, Garima Bhardwaj and Gulzar Ahmad Nayik

Published by the Royal Society of Chemistry, www.rsc.org

Table 14.1 Russula mushrooms: established history of the uses of medicinal mushrooms.

Russula mushrooms	Uses
R. cyanoxantha	Fever
R. delica	Gastritis
R. luteotacta	Wound healing
R. nobilis	Fever
R. luteotacta	Sleep promoting agent
R. acrifolia	Skin cancer
R. parazurea	Gastritis

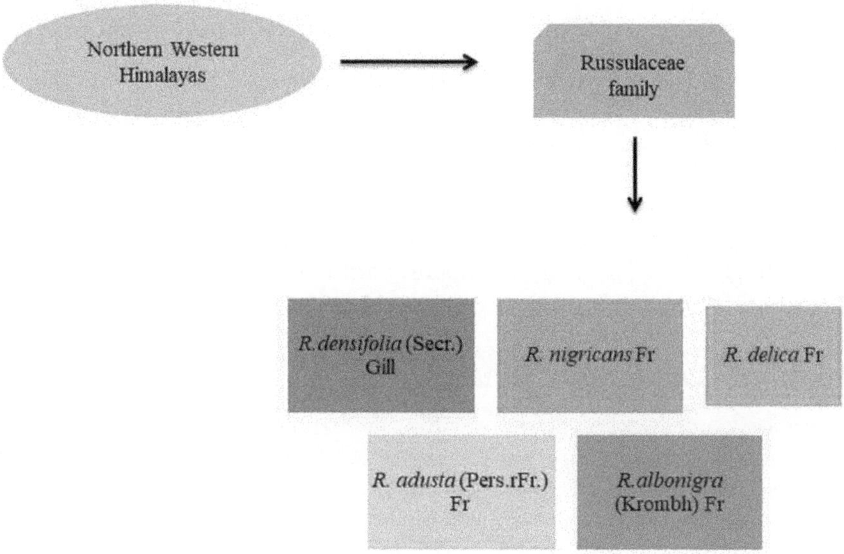

Figure 14.1 Russula species discovered in the northern Himalayan region.

and uneven lamellae. In general, the basidiocarps of members of this group are whitish, brownish or even blackish in color. Only five species (Figure 14.1), *R. delica* Fr.; *R. nigricans* Fr.; *R. densifolia* (Secr.) Gil.; *R. adusta* (Pers.rFr.) Fr. and *R. albonigra* (Krombh.) Fr., have been discovered in India during a fungal survey of the northwestern Himalayas.[2]

In the order Agaricales is family Russulaceae, one of the largest ectomycorrhizal families (Figure 14.2). The subtropical and temperate forests of the northwestern Uttarakhand Himalayas are a major terrestrial carbon sink and one of the planet's most vulnerable vital systems. Knowing the community- and species-level diversity of macrofungi in these forests is crucial for establishing and understanding their overall ecology, as well as for assessing the critical niche interactions of the various components that make up the forest biota, including the macrofungi themselves. The characterization of

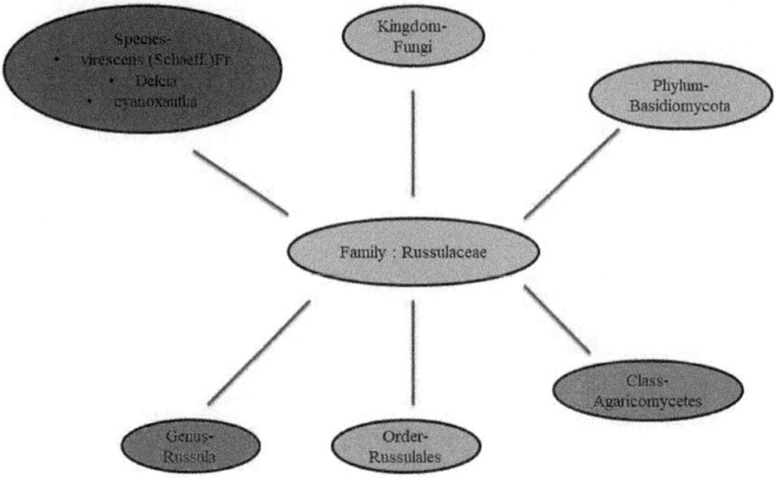

Figure 14.2 Overall diagrammatic representation of the taxonomy of the Russulaceae family.

both the systematics as well as the ecology of the accumulation of existing macrofungi is crucial for an adequate planning and management of these forests as well as for the conservation of their biodiversity.[3]

14.2 *R. virescens* (Schaeff.) Fr.

R. virescens has a green cap, spores that lack an amyloid notch, and basidia that are long and slender. It is characterized by the silky, dry surface of its cap. The thick cracks in the middle of the cap, become wider and coarser towards the rim. The trunk is sturdy but weak and can sometimes be twisted. *R. virescens*, a basidiomycete fungus of the genus Russula, is often referred to as quilted green Russula, Green Knackrussula or Green Brittlegill; it grows on pine roots and has long been used as a traditional folk treatment in Chinese medicine.

14.2.1 Distribution

It is widely distributed throughout Dhobighat, Chakrata, Mussoorie, Uttar Pradesh and Deoban.[4] It grows on the roots of pine trees throughout Europe, Nepal, Thailand, and China.[5]

14.2.2 Morphology

Pileus: 5.5–14 cm wide; pulvinate with a discretized margin and a little depressed disc when young; tuberculate-striate.
Cuticle: slender, viscid, glabrous marginally, glaucous centrally.
Trama: thick, spongy-brittle, odorless, yellowish-white.

Lamellae: wide, initially elastic, sharp, pale yellow.
Stipe: long, thick, flared at the apex, dry, dull, glabrous.
Spores: yellowish-white, broadly elliptic, obovate, reniform, and pip-shaped.[6]

14.2.3 Bioactive Phytochemical Constituents

Mushrooms have been valued for their taste and texture for decades. They are currently recognized as a nutrient-dense food and as an essential source of compounds that are pharmacologically active and may have medicinal value.[7] Secondary metabolites in fungi include polyketides, steroids, phenolic chemicals, and terpenes.[8] Flavoglaucin, a phenolic molecule found in fungi, has also been shown to be a potent antioxidant and quasi-synergist.[9] Scientists extracted a water-insoluble linear (1→3)β-D-glucan from fresh fruiting bodies, but it lacked anticancer potential; however, glucan sulfation increased antitumor activity.[10] Scientists also extracted a water-soluble polysaccharide from the fruiting bodies of *R. virescens* with a backbone composed of (1→6)-linked α-D-galactopyranosyl and (1→2,6)-linked D-galactopyranosyl residues.[4] The polysaccharide had a potent scavenging effect on hydroxyl radicals. Maleic acid, D-allitol, thio-acetic anhydride, and ribosidoadenine are secondary metabolites that have been isolated from *R. virescens*, as shown in Figure 14.3.[11]

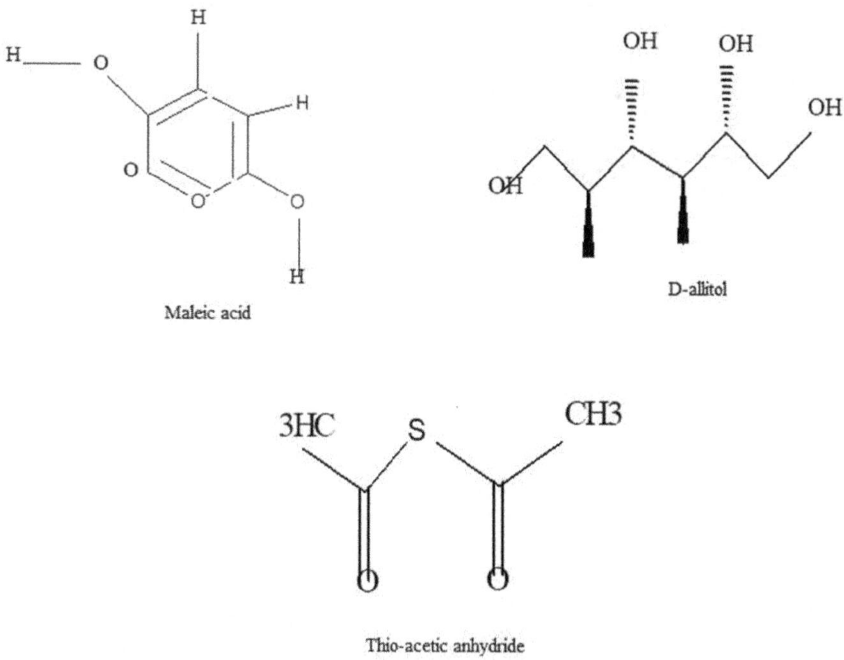

Figure 14.3 Secondary metabolites isolated from *R. virescens*.

The antioxidant activity of *R. virescens* has been studied and it has been discovered that this fungus can be used as a natural source of antioxidants.[12] *R. virescens* is rich in phenolic compounds (2–9%) and has also been found to have antioxidant capacity in the range of 10–21%, as shown in Figure 14.4.[13] Nitrite forms nitrosamines and is converted into intracellular components, diazoalkane and proteins, which can increase cancer risk.[14] The significant number of phenolic chemicals naturally occurring in plants may be responsible for their strong nitrite scavenging activity.[15]

Nitrite scavenging activity has been found to be in the range of 35–54% for water extracts and 28.5–42% for ethanol extracts of *R. virescens*. Phenolic compounds, such as catechin, camphor oil, ferulic acid, vanillic acid, luteolin, lupine, and apigenin have been detected in *R. virescens*, as shown in Table 14.2 and Figure 14.5.[13]

The hydroxyl radical generated by the interaction of O^{2-} and H_2O_2 in the presence of metal ions causes most of the oxidative damage in biological systems.[16] Flavonoids that scavenge hydroxyl radicals can prevent the Fenton reaction from occurring. The effectiveness of *R. virescens* extracts in preventing oxidative DNA damage was examined on OH^--induced DNA strand cleavage on 'pBR322' plasmid DNA prepared in a Fenton reaction. In contrast to the plasmid DNA control, DNA damage caused by hydroxyl radicals completely converted the supercoiled DNA to a linear form. Supercoiled DNA was repaired in a dose-dependent manner after exposure to water and ethanol extracts of *R. virescens*. This showed that ethanol and water-based Russula mushroom extracts had a protective effect against DNA damage.[13] Bioactive compounds such as tocopherols, carotenoids, organic acids and phenolic compounds have

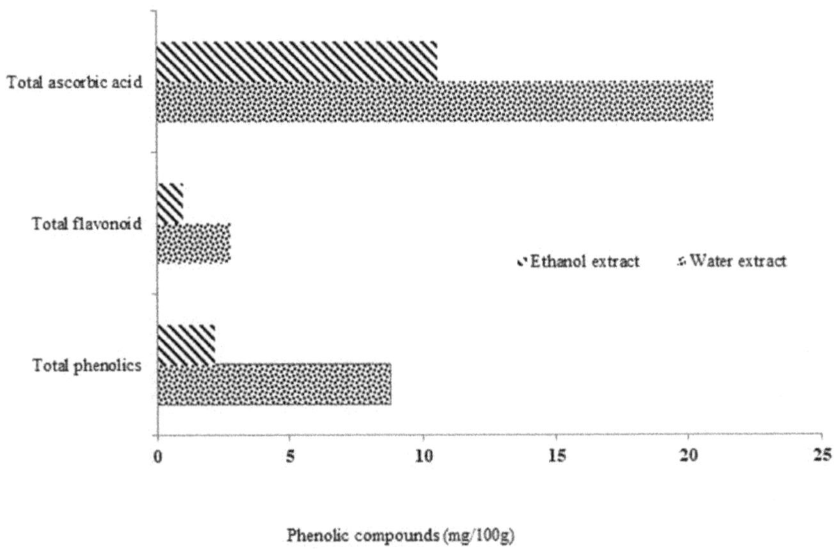

Figure 14.4 Phenolic compounds of *R. virescens*.

Table 14.2 Phenolic composition of ethanol and water extracts of *Russula virescens*.[13]

Phenolic compound	Water extract (mg ml^{-1})	Ethanol extract (mg ml^{-1})
Ferrulic acid	0.405	0.504
Luteolin	0.220	0.042
Catechin	0.151	—
Apigenin	0.019	0.047
Kaempferol	1.234	1.052
Vanillic acid	0.145	—
Lupane	0.553	0.364

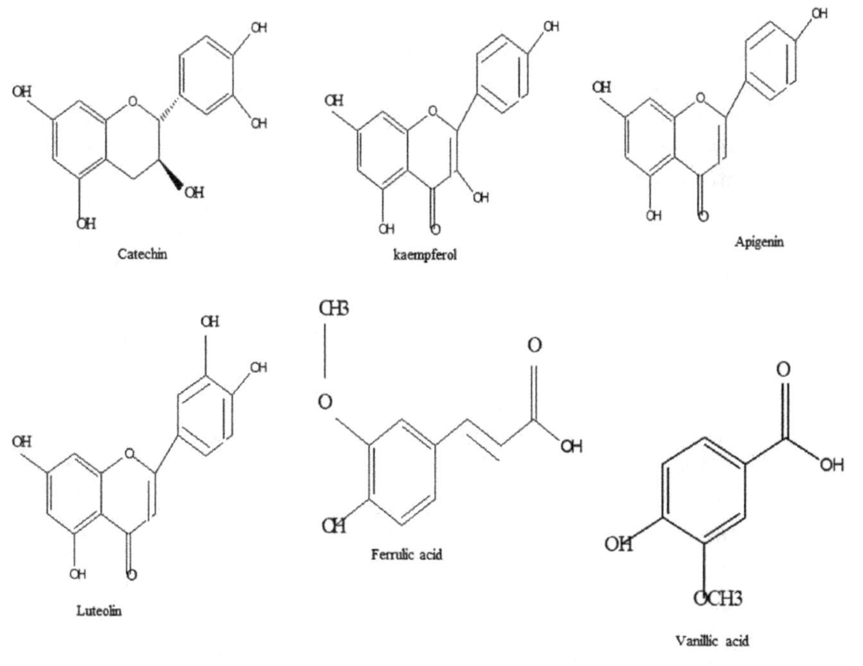

Figure 14.5 Phenolic compounds of *R. virescens*.

been identified in *R. virescens* as: α-tocopherol (20 mg per 100 g), β-tocopherol (21.30 mg per 100 g), γ-tocopherol (8 mg per 100 g), total tocopherols (49.30 mg per 100 g), lycopene (0.19 mg per 100 g), oxalic acid (0.78 mg per 100 g), malic acid (2.71 mg per 100 g), citric acid (0.55 mg per 100 g), fumaric acid (0.23 mg per 100 g), total organic acids (4.26 mg per 100 g), *p*-hydroxybenzoic acid (22.59 mg per 100 g), and cinnamic acid (15.75 mg per 100 g).[17]

14.2.4 Phytochemistry

The main component of fungi, polysaccharides, is important for maintaining biological activities.[18] In the last two decades, the study of fungal polysaccharides has made significant advances and extraordinary biological activities, such as anticancer, antioxidant, antiaging, antiviral, anti-hypotension, and immunological activity, have been observed.[19] The function of polysaccharides can be closely correlated with their structural properties. The discovery of a new fungal polysaccharide with exceptional stability and high biological activity is therefore relevant and has the potential to boost the emergence of functional foods and the bio-based pharmaceutical industry. Biological and antioxidant activities have been ascribed to the polysaccharides of *R. virescens*.[20]

Polysaccharides are natural polymers derived from plants, animals and microorganisms. Over the last few decades, polysaccharides have been extensively studied because of their diverse functions and biological activities, such as antioxidant, anticoagulant, hypoglycemic, immunomodulatory and antitumor activities, which are relevant the fields of food and biomedicine. The physical properties, chemical and complex structures of polysaccharides, such as their solubility, primary structure, and conformation in solution, affect their biological activity in different ways.[20] It has been shown that chemically altering polysaccharides using chemical modification methods such as sulfation, acetylation, phosphorylation, selenization, and carboxymethylation can enhance their inherent bioactivities or render them with new functional properties. Among the chemical modification methods, sulfate modification is considered to be an effective and easy way to modify the biological activities of polysaccharides by partially substituting sulfate groups for hydroxyl groups. Sulfated polysaccharides from various sources could serve as potential pharmacological and nutraceutical agents with potential therapeutic use for humans. According to monosaccharide composition analysis, *R. virescens* polysaccharides (RVP) are typical heteropolysaccharides consisting of xylose (1 mol/mol); mannose (15.04 mol/mol); glucose (216.89 mol/mol); galactose (58.41 mol/mol); arabinose (6.42 mol/mol); and fucose (10.66 mol/mol). Glucose and galactose were found to be the dominant monosaccharides among all the polysaccharides.[20]

Li *et al.* found that the sulfated polysaccharides of *R. virescens* exhibit high levels of ABTS and hydroxyl radical scavenging activity in addition to anticoagulant activity.[21] Using DEAE ion column chromatography, the yield of crude RVP was 10%, and the water eluting fraction RVP-1 (10.32% yield from RVP) was also prepared. The compositions of the monosaccharides of RVP-1 and RVP-2 were investigated by comparing standard spectra.[22] Fucose, mannose, glucose, and galactose were all found to be present in equal amounts in both RVP-1 and RVP-2 with molar ratios of 1 : 2.24 : 3.80 : 4.30 and 1 : 2.49 : 36.78 : 4.91, respectively. The main components of RVP-1 and RVP-2 are glucose and galactose, with trace amounts of mannose and fructose. RVP-1 and RVP-2 have also been found in *Ganoderma atrum*, *Flammulina velutipes*, *Ganoderma sinense*, and *Phellinus baumii* with the same monosaccharide compositions.[23] There is no starch or nucleic acids in either RVP-1 or RVP-2; they consist mostly of neutral carbohydrates and very small amounts of protein.[24]

14.2.5 Pharmacology

14.2.5.1 *Hypoglycemic Activity*

Hyperglycemia and disorders of nutrient metabolism are features of the chronic disease diabetes that can have long-term effects.[25] α-Amylase, released by the pancreas and salivary glands, breaks down dietary carbohydrates into oligosaccharide fragments, which are then converted to monosaccharides by α-glucosidase. Once the glucose produced is in the bloodstream, it raises blood sugar levels.[26] As a result, inhibiting digestive enzymes can help control the progression of diabetes by slowing the rise in blood sugar. On α-glucosidase (77.59%) and α-amylase (77.41%), RVP-1 and RVP-2 exhibit a high inhibitory effect. The complex structures of the polysaccharides and their indigestibility can impair enzyme function and inhibit the absorption of glucose. Therefore, the structural composition and molecular weight distribution of RVP-1 and RVP-2 may be related to their ability to inhibit α-glucosidase and α-amylase. The research results imply that RVP-1 and RVP-2 could be potential inhibitors of α-glucosidase and α-amylase, expanding the range of applications of *R. virescens* in the food industry.[23]

14.2.5.2 *Anti-cancer Activity*

Both RVP-1 and RVP-2 significantly reduced MCF-7 cell proliferation and at 1200 μg ml^{-1} there was no discernible difference in the rate of tumor inhibition between RVP-2 treatment and the positive control. *Ganoderma applanatum* polysaccharides inhibited MCF-7 cell proliferation and migration in a time-dose dependent manner with a maximum inhibition rate of 50%, similar to RVP-1. The majority of anticancer polysaccharides were found to contain different monosaccharides with different molar ratios of rhamnose, mannose, xylose, ribose, and fucose.[27] Some of them contained proteins, uronic acids, sulfated form of sugars, and glucosamine with structures similar to RVP-1 and RVP-2. Cell cycle arrest, immunomodulation, mitochondrial membrane depolarization, and the nitric oxide pathway have all been identified as anticancer pathways.[28]

14.2.5.3 *Immune Activity*

As a macromolecule that macrophages can internalize into cells through endocytosis, neutral red can be used to measure the phagocytosis of macrophages.[29] During the process of engulfing invaders *via* surface receptor proteins, macrophages can produce tumor necrosis factor, reactive nitrogen mediators, and interleukin substance to promote platelet production and increase the differentiation and maturation of megakaryocytes in the bone marrow.[30] Parameters such as molecular weight, glycoside bonding, monosaccharide, and branching characteristics are related to polysaccharide immune activity. A fungal heteroglycan containing L-fucose,

D-galactose, and D-glucose has been shown to exhibit immunostimulatory and cytotoxic effects.[23]

14.2.6 Nutritional Value of *R. virescens*

Traditionally, medicines are used to cure diseases or to alleviate the symptoms of diseases. Nutrition, on the other hand, aims primarily to prevent disease by providing the body with the optimal balance of macro and micronutrients needed for good health. The nutritional composition obtained from *R. virescens* is as follows: moisture (92.49 g per 100 g), ash (11.04 g per 100 g), carbohydrates (62.27 g per 100 g), proteins (21.85 g per 100 g), fat (1.85 g per 100 g), energy (365 kcal per 100 g), mannitol (10.90 g per 100 g), trehalose (0.2 g per 100 g), and total sugars (11.10 g per 100 g).[17] The amino acid composition (mg kg^{-1}) of *R. virescens* is as follows: lysine (850), threonine (1350), valine (1310), isoleucine (1360), leucine (1660), methionine (890), tryptophan (400), and phenylalanine (1440).[31] Moreover, *R. virescens* contains macronutrients (mg per g) such as phosphorous (5.1), potassium (27.6), sulfur (5.3), calcium (0.1), magnesium (0.8), and micronutrients (mg kg^{-1}) such as iron (283), zinc (131), manganese (18.4), copper (41.3), boron (2.1), and selenium (12.6).[32]

14.3 *Russula delica* Fr.

Russula delica, often known as milky white brittlegill, is a fungus belonging to the genus Russula, to which all brittlegills belong. It has a strong trunk and is predominantly white with ocher or tan cap patterns. This fungus grows in conifers in deciduous or mixed forests and is edible but has a bad taste.

14.3.1 Distribution

Ectomyccorhizal on Sal. It is widely distributed in Gidhani and Jhargram; West Bengal and Chakrata; and Deoban and Naina peak in Uttarakhand.[2]

14.3.2 Morphology

Russula delica basidiocarps (fruiting bodies) seem reluctant to leave the ground and are often seen semi-buried or growing hypogeal. As a result, the rough surfaces of the caps often trap leaf debris and soil. The diameter of the plate can be up to 16 cm. It is white, usually tinged with ocher or brown, and has a curled edge that remains white most of the time.[33] It starts out convex, but then flattens out and becomes funnel-shaped. The gills are decurrent and relatively close together at first. The flesh is white and does not change color when punctured. In the initial stages, a fruity odor is detected, but on maturity this changes to a slightly unpleasant odor.

14.3.3 Bioactive Constituents

Fungi accumulate a variety of secondary metabolites including phenolic compounds, polyketides, terpenes and steroids. The antioxidant properties of phenolic compounds, particularly gallic acid, catechin, caffeic acid, rutin, quercetin, ellagic acid, and *p*-coumaric acid are well known. The quality of a dietary supplement depends on the chemical composition of the fruiting body, especially with regard to the particularly high content of phenols. It has been found that phenolic compounds exhibit antioxidant activity in inhibiting the oxidation of low density lipoproteins (LDL). Phenols are one of the major groups of nonessential dietary components that have been implicated in inhibiting atherosclerosis and cancer.[33] Almost all organisms are well protected from free radical damage by enzymes, such as superoxide dismutase and catalase, or compounds such as ascorbic acid, tocopherols, and glutathione. When the mechanism of antioxidant protection is thrown off balance by factors such as aging, physiological functions can deteriorate, leading to disease and accelerated aging. Mushrooms have a variety of nutritional and nutraceutical properties and are a good source of bioactive compounds. The raw content of biochemical components of *R. delica* is shown in Figure 14.6.[33]

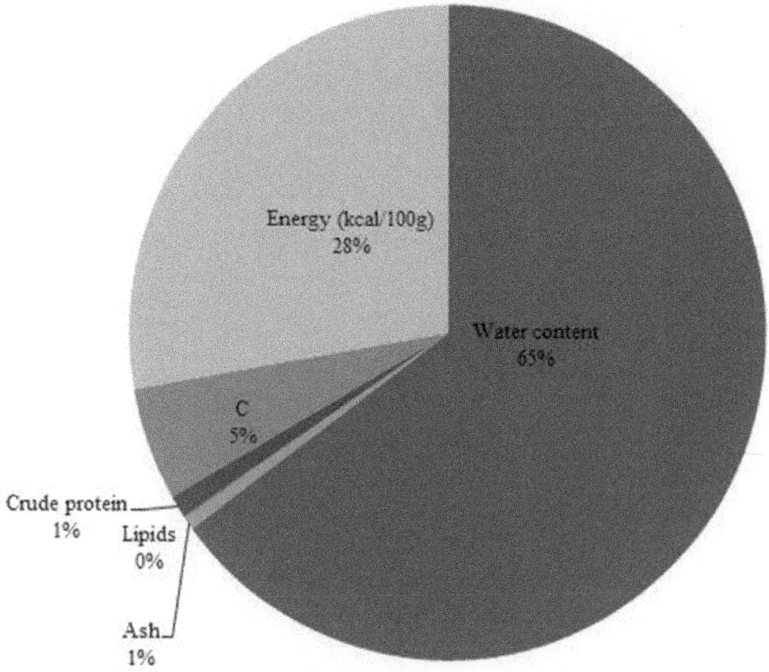

Figure 14.6 Representation of the biochemical components of *R. delica*.

14.3.4 Fatty Acids

Table 14.3 shows the fatty acid content of the *R. delica*, which includes total saturated fatty acids (SFAs), monounsaturated fatty acids (MUFAs), polyunsaturated fatty acids (PUFAs), and *trans*, n-3, and n-6 fatty acids.

14.3.5 Sterols

Ergosterol is a substance that makes up the cell membranes of yeast and fungi and serves the same purpose as cholesterol in animal cells, although it is not present in either plant or animal cells. Lanosterol and zymosterol have been implicated in the ergosterol synthesis process (Figure 14.7). The amount of ergosterol in *R. delica* was determined to be 12.51 mg per 100 g of body weight.[33]

14.3.6 Total Polyphenols and Terpenic Acid

Catechin belongs to the interesting class of flavonoids found ubiquitously, which have versatile biological effects, such as anticancer, anti-allergy and antioxidant properties. Rutin, a natural flavone derivative, quercetin-3-rhamnosylglucoside, is an important anti-lipoperoxidant that has also been shown to be a potent scavenger of hydroxyl and superoxide radicals.[33] Gallic acid (GA) is a simple phenolic acid that features a single aromatic ring, the derivatives of which are biologically active compounds widely distributed in

Table 14.3 Fatty acid composition (%) of *Russula delica*.[31]

Fatty acid	*R. delica*
C10:0	0.14
C11:0	0.17
C12:0	0.07
C13:0	0.01
C14:0	0.13
C15:0	0.31
C16:0	7.22
C16:1n9	0.08
C16:1n7	0.45
C17:0	0.07
C18:0	27.10
C18:1 *trans*	0.07
C18:1n9	13.66
C18:2n6	44.32
C18:3n3	2.50
C20:0	0.02
C20:1n9	0.08
SFA	35.87
MUFA	14.26
PUFA	47.19

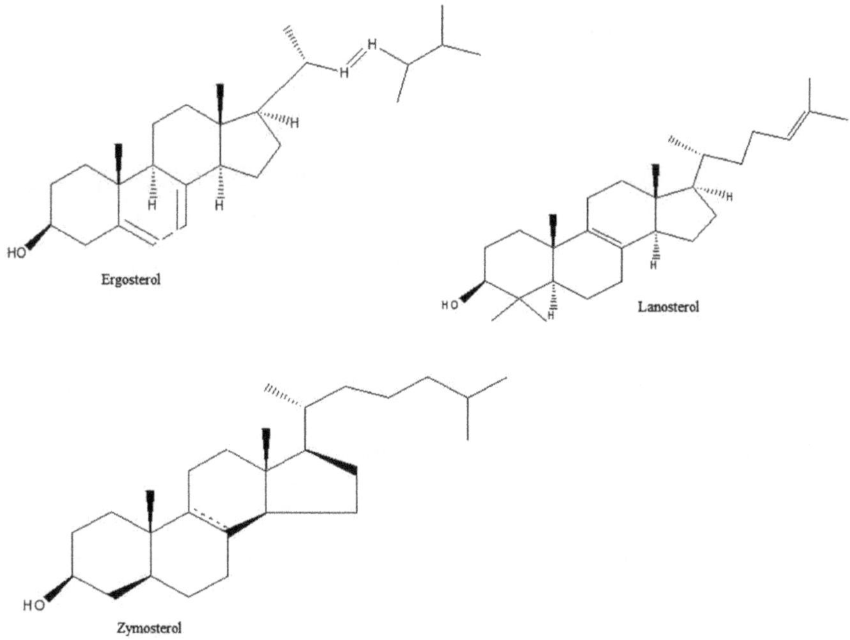

Figure 14.7 Chemical structures of *R. delica* sterols.

plants. GA is a powerful natural antioxidant and a strong chelating agent that forms highly stable complexes with Fe^{3+}. It has previously been reported to possess anti-inflammatory, antitumor, antibacterial, and antifungal properties. Catechin (5.33 $mg\,L^{-1}$) is the main component found in the ethanolic extract of *R. delica*.[33] *R. delica* extracts have been found to contain polyphenols, stilbene, resveratrol, and terpenic acids, as shown in Table 14.4 and Figure 14.8.

14.3.7 Proximate and Nutritional Composition of *R. delica*

The chemical composition (%) of the *R. delica* fungus has been determined as follows: moisture, dry matter, crude fiber, proteins, fat, carbohydrates, ash, total sugar, reducing sugars, and energy, as summarized in Figure 14.9. The carbohydrates are catabolized during mushroom growth for energy production. The protein content indicates that protein synthesis is inherent in mushroom maturity.[34] Consuming mushrooms after ripening could meet the daily soluble fiber requirements of humans, which is estimated at 25–30 g, and would help reduce the risk of hypertension, constipation, diabetes, colon cancer, and breast cancer, as soluble fiber increases absorption and decreases cholesterol and glucose in the blood.[34]

Due to their high water content and low calorific value, mushrooms are considered to be a dietary food suitable for a low-calorie diet. Minerals such as potassium and calcium, which play important roles in the human body,

Table 14.4 Phenolic compounds and terpenic acids in *Russula delica*.[33]

Phenolic acids (μg per 100 g fw)	Hydroxybenzoic acids	
	p-OH-Phenylacetic acid	0.5
	Syringic acid	1.3
	Vanillic acid	0.4
	Hydroxycinnamic acids	
	Caffeic acid	0.2
	Cinnamic acid	0.8
	Chlorogenic acid	3.2
	Ferulic acid	2.3
	o-Coumaric acid	6
	p-Coumaric acid	1.8
	Tyrosol	0.2
Phenols (μg per 100 g fw)	Vanillin	0.6
	Chrysin	46.3
Flavonoids (μg per 100 g fw)	Quercitin	0.6
	Genistein	0.1
	Oleanolic acid	9.3
Terpenic acids (μg per 100 g fw)	Ursolic acid	0.3

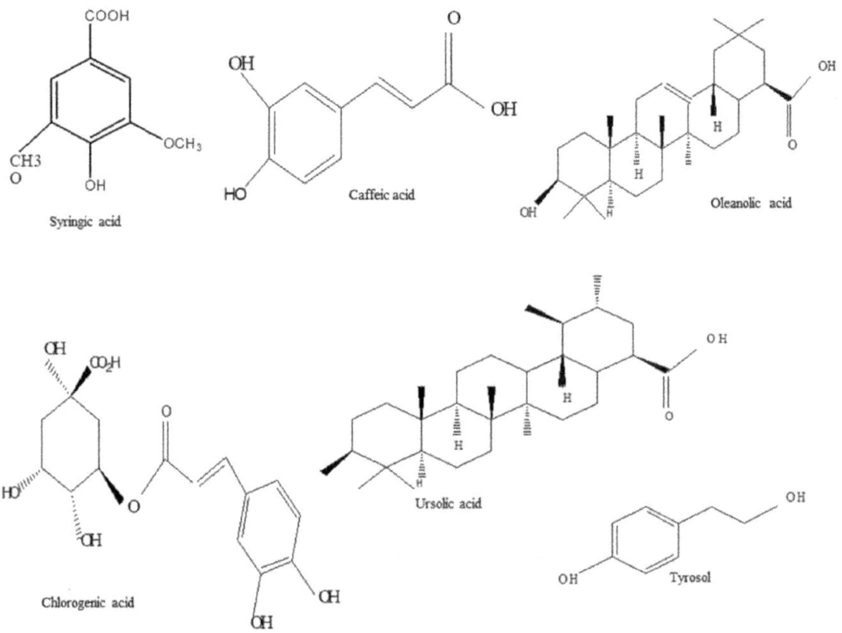

Figure 14.8 Phenolic compounds of *R. delica.*

are found in high concentrations in *R. delica*, in addition to phosphorus and magnesium. However, fungal species contain a relatively low amount of sodium; therefore, mushrooms are said to be good for patients with high

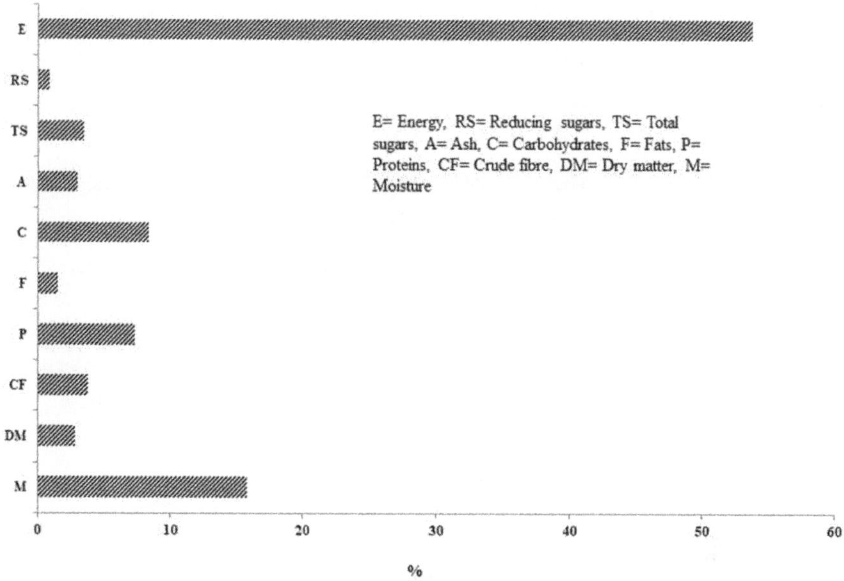

Figure 14.9 Proximate and chemical composition of *R. delica.*

blood pressure. The mineral compositions (mg per 100 g) of the *R. delica* fungus are: magnesium (27.16–34.47), potassium (76.8–92.87), phosphorous (434.24–514.06), sodium (34.95–47.89), calcium (353.31–381.09), manganese (4.69–5.04), iron (20.08–57.69), copper (1.38–2.82), and zinc (7.32–12.56), as shown in Figure 14.10.[34]

14.3.8 Pharmacology

14.3.8.1 Anti-microbial and Anti-oxidant Activity

It has been discovered that the oxidation of low-density lipoproteins (LDL) is inhibited by phenolic chemicals, which have antioxidant effects. The anti-microbial activity of an ethanolic extract of *R. delica* was tested against Gram-positive and Gram-negative bacteria. *Shigella sonnei* RSKK 8177 exhibited the highest inhibitory activity (171 mm), followed by *Bacillus cereus* RSKK 867 and *Listeria monocytogenes* ATCC 7644 (131 and 111 mm). *Pseudomonas aeruginosa* ATCC 29212 and *Salomonella enteritidis* 171 (71 and 70 mm) exhibited the worst inhibitory effect. The ethanolic extract of *R. delica* was active against several bacteria. In addition, the ethanolic extract of *R. delica* could represent a new natural antioxidant and antibacterial source.[35] The antimicrobial effect of ethanol extracts of *R. delica* were tested against various microorganisms (Table 14.5). As can be seen from the results, *R. delica* exhibits a narrow antibacterial spectrum against microorganisms. The test extract showed stronger activity against Gram-positive than against Gram-negative bacteria.

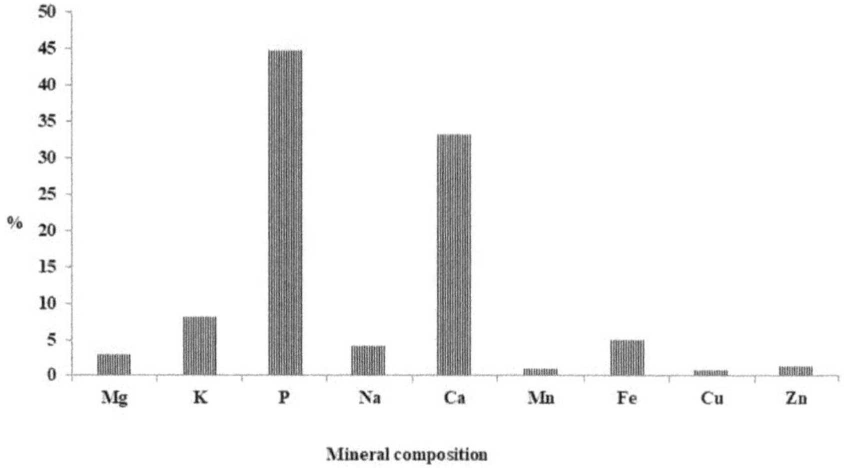

Figure 14.10 Mineral composition of *R. delica.*

The antimicrobial activities were comparable to those of commonly used antibiotics and antifungal agents against these microorganisms.[35] In general, the reference discs were more active than the ethanol extract of *R. delica.* However, over time, microorganisms become resistant to antibiotics. Therefore, *R. delica* can be used as a source of antimicrobial agents.[1]

14.3.8.2 Antiproliferative Activity

The antiproliferative effect of *R. delica* RNase is remarkable since most other fungal RNases have yet to be tested. However, some fungal ribonucleases are antiproliferative, with antifungal and HIV-1 reverse transcriptase inhibitory properties. RNase from *R. delica* has only an antiproliferative effect and none of the other activities. In addition, it lacks inhibitory effects against HIV-1 integrase and SARS coronavirus protease. This could allow the isolated ribonuclease to act as a defense protein and hydrolyze the RNA of invading pathogens. *R. delica* RNase has a different N-terminal sequence, high specific RNase activity, and antiproliferative effect against tumor cells.[36]

R. delica lectin exerts a strong antiproliferative effect on HepG2 and MCF7 cells with IC_{50} values of 0.88 M and 0.52 M, respectively. This activity is extremely strong, with IC_{50} values at nanomolar concentrations. *R. delica* lectin has significant inhibitory activity against HIV-1 reverse transcriptase with an IC_{50} value of 0.26 M. The inhibitory potency ($<$1 M) is high compared to 'proteinaceous' and 'non-proteinaceous' anti-HIV natural products. It remains to be seen whether lectin from *R. delica* can be used in the areas of cancer therapy, AIDS, and immune deficiency. The properties of *R. delica* lectin are shown in Figure 14.11. *R. delica* lectin has unique properties that infer its potential application in antitumor therapy.[35]

Table 14.5 Antimicrobial activity of the ethyl alcohol extract of *R. delica* and antibiotic sensitivity of microorganisms (zone size, mm).[1]

Microorganisms	*R. delica*	Nystatin (100 U)	Penicillin (10 U)	Ampicillin (10 μg)	Tetracycline (30 μg)	Oxacillin (1 μg)
Candida albicans	8	19	NT	NT	NT	NT
Bacillus cereus	9	NT	22	NT	NT	14
Micrococcus flavus	20		31	29	20	24
Staphylococcus aureus	8	NT	28	NT	21	18
Escherichia coli	—	NT	11	10	8	NT
Salmonella enteritidis	5	NT	NT	—	12	NT

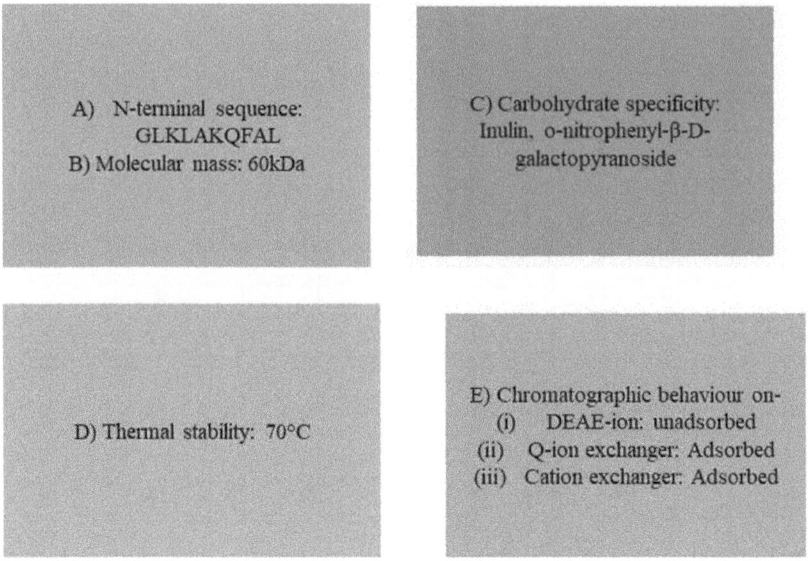

Figure 14.11 Characteristics of *R. delica* lectin.

14.4 *Russula cyanoxantha* (Charcoal Burner Mushroom)

A host plant, which is typically a deciduous tree, is required for the ectomycorrhizal fungi cyanoxantha to survive. Although it is a very tasty and high quality edible mushroom, the difficulties associated with its cultivation prevent it from being widely commercialized. This fungus shares many characteristics with other Russula species found nearby, including *R. heterophylla* and *R. virescens*. *R. cyanoxantha* can be distinguished by its flexible and soft gills and its pure white stems, devoid of brownish or yellow spots.[37]

14.4.1 Distribution

R. cyanoxantha is found in Gidhanisal forest in West Bengal, Chakrata, Kailana, Joshimath, Oli Forest, Ramgarh, Lohaghat, Champawat, Uttarakhand's Dafia Dhura, and Maitly.[38]

14.4.2 Morphology

Monopodial-pinnate mycorrhizal systems can be up to 6 mm long, with a major axis diameter of 0.4–0.5 mm, ocher to light brown in color, and with unbranched ends that are straight to slightly curved and up to 2 mm long and 0.3–0.4 mm in diameter. Rhizomorphs are uncommon, mostly found along the thicker roots of the host tree. Rhizomorphs are pressed to the root with their flattened undersides;, semicircular in cross-section, up to 0.1 mm thick and 0.2 mm wide, branched and reticulate in appearance, with thin bundles of hyphae connecting to the ectomycorrhiza along the colonized root; the surface of the rhizomorph resembles that of the mantle.[39]

14.4.3 Bioactive Compounds

R. cyanoxantha generates ergosta-4,6,8(14),22-tetraen-3-one (ergone), a bioactive steroid, shown in Figure 14.12, which exhibits antiproliferative and cytotoxic properties toward HepG2 cells. The chemical underpinnings of the cytotoxic action of ergone have been investigated by researchers.[40] Other bioactive compounds, such as ceramide, inosine, adenine, L-pyroglutamic acid, fumaric acid, and D-allitol, have been discovered in the ethanol and chloroform/methanol extracts of the fruiting bodies of the basidiomycete *R. cyanoxanotha* (Schaeff.) Fr. When HepG2 cells were treated with ergone, the following apoptotic markers were observed: nuclear fragmentation, G2/M cell cycle arrest, phosphatidylserine exposure, and chromatin condensation. Other apoptotic processes discovered included Bcl-2 downregulation and Bax upregulation, as well as PARP breakage and activation of 3,8,9 caspases. These findings have drawn more attention to the use of medicinal mushrooms in cancer treatment.[41]

The bioactive potential of *R. cyanxantha* is presented in Table 14.6. Mushroom extracts interfere with the formation of ferrous and ferrozine complexes, suggesting that they have metal-chelating activity and bind to ferrous ions before ferrozine. However, metal chelating activity is also important in the case of lipid peroxidation, where ion chelating agents promote the Fenton reaction and accelerate peroxidation by decomposing lipid hydroperoxides into peroxyl and alkoxyl radicals that can remove

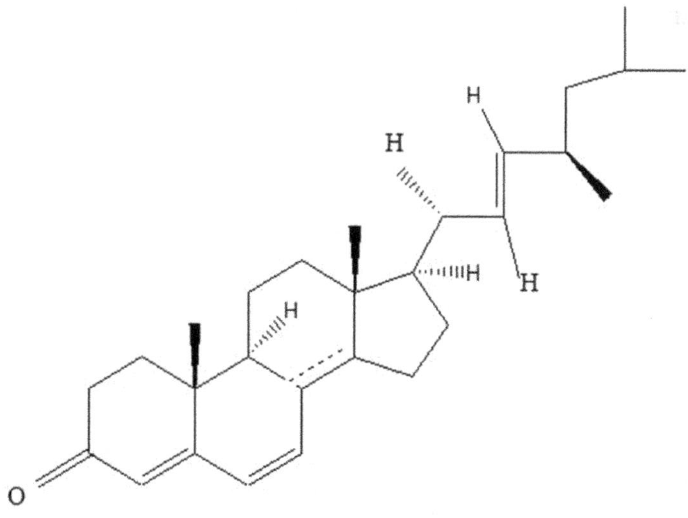

Figure 14.12 Bioactive compounds of *R. cyanoxantha.*

Table 14.6 Bioactive potentials of *Russula cyanoxantha* using three different solvent extracts (IC_{50} values in $\mu g\,ml^{-1}$).[41]

R. cyanoxantha extract	ABTS	DPPH	Metal chelating	SOD	α-Glucosidase	α-Amylase	Albumin denaturation
Ethanol	73	866	461	627	140	193	575
Chloroform	104	>1000	442	>1000	133	186	869
Water	102	>1000	125	>1000	200	97	146

hydrogen and continue the chain reaction of lipid peroxidation.[41] Therefore, the studied fungus can potentially be used as a natural metal chelator to protect against oxidative damage. The superoxide anion radical ($O^{2\bullet-}$ is one of the reactive oxygen species among the main radicals that cause oxidative damage in the human body through lipid oxidation. $O^{2\bullet-}$ scavenging activity is assessed photochemically by the riboflavin–nitro blue tetrazolium (NBT) light system, where riboflavins are reduced to generate $O^{2\bullet-}$, which then reduces NBT, resulting in the formation of blue formazan. The addition of mushroom extracts to the reaction mixture resulted in scavenging of $O^{2\bullet-}$, his resulted in the inhibition of NBT reduction and the formation of blue formazan.[41]

14.4.4 Pharmacology

14.4.4.1 Anti-diabetic Activity

An unexpected rise or fall in blood sugar levels over time is characteristic of the chronic condition of diabetes. Extracts from three different Russula species were shown to inhibit the α-glucosidase enzyme in a dose-dependent manner. It was found that as the extract concentration increased, the α-glucosidase inhibition gradually increased. Reducing post prandial glucose levels by inhibiting carbohydrate-hydrolyzing enzymes such as amylase and glucosidase, which can delay glucose absorption in the pancreas and limit glucose absorption in the small intestine, is the most efficient enzymatic treatment for type 2 diabetes.[42]

14.4.4.2 Anti-oxidant Activity

An acetone extract of *R. cyanoxantha* exhibited the highest DPPH scavenging activity ($IC_{50} = 86.272\ g\,mL^{-1}$) and was almost as potent as 'α-tocopherol'. Methanol extracts from the examined fungi showed lower DPPH radical scavenging activity than acetone extracts. For *R. cyanoxantha*, the IC_{50} for methanol extracts was 262.08 $g\,mL^{-1}$. An increasing trend in antioxidant activity was observed in the ABTS scavenger activity of the mushroom extracts. The suppression of ABTS radicals increased with concentration in all three solvent extractants (100–400 $g\,mL^{-1}$) (water, ethanol, and chloroform). The greatest scavenger effect was observed for the ethanol extract of *R. cyanoxantha*, which exhibited an IC_{50} of 73 $g\,mL^{-1}$. The maximum metal chelating activity was observed in aqueous extracts of *R. cyanoxantha*, which exhibited an IC_{50} value of 981 $g\,mL^{-1}$. Ion-chelating chemicals enhance the Fenton reaction and accelerate peroxidation by cleaving lipid hydroperoxides into peroxyl and alkoxyl radicals, which can remove hydrogen and continue the lipid peroxidation chain reaction. As a result, the studied fungi could be used to reduce oxidative damage by acting as natural metal chelators.[42]

14.4.4.3 Anti-microbial Activity

The antibacterial efficacy of water and ethanolic extracts of *R. cytoxantha* against bacterial and fungal infections has been evaluated. Significant inhibitory changes against four bacterial pathogens, *Pseudomonas aeruginosa*, *Bacillus subtilis*, *Staphylococcus aureus*, and *Escherichia coli*, were observed, with zones of inhibition in the range of 16.8–25.50 mm, whereas the conventional antibiotic ampicillin shows zones ranging from 15.8 to 32.8 mm against fungal pathogens. All pathogens examined showed lower resistance to chloroform extracts of *R. cytoxantha*. The phenolic components of the extracts could be the reason for their antibacterial effect.[42]

14.5 Conclusion

Fungi appear to be good natural sources of antioxidants and may be of importance in the therapy of human, animal, and plant diseases. High levels of protein, carbohydrates, essential minerals and low energy content make many wild mushrooms a good food for the consumer, comparable to meat, eggs and milk. The potential medicinal value of mushrooms is also high, including strengthening the immune system, controlling blood lipids, anti-tumor function, and so on. It is desirable to conduct further studies on the isolation and characterization of new compounds with antioxidant and antibacterial activity derived from fungi. Although these bioactive components of edible mushrooms are critical to human health, there is also an urgent need to test such compounds against emerging diseases, such as coronavirus (COVID-19). Further research is needed on the nano-biofortification of various fungi, their bioactive compounds and therapeutic properties, in addition to the health benefits of edible mushrooms for a healthy lifestyle.

Acknowledgements

I would like to thank my family for always giving me full support in whatever ventures I undertake.

References

1. A. Turkoglu and M. E. Duru, *Eurasian J. Anal. Chem.*, 2007, **2**, 54–67.
2. R. K. Verma, V. Pandro and A. Pyasi, *Int. J. Curr. Microbiol. Appl. Sci.*, 2018, 7, 3078–3103.
3. S. S. Saini, N. S. Atri and M. Bhupal, *Indian Phytopathol.*, 1988, **41**, 622–625.
4. Z. W. Sun, L. X. Zhang and T. G. Niu, *Food Chem.*, 2010, **118**, 675–680.
5. A. Chandrasrikul, P. Suwanarit, U. Sangwanit, S. Lumyong, A. Payapanon, N. Sanoamuang, C. Pukahuta, V. Petcharat, U. Sardsud and K. Duengkae *et al.*, *Checklist of Mushrooms (Basidiomycetes) in Thailand*, Office of

Natural Resources and Environmental Policy and Planning, Bangkok, Thailand, 2011.
6. M. E. Valverde, T. H. Perez and O. P. Lopez, *Int. J. Microbiol.*, 2015, **2015**, 376387.
7. A. Keles, I. Koca and H. Genccelep, *J. Food Process. Technol.*, 2011, **2**, 130.
8. F. Cateni, M. L. Gargano, G. Procida, G. Venturella, F. Cirlincione and V. Ferraro, *Phytochem. Rev.*, 2022, **21**, 339–389.
9. Z. Sun, Y. He, Z. Liang, W. Zhou and T. Niu, *Carbohydr. Polym.*, 2009, **77**, 628–633.
10. J. Tang, H. Shao and J. Liu, *Chin. Tradit. Herb. Drugs*, 2008, **39**, 1776–1778.
11. X. H. Chen, H. B. Zhou and G. Z. Qiu, *Asian J. Chem.*, 2010, **22**, 6867–6878.
12. P. Cosme, A. B. Rodriguez, J. Espino and M. Garrido, *Antioxidants*, 2020, **9**, 1263.
13. J. S. Beckman and W. H. Koppenol, *Am. J. Physiol.*, 1996, **271**, 1424–1437.
14. D. J. Choi, S. J. Lee, M. J. Kang, H. S. Cho, N. J. Sung, J. H. Shin and J. Korean, *Soc. Food Sci. Nutr.*, 2008, **37**, 465–471.
15. A. Kumar and S. Chattopadhya, *Food Chem.*, 2007, **100**, 1377–1384.
16. A. Lovegrove, C. H. Edwards, I. D. Noni, H. Patel, S. N. El, T. Grassby, C. Zielke, M. Ulmius, L. Nilsson, P. J. Butterworth, P. R. Ellis and P. R. Shewry, *Crit. Rev. Food Sci. Nutr.*, 2017, **57**, 237–253.
17. A. R. Leal, L. Barros, J. C. M. Barreira, M. J. Sousa, A. Martins, C. S. Buelga and I. C. F. R. Ferreira, *Food Res. Int.*, 2013, **50**, 1–9.
18. G. H. Wu, C. L. Lu, J. G. Jiang, Z. Y. Li and Z. L. Huang, *Food Funct.*, 2014, **5**, 337–344.
19. Y. X. Sun, J. C. Liu, X. D. Yang and J. F. Kennedy, *Process Biochem.*, 2010, **45**, 874–879.
20. H. Li, X. Wang, Q. Xiong, Y. Yu and L. Peng, *Int. J. Biol. Macromol.*, 2020, **154**, 1438–1447.
21. Y. M. Li, R. Zhong, J. Chen and Z. G. Luo, *Int. J. Biol. Macromol.*, 2021, **184**, 380–392.
22. Y. K. Leong, F. C. Yang and J. S. Chang, *Carbohydr. Polym.*, 2021, **251**, 117006.
23. S. Bisht, R. Kant and V. Kumar, *Int. J. Biol. Macromol.*, 2013, **59**, 214–220.
24. R. Khursheed, S. K. Singh, S. Wadhwa, M. Gulati, A. Awasthi, R. Kumar, A. K. Ramanunny, B. Kapoor, P. Kumar and L. Corrie, *Carbohydr. Polym.*, 2020, **250**, 116996.
25. X. Hanyu, L. Lanyue, D. Miao, F. Wentao, C. Cangran and S. Hui, *Int. J. Biol. Macromol.*, 2020, **146**, 353–362.
26. Q. Ma, J. G. Jiang, X. Yuan, K. Qiu and W. Zhu, *Food Chem. Toxicol.*, 2019, **125**, 422–429.
27. T. Khan, A. Date, H. Chawda and K. Patel, *Carbohydr. Polym.*, 2019, **210**, 412–428.
28. L. Huang, M. Shen, G. A. Morris and J. Xie, *Trends Food Sci. Technol.*, 2019, **92**, 1–11.

29. P. Mahdavi Sharif, P. Jabbari, S. Razi, M. Keshavarz-Fathi and N. Rezaei, *Cytokine*, 2020, **130**, 155066.
30. N. Kalogerpoulos, A. Yanni, K. Georgios and M. Aloupi, *Food Chem. Toxicol.*, 2013, **55**, 378–385.
31. X. M. Wang, J. Zhang, L. H. Wu, Y. L. Zhao, T. Li, J. Q. Li, Y. Z. Wang and H. G. Liu, *Food Chem.*, 2014, **151**, 279–285.
32. R. Sanmee, B. Dell, P. Lumyong, K. Izumori and S. Lumyong, *Food Chem.*, 2003, **82**, 527–532.
33. T. Yaltirak, B. Aslim, S. Ozuturk and H. Alli, *Food Chem. Toxicol.*, 2009, **47**, 2052–2056.
34. J. G. Oscar, A. K. Kouame, H. K. Kouassi and J. P. K. Eugene, *World J. Adv. Res. Rev.*, 2019, **2**, 21–30.
35. C. Y. Zhao, H. S. Li, Q. G. Zhang, X. H. Wang and B. T. Ng, *J. Microbiol. Biotechnol.*, 2010, **20**, 693–699.
36. A. R. Zsigmonda, I. Kantora, Z. May, I. Urak and K. Heberger, *Chemosphere*, 2020, **238**, 124566.
37. K. Das and J. R. Sharma, Botanical Survey of India, Ministry of Environment and Forests, Govt. of India, Kolkata, 2005, p. 255.
38. R. Agerer, R. M. Danielson, S. Egli, K. Ingleby, D. Luoma and R. Treu, *Descriptions of Ectomycorrhizae*, 2001, vol. 5, pp. 131–137.
39. Y. Y. Zhao, X. Shen, X. Chao, C. C. Ho, X. L. Cheng, Y. Zhang, R. C. Lin, K. J. Du, W. J. Luo, J. Y. Chen and W. J. Sun, *Biochem. Biophys. Acta*, 2011, **1810**, 384–390.
40. K. Thorat, L. Patil, D. Limaya and V. Kadam, *Int. J. Res. Pharm. Biomed. Sci.*, 2012, **3**, 730–732.
41. M. K. Panda, S. K. Das, S. Mohapatra, P. R. Debata, K. Tayung and H. Thatoi, *Plant Biosyst.*, 2021, **155**, 1041–1055.
42. M. Kosanic, B. Rankovic and M. Dasic, *Iran. J. Pharm. Res.*, 2012, **11**, 1095–1102.

CHAPTER 15

Stropharia rugosoannulata *Farlow ex Murrill (Wine Cap Stropharia or Garden Giant)*

KANIKA DULTA,*[a] BANLAMBHABOK KHONGTHAW,[b] SOMVIR SINGH,[c] ARTI THAKUR,[d] GÖZDE KOŞARSOY AĞÇELI[e] AND KESHAV THAKUR[f]

[a] Department of Food Technology, School of Applied and Life Sciences, Uttaranchal University, Dehradun-248007, Uttarakhand, India; [b] Faculty of Applied Sciences and Biotechnology, Shoolini University, Solan-173229, Himachal Pradesh, India; [c] Faculty of Sciences, Shoolini University, Solan-173229, Himachal Pradesh, India; [d] Department of Botany, Shoolini Institute of Life Sciences and Business Management (SILB), Solan, Himachal Pradesh, India; [e] Department of Biology/Biotechnology, Hacettepe University, Ankara, Turkey; [f] Department of Biotechnology, Dr YS Parmar University of Horticulture and Forestry, Nauni, Solan, HP, India
*Email: kanikadulta10@gmail.com

15.1 Introduction

Mushrooms are fruiting bodies produced by basidiomycetes and ascomycetes. They generate spores that mature and grow into mycelium. The mycelium finally generates a primordium, which develops into a novel complete mushroom, and the life cycle is continued. Mushrooms are essential in traditional medicine due to their therapeutic properties, in addition to their long history as a food source. Furthermore, the global population of 14 000 species of mushrooms serves as a rich source of

Phytochemistry and Nutritional Composition of Significant Wild Medicinal and Edible Mushrooms: Traditional Uses and Pharmacology
Edited by Ajay Sharma, Garima Bhardwaj and Gulzar Ahmad Nayik

Published by the Royal Society of Chemistry, www.rsc.org

bioactive chemicals.[1] Mushrooms are eukaryotic heterotrophic organisms that belong to the kingdom of fungi. Due to their nutritional and medicinal benefits, as well as their distinct flavor and texture, mushrooms have been ingested for over 2000 years.[2] Mushrooms are considered to be healthy foods because of their high carbohydrate, protein, vitamin, mineral, fiber, and low lipid content.[3] Mushrooms have been found to contain a variety of useful components that are linked to their bioactivity, including anticancer, antidiabetic, antioxidant, and antiviral properties.[4] The majority of the active compounds found in mushrooms are polysaccharides, which are reported to possess several biological functions, including immunomodulatory, antitumor, anticancer, and antioxidant functions.[5] There are a variety of mushroom polysaccharides used in various Asian countries as pharmaceuticals.[6] A growing number of researchers are experimenting with different edible mushrooms to consume them as natural antioxidants in the form of medicines and foods.[7]

Stropharia rugosoannulata (*S. rugosoannulata*) mushrooms are called wine-cap *Stropharia* in English and saketsubatake in Japanese and belong to the family *Strophariaceae*. *S. rugosoannulata* (Farlow apud Murrill) was found in Germany, Czechoslovakia, and Japan in the 1930s after being first described in the United States in 1922.

In terms of commercial non-mycorrhizal mushrooms, wine-cap Stropharia is one of the most recently cultivated. The initial domesticators of Stropharia were in Germany in the 1960s and used it for home gardening and small-scale commercial production.[8] Stropharia farming also continued to be successful using cereal straw in woody vegetable frames. According to reports, yields were high. By 1989, Europe's marketable *Stropharia* production had risen to about 1300 tons annually.[9] *S. rugosoannulata* wine cap mushroom is a higher Basidiomycete renowned for its stimulating gentle fragrance, crunchy, pleasant, and somewhat sweet taste, and is cultivated in Asia as it is a popular culinary mushroom that is also used for medicinal purposes.[10] *S. rugosoannulata* has anticipatory properties towards hyperglycemia, coronary heart diseases, ascites cancer, solid tumor S-180 cells, *etc.*[11] Compounds with a high nutritional value, such as crude fat, crude protein, minerals, vitamins, amino acids, and bioactive compounds with anticarcinogenic, antidiabetic, and antioxidant properties have been identified in Stropharia.[12] Recently, the intake of *S. rugosoannulata* has increased substantially due to its pharmacological activity and delightful flavor. *S. rugosoannulata* has been found to exhibit anti-endoplasmic reticulum (ER) stress, hemagglutinating, and osteoclast formation inhibiting effects, as well as antioxidant properties.[13]

The UN Food and Agriculture Organization recommends the export of wine-cap mushrooms to developing countries; therefore, it is one of the top ten mushrooms sold worldwide.[14] Stropharia cultivation regimes on forestland produce consistent increases in soil organic matter (OM) and available phosphorus (AP) and different raw materials can be used to grow these fungi, including tree branches, corn cobs, sawdust, straw, and rice husks.[15] They are simple to grow and can produce a high yield with careful management. Besides these characteristics, *S. rugosoannulata* is well-suited to under-forest cultivation. Cultural practices for Wine Cap have been performed in huge gardens with shrubs and trees[9] and under hardwood shade.[16] There have been numerous experiments conducted to increase mushroom production[17] that have boosted the large-scale production of *S. rugosoannulata*.

15.2 Taxonomy

15.2.1 Stropharia

Agaricales is a fungal order that includes mushroom-forming fungi or eu-agarics (Basidiomycota, Agaricomycetidae) and comprises more than half of all known homobasidiomycete species.[18] There are a further 26 families, 350 genera, and over 9000 species in this order.[19] *Nidulariaceae*, *Agaricaceae*, *Crepidotaceae*, *Cystodermateae*, *Panaeoleae*, *Cortinariaceae*, *Gymnopileae*, *Hymenogastraceae*, *Bolbitiaceae*, *Tubarieae*, *Hydnangiaceae*, *Inocybaceae*, and *Strophariaceae* are all members of the Agaricoid clade that are documented as smaller monophyletic groups.[20] Stropharia is a genus in the *Strophariaceae* family discovered by Fries.[21] It was later raised to generic status by Quelet,[22] as well as because only morphological structures were needed to execute classifications, so *Stropharia* was further separated into infrageneric groups after being recognized as a genus. *Stropharia* is derived from the Greek word strophos and refers to the stem circles of the fungi[12] generated by this taxonomic grouping. Although there is a paucity of information on this fungus genus, in the past years, the research on *Stropharia* has increased due to its application in health[23] and agriculture.[24] In spite of its significant applications in agriculture and health, evidence is scarce about its isolation and agro-ecologically valuable usages.

15.2.2 Stropharia Taxonomy

Singer and Smith presented the description of family,[20] and also identified two subfamilies: Stropharioideae Singer and Pholiotoideae Imai.[25] In northern temperature zones, the *Strophariaceae* family is widely distributed;[12] they span open fields, forests, plains, gardens, and the edges of forests on five continents. Fungi in this family act as decomposers in nature, recycling organic material, despite some parasitic representatives. According to the literature,[20] no *ectomycorrhizae* are formed by *Strophariaceae* fungi. Chrysocystis is a gloeocystis, a type of cystitis that is not found in species of Psilocybe, which may be unique to species of Stropharia of the subfamily Stropharioideae. There is an oily and amorphous aspect to this cystitis, which makes it easier to observe alkaline substances with variations in coloration. A different study was performed on cystitis by classifying it into two "subtypes". In Panaeolus, they are represented by the type "semiovatus" and Strophariaceae by the type "fasciculare". Stropharia is recognized by its membranous to glutinous annulus, which is followed by a purplish brown to violaceous spore print. The walls of the basidiospores are smooth and thicker, they contain a well-developed germ pore; chrysocystidia that are prostrate, gelatinized filamentous hyphae in the pileipellis, and there is no cellular hypodermium. Stropharia has been recommended as a replacement for Pholiotideae,[20,26] in which species were identified as having subviscid, viscid, and sticky pileus.[27] Molecular phylogenetic studies have also shed new light on the group's evolution and classification, such as analyses using ribosomal RNA, which have revealed that Strophariaceae belongs to a monophyletic clade.[28] Also, a study was carried out[29] to create a molecular technique for identifying Strophariaceae species using the polymerase chain reaction (PCR) and determining particular primers by which to identify Strophariaceae species (nested PCR). The application of molecular technologies and morphological features in this way can provide fresh insight into the categorization and identification of this fungus group. A generic attribute is used to identify species belonging to the Stropharia genus. Acanthocytes are a type of hyphae that form on the mycelium and rhizomorph.[30] These structures were identified at the first time by Pacioni and Lincoff.[31] Stropharia-exclusive hyphae with spinescent and crystalline projections, composed of calcium crystals that have thin walls and are hollow and fragile, are primarily found in basidium rhizomorphs. This trait has been used recently by a number of researchers to distinguish Stropharia taxa.[32]

15.2.3 *Stropharia rugosoannulata*

Stropharia rugosoannulata can reach a height of around 20 cm, and its reddish-brown, curved to flattened cap can measure up to 30 cm across, giving it the alternative slang moniker of "Godzilla mushroom" because of its size. The flesh of this fungus is firm and white, and its gills are initially pale, grey, and dark purple-brown (Figure 15.1).[32]

15.3 Scientific Classification

Kingdom: Fungi
Division: Basidiomycota
Class: Agaricomycetes
Order: Agaricales
Family: Strophariaceae
Genus: Stropharia
Species: *rugosoannulata*

15.3.1 Resource Availability

Stropharia rugosoannulata, a member of the Strophariaceae family, is widespread in northern climatic regions and is grown as a crop due to its relatively high nutritional value and pleasant flavor and has been recommended by The Food and Agriculture Organization of the United Nations (FAO) for farming in

Figure 15.1 *S. rugosoannulata* (young wine-cap stropharia). Photograph included with permission from the owners, https://www.messiah.edu/Oakes/fungi_on_wood/gilled%20fungi/species%20pages/Stropharia%20rugosoannulata.htm.

developing countries.[33] There has recently been a rise in the popularity of *S. rugosoannulata* in China as an edible mushroom due to it nutrient content and pharmacologically activity.[34] *Stropharia rugosoannulata* can be cultivated using various types of natural materials, such as sawdust, straw, corncob rice, and husks, which are all stress-free to use.[15] The Chinese government is now encouraging mushroom cultivation in forest lands because it makes efficient use of the vast spaces beneath the trees. As a result, *S. rugosoannulata* has been farmed in forests in several Chinese provinces, including Shandong, Fujian, Yunnan, and Henan.[35] The bamboo forest in Bo'ai County, Henan Province, has been cultivated for a thousand years and is renowned as the "living fossil of an artificial bamboo forest in northern China." The people who live here have made an effort to establish the *S. rugosoannulata* cultivation industry based on the plentiful bamboo resources in the region and have attained great financial and environmental benefits. A limited amount of information has been reported on the nutritional value and pharmacological properties of *S. rugosoannulata*, which grows beneath bamboo forests.[36]

In China, mushroom cultivation in forests, including beneath the shade of nursery stocks, has become widespread. The enormous expanses of space under nursery stocks can be efficiently utilized with this type of mushroom cultivation. In the meantime, straw by-products, which are often burnt or abandoned in fields, are consumed by mushrooms, minimizing waste and pollution. The Chinese government has responded by encouraging the production of economically important mushrooms in forestland.[37] The growing of mushrooms in nursery stocks can be a substitute option since a large amount of compost left over after harvesting mushrooms can be utilized. Furthermore, there is now only a rudimentary understanding of the effects of mushroom farming on soil nutrients. Given that the hyphae of these mushrooms can select particular bacterial taxa in the soil, it is also worth considering how mushroom cultivation affects microbial community composition.[38] There is also concern that the cultivation of *S. rugosoannulata* in forestlands may result in soil nutrient loss.[39]

15.4 Medicinal Uses

Stropharia fungi are edible and have been farmed for food, prompting research into their characteristics and biotechnological applications for commercial purposes. For *S. rugosoannulata,* these fungi have been shown to suppress osteoblast growth, have antifungal properties, suppress endoplasmic reticulum stress, have antitumor effects, have antiradical properties, and show antiviral properties. For example, according to the FAO, *S. rugosoannulata* is recommended as a mushroom for growing in developing countries due to its anti-inflammatory properties, risk reduction for coronary heart disease, hyperglycemic control, and antitumor activity.[21,40]

Stropharia is traditionally cultivated in spring, with beds prepared using fresh wet straw of 70–75% in humidity and irrigation achieved by watering three times a week overhead; the fungi are inoculated after the beds are formed. Beds are constructed in environments with favorable conditions for growth, in light shade and protected from wind drying. Beds built over the ground are preferable and, to avoid plagues as well as mycelium colonization, substrate temperatures of 25–28 °C are favored. In the environment, fungi may exist as saprophytes, yeasts, or as symbiotic partners of lichens or mycorrhizal symbionts associated with plant roots. Aside from their ectomycorrhizal and endomycorrhizal associations, these organisms can form associations that can affect the structure of soil (by entangling particles), the chemistry of soil (*e.g.*, acid excretion), and the growth of plants (by mobilizing nutrients and providing them with reciprocity for photosynthetic assimilates). Thus, they play an important role in controlling the function of soil biota. Aspects of the ecological role of Stropharia include assisting diverse vegetal residue decomposition in forest ecosystems, which maintains their nutrients.[16,41]

S. rugosoannulata is a popular edible variety of mushroom that is grown for food all over the world. Due to its wonderful flavor and pharmacological activity, the intake of *S. rugosoannulata* has increased substantially in recent years. Several investigations have discovered that *S. rugosoannulata* exhibits antioxidant, anti-ER stress, hemagglutinating, and inhibitory osteoclast formation properties.

15.5 Phytochemistry

15.5.1 Important Bioactive Secondary Metabolites

Stropharia rugosoannulata is rich in promising bioactive chemicals both in its basidiomes and fungal mycelia. *S. rugosoannulata* is a functional biological macromolecule due to its numerous health benefits, including antioxidant, anti-diabetic, anticancer, immunopotentiation, anti-inflammatory, and hypoglycemic properties.[42,43]

Extracts of the fruiting bodies of *S. rugosoannulata* exhibit various biological activities, such as osteoclast suppression, antifungal effects, and endoplasmic reticulum stress suppression.[44] Monosaccharides and polysaccharides are the principal bioactive components of basidiomes, and their properties vary depending on extraction methods and basidiome parts. Monosaccharides including glucose, galactose, mannose, ribose, and glucuronic acid have been found in fresh basidiomes, with galactose and glucose being the dominant components (3 : 1 ratio). Moreover, *S. rugosoannulata* has been used to isolate a novel polysaccharide, and it has been reported that this polysaccharide activates immune system function by promoting the development of T and B lymphocytes, and RAW264.7 cells, as well as IgA, IgG, and IgD production.[45] *S. rugosoannulata* was used to successfully purify two different forms of polysaccharides, SRP-1 and SRP-2. SRP-1 exhibits a triple helix shape and α-type and β-type glycosidic connections in its glycosyl residue. However, in SRP-2, a β-type glycosidic linkage predominates.[43]

S. rugosoannulata has been found to accumulate steroids. Six steroids (**1–6**) have been isolated from *S. rugosoannulata*: ergosterol (**1**), 5α-ergosta-7,22-dien-3β-ol (**2**), (22*E*,24*R*)-3β,5α,9α-trihydroxyergosta-7,22-dien-6-one (**3**), (24*S*)-ergosta-7-en-3β-ol (**4**), cerevisterol (**5**), and β-sitosterol (**6**),[46] as well as ceramide (**12**), daucosterol (**7**), 6-methyl-1-octadecene (**11**), dibutyl terephthalate (**13**), uracil (**14**), nicotinic acid (**15**) and quercetin (**16**).[47] In fresh basidiomes of *S. rugosoannulata*, a unique sterol with an unheard-of ether ring was discovered. Sterol A has anti-Methicillin-resistant *S. aureus* action and may protect neuronal cells.[48] The first purified protein from the mushroom species *S. rugosoannulata* was a novel lectin with a unique N-terminal sequence of the basidiomes. Investigations on *S. rugosoannulata* have resulted in the isolation of 16 compounds, including steroids (**1–6**), steroidal saponins (**7**), fatty acids (**8–10**), alkanes (**11**), ceramides (**12**), esters (**13**), pyrimidines (**14**), vitamins (**15**) and flavonoids (**16**). The isolation of steroids (**4–5**) and ceramides (**12**) might be used as chemotaxonomic markers for the genus of *Stropharia*[49,50] (Figure 15.2). However, the functional activities of these compounds are unknown and require additional research. There has been a dearth of research on the active compounds of the *Stropharia* family, though what research there has been has produced some significant chemotaxonomic information.

15.5.2 Nutrition and Dietary Composition

S. rugosoannulata contain abundant macro- and micro-nutrients. *S. rugosoannulata* contain a variety of nutrients, with the main components being ash, protein, crude fat, carbohydrates, minerals, coarse cellulose, energy, vitamins, fatty acids, and amino acids, which are most important macronutrients to sustain human life and have high nutritional value[36,51] (Tables 15.1–15.3).

15.6 Pharmacological Activities

15.6.1 Antibacterial Activity

S. rugosoannulata has anticancer, antioxidative, antimicrobial, antihyperglycemic, and cardiovascular disease prevention properties.[51] The bactericidal

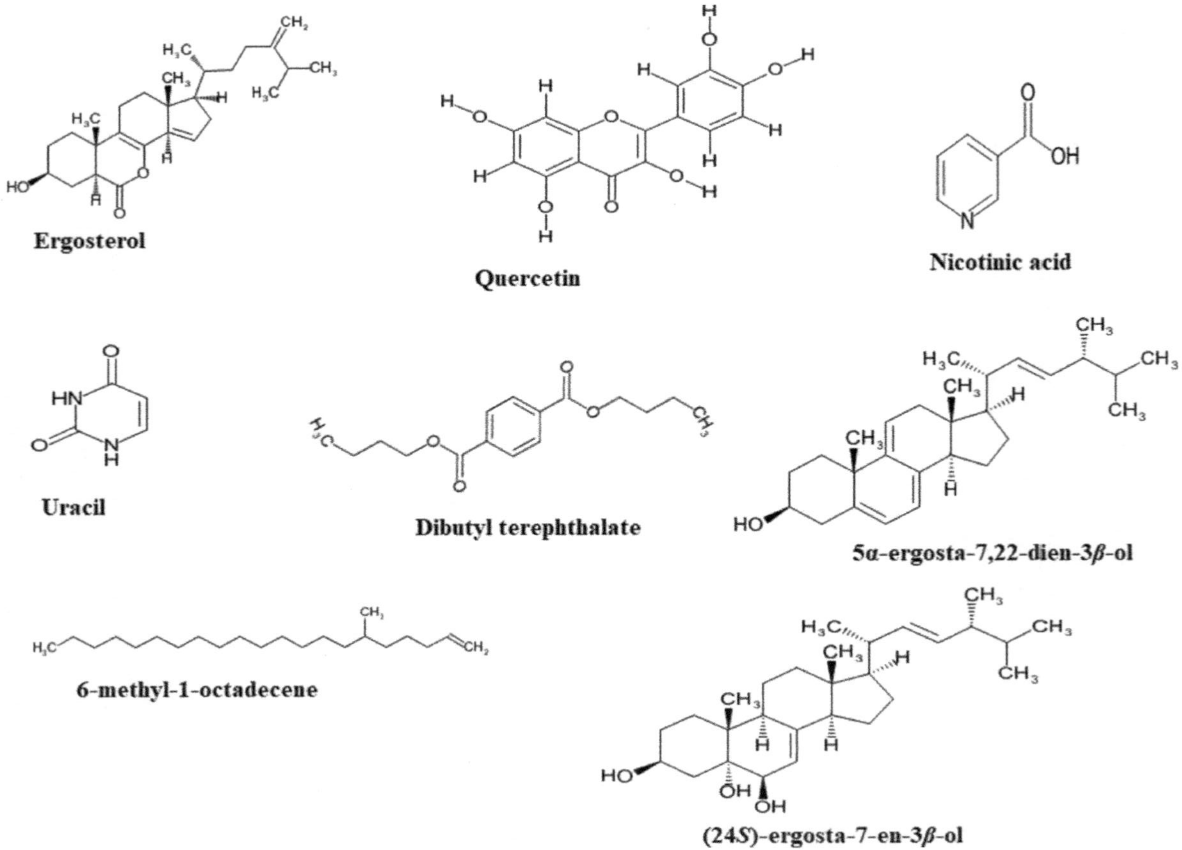

Figure 15.2 Sterols, fatty acids, and ceramide isolated from *S. rugosoannulata*.

Table 15.1 Bioactive component activity of *S. rugosoannulata*.[52]

S. no.	Components	Quantity
1	Phenolic content	5.52 ± 0.45
2	Flavonoid content	1.69 ± 0.25
3	Total polysaccharides	12.60 ± 0.57
4	Total triterpenoids	1.44 ± 0.01
5	Ascorbic acid ($mg\,g^{-1}$)	0.83 ± 0.12
6	Free-radical scavenging activity	18.21 ± 1.34
7	Ergosterol ($mg\,g^{-1}$)	7.89 ± 0.66
8	Total tocopherols ($\mu g\,g^{-1}$)	1.34 ± 0.17
9	Quercetin	13.28 ± 2.44
10	*p*-Coumaric acids	2.81 ± 0.53
11	Reducing sugars	3.55 ± 0.18

Table 15.2 Macronutrients and energy content of *S. rugosoannulata*.[52]

S. no.	Macroconstituents	Quantity
1	Moisture (%)	88.04 ± 1.37
2	Ash (g per 100 g dry weight)	6.04 ± 0.34
3	Carbohydrate (g per 100 g dry weight)	64.35 ± 0.32
4	Fibre (g per 100 g dry weight)	7.50 ± 0.08
5	Protein (g per 100 g dry weight)	25.89 ± 1.72
6	Fat (g per 100 g dry weight)	3.72 ± 0.24
7	Energy (kcal per 100 g dry weight)	1671.72 ± 2.79

Table 15.3 Mineral content of *Laetiporus sulphureus*.[52 a]

S. no.	Mineral	Quantity (mg per kg dry matter)
1	Arsenic (As)	ND
2	Calcium (Ca)	1371 ± 73
3	Copper (Cu)	29 ± 11
4	Cadmium (Cd)	1.17 ± 0.32
5	Iron (Fe)	195 ± 38
6	Potassium (K)	16 320 ± 9420
7	Manganese (Mn)	59 ± 11
8	Magnesium (Mg)	1135 ± 421
9	Sodium (Na)	411 ± 65
10	Zinc (Zn)	102 ± 37
11	Phosphorus (P)	7290 ± 1180
12	Selenium (Se)	2.21 ± 0.09

[a] ND = not detected.

activity of ethanolic and aqueous extracts exhibit low minimum inhibitory concentrations (MICs) for bamboo forest *S. rugosoannulata* (BFSR) and green house *S. rugosoannulata* (GHSR), ranging from 0.0625 to 8 $mg\,mL^{-1}$, except for the ethanolic extract of GHSR (>8 $mg\,mL^{-1}$). Amongst the strains, both extracts were most susceptible to *S. pneumoniae*, *E. coli*, and *E. faecalis*. Aqueous extracts generally displayed higher activity than ethanolic extracts.[53] Additionally, it has been reported that the minimum bactericidal concentration (MBC) values were 2 to 4 times greater than the MIC values. In antibacterial susceptibility studies, MIC values of <8 mg mL^{-1} for unpleasant solvent extracts may be regarded as therapeutically beneficial.[54,55] The MIC values, which ranged from 0.0625 to 8 mg mL^{-1}, were all <8 mg mL^{-1}, indicating that all microorganisms tested exhibited substantial action, except for GHSREe against *S. maltophilia*.[52] As a result, both BFSR and GHSR hold a lot of promise as therapeutic agents for various prevalent infectious disorders. Furthermore, Gram-negative and Gram-positive bacteria are susceptible to different antibiotics.[56,57] Only a few antibacterial medicines are effective against Gram-negative bacteria; the majority of antibacterial medicines are effective against Gram-positive bacteria. However, BFSR and GHSR showed exceptional antibacterial efficacy against *K. pneumoniae*, *E. coli*, *S. typhimurium*, *S. pneumoniae*, and *P. aeruginosa*. As a result, BFSR and GHSR can be exploited to produce new antibacterial treatments.[52]

15.6.2 Antioxidant Activity

Various trials have been conducted to estimate the antioxidant potential of aqueous and ethanolic extracts of BFSR and GHSR, which show scavenging potential for radicals of ABTS (by measuring the reduction in the absorption of ABTS after contact with scavengers of the radical), reduction in ferric power (by estimating the change of Fe^{3+} to a ferrous system), and β-carotene bleaching action (by estimating the absorption of carotene after mixing with antioxidants). All extracts showed good antioxidant potential, it was weaker than that of the positive controls (Vc and BHT). An increase in concentration was accompanied by an increase in radical scavenging power. Compared to ethanolic extracts, aqueous extracts showed greater antioxidant capacity. The strongest β-carotene bleaching inhibition was observed for BFSR aqueous extracts, which also exhibited the highest ABTS+ scavenging potential and reducing capacity.[54] FSRP-1, FSRP-2, HSRP-1 and HSRP-2 at 5 mg mL^{-1} concentration exhibited activities of 84.62%, 79%, 83.86%, and 99.7%, respectively. The findings showed that four kinds of polysaccharides showed positive ABTS scavenging capacities. SRPs and HSRP-2 exhibited the best ABTS scavenging activity, which may be due to their higher levels of β-D-Glcp compared to those of other SRPs.[58] The results revealed that samples treated with hot-air drying (HAD) had better scavenging activity than samples treated with freeze-vacuum drying (FVD). The SRP scavenging activity towards ABTS was affected by the hot-air drying procedure progress of the scavenging potentials by ABTS of the SRPs. The ferric ion reducing antioxidant power (FRAP) of HSRP-1 and HSRP-2 was estimated from their FRAP values of 28.96, 50.47, 43.66, and 75.12 mM Fe^{2+} per mg, respectively.[59] The outcomes indicated that the FRAP action of neutral polysaccharides was lower than that of acid polysaccharides. The same findings have been reported for polysaccharides of *Ganoderma lucidum*.[58] The antioxidant potential of the SRPs is a result of the various enzyme activities in *S. rugosoannulata* upon using a drying method.[60]

Polysaccharides from *S. rugosoannulata* (SRP) SRP-1 and SRP-2 showed helpful radical outcomes by ABTS assay. Compared to SRP-1 at equal dosages, SRP-2 had the most potent potential to scavenge radicals. Furthermore, at a sample concentration of 7.5 mg mL^{-1}, the SRP-1 and SRP-2 scavenging activities reached 63.2% and 100%, respectively.[61] SRP-2 showed the strongest ABTS+ radical scavenging performance compared to SRP-1,[62] which could be related to its advanced levels of α-type glycosidic linkages. Using DPPH radicals, one may assess the capacity of antioxidants to scavenge free radicals.[63] At low concentrations, the scavenging potential of SRP-1 and SRP-2 is typically improved. SRP-1 and SRP-2 showed DPPH assay scavenging activities of 46.6% and 62.5%, respectively, at 10 mg mL^{-1} sample concentration. In addition, the IC_{50} range for SRP-1 was 10.3 mg mL^{-1} and for SRP-2 was 3.1 mg mL^{-1}.[51] The findings were progressive compared to those of polysaccharides by the flowery mushroom.[60,64] Acidic polysaccharides show higher bioactivity compared to neutral polysaccharides.[65] When compared to the neutral polysaccharide SRP-1, acidic SRP-2 polysaccharides display significant scavenging action. The difference between acid and neutral polysaccharides may be interpreted as being due to the presence of additional sulphate in acidic polysaccharides. Polysaccharides from *S. rugosoannulata* (SRP) SRP-1 and SRP-2 scavenging ability in contradiction of hydroxide (OH^-) showed better scavenging activities. The OH^- activity of SRP-2 was enhanced considerably in line with the polysaccharide concentration. At a concentration of 10 mg mL^{-1}, the scavenging rates of SRP-1 and SRP-2 were 39.2% and 98.5%, respectively. Moreover, the IC_{50} of SRP-1 was 10.8 mg mL^{-1} and that of SRP-2 was 1.7 mg mL^{-1}. This different radical scavenging activity may be associated with the sulfate content, the arrangement of simple sugars, and the glycosidic linkage of the polyoses. Therefore, numerous issues have an impact on the antioxidant polysaccharide potential.[51]

15.7 Cultivation Methods of *S. rugosoannulata*

15.7.1 Culture

S. rugosoannulata can be grown in both liquid and liquid culture media. Common solid culture media include potato dextrose, yeast mannitol, and malt extract agar. Other suitable solid media can be used, such as wheat straw, which is mainly derived from agricultural waste.[62] Suitable liquid culture media are usually glucose-based and potato dextrose broth solutions.[66] Cultures are cultured for mycelium growth after being inoculated with media *via* the tissue isolation of basidiomes. The optimal incubation conditions for a solid culture of *S. rugosoannulata* are 25–30 °C, pH 5–6, and a water content of 75% for 10–30 days, and those for liquid culture are 25 °C, pH 6, and 120 rpm for 12 days.[51]

15.7.2 Production of Spawn

Commercially, *S. rugosoannulata* is grown using both cereal-based and agricultural waste-based spawn, the former being used for both small- and large-scale culture, respectively. Cereal-based spawns are created from well-cooked wheat or rice with a pH of 6.5–7 and water content of 65–70%.[67] Corn wastes, rice straw, and soybean straw are commonly used to make agricultural waste-based offspring. Large-scale spawning is usually achieved using sawdust-based spawns. Cultivation of *S. rugosoannulata*.[68] Before inoculating solid or liquid culture, spawn bags or bottles should be sanitized at 121 °C and 105 kPa for 2 h, and the ideal temperature for incubation is 25 °C.

15.7.3 Compost Preparation

Composting ingredients include rice bran, wheat bran, cereal straw, and wood sawdust, which are all derived from agricultural waste. Composting formulas often include the following: (i) sawdust from mulberry branches (100%); (ii) rice husk and straw (50%) and rice straw (50%); (iii) rice husk and wood sawdust (each 49%), gypsum, and phosphate (each 1%); (iv) spent mushroom substrate 40%, gypsum, and sugar (each 1%), rice husk powder 18%, wood sawdust 20%, cottonseed hull 10%, and wheat bran 10%. Agricultural waste should be ground (1–3 cm) and then soaked until it has absorbed enough water, followed by draining, and then placed in a compost heap, which can be up to 1.5 m wide and 1.5 m tall, with the length determined by the entire volume of the compost. The temperature of the heap increases to 60–70 °C (3 and 5 days) following heap formation.[69] The compost is rotated every three days during this stage, and it is suitable for cultivation after 26 days. Compost should contain 65–75% water at all times.[70] Composting for *S. rugosoannulata* cultivation is identical to *Agaricus subrufescens* cultivation[71] (Figure 15.3).

15.7.4 Cultivation Process

Fields, orchards, and plantations are the best places to cultivate *S. rugosoannulata.* The ideal temperature for mushroom growth is 20–30 °C, and

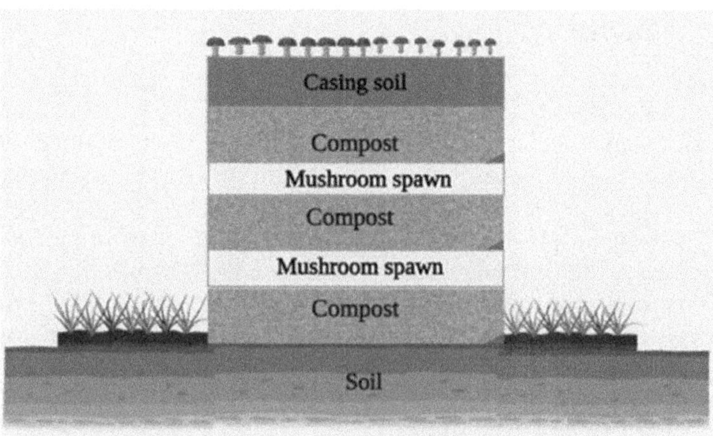

Figure 15.3 Cultivation method of *S. rugosoannulata.*

the cultivation site should be adjacent to a water source, with adequate shelter, fertile soil, drainage, and strong winds. Then, using a grid width of 0.8–1.2 m, drainage ditches should be built and land beds laid out. Before initiating spawn inoculation, quicklime should be placed on the drainage ditches and bed. Next, dibbling techniques are used to infect the spawn before adding another layer of compost to the bed. The bed is then covered with a 2–5 cm thick casing soil layer, and the overall compost layer should be 20–35 cm thick with 2–3 layers. Finally, to prevent contamination and repel insects, one layer of quicklime should be placed on the casing soil. With a spawn content of 5% (w:w), compost and spawn should be applied at a rate of 20–25 $kg\,m^{-2}$ (dry weight).[72]

15.8 Safety, Toxicity, and Regulation

Unlike most mushrooms in the genus *Stropharia*, *S. rugosoannulata* is an edible fungus that is highly regarded by researchers and is cultivated commercially, particularly in Europe. Moreover, *S. rugosoannulata* is included in Italy's list of marketable mushrooms. It is grown and sold in several countries, particularly the lutea type, but its edibility is debatable; in fact, the mushroom has been reported to have mild gastrointestinal issues, most likely as a result of their collection from contaminated areas or dietary misuse. In terms of the general public, there have been no documented negative effects or safety concerns associated with eating wine cap mushrooms. Individual adverse reactions, such as allergies and hypersensitivities, are always a possibility, but that is true of any diet. It is best not to overindulge in new foods, as a new allergy could be discovered after consuming a large quantity of the allergen. The risk of misidentification is the most likely issue linked with wine caps, especially given that they have hazardous close relatives. Another issue is that the white variation could belong to a separate species and, if so, its characteristics remain unknown; it may prove to be harmful.

15.9 Conclusions and Future Prospectives

A growing understanding of the nutritional benefits of consuming mushrooms, as well as their acceptance as a functional food, has resulted in an unprecedented increase in mushroom consumption. Consumption of wild mushrooms, like that of other wild foods, poses some potential risks, hence public safety and understanding are crucial in controlling wild mushroom collection. Unfortunately, poor taxonomy, a lack of expertise, or simply personal beliefs have resulted in a plethora of conflicting reports on which mushrooms are edible and which are not, necessitating a consensus on how to establish edibility and which species are safe to consume.

Acknowledgements

The authors are thankful to the Department of Food Technology, Uttaranchal University and Shoolini University, INDIA.

References

1. J. Wu and H. Kawagishi, *J. Antibiot.*, 2020, **73**, 657–665.
2. A. C. Ruthes, F. R. Smiderle and M. Iacomini, *Carbohydr. Polym.*, 2015, **117**, 753–761.
3. M. López-García, M. S. D. García, J. M. L. Vilariño and M. V. G. Rodríguez, *Food Chem.*, 2016, **199**, 597–604.
4. S. K. Singdevsachan, P. Auroshree, J. Mishra, B. Baliyarsingh, K. Tayung and H. Thatoi, *Bioact. Carbohydr. Diet. Fibre*, 2016, **7**, 1–14.
5. C. Liu, M. W. Choi, X. Li and P. C. K. Cheung, *J. Funct. Foods*, 2018, **41**, 90–99.
6. I. Khan, G. Huang, X. Li, W. Leong, W. Xia and W. L. W. Hsiao, *J. Funct. Foods*, 2018, **41**, 191–201.
7. Y. Liu, B. Zhang, S. A. Ibrahim, S.-S. Gao, H. Yang and W. Huang, *Carbohydr. Polym.*, 2016, **145**, 71–77.
8. S. Chang and W. A. Hayes, *The biology and cultivation of edible mushrooms*, Academic Press, 2013.
9. J. Poppe, in *Science and Cultivation of Edible Fungi: Proceedings of the 15th International Congress on the Science and Cultivation of Edible Fungi, Maastricht/Netherlands/15–19 May 2000*, A. A. Balkema, 2000, vol. 15, p. 909.
10. N. L. Huang, Classification and characterization of *Stropharia rugosoannulata, Edible Fungi*, 1995, **6**, 11.
11. N. Kodama, T. Kakuno and H. Nanba, *Mycoscience*, 2003, **44**, 257–261.
12. W. Zhang, G. Tian, X. Geng, Y. Zhao, T. B. Ng, L. Zhao and H. Wang, *Molecules*, 2014, **19**, 19880–19891.
13. J. Wu, S. Tokuyama, K. Nagai, N. Yasuda, K. Noguchi, T. Matsumoto, H. Hirai and H. Kawagishi, *Angew. Chem.*, 2012, **124**, 10978–10980.
14. D. L. Hawksworth, P. M. Kirk, B. C. Sutton and D. N. Pegler, *Rev. Inst. Med. Trop. Sao Paulo*, 1996, **38**, 272.
15. S. Gong, C. Chen, J. Zhu, G. Qi and S. Jiang, *PeerJ*, 2018, **6**, e5741.
16. J. N. Bruhn, N. Abright and J. D. Mihail, *Agrofor. Syst.*, 2010, **79**, 267–275.
17. J. Zeng, J. Zhao, B. Dong, X. Cai, J. Jiang, R. Xue, F. Yao, Y. Dong and C. Liu, *J. Nutr. Biochem.*, 2019, **66**, 70–78.
18. D. S. Hibbett and R. G. Thorn, in *Systematics and evolution*, Springer, 2001, pp. 121–168.
19. D. S. Hibbett, M. Binder, J. F. Bischoff, M. Blackwell, P. F. Cannon, O. E. Eriksson, S. Huhndorf, T. James, P. M. Kirk and R. Lücking, *Mycol. Res.*, 2007, **111**, 509–547.
20. R. Singer and A. H. Smith, *Mycologia*, 1946, **38**, 500–523.
21. N. Amaresan, M. S. Kumar, K. Annapurna, K. Kumar and A. Sankaranarayanan, *Beneficial microbes in agro-ecology: bacteria and fungi*, Academic Press, 2020.
22. S. d'Emulation (Montbéliard), *Mémoires de la Société d'émulation de Montbéliard*, 1862, vol. 1.
23. E. E. Elgorashi, N. Maekawa and H. Satoh, *Int. J. Med. Mushrooms*, 2008, **10**, 49–53.

24. K. Chuixu, Z. Chongyan, L. Jing, Z. Jun, Z. Keqin and L. Yajun, *Afr. J. Microbiol. Res.*, 2013, **7**, 5737–5741.
25. R. J. L. Singer, *The Agaricales in modern taxonomy*, University of Tucuman, Argentine, 1952, **22**, 1–832.
26. W. Xiong, X. Chen, G. Lv, D. Hu, J. Zhao and S. Li, *J. Pharm. Anal.*, 2016, **6**, 382–388.
27. R. Watling, in *Arctic and alpine mycology II*, Springer, 1987, pp. 17–45.
28. G. Gulden, Ø. Stensrud, K. Shalchian-Tabrizi and H. Kauserud, *Mycologia*, 2005, **97**, 823–837.
29. P. D. Bridge and C. Prior, *Eur. J. Soil Biol.*, 2007, **43**, 101–108.
30. V. G. Cortez, Espécies de Hypholoma (Fr.) P. Kumm. e Stropharia (Fr.) Quél. (Strophariaceae, Agaricales) no Rio Grande do Sul, Brasil, Master's Thesis, Rosa Mara Borges da Silveira, 2006.
31. G. Pacioni and G. H. Lincoff, *Simon & Schuster's guide to mushrooms*, Simon and Schuster, 1981.
32. A. Carluccio, *Complete Mushroom Book: The Quiet Hunt*, Quadrille, 2003.
33. Z. Song, L. Jia, F. Xu, F. Meng, P. Deng, K. Fan and X. Liu, *Biol. Trace Elem. Res.*, 2009, **131**, 81–89.
34. P.-S. Yan, J.-H. Jiang and W.-S. Cui, *World J. Microbiol. Biotechnol.*, 2004, **20**, 173–177.
35. F. J. Yang, C. L. Zhang, C. Xie and W. W. Li, *Mod. Agric. Sci. Technol.*, 2015, **1**, 96–97.
36. Y.-T. Liu, J. Sun, Z.-Y. Luo, S.-Q. Rao, Y.-J. Su, R.-R. Xu and Y.-J. Yang, *Food Chem. Toxicol.*, 2012, **50**, 1238–1244.
37. W. Lu, S. Yu, Y. Ma and H. Huang, *PeerJ*, 2018, **6**, e4624.
38. R. Nazir, J. A. Warmink, H. Boersma and J. D. Van Elsas, *FEMS Microbiol. Ecol.*, 2009, **71**, 169–185.
39. R. H. Socolow, *Proc. Natl. Acad. Sci. U. S. A.*, 1999, **96**, 6001–6008.
40. J. C. Chen, Y. B. Li, L. Wu, X. H. Zhou, Y. G. Duan and P. F. Lai, *J. Beijing Technol. Bus. Univ., Nat. Sci. Ed.*, 2010, **28**, 9–13.
41. H. Harms, D. Schlosser and L. Y. Wick, *Nat. Rev. Microbiol.*, 2011, **9**, 177–192.
42. W. Hong, C. Hui, Z. Jinjing and C. Mingjie, *Acta Edulis Fungi*, 2018, **25**, 115.
43. Y. Liu, C.-F. Hu, X. Feng, L. Cheng, S. A. Ibrahim, C.-T. Wang and W. Huang, *Int. J. Biol. Macromol.*, 2020, **155**, 883–889.
44. J. Wu, K. Fushimi, S. Tokuyama, M. Ohno, T. Miwa, T. Koyama, K. Yazawa, K. Nagai, T. Matsumoto and H. Hirai, *Biosci., Biotechnol., Biochem.*, 2011, **75**, 1631–1634.
45. L. Jiang, Y. Hou and X. Ding, *Lat. Am. J. Pharm.*, 2020, **39**, 1594–1604.
46. J. Wu, T. Suzuki, J.-H. Choi, N. Yasuda, K. Noguchi, H. Hirai and H. Kawagishi, *Tetrahedron Lett.*, 2013, **54**, 4900–4902.
47. S. Hu, X. Feng, W. Huang, S. A. Ibrahim and Y. Liu, *LWT*, 2020, **127**, 109428.
48. J. Wu, H. Kobori, M. Kawaide, T. Suzuki, J.-H. Choi, N. Yasuda, K. Noguchi, T. Matsumoto, H. Hirai and H. Kawagishi, *Biosci., Biotechnol., Biochem.*, 2013, **77**, 1779–1781.

49. Q.-X. Yan, M.-X. Huang, P. Sun, S. Cheng, Q. Zhang and H. Dai, *Biochem. Syst. Ecol.*, 2020, **88**, 103963.
50. F. Senatore, *Biochem. Syst. Ecol.*, 1990, **18**, 103–106.
51. D. Liu, Y.-Q. Chen, X.-W. Xiao, R.-T. Zhong, C.-F. Yang, B. Liu and C. Zhao, *Foods*, 2019, **8**, 397.
52. L. Wei, B. Jing, X. Li, Y. Hou, X. Xie, Z. Wang, Y. Liu, Y. Zhou, X. Chang and W. Wang, *J. Food Qual.*, 2021, **2021**, 1–9.
53. P. He, L. Geng, J. Wang and C. Xu, *Int. J. Med. Mushrooms.*, 2014, **16**, 319–325.
54. S. Gibbons, *Phytochem. Rev.*, 2005, **4**, 63–78.
55. W. Fabry, P. O. Okemo and R. Ansorg, *J. Ethnopharmacol.*, 1998, **60**, 79–84.
56. J. C. Meng, Q. X. Zhu and R. X. Tan, *Planta Med.*, 2000, **66**, 541–544.
57. R. M. Herrera, M. Perez, D. A. Martin-Herrera, R. Lopez-Garcia, R. M. Rabanal and A. Arias, *Phytother. Res.*, 1996, **10**, 364–366.
58. M. Shi, Z. Zhang and Y. Yang, *Carbohydr. Polym.*, 2013, **95**, 200–206.
59. S. Li, S. Zhao, C. Hu, C. Mao, L. Guo, H. Yu and H. Yu, *J. Fungi*, 2022, **8**, 99.
60. L. Chen, Y. Zhou, Z. He, Q. Liu, S. Lai and H. Yang, *Food Chem.*, 2018, **251**, 9–17.
61. Z. Xu, L. Fu, S. Feng, M. Yuan, Y. Huang, J. Liao, L. Zhou, H. Yang and C. Ding, *Molecules*, 2019, **24**, 1357.
62. Y. P. L. G. J. Jiahui and Z. Y. W. H. X. Haitao, *Acta Edulis Fungi*, 2001, **8**, 5.
63. Y. Liu, S. Ma, S. A. Ibrahim, E. Li, H. Yang and W. Huang, *Food Chem.*, 2015, **185**, 159–164.
64. D. Kothari, S. Patel and S.-K. Kim, *Biomed. Pharmacother.*, 2018, **105**, 377–394.
65. D. Gan, L. Ma, C. Jiang, R. Xu and X. Zeng, *Carbohydr. Polym.*, 2011, **84**, 997–1003.
66. R. Lucas, J. Mihály, G. M. Lowe, D. L. Graham, M. Szklenar, A. Szegedi, D. Töröcsik and R. Rühl, *Nutrients*, 2018, **10**(10), 1390.
67. Y. Yang, G. Meng, S. Ni, H. Zhang and C. Dong, *J. Fungi*, 2022, **8**, 162.
68. Z. Li, H. Han, M. Chen, Z. Lin and M. Liu, in *The Proceedings of the 5th Symposium of the National Agricultural Environmental Sciences Sponsored by the Chinese Society of Agro-Ecological Environment Protection in 2013, Jiangsu, China, 19–22 April, 2013*, Chinese Academy of Agricultural Sciences (CAAS), Scientech Documentation and Information Centre, 2013, pp. 923–928.
69. H. Han and M. Chen, *Fujian J. Agric. Sci.*, 2013, **28**, 1088–1092.
70. W. A. N. Sarentoya and B. A. U. Tolgor, *Acta Edulis Fungi*, 2005, **12**, 59.
71. K. Wisitrassameewong, S. C. Karunarathna, N. Thongklang, R. Zhao, P. Callac, S. Moukha, C. Ferandon, E. Chukeatirote and K. D. Hyde, *Saudi J. Biol. Sci.*, 2012, **19**, 131–146.
72. G. Tian, X. Xu and K. Zhang, *Edible Fungi China*, 2002, **21**, 9–16.

CHAPTER 16

Termitomyces clypeatus *and* Termitomyces heimii

AKSHYA SHARMA,[a] VARUNA SHARMA,[b] ARUP GIRI,[c] NITISH KUMAR[d] AND PUSHPENDER BHARDWAJ*[d]

[a] Department of Biochemistry, Panjab University, Chandigarh, 160014, India; [b] Department of Education, Panjab University, Chandigarh, 160014, India; [c] Department of Zoology, Baba Mastnath University, Asthal Bohar, Rohtak, Haryana-124021, India; [d] Medicinal Plant Division, Defence Institute of High Altitude Research, Near Airforce Industrial Area, Phase-II, Near 3BRD, Ram Darbar, Chandigarh, 160002, India
*Email: pbshawn06@gmail.com

16.1 Introduction

Fungi represent a wide group of nearly 50 000 species, which include mushrooms, toadstools, mold, mildew, and yeast.[1–4] Unlike plants, fungi have no chlorophyll to make their food; rather, they have many hyphae, which resemble cotton thread and are used to ingest substances. Macrofungi make up a sizeable fraction of these species and are widely distributed in dipterocarp woods, mixed forests, evergreen forests, cultivated, abandoned areas, and lawns. They digest garbage and return essential nutrients to the land as they grow in soil, on dead wood and other decomposed plant components. Thus, they are an important component of the ecosystem associated with degradation processes and serve as an essential indicator of the state of equilibrium. Moreover, mushrooms possess high nutritional and functional value as they are a source protein, enriched with essential and supplementary amino acids, low in calories and unsaturated fats, and contain vitamins and micronutrients. Therefore, they are considered to be an

Phytochemistry and Nutritional Composition of Significant Wild Medicinal and Edible Mushrooms: Traditional Uses and Pharmacology
Edited by Ajay Sharma, Garima Bhardwaj and Gulzar Ahmad Nayik

Published by the Royal Society of Chemistry, www.rsc.org

immediate source of food, medicine, and income to mankind as wild mushrooms are gathered and sold at large-scale markets. Fungi are also economically significant to the food industry as cheeses such as camembert, Roquefort, stilton, and Danish blue get their distinctive flavors from substances made by molds that are added to them to help them ripen.[3,4]

16.2 Classification and Taxonomy

16.2.1 Taxonomy of Termitomyces

The most prevailing Termitophilic fungi, allied to the paleotropical Genus Termitomyces, of the family Lyophyllaceae, and order, Agaricales, have been catalogued under the monophyletic group of gilled mushrooms. Termitomyces have symbiotic relationships with eclectic categories of termites, such as Odontotermes, Microtermes, Macrotermes, Hypotermes, Protermes, and Canthotermes, which are grouped under the subfamily *Macrotermitinae* (Isoptera). The termites promote the growth of the Termitomyces fungi, which later provide an enzymatic supplement to aid the digestion of the food of the termites.[5,6] Correspondingly, due to skepticism on the recognized range of taxonomy in the genus Termitomyces globally, their classification and identification has been disputed by microbiologists. Using current parameters, 101 distinguished species of Termitomyces have been indexed in MycoBank 2022 and Index Fungorum 2022 databases explicitly; however, only 30 species have global recognition in the Dictionary of Fungi (10th Ed.).[7]

In terms of geographical locations, myriad taxa of Termitomyces have been identified in Central and Southern Africa and Southeast Asia compared to other continents.[8] Two significant species of Termitomyces, namely *T. clypeatus* and *T. heimii,* have been found to be mainly prevalent in regions of Southeast Asia.

In India, 25 certified Termitomyces species identified in different regions of the country, particularly the Western Ghats, have been documented in "Agaricomycetous mushrooms of India". Axiomatically, India reckons to be the paramount gene pool of Termitophilic mushrooms.[9–11] In terms of the two most widely known Termitomyces species, *T. clypeatus* and *T. heimii*, *T. clypeatus* has been documented as being found in regions of Chambaghat, Poona in Maharashtra, as well as in Madras and the Kodagu region of Karnataka.[12–18] Likewise, *T. heimii* has been found in different parts of the Punjab, Karnataka, and Madras. The identification and cataloging of *T. clypeatus* and *T. heimii* have been the keys to identifying other documented species of termitophilic mushrooms.[14–20]

Among the diverse species of Termitomyces mushrooms, there is a great deal of diversity in the dimensions of their external morphological structure in terms of the color variations in umbo, disparities in their perforatorium, modified pseudorrhiza, *etc.*[21] For example, *T. heimii* and *T. clypeatus* show considerable variability in their morphological structure, microscopic structure, and ethnomycological utility.

In terms of morphological structures, the basidiocarp size of *T. heimii* ranges from 20–25 cm with a fleshy, cartilaginous, and tough consistency, which is proportionally larger than those of other Termitomyces species, including that of *T. clypeatus*.[19] Meanwhile, the divergency in basidiocarp colour exists to a lesser degree among various species of Termitomyces. For example, *T. heimii* shows a silky white creamy greyish to light brown color basiodiocarp, whereas *T. clypeatus* shows a greyish to dark brown color basidiocarps. The pileus structure spreads out, with or without a depressed center, with the shape of the umbo changing from spiniform to acute, obtuse, and even broad. In parallel to a pileus diameter of 10 cm in *T. heimii*, the shape of the umbo is broadly roundish and the color is pale yellow to greyish brown with a pronounced spiniform in *T. clypeatus*.[21] There is a pronounced diversity in the pseudorrhiza of Termitophilic mushroom species, as is the case for *T. heimii* or *T. clypeatus*, where a distinct brown-colored pseudorrhiza is observed for *T. clypeatus*.[19] Moreover, after rigorous microbiology investigations, a double-sized annular shaped veil has been observed for *T. heimii* in comparison to other species, with a lack of a veil having been observed for *T. clypeatus*.

In categorizing the species of Termitomyces, namely, *T. heimii* and *T. clypeatus*, on the basis of their morphological structures, so far, cataloging has only been conducted considering their external characteristics while failing to consider their phylogenetic study using molecular techniques. To confirm the assignment of Termitomyces species supported by phylogenetic relationship analysis, molecular markers, in particular the internal transcribed spacer (ITS), nuclear large subunit ribosomal DNA (nrLSU), and mitochondrial small subunit ribosomal DNA (mtSSU) regions are utilized extensively.[22,23] To date, rigorous literature analyses projecting the phylogenetic relationships of *Termitomyces* species using nLSU-rDNA and mitochondrial small subunit ribosomal DNA have been reported that have confirmed that *T. mammiformis*, *Sinotermitomycescavus* and *S. taiwansis* are conspecific with *T. heimii* and *T. clypeatus*. Moreover, in Nigeria, using ITS 18 species of *Termitomyces* have been discerned genetically, substantiating their monophylectic source.[21,24,25]

In a nutshell, Termitomyces, irrespective of its growth mechanism, has successfully captured the attention of mycologists, entomologists, and food technologists across the globe. As *ex situ* cultivation of Termitomyces can be demanding due to its obligate symbiotic relationship with termites of the Termitinae family, a prototypical boosting of Termitomyces species on coffee and paddyfield land under *in situ* conditions could be a sustainable approach. However, cultivators need to be cautious when making capital out of harvesting and disseminating Termitomyces in their own grounds to make the process market-worthy. Therefore, the identification and cataloging of Termitomyces taxa on the grounds of morphology and phylogenetic studies will be a major accomplishment for future productiveness.

16.3 Resource Availability

An unparalleled diverse range of organisms contribute towards the rich biodiversity of planet Earth and play a major role in meeting the requirements of the human population. Regions of South-east Asia and Africa have the most abundant wild fungal mushroom reserves, be it wild or cultivated species that exhibit both culinary and medicinal properties. Among the diverse groups of mushroom species, Termitomyces has exclusively caught the attention of the research community due to its high culinary value and being an additional source of income and nutrition for people who live in rural communities.

In 1847, Berkley, for the first time, categorized Termitomyces within the fungi kingdom in the Basidiomycota division, and Lyophyllaceae family. Later, in 1942, Heim elucidated on the Termitomyces–termite fungal mutualistic association, where the termite Isoptera was found to accommodate a nest substrate for the mushroom Termitomyces to grow in and in turn the lignocellulosic enzymes of Termitomyces were found to be consumed by termites to digest their food.[26,27] Moreover, a rare species of Termitomyces with flesh that resembles that of chicken has been found in the southern regions of China, where it is popularly known by the name of 'Jizong', which translates as 'chicken mushroom', or 'Yizong' meaning ant-planted mushroom. However, various other divisions of Termitomyces have also been indexed in Africa, Southeast Asia, and Central America.[28–33]

A major concern of researchers while dealing with Termitomyces is in its identification and isolation from other different species of mushrooms in a natural habitat. At the same time, a large undernourished population of people in tribal areas across the globe rely on Termitomyces as a food and nutraceutical source to sustain their lives. Moreover, Termitomyces are considered by vegans to be an excellent food source due to them being cholesterol free, having excellent mineral and protein content, and representing a good substitute for animal protein. Biologists have discovered 20 species globally in both wild and cultivated groups. In terms of being relevant to ethnomedicine, various Termitomyces species, such as *T. heimii*, *T. microcarpus*, *T. eurrhizus*, *T. aurantiacus*etc are useful to mankind.[34]

As an exemplary example of a study based in India, in 2016, Venkatachalapathi and Paulsamy shed light on using *T. heimii* and *T. microcarpus* in treating shivering and hot flushes as symptoms of bacterial and fungal infections. Moreover, *T. eurrhizus* has been used to reduce blood pressure and treat rheumatic disorders, and *T. heimii* has been used to aid wound healing and blood coagulation.[35,36] In addition, in India *T. aurantiacus*, *T. eurrhizus*, *T. letestui*, and *T. titanicus* have been used in tonic form to treat various alimentary mycotoxicosis, such as abdominal cramps and peptic ulcers. Furthermore, in Tanzania *T. microcarpus* is used to boost the immune systems of patients dealing with long term illness.[37]

The propagation of *Termitomyces* fruiting bodies is widely acknowledged as an up-hill battle due to its slow growth mechanism, and in this vein,

T. heimii has been classified as among the most onerous mushrooms to propagate globally.[38] However, to date, unparalleled propagation and turn out have been observed for *T. microcarpus* basidiospores.[39] Correspondingly, a dearth has been observed in artificially cultivated Termitomyces fruiting bodies and for that reason additional measures need to be in place in the process for precisely augmenting the mechanism of the artificial cultivation of Termitomyces species mycelium.

Overall, review studies have focused on the presence of bioactive components, such as phenolic compounds, polysaccharides, high protein content (27–36%), ascorbic acid (10–18 $mg\,g^{-1}$) ergosterol and important minerals such as phosphorus, potassium, calcium, copper, and iron, in the fruiting bodies and mycelia of Termitomyces mushrooms. To this aim, *T. heimii* and *T. clypeatus* species have been reviewed so far, constituting evidence in support of their nutraceutical benefits, such as antioxidant, antitumor, antibacterial, immunomodulatory, cardiovascular, liver-protective, anti-viral, anti-inflammatory, antidiabetic, and anti-fibrotic properties.[40–48]

Delving further into the structural analysis of two particular species of Termitomyces, *i.e.*, *T. heimii* and *T. clypeatus*, as well as their potential activity, water-soluble glucan isolated from *T. heimii* and water-soluble heteroglycan isolated from the fruiting bodies of *T. clypeatus* have exhibited antioxidant properties against nicotine toxicity. In addition to polysaccharides, phenolic compounds, namely cinnamic and benzoic acid derivatives, have been procured from the fruiting bodies of 23 species of edible Termitomyces mushrooms using water and methanol as solvent.[41,49,50] Here, due to their enhanced antioxidant attributes, scavenging and chelating actions have been observed for both *T. heimii* and *T. clypeatus* species of Termitomyces owing to the high amounts of phenolic compounds in their aqueous extracts, such as tannic, gallic, gentisic, protocatechuic, vanillic, syringic, coumaric, caffeic, ferulic, and cinnamic acid.

In terms of the edible utility of Termitomyces mushroom species, it has been ascertained that various species of Termitomyces dominate the food industry, including bakery houses, where the extracted enzymes from Termitomyces species are used to soften and ferment food products so as to add flavor to food stuffs. In view of this, the hemicellulase, amylase, and cellulase enzymes present in *T. clypeatus*, in addition to other components such as yeast, salt, sugar, *etc.*, facilitate the production of bread that is softer and bigger than standard bread due to better leavening. Likewise, the enzyme pectinase from *T. clypeatus* is also used to extract, filter, and simultaneously remove pectins from fruit juices and wines.[51]

Moreover, the potential activity of mixtures of various enzymes, like pectinase, cellulase, hemicellulase, arabinose, and xylanase, present in *T. clypeatus* are used to clarify apple juice.[52]

Aside from the food industry, several other remarkable enzymes have made a significant impression on the bioremediation industry. For example, the enzyme alkaline protease extracted from *T. clypeatus* can productively epilate goat skin and bird feathers from their shafts besides reducing the biochemical oxygen demand (BOD) and chemical oxygen demand (COD) of its effusions. In addition, lignocellulolytic enzymes produced by cultivating *T. clypeatus* in agro-residues of mustard have known to serve as a substrate for saccharification.[53]

As a result of their edibility and nutraceutical benefits, the cultivation of Termitomyces mushrooms has gained traction across the globe. However, the slow growth of Termitomyces mushrooms has placed constraints on their cultivation for various industrial sectors. Thus, it is necessary to advance the research on cultivation mechanisms so as to improve Termitomyces cultivation towards a sustainable future.

16.4 Medicinal and Edible Properties

Mushrooms, with their organoleptic merits and economic benefits due to their extraordinary functional, nutritional, and medicinal properties, are a far-reaching chief ingredient in bakehouses worldwide, endorsing them as 'nutraceutical foods'. Catalogued mainly as Basidiomycetes and Ascomycetes, >2000 species of mushrooms have been identified in the wild, with the exception of only 25 species that are comprehensively acknowledged for their commercial purposes. However, differentiating between medicinal and edible mushrooms can be problematic due to edible mushrooms presenting both therapeutic and medicinal properties.[54] Being cognizant of medicinal and edible properties, the discussion hinges upon an exotically rare species, namely Termitomyces, in the class *Basidiomycota*, order *Agaricales* and family *Agaricaceae*, which has promising applications as an ethnomedicine that could lead to morbidity-free survival.

16.5 Termitomyces: Significance and Implications

The term Termitomyces has been designated to the genus of edible mushrooms distributed and commonly consumed in tropical and subtropical areas from Africa to Southeast Asia that grow symbiotically in the nests of Macrotermitinae termites in the wild.[55] In addition to displaying antioxidant properties and a high content of phenolic ingredients, Termitomyces contain certain bioactive molecules that can be used to treat sarcomas and neurodegenerative disorders.[56–60] Moreover, as an ethnomedicine with purgative attributes, the affluent sources of sugars, protein, fiber, lipids, vitamins, and mineral in Termitomyces help to lower blood pressure, as well as treating rheumatism, kwashiorkor, obesity and diarrhoea.[61] Important studies have reported various species of the genus Termitomyces, namely *Termitomyces eurhizus* (Berk) Heim, *T. clypeatus* Heim, *T. striatus* (Beeli) Heim, *T. robustus* (Beeli) Heim, *T. microcapus* (Berkand Broome) Heim, and *T. heimii* Natarajan, to be edible mushrooms that have a high nutritive value and possess immunomodulatory or antibacterial activity.[62,63]

Among the various species of the aforementioned fleshy fungi, *Termitomyces clypeatus* R. Heim and *T. heimii* Natarajan belonging to the family

Table 16.1 Biological diversity, morphological structure, medicinal, and edible properties of *Termitomyces* mushrooms.

S. no.	Genus: *Termitomyces* Common name: termite mushroom	Specie: *Termitomyces clypeatus* R. Heim	Specie: *Termitomyces heimii* Natarajan
1.	Geographical distribution	• Distributed in tropical countries like Brazil, India, the Philippines and Cameroon.[66–69] • Indian geographical distribution: Chambaghat, Solan, Poona (Maharashtra), Madras, Punjab and Karnataka.[12–19]	• Distributed in South-east Asia.[55] • Indian geographical distribution: Madras, Punjab,[14–17,19] Karnataka.[18]
2.	Morphological heterogenity		
	Stipe length	• Stipe is 3–7 cm long, fibrous in nature, centrally solid with white to dirty brown fleshy attributes.[84]	• Stipe is 5 cm long and 1.5 cm wide, cylindrical and creamy and stuffed with a thick annulus.[84]
	Surface of pileus	• It is silky smooth with irregular-lobed margins.[84]	• Smooth-surfaced pileus with incurved margins.[84]
	Fruit bodies	• Produces orange-gray to grayish-orange colored fruiting bodies.[84]	• Produces cream-colored fruiting bodies.[84]
3.	Medicinal implications	• Acknowledged by ancestral naturopaths to treat abdominal infections.[84] • Cibacron blue affinity eluted protein present in the fungus exhibit anti-proliferative properties against tumor cells.[85] • Used in the treatment of pox. Paste of the fruit body is applied to the effected area.[86] • High antioxidant properties.[49] • Antibacterial activity and immunostimulatory effects.[88] • Used as a medicine against fever and measles.[89] • Paste of the fruiting body was applied to heal chicken pox by Santals.[41]	• Utilized in the treatment of fever, cold and fungal infections in parallel as a blood tonic during wound healing and blood coagulation.[36,87] • Presence of two major components, *i.e.*, ergosterol and linoleic acid made it known as a dietary calcium supplement in people suffering from hypocalcinaemia and osteoporosis along with maintenance of health through the production of prostaglandins; regulation of blood pressure, blood lipids, immune response, inflammation, and apoptosis.[42] • Paste used in the treatment of chicken pox.[41]
4.	Edible implications	• Consumed as an energy-dense food by locals of the Gadchiroli forest in Maharashtra.[90] • Consumed as a food with fish in Manipur.[91] • One of most consumed mushrooms in Jharkhand.[92]	• Identified as one of the edible mushrooms with a high nutritive value.[93] • Consumed as a highly delicious food in Jammu and Kashmir.[92] • Consumed as a delicious food and believed to reduce blood pressure in Odisha.[93] • Consumed in the form of soup, chutney, and salads in Nagaland.[94]

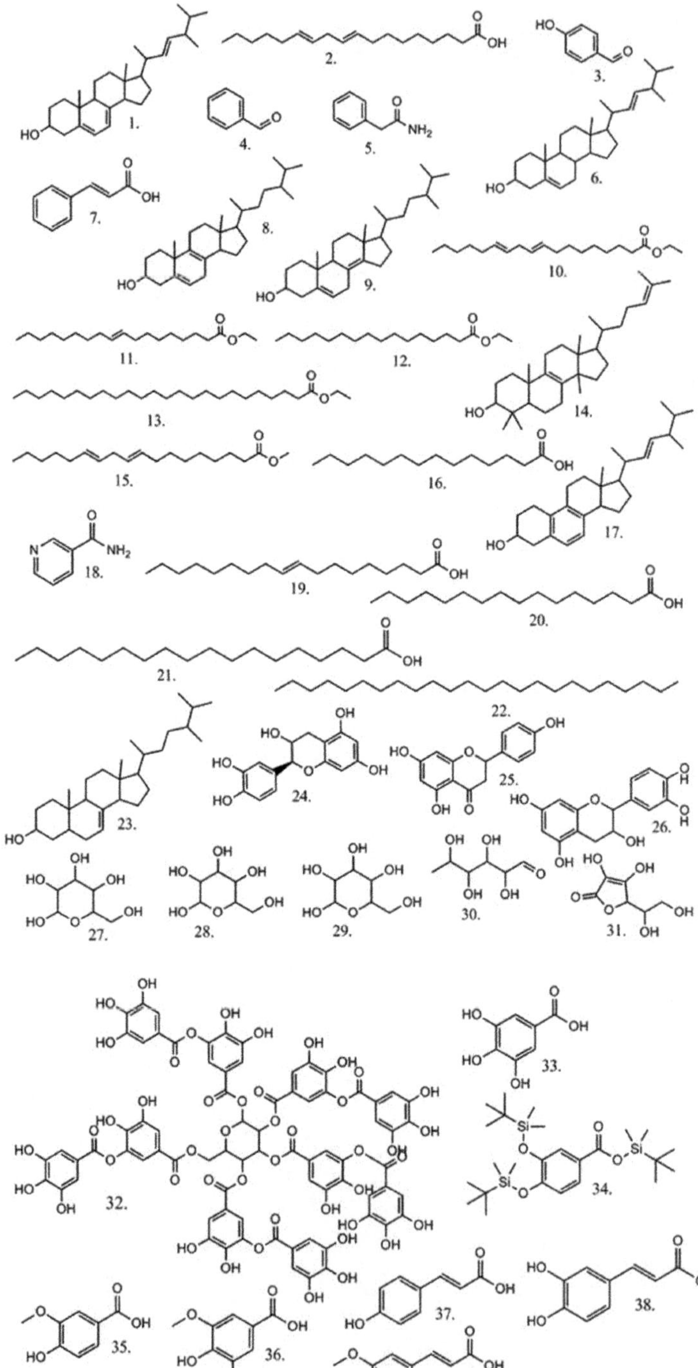
1.
2.
3.
4.
5.
6.
7.
8.
9.
10.
11.
12.
13.
14.
15.
16.
17.
18.
19.
20.
21.
22.
23.
24.
25.
26.
27.
28.
29.
30.
31.
32.
33.
34.
35.
36.
37.
38.
39.

Lyophyllaceae are the most common popular edible and medicinal mushrooms in many tropical countries, including Brazil, India, the Philippines, and Cameroon.[64–69] Not losing sight of their morphological heterogeneity, besides their edible and medicinal properties, Table 16.1 gives information on the underlying rationale for the existing diversity between *T. clypeatus* R. Heim and *T. heimii* Natarajan.

16.6 Nutritional and Dietary Composition

Mushrooms have been treated as "Food of the Gods" since ancient times due to their high content of different nutritional components. Nowadays, *Termitomyces* mushrooms are famous for having high levels of essential micronutrients and vitamins.[70] In Uganda, Kansci *et al.* (2003) reported that *Termitomyces* species have a higher water level (83.3–94.3 g per 100 g wet matter), lipid levels (2.5–5.4 g per 100 g dry weight), polyunsaturated fatty acid levels (45.1–65.1% of the total fatty acids), crude fibers (17.5–24.7 g per 100 g dry weight), protein content (15.1–19.1 g per 100 g dry weight), and ash (5.2–14.4 g dry weight) than other species grown in temperate regions. In fact, they are rich in highly digestible lysine rich proteins, vitamins, minerals such as phosphorus, calcium, copper, potassium, iron, *etc.*, are low in fat, and are low in carbohydrates.[18,70,71] In 2012, Atri *et al.* reported ample amounts of protein, carbohydrates, crude fat, crude fiber, moisture content, ash content, and mineral elements in *T. heimii* and *T. clypeatus* samples collected in northern India.[72] Other studies have indicated that these mushrooms contain sorbitol and linoleic acid (an unsaturated fatty acid) instead of starch. Moreover, they are also an excellent source of folic acid, thiamine, riboflavin, niacin, pantothenic acid, and biotin. In addition, these mushrooms are also reported to have high medicinal attributes, such as anti-ulcer, antimicrobial, immunostimulatory, and antioxidant properties.[49,50,73,74] In 2018, Hsieh and Ju reported that *Termitomyces* species have significant medicinal attributes, such as antioxidant, immunomodulatory, anticancer, and antimicrobial properties due to their high levels of bioactive compounds.[60] In 2021, Sitati *et al.* reported that *Termitomyces* mushrooms contain different bioactive

Figure 16.1 Structures of phytocomponents of *Termitomyces* mushrooms: (**1**) Ergosterol; (**2**) linoleic acid; (**3**) 4-hydroxybenzaldehyde; (**4**) Bbenzaldehyde; (**5**) benzeneacetamide; (**6**) brassicasterol; (**7**) cinnamic acid; (**8**) ergosta-5,8-dien-3-ol; (**9**) ergosta-5,8(14)-dien-3-ol; (**10**) ethyl linoleate; (**11**) ethyl oleate; (**12**) ethyl palmitate; (**13**) ethyl tetracosanoate; (**14**) lanosterol; (**15**) methyl palmitate; (**16**) myristic acid; (**17**) neoergosterol; (**18**) nicotinamide; (**19**) oleic acid; (**20**) palmitic acid; (**21**) stearic acid; (**22**) tetracosane; (**23**) γ-ergostenol; (**24**) (+)-catechin; (**25**) naringenin; (**26**) (−)-epicatechin; (**27**) D-glucose; (**28**) D-galactose; (**29**) D-mannose; (**30**) L-fucose; (**31**) ascorbic acid; (**32**) tannic acid; (**33**) gallic acid; (**34**) protocatechuic acid; (**35**) vanillic acid; (**36**) syringic acid; (**37**) P-coumaric acid; (**38**) caffeic acid; (**39**) ferulic acid.

molecules and for this reason they have the potency to reduce Gram positive and Gram negative bacterial populations (*Escherichia coli*, *Pseudomonas aeruginosa*, *Bacillus subtilis*, *Staphylococcus aureus*) and fungal growth (*Candida albicans* and *Saccharomyces cerevisiae*).[75]

16.7 Bioactive Phytochemical Constituents

T. heimii, mainly found in India, Thailand, Malaysia, and China,[31,76–79] has a symbiotic relationship with termites. These mushrooms are highly enriched with β-glucan, which is water soluble and exhibits excellent antioxidant properties.[41] Most of the polysaccharides in this species of mushroom is polybranched with two (1→6)-β-D-glucopyranosyl residues, and two (1→3)-β-D-glucopyranosyl units. This species has been found to have high ergosterol (**1**) and linoleic acid (**2**) content. Apart from these compounds *T. heimii* also contains 4-hydroxybenzaldehyde (**3**), benzaldehyde (**4**), benzeneacetamide (**5**), brassicasterol (**6**), cinnamic acid (**7**), ergosta-5,8-dien-3-ol (**8**), ergosta-5,8(14)-dien-3-ol (**9**), ethyl linoleate (**10**), ethyl oleate (**11**), ethyl palmitate (**12**), ethyl tetracosanoate (**13**), lanosterol (**14**), methyl palmitate (**15**), myristic acid (**16**), neoergosterol (**17**), nicotinamide (**18**), oleic acid (**19**), palmitic acid (**20**), stearic acid (**21**), tetracosane (**22**), and γ-ergostenol (**23**), amongst other compounds.[42,60,78,80–82] In 2017, Butkhup (2017) reported that *T. clypeatus* is highly enriched with (+)-catechin (**24**), naringenin (**25**), β-glucan and flavonoids, (−)-epicatechin (**26**), flavan-3-ols, *etc.* In other studies, it has been found that *T. clypeatus* contains D-glucose (**27**), D-galactose (**28**), D-mannose (**29**), and L-fucose (**30**), and ascorbic acid (**31**; Figure 16.1).[49,74,83]

16.8 Bioactivity Analysis

Various bioactive compounds are present in mushrooms, which are responsible for their good antioxidant properties.[41,60] *T. heimii* contains a high level of β-glucan; an antioxidant molecule that is effective in preventing nicotine toxicity.[41] In 2015, Pattanayak *et al.* reported the same type of antioxidant properties exhibited by *T. clypeatus.*[49] In 2018, Hsieh and Ju reported that *T. albuminosus* contains significant levels of exopolysaccharide (EPS), which is responsible for its antioxidant properties.[60] Research has shown that most of the antioxidant properties of mushrooms are due to their content of different phenolic compounds, such as tannic (**32**), gallic (**33**), protocatechuic (**34**), vanillic (**35**), syringic (**36**), *P*-coumaric (**37**), caffeic (**38**), and ferulic (**39**; Figure 16.1) acids. However, in 2016, Mondal *et al.* reported that treatment of a cancer cell line with *T. clypeatus* extract showed excellent antioxidant activity, exhibiting declining lipid peroxidation followed by increasing superoxide dismutase, catalase, and glutathione levels. Moreover, in 2006, Puttaraju *et al.* reported that *T. heimii* has significant flavonoid levels.[50] They also reported that the treatment of *T. heimii* effectively reduced cellular mortality after inhibiting lipid peroxidation activity.

References

1. A. Barh, B. Kumari, S. Sharma, S. K. Annepu, A. Kumar, S. Kamal and V. P. Sharma, *Smart Bioremediation Technologies*, Academic Press, Cambridge, 1st edn, 2019.
2. A. Barh, V. P. Sharma, S. K. Annepu, S. Kamal, S. Sharma and P. Bhatt, *3 Biotech*, 2019, **9**, 1.
3. S. Sharma, V. P. Sharma, S. Kumar, A. Sharma, K. Manorma and P. K. Chauhan, *Wild Mushrooms*, Taylor and Francis group, London, 1st edn, 2022.
4. V. P. Sharma, A. Barh, B. Kumari, S. K. Annepu, S. Sharma and S. Kamal, *Int. J. Med. Mushrooms*, 2020, **22**, 501.
5. S. T. Chang and T. H. Quimio, *Tropical mushrooms: biological nature and cultivation methods*, Chinese University Press, Hong Kong, 1st edn, 1982.
6. J. O. Kuja, H. I. Boga, V. Matiru and H. M. Makonde, *Int. J. Micro. Mycol.*, 2014, **2**, 29.
7. P. M. Kirk, P. F. Cannon, D. W. Minter and J. A. Stalpers, *Ainsworth & Bisby's Dictionary of the fungi*, CAB International, Wallingford, United kingdom, 10th edn, 2008.
8. R. Heim, *Termites et champignons: Les champignons termitophiles d'Afrique noire et d'Asie meridionale*, Societete Novelle des Edition, Boubee, Paris, 1977, p. 205.
9. N. C. Karun and K. R. Sridhar, *Data Brief*, 2017, **14**, 320.
10. R. C. Upadhyay, B. Verma, S. Sood, N. S. Atri, T. N. Lakhanpal and V. P. Sharma, *Documentary of Agricomycetous Mushrooms of India*, Jaya Publishing House, Delhi, India, 2017.
11. S. Mani and V. Kumaresan, *J. Threat. Taxa*, 2009, **1**, 54.
12. A. D. Sharma and R. L. Munjal, *Indian J. Mushrooms*, 1977, **3**, 12.
13. S. D. Patil, L. N. Nair and B. P. Kapandis, *J. Univ. Poona, Sci. Technol.*, 1979, **52**, 349.
14. K. Natarajan, *Curr. Sci.*, 1977, **46**, 679.
15. K. Natarajan, *Kavaka*, 1977, **5**, 35–39.
16. K. Natarajan, *Mycologia*, 1979, **71**, 853.
17. K. Natarajan, *Kavaka*, 1975, **3**, 63.
18. G. R. Janardhana and A. Pahlevanlo, *J. Adv. Lab. Res.*, 2012, **3**, 54.
19. N. S. Atri, S. S. Saini and G. Kaur, *Mushroom Res.*, 1995, **4**, 7–10.
20. K. B. Vrinda and C. K. Pradeep, *Mushroom Res.*, 2009, **18**, 33.
21. B. Kumari, V. P. Sharma, A. Barh and N. S. Atri, *Fungal Biol.*, 2022, **12**, 102.
22. Y. Taprab, M. Ohkuma, T. Johjima, Y. Maeda, S. Moriya, T. Inoue, P. Suwanarit, N. Noparatnaraporn and T. Kudo, *Biosci., Biotechnol., Biochem.*, 2002, **66**, 1159.
23. P. Sawhasan, J. Worapong and T. Vinijsanun, *Thai J. Agric. Sci.*, 2011, **44**, 183.
24. T. G. Frøslev, D. K. Aanen, T. Laessøe and S. Rosendahl, *Mycol. Res.*, 2003, **107**, 1277.
25. V. O. Oyetayo, *Sci. World J.*, 2012, 689296.

26. J. S. Sathiya Seelan, C. Shu Yee, F. She Fui, M. Dawood, Y. S. Tan, M. J. Kim, M. S. Park and Y. W. Lim, *Mycobiology*, 2020, **48**, 95.
27. M. Wisselink, D. K. Aanen and A. van 't Padje, *Insects*, 2020, **11**, 527.
28. L. P. Tang and Z. L. Yang, *Resour. Sci.*, 2014, **36**, 282.
29. R. Heim, *Etudes descriptives et expérimentales sur les agarics termitophiles d'Afrique tropicale*, Gauthier-Villars, 1940.
30. K. Natarajan and N. Raman, *South Indian Agaricales: A Preliminary Study on Some Dark Spored Species*, 1983, no. 589.222095 N3.
31. D. N. Pegler and M. Vanhaecke, *Kew Bull.*, 1994, **49**, 717.
32. N. G. A. Koné, B. Soro, L. P. L. Vanié-Léabo, S. Konaté, A. Bakayoko and D. Koné, *Mycology*, 2018, **9**, 307.
33. L. D. Gómez, *Rev. Biol. Trop.*, 1994, **42**, 439.
34. V. E. C. Gruen and H. X. Wong, *Sciences*, 1982, **57**, 269.
35. S. K. S. Sachan, J. K. Patra and H. N. Thatoi, *J. Agric. Technol.*, 2013, **9**, 403–416.
36. S. P. Chandrawati, K. Narendra and N. N. Tripathi, *Am. Int. J. Res. Formal Appl. Nat. Sci.*, 2014, **5**, 71.
37. D. D. Tibuhwa, *J. Ethnobiol. Ethnomedicine*, 2012, **8**, 1.
38. S. Mallick, A. Dutta, S. Dey, J. Ghosh, D. Mukherjee, S. S. Sultana, S. Mandal, S. Paloi, S. Khatua, K. Acharya and C. Pal, *Exp. Parasitol.*, 2014, **138**, 9.
39. D. Olila, R. Bukenya-Ziraba and D. Kamoga, *Res. J. Pharmacol.*, 2007, **1**, 56.
40. A. Z. Woldegiorgis, D. Abate, G. D. Haki and G. R. Ziegler, *Food Chem.*, 2014, **157**, 30.
41. D. K. Manna, A. K. Nandi, M. Pattanayak, P. Maity, S. Tripathy, A. K. Mandal, S. Roy, S. S. Tripathy, N. Gupta and S. S. Islam, *Carbohydr. Polym.*, 2015, **134**, 375.
42. S. N. Abd Malek, G. Kanagasabapathy, V. Sabaratnam, N. Abdullah and H. Yaacob, *Int. J. Food Prop.*, 2012, **15**(4), 809.
43. H. P. Aryal and U. Budhathoki, *Med. Plants – Int. J. Phytomed. Relat. Ind.*, 2014, 128.
44. M. Rashid, M. Rahman, R. Correll and R. Naidu, *Int. J. Environ. Res. Public Health*, 2018, **15**, 919.
45. S. P. Wasser and A. L. Weis, *Int. J. Med. Mushrooms*, 1999, **1**, 31.
46. N. K. Ishikawa, M. C. M. Kasuya and M. C. D. Vanetti, *Braz. J. Microbiol.*, 2001, **32**, 206.
47. P. Poucheret, F. Fons and S. Rapior, *Cryptogam.: Mycol.*, 2006, **27**, 311.
48. M. Zhang, S. W. Cui, P. C. K. Cheung and Q. Wang, *Trends Food Sci. Technol.*, 2007, **18**(1), 4–19.
49. M. Pattanayak, S. Samanta, P. Maity, I. K. Sen, A. K. Nandi, D. K. Manna, P. Mitra, K. Acharya and S. S. Islam, *Carbohydr. Res.*, 2015, **413**, 30.
50. N. G. Puttaraju, S. U. Venkateshaiah, S. M. Dharmesh, S. M. N. Urs and R. Somasundaram, *J. Agric. Food Chem.*, 2006, **54**, 9764.
51. D. R. Kashyap, S. Chandra, A. Kaul and R. Tewari, *World J. Microbiol. Biotechnol.*, 2000, **16**, 277.

52. K. Singha, D. K. Sahoo, A. Banerjee and P. K. Das, *Mohapatra, Biotechnology for Sustainable Utilization of Bioresources*, Daya Publishing House, New Delhi, 2019.
53. S. Pal, S. P. Banik and S. Khowala, *Ind. Crops Prod.*, 2013, **41**, 283.
54. E. Guillamón, A. García-Lafuente, M. Lozano, M. A. Rostagno, A. Villares and J. A. Martínez, *Fitoterapia*, 2010, **81**, 715.
55. K. Ono, T. Hata, T. Yoshimura and K. Kinjo, *J. Wood Sci.*, 2017, **63**, 83.
56. Y. Qu, K. Sun, L. Gao, Y. Sakagami, H. Kawagishi, M. Ojika and J. Qi, *Biosci., Biotechnol., Biochem.*, 2012, **76**, 791.
57. J. Qi, M. Ojika and Y. Sakagami, *Bioorg. Med. Chem.*, 2001, **9**, 2171.
58. L. Li, R. Yang, K. Sun, Y. Bai, Z. Zhang, L. Zhou, Z. Qi, J. Qi and L. Chen, *Int. J. Neuropsychopharmacol.*, 2012, **15**, 497.
59. Y. J. Guo, G. F. Deng, X. R. Xu, S. Wu, S. Li, E. Q. Xia, F. Li, F. Chen, W. H. Ling and H. B. Li, *Food Funct.*, 2012, **3**(11), 1195–1205.
60. H. M. Hsieh and Y. M. Ju, *Appl. Microbiol. Biotechnol.*, 2018, **102**, 4987.
61. M. M. Apetorgbor, A. K. Apetorgbor and E. Nutakor, Utilization and cultivation of edible mushrooms for rural livelihood in Southern Ghana, in *Proceedings of the 17th Commonwealth Forestry Conference*, Colombo, Sri Lanka, 2005, pp. 1–22.
62. I. Chakraborty, S. Mondal, D. Rout and S. S. Islam, *Carbohydr. Res.*, 2006, **341**, 2990.
63. K. A. Thillaimaharani, K. Sharmila, P. Thangaraju, M. Karthick and M. Kalaiselvam, *Int. J. Pharma Sci. Res.*, 2013, **4**, 1540.
64. D. K. Aanen, H. H. de Fine Licht, A. J. Debets, N. A. Kerstes, R. F. Hoekstra and J. J. Boomsma, *Science*, 2009, **326**(5956), 1103.
65. U. G. Mueller, N. M. Gerardo, D. K. Aanen, D. L. Six and T. R. Schultz, *Annu. Rev. Ecol. Evol. Syst.*, 2005, **36**, 563.
66. A. L. Njouonkou, E. De Crop, A. M. Mbenmoun, T. R. Kinge, E. H. Biyé and A. Verbeken, *Int. J. Med. Mushrooms*, 2016, **18**, 387.
67. M. E. G. De Castro, R. M. R. Dulay and M. L. D. Enriquez, *Der Pharm. Lett.*, 2016, **8**(5), 237.
68. S. D. Ghate and K. R. Sridhar, *Indian J. Nat. Prod. Resour.*, 2017, **8**, 120.
69. J. P. Tamang, K. Watanabe and W. H. Holzapfel, *Front. Microbiol.*, 2016, **7**, 377.
70. I. Nakalembe, J. D. Kabasa and D. Olila, *Comparative nutrient composition of selected wild edible mushrooms from two agro-ecological zones, Uganda*, SpringerPlus, 2015, vol. 4, p. 1.
71. B. Kumari, R. C. Upadhyay and N. S. Atri, *Int. J. Med. Mushrooms*, 2013, **15**, 191.
72. N. Atri, B. Kumari, R. C. Upadhyay and S. K. Sharma, *Int. J. Med. Mushrooms*, 2012, **14**(5), 471.
73. S. Dordević, S. Petrović, S. Dobrić, M. Milenković, D. Vucićević, S. Zizić and J. Kukić, *J. Ethnopharmacol.*, 2007, **109**, 458.
74. A. Mondal, D. Banerjee, R. Majumder, T. K. Maity and S. Khowala, *Pharm. Biol.*, 2016, **54**, 2536.

75. C. Sitati, K. O. Ogila, R. W. Waihenya and L. A. Ochola, *Evidence-Based Complementary Altern. Med.*, 2021, 3025848.
76. E. Bao, *Econ. Bot.*, 2004, **60**(1), 99–100.
77. A. Chandrasrikul, P. Suwanarit, U. Sangwanit, S. Lumyong, A. Payapanon, N. Sanoamuang, C. Pukahuta, V. Petcharat, U. Sardsud, K. Duengkae and U. Klinhom, *Checklist of mushrooms (Basidiomycetes) in Thailand*, Office of Natural Resources and Environmental Policy and Planning, Bangkok, Thailand, 2011, pp. 112–113.
78. P. Mitra, N. C. Mandal and K. Acharya, *J. Verbrauch. Lebensm.*, 2016, **11**, 25.
79. S. Schmidt, S. Kildgaard, H. Guo, C. Beemelmanns and M. Poulsen, *Nat. Prod. Rep.*, 2022, **39**, 231.
80. Z. M. Thu, K. K. Myo, H. T. Aung, M. Clericuzio, C. Armijos and G. Vidari, *Molecules*, 2020, **25**(8), 1972.
81. G. Yang, F. Ahmad, S. Liang, H. Fouad, M. Guo, H. A. Gaal and J. Mo, *Appl. Biochem. Biotechnol.*, 2020, **192**, 1270.
82. N. Rahmad, J. R. Al-Obaidi, N. M. N. Rashid, N. B. Zean, M. H. Y. M. Yusoff, N. S. Shaharuddin, N. A. M. Jamil and N. M. Saleh, *Biol. Res.*, 2014, **47**, 1.
83. R. Majumder, S. P. Banik and S. Khowala, *Int. J. Biol. Macromol.*, 2016, **91**, 198.
84. B. Srivastava, A. K. Dwivedi and V. N. Pandey, *J. Bull. Environ., Pharmacol. Life Sci.*, 2011, **1**, 54.
85. S. Maiti, S. K. Bhutia, S. K. Mallick, A. Kumar, N. Khadgi and T. K. Maiti, *Environ. Toxicol. Pharmacol.*, 2008, **26**, 187.
86. A. K. Dutta and K. Acharya, *Asian J. Pharm. Clin. Res.*, 2014, **7**, 36.
87. A. Venkatachalapathi and S. Paulsamy, *Mycosphere*, 2016, **7**, 118–130.
88. O. Mahamat, N. André-Ledoux, T. Chrisopher, A. A. Mbifu and K. Albert, *Clin. Phytosci.*, 2018, **4**, 1.
89. N. S. Atri, *Kavaka*, 2018, **51**, 11.
90. K. B. Vrinda and C. K. Pradeep, *Mushroom Res.*, 2009, **18**, 33.
91. S. Kavishree, J. Hemavathy, B. R. Lokesh, M. N. Shashirekha and S. Rajarathnam, *Food Chem.*, 2008, **106**(2), 597.
92. A. Farook, S. S. Khan and P. Manimohan, *Mycosphere*, 2013, **4**, 97.
93. E. J. M. Florence and K. Yesodharan, *Kerala Forest Research Institute Research Report*, 2000, vol. 191, p. 1.
94. C. R. Deb and A. Paul, *Bioresources and Sustainable Livelihood of Rural India*, Mittal Publications, New Delhi, India, 2021.

CHAPTER 17

Volvariella bombycina (Schaeff.) Singer (Silky Rosegill)

ANKITA SHARMA,*[a] NEHA KAUSHAL,[a] SACHIN KUMAR,[b] SHALINI SHARMA[c] AND SHAKSHI SHARMA[a]

[a] Dr Yashwant Singh Parmar University of Horticulture and Forestry, Nauni, India; [b] Dr B. R. Ambedkar National Institute of Technology, Jalandhar, India; [c] State Forensic Science Laboratory, Junga-Shimla, India
*Email: ankitas238@gmail.com

17.1 Introduction

Mushrooms have long since been regarded as a culinary wonder to mankind and have been awarded status as gourmet cuisine throughout the globe. The usage of mushrooms is regarded to be a symbol of elegance, not only due to their high nutritional and functional value but also due to their organoleptic merit, therapeutical properties, and economic value.[1,2] In ancient times, Egyptians declared mushrooms to be the plant of immortality, and thus limited their access to royal families only.[3] Approximately, 14 000 species of mushrooms are known worldwide. Of these, only 3000 species have been proven to be safe for human consumption, whereas 105 species are major edible mushrooms.[4] For many centuries there was no particular definition for mushrooms. However, with the passage of time, the need was felt to have a comprehensive definition. In 1992, Chang and Miles defined the mushroom as a macro-fungus with a unique ripening body that is visible by the naked eye and can be picked by hand.[1] These can grow either epigeously (the seed emerges out of the soil) or hypogeously (the seed remains inside the soil).[5] Furthermore, mushrooms have been classified into four categories based on their utility to humans: (1) consumable mushrooms, such as

Phytochemistry and Nutritional Composition of Significant Wild Medicinal and Edible Mushrooms: Traditional Uses and Pharmacology
Edited by Ajay Sharma, Garima Bhardwaj and Gulzar Ahmad Nayik

Published by the Royal Society of Chemistry, www.rsc.org

Lentinula edodes (Shitake) and *Agaricus bisporus* (button mushroom); (2) medicinal/therapeutic mushrooms (*Ganoderma lucidum* (Lingzhi); (3) poisonous mushrooms, such as *Amanita phalloides*/death cap; and (4) mushrooms with unknown properties.[6] Mushrooms have a long history with humans and have had significant biological and economic impact.[7] They have been found to be a rich source of proteins, including essential amino acids. Edible mushrooms have also been found to contain many therapeutically and nutraceutically-active components, such as flavonoids, ascorbic acid, carotenoids, unsaturated fatty acids, and tocopherols. In addition, mushrooms are also observed to be a rich source of different vitamins, including vitamin B1, B2, B12, D, as well as minerals, fiber, and trace elements, with little/no calories and cholesterol.[8] The multifunctional nature of mushrooms mean that they are of relevance to health, wealth, the environment, and industry, making fungi a viable strategic commodity to resuscitate and contribute considerably to the world economy. The global market for mushroom cultivation has been estimated to be 16.7 billion USD in 2020. This significant increase in the cultivation market is projected to grow further due to many factors, such as the multiple nutritional and health benefits of mushrooms, the increasing consumption of mushrooms per capita, the development of technologies for their cost-effective production in different climates, an increase in the number of health-conscious individuals in society, along with a growing need for natural, vegan food in the diet across the globe.[9] In the financial year 2020–2021, mushroom export from India was worth around 2.16 million USD. Major countries where mushroom were exported to included France, USA, Germany, Switzerland, and Nepal.[10]

17.2 *Volvariella*

'*Volvariella*' is a lesser acknowledged genus of approximately 100 species of mushrooms, which are found to be distributed mainly in tropical, subtropical, and cold climatic regions (temperate) of the eastern as well as western hemisphere.[11] Members of this genus are well known for their ability to produce pink spores and free lamellae with an annulus-free stipe. The presence of a swollen structure at its base is a characteristic of this genus (see Table 17.1). The generic name of this species is derived from the Latin word, 'volva', meaning 'to turn, wrap around'.

17.3 *Volvariella bombycina* var. *bombycina*

Volvariella bombycina, also known as the silky rosegill or silky-capped mushroom, is one of the rare species of the *Volvariella* genera. It is often found to grow singly but may appear in small clusters. It is a mildly flavored mushroom, laced with many antibacterial and antioxidants compounds; however, a proper confirmation of this mushroom is required as it resembles with poisonous death caps and destroying angels.[18] *Volvariella*

Table 17.1 Description of some species of the *Volvariella* genera.

Name of species	Habitat	Description
Volvariella canalipes	Solitary on sandy soil	Pileus is convex, whitish, ambrosial; lamellae are free, crowded, ventricose, whitish to pinkish in color, volva is large, whitish, cup-shaped with brink.[12]
Volvariella earlei	Solitary to gregarious on soil	Pileus is 3–5 cm wide, plano-convex, dry or slightly viscid. Lamellae are closed, fairly wide as well as crowded. Volva is 5–10 mm high, white to pale gray.[13]
Volvariella alabamensis	Solitary on soil	Pileus is 6–7 cm wide, thick, hard, convex, single; smooth with a silky sheen, lamellae are free, volva is white, a little fibrillose, abundant, closely clinging down, with a free, mutilate border, a portion of which is transferred to the top of pileus.[14]
Volvariella speciosa	Single to clubbable on manure or rich soil in lawns, gardens, woods, and greenhouses	Pileus is ovoid to globose when young, stretched to convex, more or less umbonate, sticky, smooth, with striped edge, white to light brown or gray, sometimes containing patches of the universal veil. Lamellae are white to brown in color and are close, crowded, wide, thrusting outward, free, with uneven edges. Volva is uncomplicated, with a wide edge and a nearly even, divided, whitish to grayish coloration.[12]
Volvariella jamaicensis	Gregarious on dead wood	Pileus is slender, with a whitish or gray surface, light brown or yellowish brown at the center, scattered stripes, slightly white to salmon-colored; volva is fragile, small, light brown, 1–2 cm long.[13]
Volvariella bakeri	Solitary to gregarious on wood	Pileus is corpulent, ovoid to convex, compressed fibrillose with the rupturing cuticle, lamellae are free, packed, not very wide, whitish to pinkish; volva is free, open, dark and fuliginous.[13]
Volvariella bombycina var. *bombycina*	Solitary to gregarious on wood	Pileus is globose to ovoid, later expanding to campanulate or convex white edges. Lamellae are congested, wide, thrusting outward, free, with irregular edges. The volva is adequate, thick, approximately even and multilateral margins and the surface is divided into smaller areas by cracks, whitish to yellowish or parchment in color on the areolae.[14]
Volvariella bombycina var. *flaviceps*	Solitary to gregarious on wood (magnolia)	Pileus is campanulate, solitary, lamellae are free, wide and circular at the margin, narrow behind, congested, slender, entire, whitish, sterile; smooth, pale-brown scales, their acute tips point upward, giving the mushroom a fetid smell during drying.[11]

Table 17.1 *(Continued)*

Name of species	Habitat	Description
Volvariella volva	Solitary to gregarious on soil or compost piles	Pileus is circular when young and expands to bell shaped, black to grayish brown smooth margin, with a tender taste along with meaty aromatic smell. Lamellae are wide, free, with fringed edges, whitish to a deep flesh color. The volva is large, with an irregular and deeply intended margin; membranaceous, flocculent, and brownish.[15]
Volvariella cubensis	Solitary on soil	Pileus is hard, corpulent, rather tough, unevenly expanded, dark ashy brown surface, rigorously covered with hair-like appendages, smooth, the disk is dark brown and smooth; lamellae are free, the volva thick and fleshy, cupulate, obscure, divided by a deep cleft into two parts and of uniform color.[11]
Volvariella taylori	Solitary to gregarious on soil	Pileus is obtusely conical to convex. Lamellae are not quite distant but not close either, adequately coarse. Depth of the volva is near to 1 cm, cupulate, with lobes of usually 3–5 in number, outwardly brownish or grayish and whitish inside.[16]
Volvariella lepiotospora	Solitary on dead roots (*Quercus virginiana*) in tropical thicket	Pileus has a rounded convex form, dry, subrimulose and scattered brownish to blackish fibrils and smooth disk with brownish black color, completely covered by a fine, pale and wooly covering. Lamellae are close, free, with laciniated edges, warm pink in color. The volva is cauldron-shaped, arachnoid, gray in color.[17]

bombycina has been reported to emerge from knotholes, typically from cracks of bark, old wounds, and other damaged/decaying plant parts, mostly at a prominent height on standing trees. Other habitats include stones covered with clay, moist timber, or basement wall, wastepaper, and compost piles, *etc.* In European countries, *Volvariella bombycina* has been observed to grow in open deciduous forests containing trees chiefly from the genera *Acer, Ailanthus, Betula, Aesculus, Castanea, Celtis, Crataegus, Carpinus, Fraxinus, Juglans, Fagus, Malus, Populus, Quercus, Pyrus, Robinia, Sorbus, Salix, Tilia, and Ulmus.* Outside of Europe, trees of the genera *Acacia, Ficus, Liquidambar, Magnolia, Mangifera,* and the dead trunks of the Arecaceae family have been observed as habitats for *V. bombycina.*[19,20] In the Indian subcontinent, this species has been reported to grow on the stumps of dead mango trees and in the northwestern Himalayas on decorticated trunks of *Picea smithiana*. This species of mushrooms is being cultivated in Asia, Europe, North and South America, Africa, and Australia. Different countries where this mushroom is cultivated on a commercial scale are India, China, Korea, Pakistan, Phillipines, Poland, Denmark, Estonia, Hungary, Norway and Sweden.

17.4 Biological Structure of *V. bombycina*

V. bombycina belongs to one of the largest groups of fungi, designated as Basidiomycota, the club fungi, which comprise some of the best-known fungi, including puffballs, shelf fungi, jelly fungi, rusts, and smuts, in addition to mushrooms that are saprophytes or mycorrhizal symbionts with the roots of plants.[21,22] All the members of this group can be characterized by the existence of a club-shaped generative organ called the basidium. The organs are presumably acquired from the ascus found in Ascomycota, and these parts have numerous attributes. For example, both arise as a dikaryotic arrangement, which serve as a site for karyogamy and meiosis. However, the basidium features spores on its exterior, whereas the ascus retains all the ascospores within its structure. The body of *V. bombycina* is divided into the following parts:

- cap
- gills
- stem
- volva
- spores

1. **Cap**: The *V. bombycina* body is covered by a white or pale cream cap, which varies in size from 5 to 20 cm. It consists of silky, long fibers, initially arranged in an oval shape followed by a bell shape later on. With ageing, the fibers change from yellow to brown. The mushroom cap acts as an umbrella and protects the pores or gills, wherein the mushroom spores are produced. The primordia of the cap arises from the egg veil. With the passage of time, it achieves a bell shape followed by flat shape at the time of maturity that is covered with whitish or yellow silky fibrils.

The upper side of the cap is covered with a number of gills, white in the beginning and pink on maturity. Gills are packed together and produce spores on maturity.[23]

2. **Gills**: Gills, also known as lamellae, papery hymenophore ribs, are developed under the cap of this mushroom species, which are mostly white in color in the beginning and change to pink on maturity. These are used by mushrooms as a tool for spore dispersal. The types and arrangement of gills are usually used as a tool for the identification and categorization of species in basidiomycetes.[23] Gills are usually packed together and not attached to the stem and produce spores. Spores are pushed horizontally from the surface of the gills through a mechanism, known as a surface tension catapult.[24,25] This tool was deciphered by John Webster and associates in the 1990s.[26] Within one millisecond of the launch, the plane segment of the path is impeded by air viscosity, followed by spores drooping vertically between the gills for a few seconds before coming into direct contact with the airflow beneath the cap.[27]
3. **Flesh**: The flesh of *V. bombycina* is usually thin, soft, and white in color with an odor similar to that of raw potatoes.[28]
4. **Stem**: These rare striking silky capped mushrooms are usually 7–15 cm long and 1–2 cm wide and are tapered toward the apex or thickened below with no ring. The stem is white in color with a smooth surface and is slightly curved. The stem holds the cap and gills high enough from the ground level and provides them with support.[29]
5. **Volva**: Also known as the universal veil, the volva is a uniform layer of protective tissue that helps immature mushrooms to grow out of the ground. It is present as a thin sheet of interwoven hyphae surrounding the globular base of the stipe. On reaching maturity, the mushroom separates itself from the bottom, leaving the veil at the base of the stalk. This results in the formation of a cup-like structure at the base of the stem. The volva type is a characteristic criterion in the identification of wild mushrooms. As these mushrooms grow from mycelium below the surface and thrust upward; they are protected by a thin layer that fissures, leaving fragments on the cap.[29] They are surrounded by a membranous volva developed throughout the stem base, by residues of the membranous universal veil which covers growing fruit bodies.[30] The volva looks like a persistent bag, turning yellowish from whitish and finally browning with age.
6. **Spores**: The spores of the mushroom are usually brownish pink to salmon in color with an elliptical to smooth surface and have dimensions of 6.5–10 × 4.5–6.5 μM.[31]

17.5 Life Cycle

Volvariella is well adapted to grow in tropical and subtropical climates.[32] Under favorable environmental conditions, it completes its life cycle in approximately 4–5 weeks.[33] In its early developmental stages, the cap arises as an egg-shaped structure, which becomes cone-like to almost flat on maturity. The vegetative body is composed of thread and cord-like mycelial strands that come out of the substrate.[1] The first stage of development of these mushroom is referred to as the 'pinhead stage' and is characterized by the formation of small clusters of interlinked thread-like hyphae. The second stage is referred to as the 'button stage', which involves the formation of a button-shaped universal veil surrounding the egg-shaped primordia. Early developmental stages of the stalk, pileus, and gills are visible at this stage if cut longitudinally. Commercial mushrooms are harvested at this stage due to their abundance of protein and other nutrients, good palatability, and increased shelf life.[34] The next stage is the elongation state, which can be characterized by rupturing of the volva by the stipe exposing the pileus and gills. The final stage is the maturity stage, which involves the formation of fully expanded pink colored gills followed by the development and dispersal of basidiospores.[33,35] Using gamma radiation as a tool of identification, it has been reported that the basidiospores of *V. bombycina* can be divided into two classes, in which one class is radioresistant and the other is radiosensitive, which makes basidiospores haploid.[36] The hymenial cell initially uninucleates and then becomes binucleate. A binucleate hymenial cell can differentiate into a multinucleate cystidium through mitotic divisions or commit to becoming a basidium.[29]

17.6 Importance

Volvariella bombycina has considerable industrial value globally. Due to its high protein content and significant therapeutic significance, it can be used as a meat substitute.[38] This mushroom exhibits different medicinal attributes, including antibacterial, antitumor, hypercholesterolemic, and antioxidant properties. Different bioactive compounds have been isolated and identified from the mycelium or pure culture of *Volvariella bombycina.* The mushroom contains a high level of nutrients, such as vitamin D, zinc, magnesium, copper, iron, and potassium (see Table 17.2). Further the fruiting bodies (basidiocarp) of this species are usually of a fine standard, appetizing, and scrumptious, which makes them delicious to eat.[39] Different compounds, such as ergosta-4,6,8,22-tetraene-3-one, ergosterol peroxide, indole-3-carboxaldehyde, and indazole, have been found in the liquid culture of *V. bombycina,* which endow the mushroom with antitumor properties.[40]

17.7 Phytochemistry of *Volvariella bombycina*

Mushrooms have shared a very long history as a source of food and medicinal components that are used by humans.[41] These are collected from their natural wild habitats or cultivated under controlled conditions that mimicking their natural environment. Recently, there has been great interest on the study of the phytochemical profile of various extracts of mushrooms to

Table 17.2 Nutritional profile of the fruiting bodies of *Volvariella bombycina*.

Nutritional parameters	Amount (dry weight)
Dry matter (%)	8.13
Ash (%)	9.37
Fat (%)	2.05
Carbohydrate (%)	46.07
Protein	34.38
Crude fibers (%)	13.51
Calcium (mg per 100 g)	25.61
Zinc (mg kg^{-1})	119.95
Iron (mg kg^{-1})	72.5
Copper (mg kg^{-1})	50.2
Magnesium (%)	0.12
Potassium (%)	4.12
Vitamin D (IU per g)	106.995

understand the biochemistry behind the various therapeutic attributes of fungi, such as their antimicrobial, antitumor, and immunomodulatory properties. Therefore, there is a deep curiosity, even in consumers, to understand the basic phytochemical profile of mushrooms apart, aside from that of professional phytochemists. The concept of phytochemicals and phytochemistry is not new. It is a well-known fact that plants have the ability to convert energy from sunlight to high-energy-yielding compounds, such as sugars, *via* photosynthesis. These compounds can be further used by various heterotrophic organisms to accomplish their energy requirements by being a part of food chain. Plants are also known to produce various chemical compounds, also known as phytochemicals, particularly secondary metabolites. The study of these compounds is known as phytochemistry, which involves the study of various classes of compounds, their structural composition, biosynthetic pathways, mechanisms of action, and their functions. Phytochemistry has been influential in the discovery of various plant products with numerous commercially marketable properties. Phytochemicals have been present in nature from the very beginning and are also responsible for sensory properties, such as taste, aroma, sight, color, and texture. For example, the pungency of ginger can be attributed to phenylalkylketones such as gingerols and shogaols, and the deep purple color of blueberries can be attributed to the pigment anthocyanin. These phytochemicals have found commercial value in the pharmaceutical, nutraceutical, cosmeceutical, clothing and textiles, food, and beverage industries.[42] The commercial value of these phytochemicals and their potential were discovered by studying their chemical compositions. The production of phytochemicals by plants can be directly related to their interaction with other species and their surrounding environment. The increased availability and discovery of new phytochemicals contribute toward research in the fields of biotechnology and biochemistry. Mushrooms being a part of the plant kingdom are no exception. They synthesize a number of phytochemicals for their survival in response to their external environmental conditions.[43] The phytochemicals of mushrooms can be classified mainly into phenolics, terpenes, terpenoids, and N, and S containing compounds. The secondary metabolites found in mushrooms include phenolic chemicals, polyketides, terpenes, and steroids. Moreover, these substances exhibit antioxidant, antiviral, antibacterial, and medicinal properties. They have uses as antibiotics, dyes, anti-cancer compounds, and flavoring and odor compounds, among others. Many phytochemicals, such as polysaccharides, peptides, protein complexes, as well as terpenoids such as lanostames, carotenoids, phenolics and flavonoids, have been extracted from edible and wild mushrooms.[44] Carotenoids and polyphenols have been reported to be efficient antioxidants and also interfere with the initiation and proliferation of cancer.[45] Polyphenolic compounds have also been shown to exhibit anti-ageing, anti-inflammatory, and neuroprotective properties.[46]

Volvariella bombycina is a wild edible mushroom in the Pluteaceae family that contains various phytochemicals that exhibit medicinal and therapeutic properties. Most of the phytochemicals with therapeutic properties are found in their fruiting bodies.

In 2003, Badalyan reported 20% antioxidant activity by the mycelia of *Volvariella bombycina* and observed great potential to inhibit free-radical peroxide oxidation. Various bioactive compounds with anticancer properties have been reported in *Volvariella bombycine,* such as methylester, hexadecanoic acid, pyrrolo[1,2-a]pyrazine-1,4-dione, hexahydro-3-[2-methylpropyl], 1-leucyl-D-leucine, and 1,1-biphenyl,3-[1,3-butadienyl]-indole.[47,48] Some of the phytochemical compounds that have been reported in *Volvariella bombycina* are discussed here with their medicinal and other properties.

17.7.1 Hexadecanoic Acid

Hexadecanoic acid, also known as palmitic acid ($(CH_2)_{14}COOH$, molecular weight = 256.42), is a long-chain fatty acid of 16 carbon atoms (see Scheme 17.1).[49] Hexadecanoic acid from *Volvariella bombycina* has been shown to exhibit anticancer properties.[48] Palmitic acids have been examined for toxicity against multiple myeloma cells and shown to have an adverse effect on their survival. The inhibitory effect of palmitic acid might be related to a number of mechanisms, including modification of the organization of lipids present in the cells and its interaction with nuclear receptors.[50] In 2009, Lee *et al.* reported the antiviral activity of palmitic acid, showing it to inhibit HIV infection by binding to the CD4 receptor.[49] Moreover, it

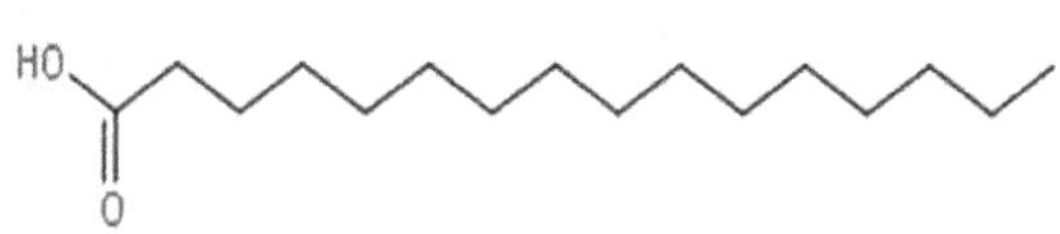

Scheme 17.1 Structure of hexadecanoic acid.[52]

also blocked the entry of HIV-1 and was thus able to prevent infection. Palmitic acid also functions as an emollient, emulsifier, surfactant, and opacifying agent, and can therefore be used to treat disruption of the skin's barrier in conditions such as acne, eczema, sensitive skin, and even signs of aging.[51]

17.7.2 Pyrrolo[1,2-*a*]Pyrazine-1,4-dione,hexahydro-3-(2-methylpropyl)

Pyrrolo[1,2-*a*]pyrazine-1,4-dione,hexahydro-3-(2-methylpropyl) ($C_{10}H_{16}N_2O_2$, molecular weight = 196.25) is a diketopiperazine compound derived from an α-amino acid. It is an organooxygen compound and an organonitrogen compound (see Scheme 17.2). This compound has been reported to have great reducing power, which suggests that it holds antioxidant potential. Pyrrolo[1,2-*a*]pyrazine-1,4-dione,hexahydro-3-(2-methylpropyl) reduces potassium ferricyanide to potassium ferrocyanide, which subsequently interacts with ferric chloride to produce a ferro-ferric complex. The compound also scavenges nitric oxide (NO) and 2,2-diphenyl-1-picrylhydrazyl (DPPH) radicals.[53] Free radicals, such as reactive oxygen species (ROS), reactive nitrogen species (RNS), and DPPH play a critical role in various diseases and also cause DNA damage.[54–56] In 2019, Kumar *et al.* reported the presence of pyrrolo[1,2-*a*]pyrazine-1,4-dione,hexahydro-3-(2-methylpropyl) in an ethanolic extract of the fruiting body of *Volvariella bombycina* through gas chromatography–mass spectrometry (GC–MS) analysis.[48] In 2015, Ser *et al.* reported the antioxidant activity of the bioactive compound pyrrolo[1,2-*a*]pyrazine-1,4-dione,hexahydro-3-(2-methylpropyl). Similar findings related to the antioxidant activity of this compound revealed that it has the ability to reduce the number of free radicals or scavenge them.[57–59] In short, an antioxidant presumably play key roles in obviating and treating chronic diseases.[60]

In 2010, Xu *et al.* reported the novel compound isodeoxyhelicobasidin ($C_{14}H_{18}O_3$), which is a benzoquinone compound with the structural composition (*S*)-5-hydroxy-3-methyl-6-(1,2,2-trimethylcyclopentyl)-1,4-benzoquinone (see Scheme 17.3).[61] The stereochemistry of the compound was assigned as an *S* configuration with negative optical rotation. Isodeoxyhelicobasidin, a human

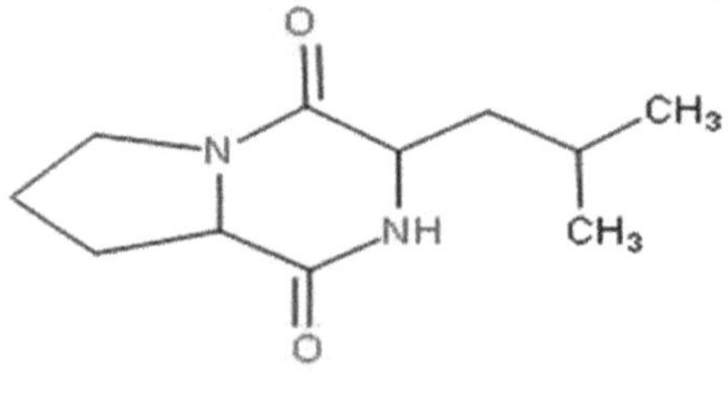

Scheme 17.2 Structure of pyrrole[1,2-*a*]pyrazine-1,4-dione, hexahydre-3-[2-methylpropyl].[53]

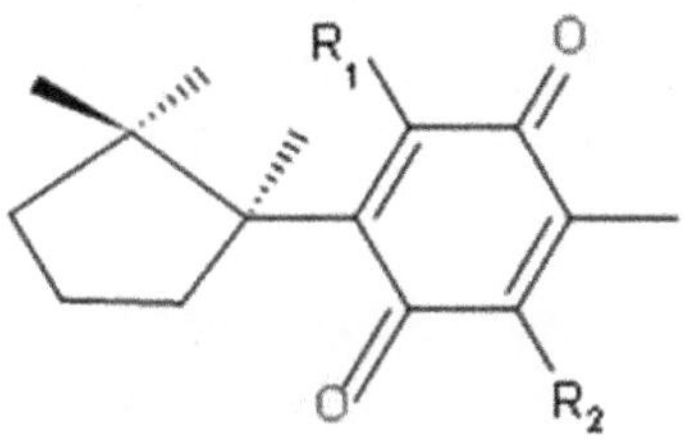

Scheme 17.3 Structure of isodeocyhelicobasidin.[61]

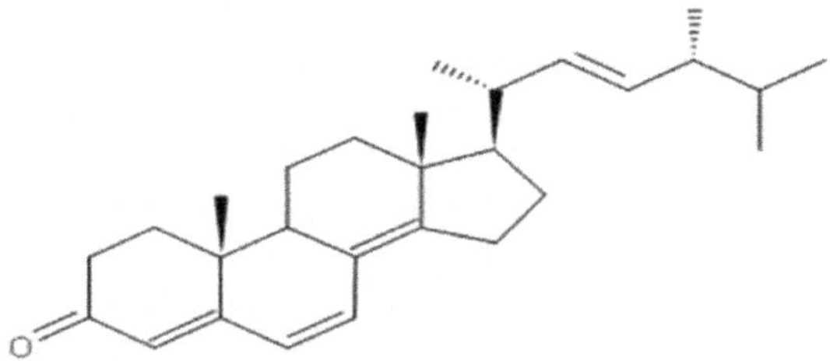

Scheme 17.4 Structure of ergesta-4,6,8(14),22-tetraen-3-one.[61]

neutrophil elastase generated by *Volvariella bombycina*, shows antibacterial action against Gram-positive pathogens such as *Staphylococcus aureus*, MRSA, QRSA, *Bacillus subtilis*, *Staphylococcus epidermis* and *Staphylococcus mutans*. The phytochemical compound, isodeoxyhelicobasidin produced by *V. bombycina* shows potential human neutrophil elastase (HNE) inhibitory activity. HNE is a stereo protease inhibitor that inhibits the elastin-a structural protein of the extracellular matrix responsible for the resilience and elasticity of skin, lungs, ligaments, *etc.* The HNE is the sole enzyme responsible for degrading ECM proteins such as elastin, collagen, fibronectin, *etc.* Therefore, by inhibiting HNE, these phytochemicals are useful for developing anti-aging cosmetics and other products.[61]

17.7.3 Ergosta-4,6,8(14),22-tetraene-3-one

Ergosta-4,6,8(14),22-tetraen-3-one is substituted by an oxo group at the 3-position, which makes it an ergostanoid and a 3-oxo-delta (4) steroid (see Scheme 17.4). It is one of the most well-known bioactive steroids and exhibits cytotoxic activity. In 2011, Zhao *et al.* observed chromatin condensation, nuclear fragmentation, and cell cycle arrest after treatment of human hepatocellular carcinoma HepG2 cells with ergosta-4,6,8(14),22-tetraen-3-one, resulting in apoptosis. The remarkable anti-proliferative activity of this bioactive compound clearly shows its potential for use in cancer treatment. The presence of ergosta-4,6,8(14),22-tetraene-3-one has also been observed in *V. bombycina,* exhibiting inhibitory activity on melanogenesis and cytotoxic activity.[61,62] Ergosta-4,6,8(14),22-tetraen-3-one has been reported to show dieresis, immunosuppression, as well as treatment of chronic kidney disease.[63]

17.7.4 Heteroglycan

The complement system is composed of a set of structurally different proteins present in an inactive form that can be activated *via* a set of pathways. One of these pathways is activated by polysaccharides through antibody independent mechanisms.[64,65] Many of these anti-complementary polysaccharides have been isolated from hot-water extracts of several medicinal plants as complement activators (see Scheme 17.5).[66] In 2008, Das *et al.* isolated a water soluble polysaccharide, a heteroglycan from a hot-water extract of *Volvariella bombycina.*[43] Acid hydrolysis and gas–liquid chromatography (GLC) analysis of the polysaccharide revealed the molar ratio of D-glucose, D-galactose, and D-mannose to be 2:1:1. The polysaccharides present in mushrooms are considered to be important antitumor agents. In 1985, Yamada *et al.* reported the anticomplementary activity of heteroglycans, showing that they could be of potential benefit for use in the treatment of complement-mediated diseases.[67]

17.7.5 Ergosterol Peroxide

Ergosterol and ergosterol peroxide are commonly-found bioactive metabolites in a diverse range of mushrooms and are known to possess various health promoting attributes, such as anti-inflammatory, anticancer, and antiviral properties (see Scheme 17.6).[68,69] The bioactive compounds identified in *Volvariella bombycina* and their biological properties are shown in Table 17.3. In 2007, Kobori *et al.* reported ergosterol peroxide to be the major antitumor sterol present in edible or medicinal mushrooms and observed suppression of inflammatory responses and growth of adenocarcinoma cells.[69] In 2010, Xu *et al.* reported the presence of ergosterol peroxide, identified as 5α,8α-epidioxy-22*E*,24*R*-ergosta-6,22-dien-3β-ol, in an extract of *Volvariella bombycina.*[61] In 2020, Jeong and Park reported the significant inhibition of triglyceride synthesis, expression of lipogenic factors, and

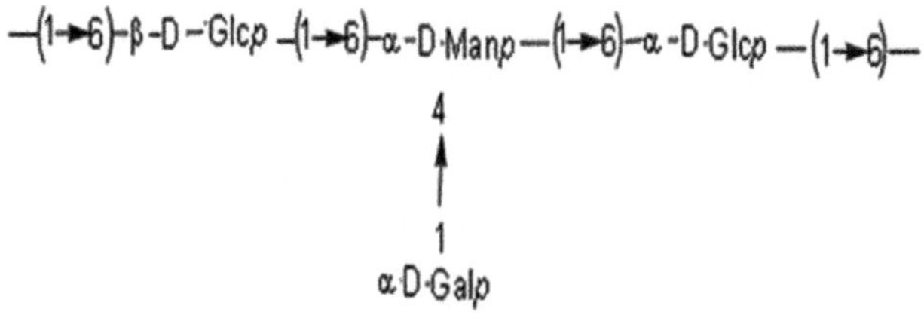

Scheme 17.5 The repeating unit of the polysacchanide.[43]

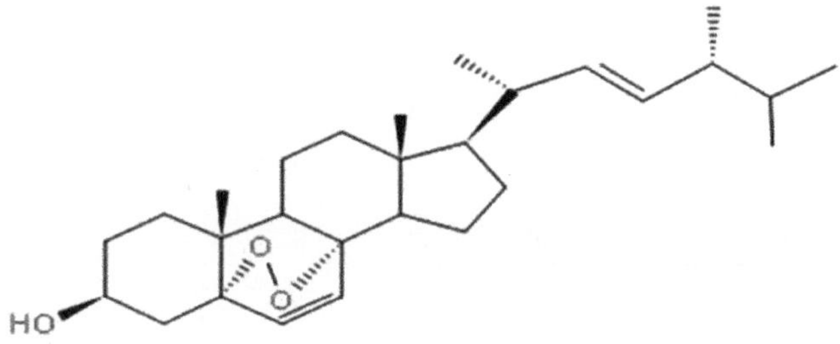

Scheme 17.6 Structure of ergesteral peroxide.[71]

Table 17.3 Bioactive compounds identified in *Volvariella bombycina* with their biological activities.

Metabolite	Biological activity
Isodeoxyhelicobasidin	HNE (human neutrophil elastase) inhibition, inhibitory effect on melanogenesis[61]
Ergosterol peroxide	Inhibition of triglyceride synthesis, expression of lipogenic factors and mitogen-activated protein kinase (MAPK) phosphorylation[70]
	Anti-tumor activity and suppression of inflammatory responses[69]
Heteroglycan	Anti-complementary activity[67]
Pyrrolo[1,2-*a*]pyrazine-1,4-dione, hexahydro-3-(2-methylpropyl)	Anti-oxidant activity[53]
Hexadecanoic acid	Anti-cancerous
	Emollient, emulsifier, surfactant and opacifying agent[51]
1,1′-Biphenyl (CAS)	Antimicrobial food preservative and an antifungal agrochemical[72]
Phthalic acid, butyl undecyl ester	Plasticizer compound, antimicrobial and antifouling[73]
13-Docosenamide, (*Z*)-	Host bacteria interaction, fluorescein quenching[74]
2,3-Dimethyl-5-propyldecane	Used in the preparation of novel S1P1 receptor agonists useful in lymphatic transport[75]

mitogen-activated protein kinase (MAPK) phosphorylation of an extract of *Volvariella bombycina*, proving it to be a potential therapeutic agent for obesity and other associated metabolic diseases.[70]

In 2022, Nowak *et al.* suggested that mushrooms represent a great source of ergosterol peroxide in the daily diet and can be utilized for the preparation of nutraceuticals and functional foods.[69] In 2010, Jegadeesh *et al.* examined the nutritional and composition of the mycelia and fruiting body of *Volvariella bombycina* (see Table 17.4). The nutritional composition of *V. bombycina*, with rich protein and low lipid content, proves it to be an ideal food for good health.[76]

Table 17.4 Nutritional analysis of the edible mushroom *Volvariella bombycina*.

	Carbohydrate (%)	Crude fiber (%)	Lipid (%)	Moisture content (%)	Protein (%)	Total ash (%)
Fruiting body	38.90	24.60	2.72	9.68	28.30	10.90
Mycelia	34.75	31.80	1.15	8.5	25.50	9.03

Table 17.5 Antagonistic property of *Volvariella bombycina*.

Compound/extract	Bacterial pathogens
Isodeoxyhelicobasidin, ethyl acetate extract	*Bacillus subtilis* ATCC 6633, *Bacillus subtilis 1021*[76]
Isodeoxyhelicobasidin, ethyl acetate, hexane, chloroform, methanol extract	*Staphylococcus aureus* ATCC 12600, *S. aureus* 503, *methicillin-resistant S. aureus* CCARM3167 (MRSA), *quinolone-resistant S. aureus* CCARM 3505 (QRSA)[61]
Hexane, chloroform, methanol extract	*Escherichia coli* ATCC 11675[76]
Ethyl acetate, methanol extract	*Klebsiella pneumonia* ATC 13883[76]
Chloroform, methanol extract	*Pseudomonas aeruginosa* ATCC 10145[76]
Isodeoxyhelicobasidin	*Staphylococcus epidermidis* 3958[61]
Isodeoxyhelicobasidin	*Streptococcus mutans* 3065[61]

Different extracts of *V. bombycina* have been tested for antibacterial activity against clinically important bacterial strains, such as *Bacillus subtilis*, *Staphylococcus aureus*, *Escherichia coli*, *Klebsiella pneumoniae*, and *Pseudomonas aeruginosa*. Phytochemicals found in *V. bombycina* have shown more antibacterial activity against Gram-positive bacteria than Gram-negative bacteria (see Table 17.5). This can be related to the mechanisms involving the inhibition of cellular processes, such as increasing plasma membrane permeability and ion leakage. The activity of these compounds can be related to the solubility of the active compounds in different solvents and their polarity.[76] Phytochemicals, *viz.*, ergosta-4,6,8(14),ergosterol peroxide,22-tetraene-3-one, indazole and, indole-3-carboxaldehyde have been reported to be produced by *V. bombycina* and show inhibitory activity of melanogenesis and cytotoxic activity.[61]

Despite its good quality, excellent taste, and pleasant flavor, the edible wild mushroom *Volvariella bombycina* is a rich source of a large number of bioactive compounds that have numerous medicinal and nutritive properties. An extract of *V. bombycina* has been shown to exhibit antimicrobial, antioxidant, anti-proliferative, cytotoxic, and anticancer properties, which makes this fungus a potential source of medicinal remedies. *V. bombycina* can also be used as a functional food due to its high protein and fiber content and low lipid content.

17.8 Cultivation Technology of *Volvariella bombycina*

Volvariella volvacea, also named the paddy straw mushroom, is a mushroom that grows in tropical climates below tree shade conditions in the temperature range of 28–35 °C. It is a rapid-growing mushroom with the ability to produce 7–8 crops under growing conditions from March to October only. Moreover, a short shelf life is the major constraint on the cultivation of this species. The cultivation of *Volvariella bombycina* is one of the alternative approaches by which to produce paddy straw mushrooms throughout the year. The three distinguishing characteristics of silver stalk mushrooms, *viz.*, the growth and development of its fruiting bodies at low temperature, superior shelf life of its fruiting bodies on refrigeration, and slow opening of fruiting bodies on maturity, have made them an ideal choice to grow in paddy straw cultivation regions across the globe. Ready to mix spawns of *V. bombycina* can be prepared by incubation at 28–30 °C spawn run temperature for 10 days using wheat seeds or chopped paddy straw as a substrate. After completion of mycelial colonization, bags carrying inoculums are shifted to temperature conditions of 23–25 °C with the relative humidity maintained at 80–85%. Under such conditions, the pinhead starts developing within 3 to 4 days, followed by initiation of the full matured fruiting body in the next 4 to 5 days. First harvesting of the crop can be undertaken within 16–18 days of spawn sowing. The first flush remains for 4 to 5 days, giving a yield of 60% of total fruit bodies. The next flush follows after a 4 to 5 day interval with a continuous staying duration of 4 to 5 days. The entire cropping cycle from initiation of spawn mixing to completion is almost 30 days. The average fruiting body weighs 40–60 kg. In 2016, Ahlawat *et al.* reported the cultivation of this species of mushrooms using compost composed of paddy straw mixed with cotton ginning mill waste in a ratio of 1 : 1 (w/w).[39] Admissible quality was observed up to 7 days under storage conditions of 4 ± 2 °C with minute variations in weight (−5.80%) as well as in nutritional structure (±4.17 to 6.95%), making it an ideal candidate for large-scale cultivation.

17.9 Threat

V. bombycina is consumed in many regions of the world, mainly in Asia and Africa. Although it has been reported to be distributed widely across the world, some European countries have categorized this species on the red data list. Its existence has been placed in danger due to the removal of its natural habitat, *i.e.*, standing or lying deadwood matter. Moreover, attempts have been made to cultivate this fungus.[19]

17.10 Conclusions

Mushrooms represent a valuable class of healthy dietary food that can be consumed by people of all age groups due to their abundance of nutrients.

Further, the richness of the different phytochemicals of mushrooms, such as terpenoids, polyphenols, proteins, *etc.*, making them an ideal candidate for the development of functional foods. These functional foods can play an important role in increasing immune strength, lowering blood cholesterol levels, inhibiting tumor cell proliferation, *etc.* Moreover, the prebiotic effects of mushrooms have led scientists to find and study novel species to understand the correlation between the regulation of the gut microbiome and consumption of mushrooms. *Volvariella bombycina* is one of less explored species of the *Volvariella* genus that can be grown in subtropical or temperate climates. The fruiting bodies of this species are a rich source of carbohydrates, proteins, vitamins, and different minerals, such as Zn and Ca.

References

1. S. T. Chang and P. G. Miles, *Mushrooms: cultivation, nutritional value, medicinal effect, and environmental impact*, CRC Press, Boca Raton, Fla, USA, 2nd edn, 2008.
2. P. G. Ergönül, I. Akata, F. Kalyoncu and B. Ergönül, *Sci. World J.*, 2013, 163964.
3. S. T. Chang, *Int. J. Med. Mushrooms*, 2006, **8**, 297–314.
4. S. Kamal, J. Pandey, S. Ghignone and A. Varma, in *A Textbook of Molecular Biotechnology*, ed. A. K. Chauhan and A. I. K. Varma, International Publishing House Pvt. Ltd., New Delhi, India, 3rd edn, 2009, pp. 573–628.
5. S. T. Chang and P. G. Miles, *Mycologist*, 1992, **6**, 64–65.
6. S. Zafar, F. Jabeen, M. Akram, Z. Riaz and N. Munir, in *Poisonous plants and phytochemicals in drug discovery*, ed. A. G. Mtewa, C. Egbuna and G. N. Rao, Wiley John Wiley & Sons, USA, 2020, DOI: 10.1002/9781119650034.ch8.
7. K. Das, *NeBIO*, 2010, **1**(2), 1–13.
8. H. Thatoi and S. K. Singdevsachan, *J. Biotechnol.*, 2014, **13**(4), 523–545.
9. (https://www.globenewswire.com/news-release/2020/04/27/2022477/0/en/GlobalMushroom-Cultivation-Industry-2020-to-2025-Economic-Viability-of-Mushroom-Cultivation-and-Trade-by-Developing-Countries-Presents-Opportunities.html) (online accessed: 10/04/2022).
10. (https://connect2india.com/global/export-Mushroom-fromindia#:~:text=Summary%20for%20the%20Indian%20Mushroom%20export&text=It%20was%20around%2024.48%25%20of,)%2C%20Nepal(0.06).
11. R. L. Shaffer, *Mycologia*, 1957, **57**, 542–579.
12. W. A. Murrill, *Mycologia*, 1938, **30**, 359–371.
13. W. A. Murrill, *Mycologia*, 1911, **3**, 271–282.
14. W. A. Murrill, *Mycologia*, 1917, **10**, 140–145.
15. W. C. Coker, *J. Elisha Mitchell Sci. Soc.*, 1947, **63**, 220–230.
16. W. A. Murrill, *Lloydia*, 1942, **5**, 136–157.
17. W. A. Murrill, *Lloydia*, 1951, **14**, 194–200.
18. (https://www.press.uillinois.edu/wordpress/mushroom-monday-volvariella-bombycina).

19. A. Szczepkowski, A. Kujawa and M. Halama, *Pol. J. Environ. Stud.*, 2013, **22**, 41–51.
20. M. Ajana, A. Ouabbou, S. El-kholfy, A. Nmichi, T. A. Ouazzani, R. Benkirane and A. Douira, *Int. J. Environ. Agric. Biotechnol.*, 2017, **2**, 940–945.
21. M. P. Kirk, P. F. Cannon, D. W. Minter and J. A. Stalpers, *Dictionary of the Fungi*, CABI, Oxon, United Kingdom, 10th edn, 2008.
22. D. S. Hibbett, *Mycologia*, 2006, **98**, 917–925.
23. S. J. Seok, Y. S. Kim and K. J. Lee, *Kor J. Mycol.*, 1994, **22**, 209–215.
24. (https://www.messiah.edu/Oakes/fungi_on_wood/gilled%20fungi/species%20pages/Volvariella%20bombycina.html) (online accesses: 10/04/2022).
25. J. C. R. Turner and J. Webster, *Chem. Eng. Sci.*, 1991, **46**, 1145–1149.
26. N. P. Money, *Mycologia*, 1998, **90**, 547–558.
27. M. F. W. Fischer and N. P. Money, *Fungal Biol.*, 2010, **114**, 57–63.
28. P. Roberts and S. Evans, *The Book of Fungi*, Illinois University of Chicago Press, Chicago, 2011, p. 320.
29. S. W. Chiu and D. Moore, *Mycol. Res.*, 1990, **94**, 327–337.
30. M. Kuo, Volvariellabombycina. Retrieved from the MushroomExpert.Com. Website: http://www.mushroomexpert.com/volvariella_bombycina.html (online access: 2020).
31. D. Arora, *Mushrooms demystified: a comprehensive guide to the fleshy fungi*, Ten Speed Press, Berkeley, California, 1986, vol. 903, pp. 261–262.
32. M. Obodai and G. T. Odamtten, *J. Basic Appl. Mycol.*, 2012, **3**, 21–26.
33. M. K. Biswas, *Intl. J. Econ. Plants*, 2014, **1**, 43–47.
34. V. Thuc, R. G. Corales, J. T. Sajor, N. T. T. Truc, P. H. Hien, R. E. Ramos, E. Bautista, C. J. M. Tado, V. Ompad, D. T. Son and N. V. Hung, in *Sustainable rice straw management*, ed. M. Gummert, N. Hung, P. Chivenge and B. Douthwaite, Springer, Cham, Denmark, 2020, pp. 93–109.
35. D. Bao, M. Gong, H. Zheng, M. Chen, L. Zhang, H. Wang, J. Jianping, W. Lin, Z. Yongqiang, Z. Gang, Z. Yan, L. Chuanhua, W. Shengyue, Z. Yan, Z. Guoping and T. Qi, *PLoS One*, 2015, **8**, 1.
36. J. R. Quaye, *Mushroom J. Trop.*, 1987, **7**, 39–44.
37. https://commons.wikimedia.org/wiki/File:Volvariella_bombycina_growth_stages.jpg (Accesses: 10/04/2022).
38. P. K. Chaudhary, M. Shrestha, B. H. Poudel and M. K. Adhikari, *Nepal J. Biotechnol.*, 2017, **5**, 27–31.
39. O. P. Ahlawat, P. Sinha and M. Singh, *J. Sci. Ind. Res.*, 2016, **75**, 562–569.
40. G. H. Xu, S. J. Choo, Y. H. Kim, I. J. Ryoo, S. J. Seok, J. S. Ahn and I. D. Yoo, *J. Microbiol. Biotechnol.*, 2010, **20**, 78–81.
41. Z. M. Thu, K. K. Myo, H. T. Aung, M. Clericuzio, C. Armijos and G. Vidari, *Molecules*, 2020, **1972**, 1–42.
42. C. Egbuna, J. C. Ifemeje, T. L. Kryeziu, M. Mukherjee, H. Shah, G. M. N. Rao, L. J. F. J. Gido and H. Tijjani, in *Phytochemistry, Fundamentals, modern techniques and applications*, Apple Academic Press, USA, 2018, vol. 1, pp. 3–35.

43. D. Das, D. Maiti, K. Chandra, S. Mondal, A. K. Ojha, S. K. Roy, K. Ghosh and S. S. Islam, *Carbohydr. Res.*, 2008, **343**, 2258–2265.
44. H. Tong, F. Xia, K. Feng, G. Sun, X. Gao, L. Sun, R. Jiang, D. Tian and X. Sun, *Bioresour. Technol.*, 2009, **100**, 1682–1686.
45. L. Barros, M. J. Ferreira, B. Queiros, I. C. F. R. Ferreira and P. Baptista, *Food Chem.*, 2007, **103**, 413–419.
46. S. Rossi, V. De-Chiara, A. Musella, H. Kusayanagi, G. Mataluni, G. Bernardi, A. Usiello and D. Centonze, *J. Neurol. Sci.*, 2008, **28**, 7284–7292.
47. S. M. Badalyan, *Int. J. Med. Mushrooms*, 2003, **5**, 153–162.
48. S. G. Kumar, M. Karnan and A. Panneerselvam, *Int. J. Pharm. Biol. Sci.*, 2019, **9**, 344–347.
49. D. Y. Lee, X. Lin, E. E. Paskaleva, Y. Liu and S. S. Puttamadappa, *et al.*, *AIDS Res. Hum. Retroviruses*, 2009, **12**, 1231–1241.
50. Y. Nagata, I. Ishizaki, M. Waki, Y. Ide, M. A. Hossen, K. Ohnishi, T. Miyayama and M. Setou, *Leuk. Res.*, 2015, **39**, 638–645.
51. A. Mieremet, R. Helder, A. Nadaban, G. Gooris, W. Boiten, A. E. Ghalbzouri and J. A. Bouwstra, *Int. J. Mol. Sci.*, 2019, **20**, 6069.
52. National Library of Medicine, PubChem, 2022, https://pubchem.ncbi.nlm.nih.gov/compound/hexadecanoicacid (online accessed: 10/04/22).
53. M. Manimaran and K. Kannabiran, *Open Bioact. Compd. J.*, 2017, **5**, 23–30.
54. A. F. Chen, D. D. Chen and A. Daiber, *et al.*, *Clin. Sci.*, 2012, **123**, 73–91.
55. S. Reuter, S. C. Gupta, M. M. Chaturvedi and B. B. Aggarwal, *Free Radical Biol. Med.*, 2010, **49**, 1603–1616.
56. G. Yen, H. Lai and H. Chou, *Food Chem.*, 2001, **74**, 471–478.
57. H. L. Ser, U. D. Palanisamy, W. F. Yin, S. N. A. Malek, K. G. Chan, B. H. Goh and L. H. Lee, *Front. Microbiol.*, 2015, **6**, 1–11.
58. M. Gopi, N. B. Dhayanithi, K. N. Devi and T. T. A. Kumar, *J. Coastal Life Med.*, 2014, **2**, 632–637.
59. D. Balakrishnan, A. Bibiana, A. Vijayakumar, R. Santhosh, K. Dhevendaran and P. Nithyanand, *Indian J. Microbiol.*, 2015, **55**, 13–18.
60. J. A. Morales-González, *Oxidative Stress and Chronic Degenerative Diseases - a Role for Antioxidants*, IntechOpen London, 2013, DOI: 10.5772/4572.
61. G. H. Xu, S. J. Choo, Y. H. Kim, I. J. Ryoo, S. J. Seok, J. S. Ahn and I. D. Yoo, *J. Microbiol. Biotechnol.*, 2010, **20**, 78–81.
62. Y. Y. Zhao, X. Shen, X. Chao, C. Chao, X. L. Cheng, Y. Zhang, R. C. Lin, K. J. Du, W. J. Luo, J. Y. Chen and W. J. Sun, *Biochem. Biophys. Acta*, 2011, **1810**, 384–390.
63. A. F. Chen, D. D. Chen and A. Daiber, *et al.*, *Clin. Sci.*, 2012, **123**, 73–91.
64. M. C. Carroll, *Nat. Immunol.*, 2004, **5**, 981–986.
65. M. W. Turner, *Immunol. Today*, 1996, **17**, 532–540.
66. H. Yamada and H. Kiyohara, in *Immunomodulatory agents from plants*, ed. H. Wagner, Birkhäuser Publishing: Basel, Switzerland, 1999, pp. 161–202.
67. H. Yamada, T. Nagai, J. C. Cyong, Y. Otsuka and M. Tomoda, *et al.*, *Carbohydr. Res.*, 1985, **144**, 101–111.

68. M. Kobori, M. Yoshida, M. O. Kameyama and H. Shinmoto, *Br. J. Pharmacol.*, 2007, **150**, 209–219.
69. R. Nowak, N. N. Jechalke, W. Pietrzak and U. G. Dziki, *Food Chem.*, 2022, **369**, 130927.
70. Y. U. Jeong and Y. J. Park, *Int. J. Mol. Sci.*, 2020, **21**, 460.
71. H. Y. Wu, F. L. Yang, L. H. Li, Y. K. Rao, T. C. Ju, W. T. Wong, C. Y. Hsieh, M. V. Pivkin, K. F. Hua and S. H. Wu, *Sci. Rep.*, 2018, **8**, 17956.
72. National Library of Medicine, https://pubchem.ncbi.nlm.nih.gov/compound/Biphenyl (online access: 12/4/2022).
73. G. P. Beulah, P. Tresina Soris and V. R. Mohan, GC–MS Determination of Bioactive Compounds of *Dendrophthoefalcata* (L.F) Ettingsh: An Epiphytic Plant, *Int. J. Health Sci. Res.*, 2018, **8**, 261–269.
74. E. Tamilmani, R. Radhakrishnan and K. Sankaran, 13-Docosenamide release by bacteria in response to glucose during growth-fluorescein quenching and clinical application, *Appl. Microbiol. Biotechnol.*, 2018, **102**, 6673–6685.
75. M. Bolli, *et al.*, *J. Med. Chem.*, 2014, **57**, 78.
76. R. Jegadeesh, N. Raaman, K. Periyasamy, L. Hariprasath, R. Thangaraj, R. Srikumar and S. R. Ayyappan, *Int. J. Microbiol. Res.*, 2010, **1**, 110–113.

CHAPTER 18

Lesser-known Wild Medicinal and Edible Mushrooms

RAMANDEEP KAUR,*[a] URVASHI,[a] PRATIBHA VYAS[b] AND TANVI SAHNI[a]

[a] Department of Chemistry, Punjab Agricultural University, Ludhiana-141004, India; [b] Department of Microbiology, Punjab Agricultural University, Ludhiana-141004, India
*Email: ramansaini@pau.edu

18.1 Introduction

Mushrooms, a soft, spongy, and flesh-like fungal species that bear fruiting spores, can be found on decaying wood, in soil, and other materials. These fungi have attained the status of vegetable despite not being plants, and are self-growing. Due to their highly nutritious value they have been a part of the diet of humans for a long time. Most edible mushroom varieties are cultivated, among which 'Button mushrooms' are at the top due to their market demand. Apart from their use in human diet, mushrooms have long been used in traditional medicines. However, these medicinal mushrooms are not becoming part of our diet due to their therapeutic potential and are a new source for dietary supplements. Due to their remedial properties, a new class of drugs, 'mushroom pharmaceuticals', is emerging. In addition to their role in the human diet, mushrooms are also able to control several agricultural pests and also serve as plant protection agents.[1] Moreover, some mushrooms are used in the cosmetic industry.[2] Like plants, medicinal mushrooms are also used in powder or extract form for the treatment of diseases. Synonymous to 'herbal drugs', medicinal mushrooms, their spores or other dried parts, are termed as 'mushroom drugs'. Most of the research in the last

Phytochemistry and Nutritional Composition of Significant Wild Medicinal and Edible Mushrooms: Traditional Uses and Pharmacology
Edited by Ajay Sharma, Garima Bhardwaj and Gulzar Ahmad Nayik

Published by the Royal Society of Chemistry, www.rsc.org

years has been focused on the nutritional properties of fungi, yet more is still under progress to explore their medicinal potential. Fungi can be a rich source of medicinal molecules that have different modes of action against particular diseases; different classes of mushrooms are source of different categories of molecules. In this chapter, we have compiled information on the lesser known mushrooms that are edible and possess medicinal properties but are still not well known due to the limited literature available on them.

18.2 *Pholiota* spp.

18.2.1 Taxonomy

The genus *Pholiota*[3] comprises small or medium sized mushrooms that are fleshy in texture and bear a scaly or glutinous to dry cap. This genus constitutes a large number of species and to date approximately 157 species have been recorded that belong to this genus.[4] The name Pholiota is derived from the Greek word *pholis*, which means 'scales'; hence, the basic characteristic feature of this species is that they bear scales on their caps. The caps contain smooth surface spores that are mostly brownish in color.[5] These spores feature germ pores that are somewhat narrow in most species. The cystidia are generally arranged towards the outer edge (pleurocystidia), are mostly yellow in color (chrysocystidia).

Overall, the main distinguishing features of *Pholiota* species are their cystidia, color and thickness of their walls. However, the distinction between different species cannot be made based on the above morphological features only as the outer physiology of the species is often dependent upon the environmental conditions. Also, for example, a few species feature a gelatinous layer during their early growth stages and others feature this layer during their mature stages. Sometimes within the same species, diverse morphological characteristics can be observed depending upon the environmental conditions.[5]

Out of a large number of species of the genus *Pholiota*, a small group of species mainly grow on wood, which often results in the formation of a white rot, whereas there are other species in this genus that are able to grow even on burnt ground and forest soil; however, no single species has been found to be mycorrhizal.[6] Another distinguish feature of a few species is the formation of partial veil along the stripes forming the annular rings.

18.2.2 Resource Availability, Sustainability and Distribution

Pholiota species grow well in temperate climates.[7] They are active participants in aiding the decomposition of wood as well as organic matter in soil. As most of the species of this genus grow on wood they are known to cause heartwood rot, hence they are commonly called as 'wood decomposers'.[5]

Pholiota species are most commonly found in low-temperature conditions; they can be found on mountains during the summer and on plains only during the winter.

18.2.3 Identification

The common identification feature of *Pholiota* is the presence of brown-colored (rusty or cinnamon brown) spore prints. Gills are found on their stem that do not reach the ring zone. The most common feature of *Pholiota* is the presence of a scaly to slimy cap. However, as discussed earlier, the species cannot be distinguished only on the basis of morphology characteristics; proper identification at the microscopic level is required.

The written record regarding the identification of these species is documented in North American monographs,[8,9] which detail only their physical aspects. A few other reports are available that give distinguished features based on reproduction studies, *etc.*[10–13] Proper identification features based on genetic data need to be investigated and documented.

18.2.4 Pharmacological Potential, Bioactive Secondary Metabolites and Medicinal Uses

The Pholiota genus comprises a wide range of bioactive molecules that exhibit anticancer[14] and antioxidant properties.[15,16] Gas chromatography (GC) analysis of six species of *Pholiota* (*P. adiposa, P. aurivella, P. highlandensis, P. nameko, P. squarrosa* and *Pholiota* sp.) has revealed the presence of campesterol, stigmasterol, and β-sitosterol as the major phytosterols in their fruiting bodies[17] (Figure 18.1).

Four new lanostane diesters, named pholiols A–D (1–4), together with an acyclic triterpene, (3*S*,6*E*,10*E*,14*E*,18*E*,22*S*)-2,3,22,23-tetrahydroxy-2,6,10,15,19,23-hexamethyl-6,10,14,18-tetracosatetraene (5), ergosterol (6), and 3β-hydroxyergosta-7,22-diene (7), have been isolated from the *n*-hexane extract of *Pholiota populnea* (Figure 18.1). The isolated metabolites (1–6) were investigated for cytotoxic activity against Colo205 and Colo320 colon adenocarcinoma and nontumoral MRC-5 cell lines, with the results indicating that *P. populnea* is a promising source for finding new triterpenes with significant chemosensitizing activity toward cancer cells.[18]

The activity and application of lignocellulase from *P. adiposa* have been reported in several studies.[19,20] Some Pholiota species are edible—*e.g.* *P. microspora* is well-known for its culinary usage in Asian countries,[21] whereas *P. squarrosa* is poisonous. The antioxidant, antibacterial and antimicrobial properties of a few species of *Pholiota* (*Pholiota adiposa* (Batsch) P. Kumm, *P. lubrica* (Pers.) Singer, and *P. squarrosa* (Vahl) *P. Kumm*) have been reported and their use was recommended as natural antioxidant and antimicrobial agents.[22]

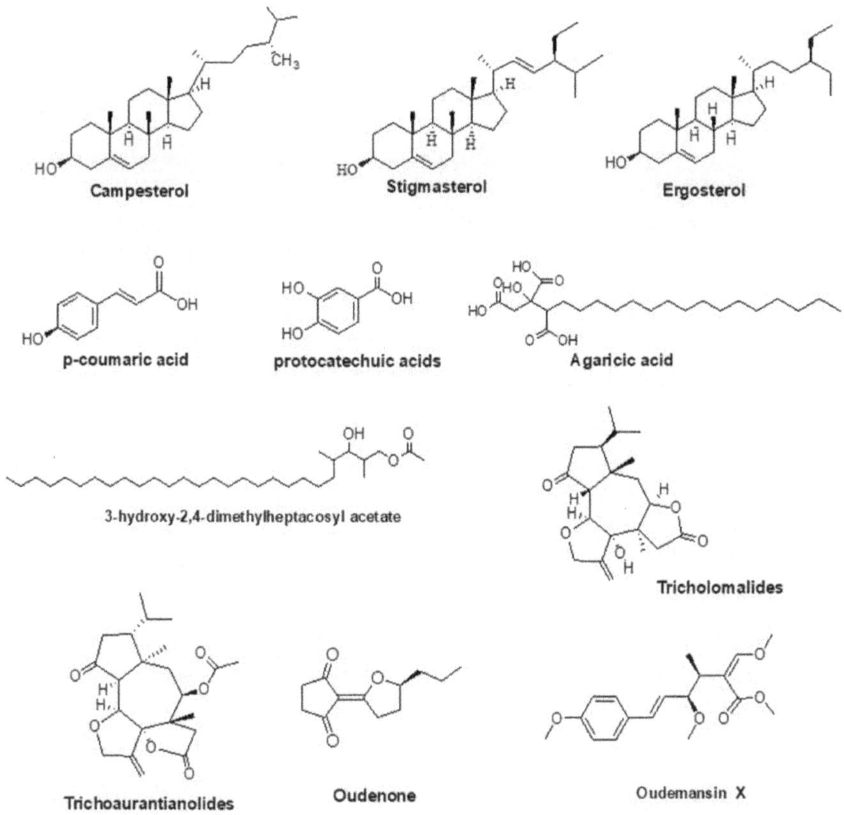

Figure 18.1 Chemical structures of the major compounds present in mushrooms.

18.3 *Pholiota aurivella* (Batch) P. Kumm – Golden Scalycap

P. aurivella (Golden scalycap) is a mushroom belonging to the family Strophariaceae. The name *Pholiota aurivella* is derived from two words: *Pholiota* meaning scaly, and *aurivella* meaning golden fleece. The morphological distinction of this species is not as easy as other genus as most of the species in the Pholiota genus bear a greasy or slimy cap like *P. alnicola* (grows on alders) and *P. gummosa* (grows in grass beside sick/deciduous trees). *P. aurivella* is one the largest mushrooms in the group. As already discussed earlier, due to the overlapping morphology features among different Pholiota species, proper identification at the microscopic level is necessary. For example, *P. aurivella* shares a common physical appearance to *P. limonella*, with only a difference in spore size (smaller than *P. limonella*).

18.3.1 Distribution

P. aurivella is native to the northern temperate and boreal forests of Canada (Southern), New Zealand and the United States and is commonly found in west and south-west America during the end of summer and autumn. It is also commonly found in the south-eastern regions of England and Wales, where beech trees are mature and are found in abundance; however, in other areas it is not so common. In addition to Britain and Ireland, this species is also abundantly found in the central and northern regions of Europe and Asia.

18.3.2 Occurrence

P. aurivella is greasy or slippery in appearance and grows as a parasite (in clusters) on live or dead coniferous trees, most commonly, maple, elm, beech and birch.[23]

18.3.3 Edibility

P. aurivella has been categorized as edible; however, a few studies have reported gastric issues in people who consumed it. Moreover, it is not very palatable as it tastes like unsweetened marshmallows.[24]

18.3.4 Morphology

18.3.4.1 Cap

The cap of *P. aurivella* is usually 2 to 6 inches wide with a bell shape bearing a broad knob. The texture of the cap is slippery and is yellow or yellowish orange in color, and is covered with a sticky liquid in its early growth stages. The scales are dark and sticky during the early growth stages and later grow bigger and become flattened and red-wine in color compared to in the young stages.

18.3.4.2 Gills

P. aurivella features broad gills at its base that are yellowish to rusty-brown in color.

18.3.4.3 Spores

The spores of *P. aurivella* are brown colored and smooth. The apical pore is basically oblong or vaguely bean-like in appearance and is broadly elliptical at the apex. The thickness of the wall is around 0.25 μm.

18.3.4.4 Stalk/Stipe

The yellowish brown stalk is solid and thick with a dry texture and is 5 to 10 cm in length and 0.5 to 1.5 cm in diameter. It can vary from centric to eccentric and bears scales under the ring; it is covered with soft white wool above the ring. The annulus is covered with a woolly coating on top and fibrils below, which become scaly on moving downwards.

18.3.4.5 Veil

The veil of *P. aurivella* is also covered with fibers that are white to off-white in color, where half of the veil toward the upper end shows a decrease in the number of fibers.

18.3.5 Taxonomy

Kingdom: Fungi
Phylum: Basidiomycota
Class: Agaricomycete
Order: Agaricales
Family: Strophariaceae
Genus: Pholiota
Species: *P. aurivella*

18.3.6 Cultivation

P. aurivella exists either as a parasite on living trees or as a saprophyte on dead wood, mainly on Salix species. Usually, *P. aurivella* grows above ground most commonly on branches or stems and is found growing in dense bunches.[25] A few lab experiments have reported the effect of light on the production of sporophores of *P. aurivella*, where light showed a positive correlation with sporophore production as well as the length of time of the light stimulus within the first 14 days of mycelium plantation. The nutrient supply also had a positive impact on the number and size of the spores, where 7% concentration of malt in the solid medium was found to have the maximum effect. The size of the container as well as proper aeration were considerations required for the production of normal sized spores, with airtight small containers preferential over cotton-plugged large containers.[26]

18.3.7 Chemical Composition

The chemical and mineral analyses of *P. aurivella* extract revealed that its extract contained phenolic (*p*-coumaric and protocatechuic acids) and fatty acids (linoleic, oleic, and palmitic acids); also, among the minerals in this fungus, arsenic was found to be present in the highest quantity[27]

(Figure 18.1). A lectin named *Pholiota aurivella agglutinin* (PAA) was isolated from the fruiting bodies of *P. aurivella* using preparative PAGE.[28]

18.3.8 Pharmacological/Medicinal Properties

The effect of *P. aurivella* extract on CCl_4-induced stress in rats and some biochemical parameters was studied. It was found that the extract decreased the antioxidant defense system (ADS) levels and increased the aspartate aminotransferase (AST), alanine aminotransferase (ALT), lactate dehydrogenase (LDH) in serum, and malondialdehyde (MDA) levels. Moreover, severe histopathological changes were observed in the liver and kidney. Overall, *P. aurivella* extract was found to be toxic when consumed orally; hence, this fungus is not recommended for direct consumption.[27] *P. aurivella* and a few other edible mushrooms produce sufficient quantities of the phenoloxidase enzymes, which exhibit maximum activity in acidic pH environments.[29]

18.4 *Pholiota squarrosa* (Vahl) P. Kumm (Scaly Pholiota)

18.4.1 Synonym: *Agaricus squarrosus* Vahl

The Latin word *squarrosa* means "with upright scales, rough, scurfy". Georg Christian Oeder initially documented *P. squarrosa* as *Agaricus squarrosus* and the scientific name was sanctioned by Elias Magnus Fries in the Systema Mycologicum. Moreover, it was included in the genus Pholiota by the German scientist Paul Kummer. The common names used for this mushroom are *scaly* or *shaggy Pholiota* and *shaggy scalycap*. Both the cap and the stem are covered with downward- and backward-facing small scales. This fungus does not have any particular smell – it may have a strong and unpleasant smell like garlic, radish, onion, or lemon. The mushroom secretes chemicals that help it to neutralize the defense mechanism of plants and infect them. The only difference between *P. squarrosa* and *P. squarrosoides* is the stickiness between the scales of a paler cap and smaller spores in the latter.

18.4.2 Edibility

P. squarrosa is considered to be an edible mushroom. However, few reports related to its poisonous nature have been documented when the mushroom is consumed with alcohol. The main symptoms of poisoning are vomiting and diarrhea, which can last up to ten hours in infected persons.[30] The mishmash combination of mushrooms and alcohol might be the reason for the toxicity. Hence, it is not usually recommended to consume mushrooms while drinking alcohol.

18.4.3 Ecology and Habitat

P. squarrosa is considered to be a white rot fungus that has the tendency to convert lignin present in wood to CO_2, as it utilizes cellulose as a source of carbon. *P. squarrosa* usually affects a number of deciduous and coniferous trees, such as sugar/red maple, yellow/paper birch, American beech, spruce, and white ash.[31] This mushroom usually develops in clusters on the lower part of trees. It is a secondary as well as opportunistic parasite and feeds on trees that have already been injured, damaged, and infected by micro-organism fungi.[32] Being rich in proteins, this mushroom is a primarily consumed as food by the red squirrel *Sciurus vulgaris* and other *Sciurus* spp.,[33] whereas decaying mushrooms are consumed by fruit flies of the *Drosophila* genus.[34]

18.4.4 Distribution

P. squarrosa is distributed mainly in North America (extending from Canada to Mexico) and Europe,[35] where it is limited to coniferous forests. In the Netherlands, its fruiting bodies are mostly found on ancient timber wharves.

18.4.5 Taxonomy

Phylum: Basidiomycota
Family: Strophariaceae
Order: Agaricales
Family: Agaricales
Genus: *Pholiota*
Species: *P. squarrosa*

18.4.6 Description

18.4.6.1 Dimensions

Caps are approximately 1 to 4 inches; stipes 2 to 4 inches tall and 3–15 mm in thickness.

18.4.6.2 Cap

Its diameter is about 4–12 cm, yellowish-orange, and covered with upturned triangle-shaped brown scales arranged in concentric rings. Initially, the cap is convex and flattens with maturity, with margins rolled toward its inside. The cap flesh is dry, firm, light yellowish, and covered with pink or brown scales.

18.4.6.3 Gills

The gills are properly attached to this fungus. The color of the gills changes with maturity from yellowish to greenish and finally a rusty brown. The young gills are covered by a thin web-like veil.

18.4.6.4 Stem

It has dimensions of 7–15 cm in diameter and 6–15 cm in length. The lower portion has a distinct scaly ring that is covered with a scale-like cap, while the upper portion is smooth and yellowish in color.

18.4.6.5 Stipe

Dry, and bearing scales like its cap.

18.4.6.6 Veil

Yellow, partial veil that leaves a continual membranous ring or occasionally a ring zone on the upper stalk.

18.4.6.7 Spore Print

Rusty Brown.

18.5 *Polyporus* spp.

18.5.1 Taxonomy

The genus *Polyporus* was reported in 1729 by the Italian botanist Pier Antonio Micheli to include 14 species with the characteristic fruiting bodies having centrally-placed stipes and the underside of the cap containing pores.[36] The genus has been named *Polyporus* (Gr. *Poly* meaning many and *poros* meaning pores) due to the occurrence of numerous minute pores beneath the surface of the fruit bodies, and belongs to the family *Polyporaceae*, order *Polyporales,* and the morphological group of basidiomycete-like gilled mushrooms. The members of this family are often called 'Bracket or Shelf fungi' because they bear shelf or bracket like fruiting bodies. The fruiting bodies are known as 'conks' and are present either separate or in interconnected horizontal rows. The brackets range from a single row containing a few caps to multiple rows of caps weighing several hundred pounds.

In 1855, the genus was divided into three subgenera by Elias Fries in his book *Novae Symbol Mycologici*: *Eupolyporus*, *Fomes*, and *Poria*.[37] In a 1995 monograph, Maria Núñez and Leif Ryvarden placed 32 *Polyporus* species into six morphologically distinct groups: *Polyporus*, *Polyporellus*, *Admirabilis*,

Favolus, *Dendropolyporus*, and *Melanopus sensustricto*.[38] The species *Badius* was transferred to the genus *Polyporus* and later in 1801 was given its widely-accepted scientific name *Polyporus durus*. *Polyporus durus* is also known as *Polyporus badius* (Pers.), *Polyporus picipes* Fr., *Boletus badius* Pers., *Polyporellus picipes* (Fr.), P. Karst., *Boletus durus* Timmerm. and *Grifola badia* (Pers.) Gray, Schwein.

18.5.2 Resource Availability (Wild or Cultivated): Sustainability and Distribution

The majority of *Polypores* inhabit the trunks or branches of trees and consume their wood, but some can form mycorrhizal associations with trees. They play an important role in nutrient cycling and CO_2 production owing to their decaying nature.

Polyporus badius, previously known as *Polyporus durus*, grows well on the fallen branches of deciduous trees. Due to its brown colored cap, this species is referred to as a Bay Polypore. As they are well camouflaged by fallen brown leaves, they are very difficult to spot. The pores cannot be separated from the upper layer of the cap. The species is inedible due to its tough nature; however, their dried caps are used for as table decorations. This species is commonly found across Britain and Ireland and throughout Europe. It is also found in many different parts of North America and Asia.

18.5.3 Identification

The upper surface of the cap of *P. badius* is smooth and reddish-brown in color with a dark center without any zones. When fully developed, the size varies from 5 to 20 cm, and is irregular and lobed. The thin caps are funnel-shaped with wavy margins when mature.

The stem is 1.3–5 cm long and 0.5–1.5 cm wide, although there is no clear distinction between the cap and stem. The stem is grayish in color but sometimes is black in appearance near the base and tapers towards it.

Minute white tubes are packed beneath the cap at a density of 5 to 8 per mm, which are not visible to the naked eye. These tubes are 0.5–2.5 mm deep and end in whitish pores that upon aging turn yellow in color.

The spores are smooth, cylindrical or sausage-shaped with white spore print, and measure about 5–9 × 3–4 μm in size. The odor is mushroomy but not distinctive. The taste is mild.

18.5.4 Nutritional and Dietary Composition

The species of the genus *Polyporus* play an important role in the degradation of organic compounds. The compounds isolated from *Polyporus fomentarius* have been identified as polysaccharides with α 1–6 linked D-galactopyranosyl residues. About 30–40% of the residues are substituted by L-fucopyranosyl residues, D-galactopyranosyl or 3-*O*-α-D-mannopyranosyl-L-fucopyranosyl residues in the 2-position.[39] The compositions of *Polyporus squamosus* from Portugal and Serbia have shown to have 17.14 g per 100 g of protein, 2.69 g per 100 g of fat, 3.15 g per 100 g of ash and 77.02 g per 100 g of carbohydrates.[40]

18.5.5 Pharmacological Potential, Bioactive Secondary Metabolites and Medicinal Uses

The genus *Polyporus* has medicinal importance and different industrial applications. The species of the genus are used in traditional medicine and are known to possess hepatoprotective, nephroprotective, diuretic, antioxidative, antimicrobial, anti-inflammatory, cytotoxic, immuno-enhancing and hair-growing activities (Table 18.1). Moreover, agaric acid or laricic acid isolated from *P. officinalis* has been used in the treatment of tuberculosis (Figure 18.1). Crude extracts and compounds isolated from *P. umbellatus* have been shown to have broad-spectrum pharmacological potential, including anti-inflammatory, hepatoprotective, antioxidative, diuretic, nephroprotective, anti-cancer and immuno-enhancing activities.[41] Although different species of *Polyporus* have been studied for their medicinal importance, very few reports are available on the use of *P. badius* in the treatment of diseases.

Extracts of *Polyporus badius* (Gray) Schwein, *P. betulinus, T. versicolor L. vellereus* (Fr.) Fr., and *Heterobasidion annosum* (Fr.) Bres. have been shown to inhibit the binding of lipopolysaccharides (LPS) to receptors, suggesting that the extracts from these fungi contain structures for drug discovery against LPS-mediated septic shock.[42] The inhibition in acetylcholinesterase activity of ethanolic extracts of *Polyporus* spp., including *P. badius, P. fomentarius, P. annosus*, *P. gilvus*, *P. pinicola*, *P. sulphureus*, *P. radiatus*, *P. volvatus* and *P. stevenii* grown in Turkey was evaluated.[43] The highest activity of about 31% inhibition was shown by *P. sulphureus*, while extracts from all other fungi showed low ability in the chelation of ferrous ions and reduction of ferric ions. The lowest scavenging activity was shown by *Polyporus badius* in view of its low phenolic content of around 4.27 mg gallic acid equivalents (GAE) per g of extract.

Polyporus tenuiculus grown on wheat straw has been shown to contain 39 volatile compounds, including alcohols, acids, hydrocarbons, ketones, esters, and aldehydes, with its main volatile compound being 1-octen-3-ol. The fungus grown on supplemented wheat straw was found to have a higher protein content and lower fat and carbohydrate contents. The substrates were shown to have significant effects on the fiber and ash contents. The high fiber and protein contents of *P. tenuiculus* suggest its suitability for use in the food industry.[44]

Table 18.1 *Polyporus* spp. in the treatment of various diseases and their bioactive compounds.

Polyporus sp.	Activity	Activity against	Bioactive compound	Ref.
P. umbellatus	Anti-protozoal	*Plasmodium falciparum*		117
P. umbellatus	Antibacterial and antifungal	*Bacillus subtilis*, *Escherichia coli*, *Staphylococcus aureus*, *Chlamydia trachomatis*, *Saccharomyces cerevisae*, *Candida tropicalis* and Fusarium *graminearum*	Non-water-soluble type II antibiotics	118 and 119
P. badius	Inhibition of LPS binding to receptors	Gram negative bacteria	—	42
P. sclerotium	Anti-aldosteronic diuretic	—	Ergone (Ergosta-4,6,8(14), 22-tetraen-3-one)	120
P. umbellatus	Immune system enhancement	—	Polysaccharide	121
P. umbellatus	Hepatitis B infection	Hepatitis B virus	Polysaccharides	122 and 123
P. squamosus	Antibacterial and antioxidant	*Pseudomonas aeruginosa*	β-Tocopherol, polyunsaturated fatty acids	124
P. squamosus	Antibacterial activity	*Pseudomonas aeruginosa*	Tocopherols and phenolics	40
P. umbellatus	Antioxidant		Exopolysaccharides	125
P. badius, *P. squamosus*	Antimicrobial	*Bacillus subtilis, Micrococcus luteus, Pseudomonas fluorescens, P. aeruginosa*, and *Staphylococcus epidermidis*		56

18.5.6 Processing in the Food Industry

Polyporus badius has long since been used in the gelation of ewe's milk, leading to the formation of a curd. The product is consumed with rice and called "halbi" or "halvi". The product is soft and spongy and contains high levels of fat and nutritional content. In a study, enzymes from *Polyporus badius* along with ficin were used to develop a protocol for the coagulation of milk proteins for providing novel structures in foods.[45]

18.6 *Rhizopogon luteolus* Fr.

Rhizopogon luteolus is an ectomycorrhizal fungus used as a soil inoculant in agriculture and horticulture. To improve tree growth, it was purposely introduced into *Pinus radiata* plantations in Western Australia.[46] In 1935, the seedlings of *Pinus radiata*, *P. taeda* and *P. caribaea* grown in nursery beds were found to have ectomycorrhizal association with Rhizopogon luteolus at Banyabba in Northern New South Wales.[47]

18.6.1 Taxonomy

In 1817, this fungus was first described and named by the Swedish mycologists Elias Magnus Fries and Johan Nordholm.[48] The genus *Rhizopogon* was treated as ascomycetous rather than basidiomycetous by Carlo Vittadini. The genus comprises two words, *Rhiz* meaning root and *pogon* meaning beard. Thus, from its name, it is expected that the genus false truffles has root-resembling extremities. They are permanently fixed to pine tree roots by pale mycelial strings. The name of the species *luteolus* means 'false truffles' and as they are yellow in color they are also referred to 'the yellow truffle'. In Western USA and Britain, approximately 100 and 6 *Rhizopogon* species have been found, respectively.[49] Their size is typically 1.5–4.5 cm in size and they can be ovoid, ellipsoid, and oblate spheroid. They have no stem-like structure, and their interior tissue is softer than the outer skin. During their initial growth phase, their outer tissue is cream and eventually changes to ochre and olive brown during maturation. As the fruiting body expands, the outer surface cracks irregularly and is arbitrarily highlighted with ochre mycelial strands which furnish it a vague outer skin. The fruiting body core comprises numerous petite chambers creased with basidia and spores developing from it.[48] *Rhizopogon* belongs to the phylum Basidiomycota, class Agaricomycetes, order *Boletales*, family *Rhizopogonaceae* and genus *Rhizopogon.*

R. luteolus is symbiotic or parasitic on organisms, some are mycorrhizal or saprotrophic, and its cell walls are composed of chitin, polysaccharides (*e.g.* glucans) or both; it is mushroom-forming, filamentous; its hyphae are septate, its basidia may be branched or unbranched; dolipore septa and parenthesomes are present; it bears bridge-like clamp links over the septa; and the species includes mushrooms, bracket fungi, and puffballs. Asexual reproduction occurs *via* fragmentation, oidia or conidia; sexual reproduction by somatogamy, oidization or spermatization, resulting in dikaryotic hyphae that eventually give rise to basidia. The dikaryotic hyphae produce teleutospores (thick-walled resting spores).

18.6.2 Resource Availability (Wild or Cultivated): Sustainability and Distribution

Rhizopogon luteolus are abundant in Europe, as well as in North and South America, different parts of Africa, Australia and New Zealand.[50] In the western USA and Britain, approximately 100 and 6 Rhizopogon species have been found, respectively.[49] Spores covered in fruiting body, interior is soft during initial time, but become dry and powdery at maturity and are debarred into the atmosphere; includes several edible boletes, such as king boletes, butter boletes and queen boletes as well as pigskin poison puffballs.

The order of germination and growth of the germ tubes of basidiospores on seedlings raised in DW and placed in DW, DWP, PNS or PNS-P after inoculation was DW (64%) > PNS-P (36%) > DWP (25%) = PNS (24%).[51]

{DW: only distilled water; DWP: distilled water and P; PNS: plant nutrient solution along with P;[52] PNSP: plant nutrient solution without P}.

18.6.3 Medicinal Uses

Heteropolysaccharides, isolated from *Rhizopogon luteolus*, restrain innate methylated and acid monosaccharides, which makes it complicated to distil and elucidate their structures.[53,54] Due to their mixed and complex structures, they are considered to be a resource for medical and industrial applications. Ruthes *et al.*[55] investigated the antinociceptive, immunomodulatory, antitumor, anti-inflammatory, antioxidant, anti-sepsis and cytotoxic activities of mushroom heteropolysaccharides. Galactomannans, the major heteropolysaccharides of Ascomycetes, are effective in mammals as immunostimulants against tumors and infections.[56] More than 20 million people globally are affected by neurodegenerative Alzheimer's disease (AD). Aulitto *et al.*[57] reported that galactomannan was found to be a weak inhibitor against acetylcholinesterase.

18.6.4 Phytochemistry

18.6.4.1 Important Bioactive Secondary Metabolites/Natural Products

Among the isolated bioactive compounds, polysaccharides exhibit anti-hepatoprotective, antioxidant, anti-inflammatory, antiviral, antitumor, anti-hypoglycemic, immunomodulatory and antinociceptive properties. Homo- and hetero-polysaccharides are two classes of polysaccharides. Ruthes *et al.*[55] reported the β-glucans to be the most studied mushroom polysaccharides. Glucomannan and galactomannan are major heteropolysaccharides of Ascomycetes, which include D-mannan leading chains.[58] Galactomannans are composed of various ratios of D-mannose (Man) and D-galactose (Gal) monomers; their structure comprises a β-(1,4)-mannose linkage with α-(1,6)-galactose as side groups.

Flavonoids and phenolics exhibit a wide range of bio-activities, including anti-mutagenic, anti-oxidant, anti-bacterial, anti-carcinogenic and cytotoxic properties.[59,60] Fumaric, gallic and protocatechuic acids are major compounds found in *R. luteolus*. The total flavonoid and phenolic contents of extracts is in the order of: hexane > ethyl acetate > methanol > water.[61,62] A total of eight fatty acids, ranging from C15:0 to C24:0 have been detected in *R. luteolus.* Linoleic acid (~46%), oleic acid (~16%), and stearic acid (~24%) are the major fatty acids in this fungus. Heleno *et al.*[63] reported that linolenic acid is the major aromatic compound that gives the mushroom flavor. Other fatty acids, such as C15:0, C16:1, C16:0, C20:4, and C24:0 were found to total <7%. The total unsaturation was 68.2%, directly proportional to the nutritional value of mushroom. From a chemotaxonomic viewpoint, the linoleic : oleic acid ratio (L : O) was 2.80. Phytochemical investigation of *Rhizopogon luteolus* Fr. led to the isolation of 3-hydroxy-2,4-dimethylheptacosyl acetate together with tetracosanoic acid and ergosterol (Figure 18.1). The compound ergosterol exhibited privileged anti-oxidant and anti-cholinesterase activities. Thus, *R. luteolus* can be used in the cosmetic, food and pharmaceutical industries.

Sadi *et al.*[64] reported anti-oxidant activity using metal chelating, DPPH radical scavenging and reducing power assays of the various extracts of *R. luteolus*.

18.6.4.2 Nutritional and Dietary Composition

Truffle fruit-bodies serve as a major food source for many forest-dwelling mammals. The sporocarp of *R. luteolus* has no distinct peridium that can be removed easily.[65] Therefore, small mammals and deer feed on whole sporocarps of *R. luteolus*. In the sporocarps, approximately 76% digestible fiber is present.

Although the total nitrogen (N) content of the inner core of *R. luteolus* sporocarps was high, most of the N was in non-protein form and was allied with the cell wall components. The fungal sporocarps has total nitrogen, protein nitrogen, cell content nitrogen, non-protein nitrogen, gross energy, cell wall nitrogen, ash, water and total cell wall contents of 15.98, 5.05, 12.82, 7.77, 19.86, 3.16, 20.40, 76.45 and 189.30 mg per g fresh weight, respectively.[66]

18.6.5 Pharmacological Potential (Biological Activities Related to Medicinal Uses or Clinical Evidence)

Fumaric acid has acidifying properties[61] and is also broadly used in food and other industries. Valentão *et al.*,[62] reported fumaric acid to be the chief organic acid in a few edible mushrooms.

18.6.6 Cultivation and Harvesting

In regional restaurants, a few truffle species are commercially harvested for consumption by connoisseurs.

18.6.7 Safety, Toxicity and Regulation

Boer *et al.*[67] reported that Ascomycota fungus damages wood by causing changes in cell wall constituents and anatomy features. It can cause white rot, brown rot and soft rot decay. In white rot decay, the cell walls are attacked progressively from lumen and hemicellulose to lignin. In brown rot decay, the entire wall is attacked and all of the cell wall constituents except for lignin are used. In soft rot decay, longitudinal cavities develop and the walls are attacked from the lumen.

18.7 *Tricholoma terreum*

18.7.1 Taxonomy and Distribution

Tricholoma, a genus of Basidiomycete fungi (Basidiomycota), order Agricales and family Tricholomataceae, usually comprises gilled mushrooms. It grows in the temperate regions of both hemispheres with >250 known species. *Tricholoma* form ectomycorrhizal associations, such as arbutoid mycorrhizal relationships, with some herbaceous plants, some shrubs, and a number of trees. *Tricholoma terreum* first takes the form of a pinhead followed by globos, campanulate, convex, and humbified forms before finally becoming flat. It is covered with numerous scales that are silky and gray-blackish in color. The lamellae are not very dense but have a smeared denticulate margin with a color that varies from white to gray-ash. The stem is cylindrical (4–8 cm), fistulous, and hollow, with a varying diameter of 1–2.5 cm.[68]

18.7.2 Nutritional and Dietary Composition

A study conducted on the nutritional aspects of two widely consumed mushrooms in Jordan showed that the protein content of *Tricholoma terreum* was around 17.5% compared to the protein content of *Pleurotus eryngi* of 30.6%. It was also observed that 72–83% of the total nitrogen was in digestible protein form. The protein content is also affected by the mushroom type, nitrogen availability, and habitat. *Tricholoma terreum* was found to contain a higher ash percentage (12.3%) than *Pleurotus eryngi* mushrooms. However, the amount of crude fat present was found to be the same for both mushrooms. The data on the mineral analysis of the mushrooms showed significant differences between the two mushrooms. A higher potassium content was observed in *Tricholoma terreum* compared to *Pleurotus eryngi*. *Tricholoma terreum* had the highest carbohydrate content (50.8%) compared to *Pleurotus erygni*, with a higher dietary fiber content (45% of dry matter) in two *Tricholoma* species.[69]

18.7.3 Cultivation and Harvesting

The process of collecting wild edible fungi is usually compared with picking fruit from a tree. The future harvests of a tree are not affected by removing all the fruits, unless the is damaged, but may have an impact on its regeneration. This process can be followed for wild edible fungi but keeping in mind some limitations that the removal of unopened fruiting bodies inhibits the spread of spores. In a study conducted in Switzerland on 15 species of macro-fungi over a ten-year period, it was observed that harvesting of all the fruiting bodies from these species of mushroom had no significant effect on production. The production was affected by soil compaction, leaf litter layers, and indiscriminate digging for truffles. A number of edible fungi have to be picked without causing any damage to the fruiting bodies and edible parts as they are all above the ground. Proper management of forests is required to enhance the production of mushrooms. The following factors can be kept in mind for increasing the production of wild mushrooms:

- A clear-cut harvesting inhibits the formation of a number of edible ectomycorrhizal fungi for ten or more years. Recovery is only observed through re-establishment of fungi on trees that have the necessary nutrients to promote their growth.
- Trees should be selectively removed to promote the growth of other trees and inhibit the growth of weak specimens. This introduces a higher rate of rain as well as sunshine for rapid wetting and drying of the forest floor to maintain fungal productivity.
- Logging operations cause compaction of soil, thereby reducing productivity. It is also observed that the removal of larger branches helps to easily find wild fungi without increasing the base productivity.[70]

18.7.4 Processing in the Food Industry

The mushroom *T. terreum* grows widely and naturally in the geographical habitat of Turkey. It is known to be a late-season mushroom that forms huge groups in cold spells. Mushroom hunters with a low income visit the habitat regularly during the rainy season with the motive of collecting large quantities of *T. terreum* and selling them at the local market. Thus, this mushroom is an important source of income. *T. terreum* is considered to be a valuable food for rural populations due to its nutritional properties. It is known to contain ash (8.2%), protein (18.0%), fats (8.8%), cellulose (19.8%), carbohydrates (60.0%) and common sugars (2.4%). It was reported that in every 100 g of dry weight of the mushroom that there are around 730 mg of Fe, 139 mg of Zn, 19.2 mg of Cu and 15.7 mg of Mn.[71]

A high protein content and quality was reported in *T. terreum* compared with other edible mushrooms, vegetables and wild plants.[72]

18.7.5 Important Bioactive Secondary Metabolites/Natural Products

It becomes evident to note that a number of metabolites obtained from *Tricholoma* species have shown extensive biological activities, such as cancer cell cytotoxicity, leukemia cell growth inhibition, *etc.* A number of secondary metabolites have been isolated from *Tricholoma* fruiting bodies and have been studied according to their presumed biosynthetic origin as follows:

18.7.5.1 Terpenoids

These secondary metabolites most commonly occur in *Basidiomycota* and *Tricholoma* species. C-30 terpenoids are sesquiterpenoid dimers that are derived from the coupling of two C-15 farnesyl pyrophosphate units. It is also reported that *Tricholoma* species also contain some diterpenes, which are rarely found in *Basidiomycota.*

The diterpenes found in *Tricholoma* species have an unprecedented terpenoid skeleton. Another two types of secondary metabolites, tricholomalides and trichoaurantianolides, are also found in Tricholoma species, with an epimeric structure at C8 (Figure 18.1). It has also been reported that the seconeodolastane diterpene trichoaurantianolide A exhibits antibiotic activity against *Bacillus subtilis* and *Staphylococcus aureus.*[73]

18.7.5.2 C-30 Terpenoids

Another class of terpenoids is formed by saponaceolides, which were first obtained from the fruiting bodies of *T. saponaceum* growing in the forests of Italy; these were extracted using ethyl acetate (EtOAc) at −20 °C.[74]

Saponaceolides exhibit a number of bioactivities, with the crude EtOAc extract exhibit a strong inhibitory effect on the growth of leukemia cells of P388 mice. Saponaceolide A was reported to have some cytotoxic effects on human colon adenocarcinoma cells.[75] It was observed that when the same cytotoxicity tests were conducted, saponaceolide B was found to be more active than saponaceolide A, showing a strong inhibitory effect against KB and P388 tumor cells.[76] The National Cancer Institute selected saponaceolides A and B to conduct tests against a panel of around 60 human disease-oriented cancer cell lines.[77] The studies conducted showed the presence of all important nutrients, with a reported calorific value of 19.8 kcal per 100 g of fresh mushrooms.[78]

18.8 *Xerula radicata* (Relhan) Dörfelt

18.8.1 Taxonomy

Kingdom: Fungi
Phylum: Basidiomycetes
Class: Agaricomycetes
Order: Agaricales
Family: Physalacriaceae
Genus: Hymenopellis R. H. Petersen
Species (two accepted names): Hymenopellisradicata (Relhan) R.H. Petersen; *Xerula radicata* (Relhan) Dörfelt

Xerula radicata is more often reported as *Hymenopellis radicata* and *Oudemansiella radicata* under the genus *Oudemansiella*. These are popular as deep-root mushrooms, Beech rooters, or rooting shanks.

18.8.2 History

The genus *Xeruloids*, a common variety of mushrooms, have been distributed under the genera *Collybia*, *Oudemansiella*, and Xerula. Later on, in monographs of *Xeruloids*/*Oudemansiella* and allied genera, these species were segregated into three separate genera: *Hymenopellis*, *Paeaxeula*, and *Xerula* on the basis of characterization.[79] In 1785, *X. radicata* was named *Agaricus radicatus* by the botanist/mycologist Richard Relhan. In 1995, this fungus was classified as *Xerula radicata* by the German mycologist Heinrich Dörfelt; however, in 2010, it was again reclassified under the genus *Hymenopellis radicata* by the American mycologist Ron Peterson. *Xerula* is represented by 17 taxa. These fungus is reported with various synonyms, such as *Xerula radicata* var. alba Dörfelt, *Agaricus radicatus* Relhan, *Collybia radicata* (Relhan ex Fr.) Quel., *Gymnopus radicatus* (Relhn) Gray, *Collybia radicans* P. Kumm., *Collybia radicata* (Relhan) Singer, *Oudemansiella radicata* var. marginata (Konrad & Maubl.) Bon & Dennis, *Oudemansiella radicata* (Relhan ex Fr.) Singer, and *Xerula radicata* (Relhan Fr.) Dörfelt.

18.8.3 Distribution

It is found in North America and India,[80–83] among other countries. The mushroom is commonly found on stumps and trees. Saprophytic association has been observed between the mushroom and the wood of dead and dying tree roots, but it appears to be terrestrial. Identification of the mushroom has been reported in mating studies[84,85] by DNA analysis.[79]

18.8.4 Morphology

This edible fungus is tall and slender in structure. It has white spore prints and stems that are thin and lack a ring or volva with no partial and universal veins in

North American species. The cap is yellow to olive-brown in color with a diameter in the range of 3–10 cm. The shape of the fungus varies from convex or campanulate in the earlier growth stage while becoming plano-complex with a viscid surface or radially wrinkled in its mature stages. It has uplifted margins. Its gills are white in color, thick, broad, and distant. The stem of the mushroom is deeply rooted to touch the wood of the tree for nourishment, is 5–20 cm tall and 0.3–1 cm in diameter, with a color varying from white at the apex to brownish-gray at the roots. The spores of the fungus are smooth, ellipsoidal, hyaline, and non-amyloid, with dimensions of 12–16 × 9–12 μm.

18.8.5 Nutritional Content

Yilmaz *et al.*[82] analyzed the lipid profile of this mushroom. A total of 21 fatty acids were found in a chloroform : methanol (2 : 1) extract of the stem and fruiting body of this fungus. These were further categorized into saturated, monosaturated, and polyunsaturated fatty acids. Its cap consists of 42.7% polyunsaturated fatty acids, with linolenic acid constituting a major percentage. It is recommended as an edible fungus that holds good nutritional importance for cholesterol patients.

18.8.6 Biological Activity

Mushrooms have economic importance due to their biological activities, such as immunomodulatory, antibiotic,[86] anti-inflammatory, antioxidant,[87] antifungal, antitumor, lung-protective, and hepatoprotective properties.[88] Hamao *et al.*,[89] reported the presence of exopolysaccharides and oudenone in *Oudmansiella radicata,* responsible for anticancer and anti-hypertensive properties, respectively.[90] Ono *et al.*[91] reported the structure and synthesis of oudenone (Figure 18.1); its mechanism of action involves the inhibition of the enzymes phenylalanine and tyrosine hydroxylase in the adrenal glands and the catecholamine levels in the adrenal glands, brain, and heart. This results in a decrease in blood pressure. Oudemansin X, an antifungal agent (Figure 18.1), along with Strobilurin A, were extracted from *O. radicata.*[92] The polysaccharides of *O. radicata* have been found to increase the immunity, appetite, growth, and disease resistance of meat livestock. Moreover, *O. radicata* also improves the quality of poultry eggs. *O. radicata* accumulates heavy metals; its fruiting body shows good lead accumulation potential.[93] Another study[94] reported the metal tolerance and detoxification potential of *O. radicata* as a copper metal bio-accumulator. Moreover, the production of economically important cellulose and laccase enzymes was reported by *O. radicata* utilizing agro waste.

18.9 *Verpa bohemica* (Krombh) J. Schrot

18.9.1 Taxonomy

Kingdom: fungi
Phylum: Ascomycota

Class: Pezizomycetes
Order: Pezizales
Genus: *Verpa* Sw.
Species: *Verpa bohemica* (Krombh) J. Schrot

18.9.2 History

Verpa bohemica is a variety of mushroom that was first reported in 1823 as *Morella bohemica* by the Czech physician and mycologist Julius Vincenz von Krombholz, which was further modified to the genus *Verpa* in 1893 by Joseph Schroter. This species is popular, having been labeled with various names, such as *Ptychoverpa bohemica*, which was given to *Verpa bohemica* by Frenchman Jean Louis Émile Boudier in 1907. It is recognized as a new species due to its curved ascospores and short paraphyses. Further, it is classified as ptychoverpa due to its thick longitudinal ridge on its cap. In Canada, the species was first discovered in 1910 by Alfred Brooker Klugh and given the name *Morchella bispora*. It has been found that this species has been reported in the literature with various obsolete names, such as *Morchella bohemica* var. bispora, *Morchella gigaspora, Morchella bohemica, Ptychoverpa bohemica, Phalobolltus bisporus, Verpa bohemica* var. pallida, *Morchella bispora* var. truncata, *Morchella bispora* var. bispora, *Verpa bispora, Verpa bohemica* var. bohemica, *Morchella bispora, Ptychoverpa bohemica* var. pallida, *Phalloboletus bohemicus, Ptychoverpa bohemicus, Mitrophora bohemicus, Morchella speciosa* var. bohemica, *Verpa bohemica* var. bispora, and *Morchella speciosa* var. gigaspora.

18.9.3 Distribution

The genus *Verpa* consists of five species, among which the species *V. bohemica* is found on continents such as Asia, Europe, and North America (including Canada (British Columbia)). In India, it is found in the Apharwhat range, the Kashmir Himalayas, Jammu, and Kashmir,[95] and in Turkey it has been found in Mut-dagpazi village in the black pine forest[96] on the continent of Asia. This fungus is also found in the European countries Ukraine,[97] Sweden,[98] Spain (1907), Slovenia (1907), Russia,[99] Poland,[100] Czech republic,[101] Denmark (1907), Austria (1907), Finland,[102] and Germany,[103] while in North America[104] is found at a range of great lakes in the midwestern states from the south to the north to California on the west coast.[105]

18.9.4 Morphology

V. bohemica is commonly known as the wrinkled thimble-cap, a name derived from its morphological characteristics of a pale yellow to tarnished brown bell-shaped cap (apothecium) attached at a single point to the stem. The surface of its apothecium features brain-like wrinkles. Its length varies from 2 to 5 cm and its diameter varies from 2 to 4 cm. Its stem is white and brittle, thick at the base and thin at the top, with a length of 6–12 cm and a thickness of almost 2.5 cm.[106] The stem is filled with hyphae at the early stage, which disappear on maturity.[107] *V. bohemica* usually bears 60–80 spores with a distribution of 2–3 spores per ascus.[108,109] The asci are 275–350 μm in length and 16–23 μm wide.[105] The spore are translucent to yellow in appearance[110] and 15–18 μm in size, with a shape varying from smooth and curved, to elliptical.[110]

18.9.5 Toxicity

The toxicity of *V. bohemica* is still a matter of concern for scientists. According to a study by Langrange and Vernoux,[111] the false morel (*V. bohemica*) is considered to be inedible due to it inducing severe gastrointestinal symptoms and cerebellar syndrome due to its mycotoxin content. In Russia, *V. bohemica* is sold as a frozen commercial species. No clinical cases of intoxications have been reported. In another study in California,[112] it was found that consuming large portions of mushrooms on successive days caused food poisoning. Like other toxic mushrooms, it is hypothesized that *V. bohemica* synthesizes low levels of toxins, but there have been no reports on this to date. To avoid the harmful effects of *V. bohemica*, it is advised to parboil the fungus with water and dry specimens before consuming them.[113] Moreover, the consumption of this fungus should be limited.[108] Another piece of advice is to eat only the cap and discard the stem.[114] The usage of *V. bohemica* as a dietary material was supported by analysis of its proximate composition by Sheikh *et al.*,[115] who investigated the *V. bohemica* of Kashmir and reported a crude protein content of 34.37%, total carbohydrate content of 49.67%, fat content of 1.60%, ash content of 10.34%, and moisture content of 8.54%, while other minerals and elements, such as phosphorous (410 $mg\,g^{-1}$), potassium (2230 $mg\,g^{-1}$), calcium (10 $mg\,g^{-1}$), magnesium (40 $mg\,g^{-1}$), iron (5.15 $mg\,g^{-1}$), magnesium (1.29 $mg\,g^{-1}$), zinc (5.730 $mg\,g^{-1}$), copper (2.16 $mg\,g^{-1}$), sodium (76 $mg\,g^{-1}$), nickel (0.55 $mg\,g^{-1}$), lead (0.112 $mg\,g^{-1}$), and cadmium (0.089 $mg\,g^{-1}$) were also present in the fungus. It was concluded that mushrooms with as low as possible poison content will attain greater acceptability for consumption; although, the levels are dependent on environmental factors.

18.9.6 Biological Activity

An antimicrobial evaluation of *V. bohemica* extract was conducted against *Escherichia coli*, *Staphylococcus aureus*, *Pseudomonas aeruginosa*, *Bacillus subtilis*, *Proteus vulgaris*, and fungal strains like *Aspergillus niger*, *Candida albicans*, *Candida kruesis*, *Candida paralopsis*, and *Aspergillus fumigates* were investigated using disc dilution and microbroth dilution methods.[95] Extracted compounds from different solvent extracts of *V. bohemica* were found to be effective against *E. coli*.

The anti-proliferative activities of *V. bohemica*, *Amanita rubescens*, and *Pleurotus ostreatus* from hexane extracts of caps were analyzed and examined in a HT29 cell line using BrdU cell proliferation ELISA.[116] The evaluation of extracts of *V. bohemica* as a dietary supplement that protects against DNA damage using hydroxyl radicals was proposed. It was found that among six different extracts, ethyl acetate extract was more promising against a DPPH assay. Ethyl acetate and butanol extracts were effective against a lipid peroxidation assay. The ethanol extract was effective against FTC and TBA assays. These extracts exhibited better activity than ascorbic acid and BHT standards. The antioxidant potential was assessed by DPPH assay, where the ethanol extract showed 83.60% inhibition, methanol extracts showed 85.68% inhibition, and the distilled water extract showed 83.00% inhibition.

18.10 Conclusions

Mushroom have been a part of the human diet since time immemorial. With advances in medicinal research, many mushroom species have been found to cure certain diseases and this has driven the research to determine their remedial properties rather than only their nutritional aspects. In this chapter, a few species of wild and medicinal mushrooms that have not been well explored have been compiled and details about their taxonomical data, distribution, biological activities, and bioactive molecules have been presented. Although a few species have been shown to exhibit therapeutical potential, the fungi described are toxic to consume directly, *e.g.*, *Pholiota aurivella* is toxic and causes stomach discomfort when consumed alongside alcohol. In addition, the bioactive components present in these species and their chemistries are still unexplored. Further clinical trials are thus required to confirm their medicinal potential.

References

1. S. Sivanandhan, A. Khusro, M. G. Paulraj, S. Ignacimuthu and N. A. AL-Dhabi, *J. Fungi*, 2017, **3**, 2.
2. Y. Wu, M. H. Choi, J. Li, H. Yang and H. J. Shin, *Cosmetics*, 2016, **3**, 22.
3. P. Kummer, *Der Führer in die Pilzkunde (in German)*, 1st edn, 1871, p. 84.
4. M. Q. He, R. L. Zhao and K. D. Hyde, *et al.*, *Fungal Divers*, 2019, **99**(1), 105.
5. J. Holec, *The genus Pholiota in central and western Europe*, IHW-Verlag und Verlagsbuchhandlung, Eching, Germany, 2001.
6. J. M. Moncalvo, R. Vilgalys, S. A. Redhead, J. E. Johnson, T. Y. James, A. M. Catherine, V. Hofstetter, S. J. Verduin, E. Larsson, T. J. Baroni, T. R. Greg, S. Jacobsson, H. Clémençon and O. K. Miller, *Mol. Phylogenet. Evol.*, 2002, **23**(3), 357, PMID 12099793.
7. J. W. Lee, M. S. Park, J. H. Park, Y. Cho, C. Kim, C. S. Kim, J. W. Jo and Y. W. Lim, *Mycobiology*, 2020, **48**(6), 476.
8. L. O. Overholts, *Ann. Mo. Bot. Gard.*, 1927, **14**, 87.

9. A. H. Smith and L. R. Hesler, *The North American Species of Pholiota*, Hafner, New York, 1968, p. 402.
10. E. R. Farr, O. K. Miller, Jr. and D. F. Farr, *Can. J. Bot.*, 1977, **55**, 1167.
11. S. Jacobsson, *Mycotaxon*, 1987, **17**, 1.
12. S. Jacobsson, *Mycotaxon*, 1989, **36**, 95.
13. J. Klan, D. Baudisova and I. Rulfova, *Mycotaxon*, 1989, **36**, 249.
14. Y. Zhang, Y. Zhang, W. Gao, R. Zhou, F. Liu and T. B. Ng, *Int. J. Biol. Macromol.*, 2020, **164**, 3171, PMID: 32858105.
15. D. Gan, L. Ma, C. Jiang and Z. Xiaoxiong, *Carbohydr. Polym.*, 2011, **84**(3), 997.
16. Q. Hu, H. Wang and T. B. Ng, *Int. J. Med. Mushr.*, 2012, **14**(3), 271.
17. S. K. Cho, G. S. Seo, M. K. Kim and J. S. Lee, *Korean J. Med. Mycol.*, 2009, **37**(2), 195.
18. M. Yazdani, Z. Béni, M. Dékány, N. Szemerédi, G. Spengler, J. Hohmann and A. Ványolós, *J. Nat. Prod.*, 2022, **85**(4), 910.
19. S. S. Dhiman, S. S. Jagtap, M. Jeya, J. R. Haw, Y. C. Kang and J. K. Lee, *Biotechnol. Lett.*, 2012, **34**(7), 1307.
20. S. S. Jagtap, S. S. Dhiman, M. Jeya, J. R. Haw, Y. C. Kang and J. K. Lee, *Bioresour. Technol.*, 2012, **120**, 26.
21. H. Neda, *Mycoscience*, 2008, **49**(1), 88.
22. C. Islek, B. G. O. Saridogan, M. Sevindik and I. Akata, *Fresenius Environ. Bull.*, 2021, **30**(6), 6109.
23. A. E. Bessette, *Mushrooms of Northeastern North America*, Syracuse University Press, Syracuse, New York, 1997.
24. S. Trudell and J. Ammirati, *Mushrooms of the Pacific Northwest*, Timber Press Field Guides, Timber Press, Portland, OR, 2009, p. 174.
25. J. Holec, *Czech Mycol.*, 1998, **50**, 201.
26. A. Lavallée and M. Lortie, Notes sur la production, *in vitro*, de sporophores par *Pholiotaaurivella*, *Can. J. Bot.*, 1971, **49**(4), 567.
27. A. Dogan, A. Dalar, A. Battal, A. Uyar, C. Sadullahoglu, I. Celik and Y. Uzun, Evaluation of the *in vitro* and *in vivo* effects of the ethanolic lyophilized extract obtained from *Pholiotaaurivella* mushroom on CCl4-induced toxicity in rats, *Fresenius Environ. Bull.*, 2019, **28**, 9503.
28. H. Kawagishi, Y. Abe, T. Nagata, A. Kimura and S. Chiba, A Lectin from the Mushroom *Pholiotaaurivella*, *Agric. Bioi. Chern.*, 1991, **55**(10), 2485.
29. K. Morisaki, T. Fuhimi, S. Kaneko, I. Kusakabe and H. Kobayashi, Screening for phenoloxidases from edible mushrooms, *Biosci., Biotechnol., Biochem.*, 2001, **65**(10), 2334.
30. R. M. Davis, R. Sommer and J. A. Menge, *Field Guide to Mushrooms of Western North America*, University of California Press, Berkeley, 2012, p. 237.
31. A. L. Shigo, Growth of Polyporusglomeratus, Poriaobliqua, Fomesigniarius, and Pholiotasquarrosa-adiposa in media amended with manganese, calcium, zinc, and iron, *Mycologia*, 1970, **62**(3), 604.
32. M. Kuo, *Pholiotasquarrosa*, 2007, Retrieved from the *MushroomExpert.Com*, Web site: http://www.mushroomexpert.com/pholiota squarrosa.html.

33. O. Grönwall and A. Pehrson, Nutrient content in fungi as a primary food of the red squirrel Sciurus vulgaris L, *Oecologia*, 1984, **64**(2), 230.
34. M. T. Kimura, Evolution of food preferences in fungus-feeding Drosophila: an ecological study, *Evolution*, 1980, **34**(5), 1009.
35. F. W. M. R. Schwarze, J. Engels and C. Mattheck, *Fungal Strategies of Wood Decay in Trees*, Springer, Springer-Verlag Berlin Heidelberg, New York, 2000, p. 138.
36. P. A. Micheli, *Nova Plantarum Genera*, 1729, p. 70.
37. E. M. Fries, *Nova Acta Regiae Societatis Scientiarum Upsaliensis*, 1855, p. 17.
38. M. Núñez, *Synop. Fungorum*, 1995, **10**, 1.
39. H. Björndal and B. Lindberg, *Carbohydr. Res.*, 1970, **12**, 29.
40. Â. Fernandes, J. Petrović, D. Stojković, L. Barros, J. Glamočlija, M. Soković, A. Martins and I. C. Ferreira, *LWT-Food Sci. Technol.*, 2016, **69**, 91.
41. Y. Y. Zhao, *J. Ethnopharmacol.*, 2013, **149**, 35.
42. J. Koch, S. Witt and U. Lindequist, *Int. J. Med. Mushrooms*, 2002, **4**, 3.
43. I. Orhan and O. Üstün, *J. Food Composition Anal.*, 2011, **24**, 386.
44. A. Omarini, C. Henning, J. Ringuelet, J. A. Zygadlo and E. Albertó, *Internat. J. Food Sci. Technol.*, 2010, **45**, 1603.
45. R. Shabani, S. A. Shahidi and A. Rafe, *Food Sci. Nutr.*, 2018, **6**, 287–294.
46. S. L. Kessel, *Emp. For. Rev.*, 1927, **6**, 70.
47. H. E. Young, *Int. J. Forest Res.*, 1937, **11**, 30.
48. M. Binder and D. S. Hibbett, *Mycologia*, 2006, **98**, 971.
49. A. H. Smith and S. M. Zeller, *Memoirs New York Botanic Garden*, 1966, **14**, p. 1.
50. M. A. Sulzbacher, T. Grebenc, M. Á. García, B. D. Silva, A. Silveira, Z. I. Antoniolli, P. Marinho, B. Münzenberger, M. T. Telleria, I. G. Baseia and M. P. Martín, *Mycorrhiza*, 2016, **26**, 377.
51. C. Theodorou, *Soil Biol. Biochem.*, 1993, **25**, 647.
52. D. R. Hoagland and D. I. Amon, *California Agri. Exp. Station Cir.*, 1950, **347**, 1.
53. F. R. Smiderle, E. R. Carbonero, G. L. Sassaki, P. A. Gorin and M. Iacomini, *Food Chem.*, 2008, **108**, 329.
54. G. Tel-Çayan, A. Muhammad, E. Deveci, M. E. Duru and M. Öztürk, *Int. J. Biol. Macromol.*, 2020, **165**, 2395.
55. A. C. Ruthes, F. R. Smiderle and M. Iacomini, *Carbohydr. Polym.*, 2016, **136**, 358.
56. F. Hassan, S. Ni, T. L. Becker, C. M. Kinstedt, J. L. Abdul-Samad, L. A. Actis and M. A. Kennedy, *Int. J. Med. Mushrooms*, 2019, **21**(2), 131–141.
57. M. Aulitto, S. Fusco, D. Limauro, G. Fiorentino, S. Bartolucci and P. Contursi, *World J. Microbiol. Biotechnol.*, 2019, **35**, 1.
58. E. Barreto-Bergter and P. A. Gorin, *Advances in carbohydrate chemistry and biochemistry*, Academic Press, 1983, vol. 1, pp. 67–103.

59. K. E. Heim, A. R. Tagliaferro and D. J. Bobilya, *J. Nutr. Biochem.*, 2002, **13**, 572.
60. B. Modak, M. L. Contreras, F. González-Nilo and R. Torres, *Bioorg. Med. Chem. Lett.*, 2005, **15**, 309.
61. B. Ribeiro, P. B. Andrade, P. Baptista, L. Barros, I. C. Ferreira, R. M. Seabra and P. Valentao, *J. Agric. Food Chem.*, 2008, **56**, 4769.
62. P. Valentão, G. Lopes, M. Valente, P. Barbosa, P. B. Andrade, B. M. Silva, P. Baptista and R. M. Seabra, *J. Agric. Food Chem.*, 2005, **53**, 3626.
63. S. A. Heleno, L. Barros, M. J. Sousa, A. Martins and I. C. Ferreira, *Microchem. J.*, 2009, **9**, 195.
64. G. Sadi, B. Emsen, A. Kaya, A. Kocabaş, S. Çınar and D. I. Kartal, *Pharmacogn. Mag.*, 2015, **11**, 6.
65. S. L. Miller, *Mycotaxon*, 1986, **112**, 193.
66. A. W. Claridge and S. J. Cork, *Aust. J. Zool.*, 1994, **42**, 701.
67. W. D. Boer, L. B. Folman, R. C. Summerbell and L. Boddy, *FEMS Microbiol. Rev.*, 2005, **29**, 795.
68. M. Clericuzio, G. G. Mellerio, P. V. Finzi and G. Vidari, *Nat. Prod. Commun.*, 2018, **13**, 1213–1224.
69. A. M. Almomany, *J. Life Sci.*, 2013, **7**, 653.
70. E. R. Boa, *Wild edible fungi: a global overview of their use and importance to people*, Rome Italy, 2004, p. 31.
71. B. Kibar and A. Peksen, *Agriculture*, 2011, **98**, 409–414.
72. V. A. Dıez and A. Alvarez, *Food Chem.*, 2001, **75**, 417.
73. A. G. Invernizzi, G. Vidari and P. Vita-Finzi, *Tetrahedron Lett.*, 1995, **36**, 1905.
74. S. Tsukamoto, A. D. Macabalang, K. Nakatani, Y. Obara, N. Nakahata and T. Ohta, *J. Nat. Prod.*, 2003, **66**, 1578.
75. M. De Bernardi, L. Garlaschelli, G. Gattl, G. Vidari and P. V. Finzi, *Tetrahedron*, 1988, **44**, 235.
76. C. Geraci, M. Piattelli and C. Tringali, *Magn. Reson. Chem.*, 1991, **29**, 603.
77. G. Vidari, G. Lanfranchi, P. Sartori and S. Serra, *Tetrahedron*, 1995, **6**, 2977.
78. V. E. Prakasam, B. A. Karthikayani, G. U. Thiribhuvanamala, G. O. Chandrasekar, S. U. Veeralakshmi, P. A. Ahila, K. R. Sakthivel and B. A. Malarkodi, *Proceedings of the 7th International Conference on Mushroom Biology and Mushroom Products*, 2011, p. 438.
79. R. H. Petersen and K. W. Hughes, *The Xerula/Oudemansiella Complex (Agaricales)*, J. Cramer, Stuttgart, 2010, p. 625.
80. N. A. Wani, M. K. Saini and N. A. Malik, *Xerula radicata* var. setosa var. nov. and three new records of the family Physalacriaceae, Agaricales from the Indian subcontinent, *Curr. Res. Environ. Appl. Mycol.*, 2021, **11**(1), 248.
81. Y. Liu, Y. Li, Y. Ke, C. Li, Z. Zhang, Y. Wu and W. Wu, *In vitro* saliva-gastrointestinal digestion and fecal fermentation *of Oudemansiellar-adicata* polysaccharides reveal its digestion profile and effect on the

modulation of the gut microbiota, *Carbohydr. Polym.*, 2021, **251**, 117041.
82. N. Yilmaz, M. Solmaz, I. Türkekul and M. Elmastaş, Fatty acid composition in some wild edible mushrooms growing in the middle Black Sea region of Turkey, *Food Chem.*, 2006, **99**(1), 168.
83. A. Ronikier, Revision of the genus *Xerula maire* [Basidiomycetes, Agaricales] in Poland, *Acta Soc. Bot. Pol.*, 2003, **72**(4), 339.
84. R. H. Petersen and A. S. Methven, Mating systems in the Xerulaceae: Xerula, *Can. J. Bot.*, 1994, **72**(8), 1151.
85. R. H. Petersen and K. W. Hughes, Une nouvelle espece de *Xerula*, *Xerulalimonispora. Bull. Trimest. Soc. Mycol. Fr.*, 2004, **120**, 37.
86. B. Gao, R. Wang, Y. Peng and X. Li, Effects of a homogeneous polysaccharide from Sijunzi decoction on human intestinal microbes and short chain fatty acids *in vitro*, *J. Ethnopharmacol.*, 2018, **224**, 465.
87. Z. Gao, X. Liu, W. Wang, Q. Yang, Y. Dong, N. Xu and L. Jia, Characteristic anti-inflammatory and antioxidative effects of enzymatic- and acidic-hydrolysed mycelium polysaccharides by Oudemansiellaradicata on LPS-induced lung injury, *Carbohydr. Polym.*, 2019, **204**, 142.
88. Q. Liu, M. Zhu, X. Geng, H. Wang and T. B. Ng, Characterization of polysaccharides with antioxidant and hepatoprotective activities from the edible mushroom *Oudemansiellaradicata*, *Molecules*, 2017, **22**(2), 234.
89. U. Hamao, T. Osamn and T. Takenchi, Hypotensive agent, oudenone, its salts and process for production and preparation thereof, *US Pat.*, 3835170, 1974.
90. Y. S. Tsantrizos and F. Zhou, Biosynthesis of the hypotensive metabolite oudenone by *Oudemansiellaradicata*, *J. Org. Chem.*, 1995, **60**, 6922.
91. M. Ono, M. Okamoto, N. Kawabe, H. Umezawa, T. Takeuchi, H. Iinuma and S. Takahashi, Oudenone, a novel tyrosine hydroxylase inhibitor from microbial origin, *J. Amer. Chem. Soc.*, 1971, **93**(5), 1285.
92. T. Anke, A. Werle, M. Bross and W. Steglich, Antibiotics from basidiomycetes XXXXIII oudemansin X, A new antifungal e-β-methoxyacrylate from *Oudemansiella radicata* (Relhan ex fr.) Sing, *J. Antibiot.*, 1990, **43**(8), 1010.
93. W. Zhang, Y. Hu, Y. Cao, F. Huang and H. Xu, Tolerance of lead by the fruiting body of Oudemansiellaradicata, *Chemosphere*, 2012, **88**(4), 467.
94. J. Jiang, C. Qin, X. Shu, R. Chen, H. Song, Q. Li and H. Xu, Effects of copper on induction of thiol-compounds and antioxidant enzymes by the fruiting body of Oudemansiellaradicata, *Ecotoxicol. Environ. Saf.*, 2015, **111**, 60.
95. N. Shameem, A. N. Kamili, M. Ahmad, F. A. Masoodi and J. A. Parray, Radical scavenging potential and DNA damage protection of wild edible mushrooms of Kashmir Himalaya, *J. Saudi Soc. Agric. Sci.*, 2017, **16**(4), 314.
96. H. H. Doğan, C. Öztürk, G. Kaşık and S. Aktaş, Macrofungi distribution of Mut province in Turkey, *Pak. J. Bot.*, 2007, **38**(1), 293.

97. D. W. Minter, V. P. Hayova, T. J. Minter and Y. Y. Tykhonenko, *Verpabohemica (Krombh.) J. Schröt. Electronic Distribution Maps of Ukrainian Fungi*, Archived from the original on 2011-09-27. Retrieved 2011-05-06.
98. S. Ryman, Sweedish Pezizales of spring and early summer, Svensk, *Bot. Tidskr.*, 1978, **72**(4), 327.
99. A. A. Skryabina, Fructification of some species of edible fungi from the family Helvellaceae in the Slobodsky Roan of the Kirov Oblast, *Rast. Resur.*, 1975, **11**(4), 552.
100. A. Skirgiello, Discomycètes de printemps de Bialowieza [Spring Discomycetes of Bialowieza], *Monogr. Bot.*, 1960, **10**(2), 3.
101. M. Svrček, Katalogoperkulátníchdiskomycetů (Pezizales) Československa II. (O-W) [List of operculate discomycetes Pezizales recorded from Czechoslovakia 2. O-W], *Czech. Mycol.*, 1981, **35**(2), 64.
102. A. Granmo, O. Skifte and A. C. Nilssen, *Ptychoverpa bohemica* Pezizales in Norway and Finland, *Karstenia*, 1982, **22**(2), 43.
103. K. O'Donnell, E. Cigelnik, N. S. Weber and J. M. Trappe, Phylogenetic relationships among ascomycetous truffles and the true and false morels inferred from 18S and 28S ribosomal DNA sequence analysis, *Mycologia*, 1997, **89**(1), 48.
104. M. Kuo, *Verpabohemica. MushroomExpert.Com.* 2005a, Retrieved 2011-05-03.
105. M. Kuo, *Morels*, The University of Michigan Press, Ann Arbor, Michigan, 2005, p. 184.
106. R. Phillips, *Verpabohemica*, Rogers Plants Ltd, Retrieved 2011-05-03.
107. H. M. E. Schalkwijk-Barendsen, *Mushrooms of Western Canada*, Lone Pine Publishing, Edmonton, Alberta, 1994, p. 178.
108. V. B. McKnight and K. H. McKnight, *A Field Guide to Mushrooms, North America*, Houghton Mifflin, Boston, Massachusetts, 1987, p. 42.
109. B. O. Dodge, Spore formation in asci with fewer than eight spores, *Mycologia*, 1928, **20**(1), 18.
110. R. A. Healy, D. R. Huffman, L. H. Tiffany and G. Knaphaus, *Mushrooms and other fungi of the midcontinental United States (Bur Oak Guide)*, University of Iowa Press, Iowa City, Iowa, 2008, p. 295.
111. E. Lagrange and J. P. Vernoux, Warning on false or true morels and button mushrooms with potential toxicity linked to hydrazinic toxins: An update, *Toxins*, 2020, **12**, 482.
112. D. B. Orr and R. T. Orr, *Mushrooms of Western North America*, University of California Press, Berkeley, California, 1979, p. 36.
113. J. D. Sept, *Common Mushrooms of the Northwest: Alaska, Western Canada & the Northwestern United States*, Calypso Publishing, Sechelt, British Columbia, 2006, p. 82.
114. M. Pelouch and L. Pelouch, *How to Find Morels*, University of Michigan Press, Ann Arbor, Michigan, 2008, p. 71.
115. P. A. Sheikh, G. H. Dar, W. A. Dar, S. Shah, K. A. Bhat and S. Kousar, Chemical composition and antioxidant activities of some edible mushrooms of Western Himalayas of India, *Vegetos*, 2015, **28**, 124.

116. Ö. Y. Sahin, M. Dumanlı, I. Türkekul and S. Tekin, *Amanita rubescens, Pleurotusostreatus* and *Verpabohemica*: Anticancer activities against HT29 cell line, *J. Biotechnol.*, 2014, **185**, S87.
117. A. Lovy, B. Knowles, R. Labbe and L. Nolan, *Herbs, Spices Med. Plants*, 2000, **6**, 49.
118. J. J. Li, Y. Y. Tu, J. Z. Tong and P. T. Wang, *China J. Chin. Mater. Med.*, 2000, **25**, 628.
119. Y. Sun and X. Zhou, *Food Sci. Human Wellness*, 2014, **3**, 73.
120. D. Yuan, J. Mori, K. I. Komatsu, T. Makino and Y. Kano, *Biol. Pharm. Bull.*, 2004, **27**, 867.
121. X. Li and W. Xu, *J. Ethnopharmacol.*, 2011, **135**, 1.
122. P. Kang, S. L. Lv, X. Y. Wang, L. J. Sun, J. W. Yu, Y. H. Zhao and S. C. Li, *Afr. J. Pharm. Pharmacol.*, 2012, **6**, 2686.
123. Z. Guo, Y. Zang and L. Zhang, *Prog. Mol. Biol. Transl. Sci.*, 2019, **163**, 29.
124. A. Mocan, Â. Fernandes, L. Barros, G. Crişan, M. Smiljković, M. Soković and I. C. Ferreira, *Food Func.*, 2018, **9**, 160.
125. G. K. Liu, N. Li, S. Y. Song, Y. J. Zhang and J. R. Wang, *Int. J. Biol. Macromol.*, 2019, **132**, 629.

Subject Index